Introduction to Ore-Forming Processes

Introduction to Ore-Forming Processes

Laurence Robb
University of Oxford
Oxford, UK

Second Edition

WILEY Blackwell

Registered Offices
John Wiley & Sons, Inc., 111 River Street, Hoboken, NJ 07030, USA
John Wiley & Sons Ltd, The Atrium, Southern Gate, Chichester, West Sussex, PO19 8SQ, UK

Editorial Office
9600 Garsington Road, Oxford, OX4 2DQ, UK

For details of our global editorial offices, customer services, and more information about Wiley products visit us at www.wiley.com.

Wiley also publishes its books in a variety of electronic formats and by print-on-demand. Some content that appears in standard print versions of this book may not be available in other formats.

Library of Congress Cataloging-in-Publication Data

Name: Robb, L., author.
Title: Introduction to ore-forming processes / Laurence Robb, Department of
    Earth Sciences, University of Oxford, Oxford, UK.
Description: Second edition. | Hoboken, NJ : Wiley-Blackwell, 2020. |
    Includes index.
Identifiers: LCCN 2019058286 (print) | LCCN 2019058287 (ebook) | ISBN
    9781119967507 (paperback) | ISBN 9781119232391 (adobe pdf) | ISBN
    9781119232384 (epub)
Subjects: LCSH: Ores.
Classification: LCC QE390 .R32 2020 (print) | LCC QE390 (ebook) | DDC
    553/.1–dc23
LC record available at https://lccn.loc.gov/2019058286
LC ebook record available at https://lccn.loc.gov/2019058287

Cover Design: Wiley
Cover Images: © Laurence Robb and Lawrence Minter

Set in 9.5/12.5pt STIXTwoText by SPi Global, Chennai, India
Printed and bound in Singapore by Markono Print Media Pte Ltd

10 9 8 7 6 5 4 3 2 1

Contents

Preface to the 2nd Edition

It is now more than a decade since the first edition of this book appeared, during which time a great deal has happened in furthering the knowledge of metallogeny and earth system science. Our understanding of global tectonic processes and the nature of crustal evolution continues to influence the practice of economic geology and assists in maintaining the supply of natural resources in a responsible and sustainable way. The economies of developing nations continue to grow so that a greater proportion of people than ever before enjoy the benefits of a lifestyle that befits the twenty-first century. However, the global economy, and the natural resources industry in particular, remain cyclical in that security of supply of strategically important commodities has become a major problem – one result of this is the identification of "critical metals" whose supply cannot be regarded as sustainable in the short or medium term. Despite the fact that new metallotects are still being discovered – and even exploration for metals in the deep ocean and outer space mooted – the replenishment of depleted natural resources is becoming more difficult and environmentally demanding. In order to mitigate these problems, the responsible custodianship of natural resources is more necessary than ever before and there is a continuing need for all earth scientists to understand metallogeny and the resource cycle.

The 2nd edition of *Introduction to Ore-Forming Processes* has been updated to play a role in meeting these demands. The book is still introductory in nature and the basic structure and layout remain unchanged – all sections have, however, been updated and expanded with respect to research undertaken since it first appeared. My grateful thanks are to Brian Skinner, Steve Kesler, Charlie Moon, Michael Meyer, and Judith Kinnaird who provided valuable commentary on the revised content for the 2nd edition. My own development as an economic geologist has benefited over the past decade or more by collaboration with geoscientists that include Mike Searle, Dave Waters, Chris Hawkesworth, Nick Gardiner, Judith Kinnaird, and Paul Nex.

This book was originally conceived in a very different format. The 2nd edition is dedicated to Professor John Moore (1946–2011), Rhodes University, whose perceptive suggestions led to the process-related approach of the present content, and which contributed in no small measure to its success.

Laurence Robb
Oxford

Preface to the 1st Edition

There are many excellent texts, available at both introductory and advanced levels, that describe the Earth's mineral deposits. Several describe the deposits themselves and others do so in combination with explanations that provide an understanding of how such mineral occurrences form. Few are dedicated entirely to the multitude of processes that give rise to the ore deposits of the world. The main purpose of this book is to provide a better understanding of the processes, as well as the nature and origin, of mineral occurrences and how they fit into the Earth system. It is intended for use at a senior undergraduate level (third and fourth year levels), or graduate level (North America), and assumes a basic knowledge in a wide range of core earth science disciplines, as well as in chemistry and physics. Although meant to be introductory, it is reasonably comprehensive in its treatment of topics, and it is hoped that practicing geologists in the minerals and related industries will also find the book useful as a summary and update of ore-forming processes. To this end the text is punctuated by a number of boxed case studies in which actual ore deposits, selected as classic examples from around the world, are briefly described to give context and relevance to processes being discussed in the main text.

Metallogeny, or the study of the genesis of ore deposits in relation to the global tectonic paradigm, is a topic that traditionally has been, and should remain, a core component of the university earth science curriculum. It is also the discipline that underpins the training of professional earth scientists working in the minerals and related industries of the world. A tendency in the past has been to treat economic geology as a vocational topic and to provide instruction only to those individuals who wished to specialize in the discipline or to follow a career in the minerals industries. In more recent years, changes in earth science curricula have resulted in a trend, at least in a good many parts of the world, in which economic geology has been sidelined. A more holistic, process-orientated approach (earth systems science) has led to a wider appreciation of the Earth as a complex interrelated system. Another aim of this book, therefore, is to emphasize the range of processes responsible for the formation of the enormously diverse ore deposit types found on Earth and to integrate these into a description of Earth evolution and global tectonics. In so doing it is hoped that metallogenic studies will increasingly be reintegrated into the university earth science curricula. Teaching the processes involved in the formation of the world's diminishing resource inventory is necessary, not only because of its practical relevance to the real world, but also because such processes form an integral and informative part of the Earth system.

This book was written mainly while on a protracted sabbatical in the Department of Earth Sciences at the University of Oxford. I am very grateful to John Woodhouse and the departmental staff who accommodated me and helped to provide the combination of academic rigor and quietude that made writing this book such a pleasure. In particular Jenny Colls, Earth Science Librarian, was a tower of support in locating reference material. The "tea club" at the Banbury Road annexe provided both stimulation and the requisite libations to break the monotony. The staff at Blackwell managed to combine being really nice people with a truly professional attitude, and Ian Francis, Delia Sandford, Rosie Hayden, and Cee Pike were all a pleasure to work with. Dave Coles drafted all the diagrams and I am extremely grateful for his forebearance in dealing amiably with a list of figures that seemingly did not end. Several people took time to read through the manuscript for me and in so doing greatly improved both the style and content. They include John Taylor (copyediting), Judith Kinnaird and Dave Waters (Introduction), Grant Cawthorn (Chapter 1), Philip Candela (Chapter 2), Franco Pirajno (Chapter 3), Michael Meyer (Chapter 4), John Parnell and Harold Reading (Chapter 5), and Mark Barley, Kevin Burke, and John Dewey (Chapter 6). The deficiencies that remain, though, are entirely my own. A particularly debt of gratitude is owed to David Rickard, who undertook the onerous task of reviewing the entire manuscript; his lucid comments helped to eliminate a number of flaws and omissions. Financial support for this project came from BHP Billiton in London and the Geological Society of South Africa Trust. My colleagues at Wits were extremely supportive during my long absences, and I am very grateful to Spike McCarthy, Paul Dirks, Carl Anhauesser, Johan Kruger, and Judith Kinnaird for their input in so many ways. Finally, my family, Vicki, Nicole, and Brendan, were subjected to a lifestyle that involved making personal sacrifices for the fruition of this project – there is no way of saying thank you and it is to them that I dedicate this book.

Laurence Robb
Johannesburg

Introduction: Mineral Resources

TOPICS

General introduction and aims of the book
A simple classification scheme for mineral deposits
Some important definitions
   metallogeny, syngenetic, epigenetic, mesothermal, epithermal, supergene, hypogene, etc.
Some relevant compilations
   periodic table of the elements
   tables of the main ore and gangue minerals
   geological time scale
Factors that make a viable mineral deposit
   enrichment factors required to make ore deposits
   how are mineral resources and ore reserves defined?
Natural resources and their future exploitation
   sustainability
   environmental responsibility

Introduction and Aims

With a global population in 2019 of close to eight billion people, and this figure set to increase to some ten billion by 2050, it is apparent that the world's economies are under growing pressure to meet the demands of an increasingly materialistic lifestyle. The unprecedented growth of human population over the past century has resulted in a dramatic increase in demand for, and production of, natural resources – it is therefore evident that understanding the nature, origin, and distribution of the world's mineral deposits remains a vital and strategic topic. The discipline of "economic geology," which covers all aspects pertaining to the description and understanding of mineral resources, is, therefore, one which traditionally has been,

and should remain, a core component of the university earth science curriculum. It is also the discipline that underpins the training of professional earth scientists working in the minerals and related industries of the world. Unfortunately, a tendency at many universities in the recent past has been to treat economic geology as a vocational topic, and to provide instruction only to those individuals who wished to specialize in the discipline or to follow a career in the minerals industry. There has been a trend, at least in many parts of the world, to sideline economic geology both as a taught discipline and a research topic.

Developments in the early twenty-first century have indicated how problematic institutional and governmental neglect can be when the security of supply of strategic metals is brought into question. Global demands to reduce greenhouse gas emissions, and to provide a framework for the responsible and sustainable supply of natural resources, have resulted in the realization that all earth scientists need to understand the resource cycle in order to properly advise the public at large, and to manage future programs aimed at the responsible custodianship of the world's finite resources. The conceptual development of earth systems science, a feature of the latter years of the twentieth century, has led to changes in the way in which the earth sciences are taught. A more holistic, process-orientated approach has led to a much wider appreciation of the Earth as a complex, interrelated system. The understanding of feedback mechanisms has created an awareness that the solid Earth, its oceans and atmosphere, and the organic life forms that occupy niches above, at and below its surface, are intimately connected and can only be understood properly in terms of an interplay of processes. Examples include the links between global tectonics and climate patterns, and also between the evolution of unicellular organisms and the formation of certain types of ore deposits. In this context the teaching of many of the traditional geological disciplines assumes new relevance and the challenge to successfully teaching earth system science is how best to integrate the wide range of topics into a curriculum that provides understanding of the entity. Understanding the processes involved in the formation of the enormously diverse ore deposit types found on Earth is necessary, not only because of its practical relevance to the real world, but also because such processes form an integral and informative part of the Earth's evolution.

The purpose of this process-orientated book is to provide a better understanding of the nature and origin of mineral occurrences and how they fit into the Earth system. It is intended for use at a senior undergraduate level, or at a graduate level, and assumes a basic knowledge in a wide range of earth science disciplines, as well as in chemistry and physics. It is also hoped that practicing geologists in the minerals and related industries will find the book useful as a summary and update of ore-forming processes. To this end the text is punctuated by a number of boxed case studies in which actual ore deposits, selected as classic examples from around the world, are briefly described to give context and relevance to processes being discussed in the main text.

A Classification Scheme for Ore Deposits

There are many different ways of categorizing ore deposits. Most people who have written about and described ore deposits have either unwittingly or deliberately been involved in their classification. This is especially true of textbooks where the task of providing order and structure to a set of descriptions invariably involves some form of classification. The best classification schemes are probably those that remain as independent of genetic linkages as possible, thereby minimizing the scope for mistakes and controversy. Nevertheless, genetic classification schemes are ultimately desirable, as there is considerable advantage to having processes of ore formation reflected in a set of descriptive categories. Guilbert and Park (1986) discuss the problem of ore deposit classification at some length in chapters 1 and 9 of their seminal book on the geology of ore deposits. They show how classification schemes reflect the development of theory and techniques, as well as the level of understanding, in the discipline. Given the dramatic improvements in the level of understanding in economic geology over recent years, the Guilbert and Park (1986) classification scheme, modified after Lindgren's (1933) scheme, is both detailed and complex, and befits the comprehensive coverage of the subject matter provided by their book. In a more recent, but equally comprehensive, coverage of ore deposits, Misra (2000) has opted for a categorization based essentially on genetic type and rock association, similar to a scheme by Meyer

(1981). It is the association between ore deposit and host rock that is particularly appealing for its simplicity, and that has been selected as the framework within which the processes described in this book are placed.

Rocks are classified universally in terms of a threefold subdivision, namely igneous, sedimentary, and metamorphic, that reflects the fundamental processes active in the Earth's crust (Figure 1a). The scheme is universal

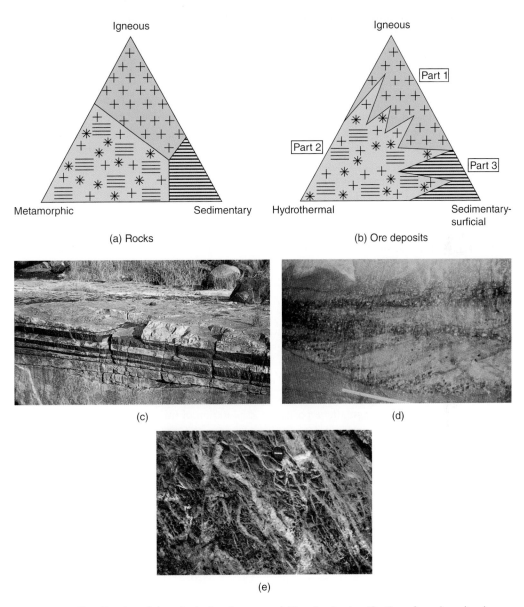

Figure 1 (a) Classification of the principal rock types and (b) a simple classification of ore deposits also based on host rock type – Parts 1, 2, and 3 represent the breakdown of sections in this book. Photographs illustrate examples representing the main ore forming processes. (c) Igneous: magmatic layering and chromitite seams, Critical Zone, Bushveld Complex, South Africa. (d) Sedimentary: Au- and U-bearing conglomerate from the Witwatersrand Basin, South Africa. (e) Hydrothermal: quartz-carbonate vein network in metasedimentary host rocks of the Lily gold mine, Barberton greenstone belt, South Africa.

because rocks are recognizably either igneous or sedimentary (generally!), or, in the case of both precursors, have been substantially modified to form a metamorphic rock. Likewise, ores are rocks and can often be relatively easily attributed to an igneous or sedimentary/surficial origin, a feature that represents a good basis for classification. Such a classification also reflects the genetic process involved in ore formation, since igneous and sedimentary deposits are typically syngenetic and formed at the same time as the host rock itself. Although many ores are metamorphosed, and whereas pressure and temperature increases can substantially modify the original nature of ore deposits, it is evident that metamorphism does not itself represent a fundamental process whereby ore deposits are created. Hydrothermal processes, however, are a metallogenic analogue for metamorphism and also involve modification of pre-existing protoliths, as well as heat (and mass) transfer and pressure fluctuation. A very simple classification of ores is, therefore, achieved on the basis of igneous, sedimentary/surficial, and hydrothermal categories (Figure 1b), and this forms the basis for the structure and layout of this book. This subdivision is very similar to one used by Einaudi (2000), who stated that all mineral deposits can be classified into three types based on process, namely magmatic deposits, hydrothermal deposits, and surficial deposits formed by surface and groundwaters. One drawback of this type of classification, however, is that ore-forming processes are complex and episodic. Ore formation also involves processes that evolve, sometimes over significant periods of geologic time. For example, igneous processes become magmatic-hydrothermal as the intrusion cools and crystallizes, and sediments undergo diagenesis and metamorphism as they are progressively buried, with accompanying fluid flow and alteration. In addition, deformation of the Earth's crust introduces new conduits that also facilitate fluid flow and promote the potential for mineralization in virtually any rock type. Ore-forming processes

can, therefore, span more than one of the three categories, and there is considerable overlap between igneous and hydrothermal and between sedimentary and hydrothermal, as illustrated diagrammatically in Figure 1b.

The main part of this book is subdivided into three sections termed Igneous (Part I), Hydrothermal (Part II), and Sedimentary/Surficial (Part III) (Figure 1a–e). Part I comprises Chapters 1 and 2, which deal with igneous and magmatic-hydrothermal ore-forming processes respectively. Part II contains Chapter 3 and covers the large and diverse range of hydrothermal processes not covered in Part I. Part III comprises Chapter 4 on surficial and supergene processes, as well as Chapter 5, which covers sedimentary ore deposits, including a section on the fossil fuels. The final chapter of the book, Chapter 6, is effectively an addendum to this threefold subdivision and is an attempt to describe the distribution of ore deposits, both spatially in the context of global tectonics and temporally in terms of crustal evolution, through Earth history. This chapter is relevant because the plate tectonic paradigm, which has so pervasively influenced geological thought since the early 1970s, provides another conceptual basis within which to classify ore deposits. In fact, modern economic geology, and the scientific exploration of mineral deposits, is now firmly cast into the frame of global tectonics and crustal evolution. Although there is still a great deal to be learnt, the links between plate tectonics and ore genesis are now sufficiently well established that studies of ore deposits are starting to contribute to a better understanding of the Earth system.

What Makes a Viable Mineral Deposit?

Ore deposits form when a useful commodity is sufficiently concentrated in an accessible part of the Earth's crust so that it can be profitably extracted. The processes by which this

Table 1 Average crustal abundances for selected metals and typical concentration factors that need to be achieved in order to produce a viable ore deposit.

	Average crustal abundance	Typical exploitable grade	Approximate concentration factor
Al	8.4%	30%	×4
Fe	5.2%	50%	×9
Cu	27 ppm	1%	×370
Ni	59 ppm	1%	×170
Zn	72 ppm	5%	×700
Sn	1.7 ppm	0.5%	×2900
Au	1.3 ppb	2 g t^{-1}	×1500
Pt	1.5 ppb	5 g t^{-1}	×3300

Note: 1 ppm is the same as 1 g t^{-1}.
Source: Average crustal abundances from Rudnick and Gao (2014). Reproduced with permission of Elsevier.

concentration occurs are the topic of this book. As an introduction it is pertinent to consider the range of concentration factors that characterize the formation of different ore deposit types. Some of the strategically important metals, such as Fe, Al, Mg, Ti, and Mn, are abundantly distributed in the Earth's crust (i.e. between about 0.5% and 10%) and only require a relatively small degree of enrichment in order to make a viable deposit. Table 1 shows that Fe

and Al, for example, need to be concentrated by factors of 9 and 4 respectively, relative to average crustal abundances, in order to form potentially viable deposits.

By contrast, base metals such as Cu, Zn, and Ni are much more sparsely distributed and average crustal abundances are only in the range 30–70 parts per million (ppm). The economics of mining dictate that these metals need to be concentrated by factors in the hundreds in order to form potentially viable deposits – degrees of enrichment that are an order of magnitude higher than those applicable to the more abundant metals. The degree of concentration required for the precious metals is even more demanding, where the required enrichment factors are in the thousands. Table 1 shows that average crustal abundances for Au and Pt are in the range 1–2 parts per billion (ppb) and even though mines routinely extract these metals at grades of around 1–5 g t^{-1}, the enrichment factors involved are between 1000 and 3000 times.

Another useful way to distinguish between the geochemically abundant and scarce metals is to plot average crustal abundances against production estimates. This type of analysis was first carried out by Skinner (1976), who used a plot like that in Figure 2 to confirm that

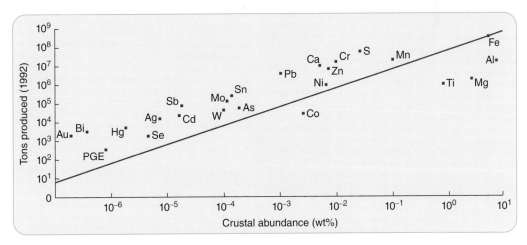

Figure 2 Plot of global production against crustal abundances for a number of metal commodities. The line through Fe can be regarded as a datum against which the rates of production of the other metals can be compared in the context of crustal abundances. Source: After Einaudi (2000).

crustal abundance is a reasonable measure of the availability of a given resource. It is by design and of necessity that we use more of the geochemically abundant metals than we do the scarce ones. The nature of our technologies and the materials we use to manufacture mechanical items depend in large measure on the availability of raw materials. As an example, the technologies (geological and metallurgical) that resulted in a dramatic increase in global aluminum production over the latter part of the twentieth century, allowed iron to be replaced by aluminum in many products such as motor vehicles. More importantly, though, Figure 2 allows estimates to be made of the relative rates of depletion of certain metals relative to others. These trends are discussed again below.

Mineral Resources and Ore Reserves

Throughout this book reference is made to the term "ore deposit" with little or no consideration of whether such occurrences might be economically viable. Although such considerations might seem irrelevant in the present context, it is necessary to emphasize that professional institutions now insist on the correct definition and usage of terminology pertaining to exploration results, mineral resources, and ore reserves. Such terminology should be widely used and applied, as it helps to reduce the incorrect, and sometimes irresponsible, usage of terminology in reports on which, for example, investment decisions might be based. Correct terminology can also assist in the description and identification of genuine ore deposits from zones of marginal economic interest or simply anomalous concentrations of a given commodity.

Although the legislation that governs the public reporting of mineral occurrences varies from one country to another, there is now reasonable agreement globally on a definition of terms. It is widely accepted that different terms should apply to mineral occurrences depending on the level of knowledge and degree of confidence that is associated with their quantification in terms of grade and mass/volume. Figure 3 is a matrix that reflects the terminology associated with an increased level of geological knowledge and confidence,

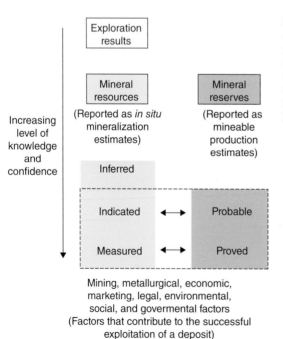

Figure 3 Simplified scheme illustrating the conceptual difference between mineral resources and ore reserves as applied to mineral occurrences. The scheme forms the basis for a more unified description of ore deposits as now required in terms of legislation that has been passed in most major mineral producing jurisdictions.

and modifying factors such as those related to mining techniques, metallurgical extraction, marketing, and environmental reclamation. Exploration results can be translated into a *mineral resource* once it is clear that an occurrence of intrinsic economic interest exists in such form and quantity that there are reasonable prospects for its eventual exploitation. Such a resource can only be referred to as an *ore reserve* if it is a part of an economically extractable measured or indicated mineral resource. One problem with this terminology is that an economically extractable ore deposit in a developing world artisanal operation may not be viable in a technically developed economy, and vice versa. The term "ore deposit" has no significance in the professional description of a mineral occurrence and is best used simply as a descriptive or generic term.

Some Useful Definitions and Compilations

General Definitions

This section is not intended to provide a comprehensive glossary of terms used in this book. There are, however, several terms that are used throughout the text where a definition is either useful or necessary in order to avoid ambiguity. The following definitions are consistent with those provided in the *Glossary of Geology* (Bates and Jackson 1987) and *The Encyclopedia of the Solid Earth Sciences* (Kearey 1993).

- *Ore*: any naturally occurring material from which a mineral or aggregate of value can be extracted at a profit. In this book the concept extends to coal (a combustible rock comprising more than 50% by weight carbonaceous material) and petroleum (naturally occurring hydrocarbon in gaseous, liquid, or solid state).
- *Syngenetic*: refers to ore deposits that form at the same time as their host rocks. In this book this includes deposits that form during the early stages of sediment diagenesis.

- *Epigenetic*: refers to ore deposits that form after their host rocks.
- *Hypogene*: refers to mineralization caused by ascending hydrothermal solutions.
- *Supergene*: refers to mineralization caused by descending solutions. The term generally refers to the enrichment processes accompanying the weathering and oxidation of sulfide and oxide ores at or near the surface.
- *Metallogeny*: the study of the genesis of mineral deposits, with emphasis on their relationships in space and time to geological features of the Earth's crust.
- *Metallotect*: any geological, tectonic, lithological, or geochemical feature that has played a role in the concentration of one or more elements in the Earth's crust.
- *Metallogenic Epoch*: a unit of geologic time favorable for the deposition of ores or characterized by a particular assemblage of deposit types.
- *Metallogenic Province*: a region characterized by a particular assemblage of mineral deposit types.
- *Epithermal*: hydrothermal ore deposits formed at shallow depths (less than 1500 m) and fairly low temperatures (50–200 °C).
- *Mesothermal*: hydrothermal ore deposits formed at intermediate depths (1500–5000 m) and temperatures (200–400 °C).
- *Hypothermal*: hydrothermal ore deposits formed at substantial depths (greater than 5000 m) and elevated temperatures (400–600 °C).

Periodic Table of the Elements

The question of the number of elements present on Earth is a difficult one to answer. There are 94 primordial nuclides present on Earth, these being the elements that formed during nucleosynthesis of the material that makes up the solar system. Most of the element compilations relevant to the earth sciences show 92 elements, the majority of which occur in readily detectable amounts in the Earth's crust. Figure 4 shows a periodic table in which these elements are presented in ascending atomic number and also categorized into groupings that are relevant to metallogenesis. There are in fact as many as 118 elements known to man, but those with atomic numbers greater than 92 (U: uranium) either occur in vanishingly small amounts as unstable isotopes that are the products of various natural radioactive decay reactions, or are synthetically created in nuclear reactors. The heaviest known element, originally called ununoctium (Uuo, atomic number 118), was only fleetingly detected in a nuclear reactor – its existence has now been confirmed and officially named "oganesson" (Symbol Og) after the Russian nuclear physicist, Yuri Oganessian. Some of the heavy, unstable elements are, however, manufactured synthetically and serve a variety of uses. Plutonium (Pu, atomic number 94), for example, is manufactured in fast breeder reactors and is used as a nuclear fuel and in weapons manufacture. Americium (Am, atomic number 95) is also extracted from spent reactor fuel and is widely used as the active agent in smoke detectors.

Of the 92 elements shown in Figure 4, almost all have some use in our modern, technologically-driven societies. Some of the elements (iron and aluminum) are required in copious quantities as raw materials for the manufacture of vehicles and in construction, whereas others (the rare earth elements, for example) are needed in very much smaller amounts for use in the alloys and electronics industries. Only two elements appear at the present time to have little or no commercial use at all (Figure 4). These are francium (Fr, atomic number 87), and protactinium (Pa, atomic number 91). Francium is radioactive and so short-lived that only some 20–30 g exists in the entire Earth's crust at any one time! Astatine (At, atomic number 85) is another very unstable element that exists in vanishingly small amounts in the crust as a decay chain by-product or is manufactured synthetically. Astatine has been manufactured in particle accelerators and is occasionally used in various nuclear medical applications.

The useful elements can be subdivided in a number of different ways. Most of the

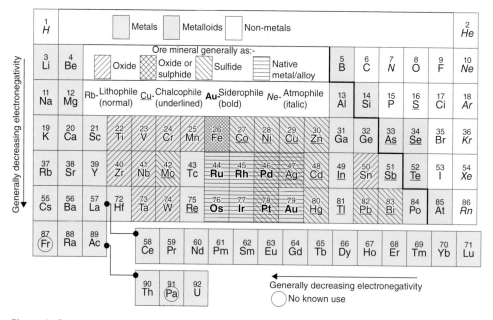

Figure 4 Periodic table showing the elements with atomic numbers from 1 to 92; classified on the basis of their rock and mineral associations.

elements can be classified as metals (Figure 4), with a smaller fraction being non-metals. The elements B, Si, As, Se, Te, and At have intermediate properties and are referred to as metalloids. Another classification of elements, attributed to the pioneering geochemist Goldschmidt, is based on their rock associations and forms the basis for distinguishing between lithophile (associated with silicates and concentrated in the crust), chalcophile (associated with sulfides), siderophile (occur as the native metal and concentrated in the core), and atmophile (occur as gases in the atmosphere) elements. It is also useful to consider elements in terms of their ore mineral associations, with some preferentially occurring as sulfides and others as oxides (see Figure 4). Some elements have properties that enable them to be classified in more than one way – iron is a good example, in that it occurs readily as both an oxide and sulfide.

Common Ore and Gangue Minerals

It is estimated that there are about 3800 known minerals that have been identified and classified (Battey and Pring 1997). Only a very small proportion of these make up the bulk of the rocks of the Earth's crust, as the common rock forming minerals. Likewise, a relatively small number of minerals make up most of the economically viable ore deposits of the world. The following compilation is a breakdown of the more common ore minerals in terms of chemical classes based essentially on the anionic part of the mineral formula. Also included are some of the more common "gangue" minerals, which are those that form part of the ore body, but do not contribute to the economically extractable part of the deposit. Most of these are alteration assemblages formed during hydrothermal processes. The compilation, including ideal chemical formulae, is subdivided into six sections, namely native elements, halides, sulfides and sulfosalts, oxides and hydroxides, oxysalts (such as carbonates, phosphates, tungstates, sulfates),

and silicates. More detailed descriptions of both ore and gangue minerals can be found in a variety of mineralogical texts, such as Deer et al. (1982), Berry et al. (1983), Battey and Pring (1997), and Wenk and Bulakh (2017). More information on ore mineral textures and occurrences can be found in Craig and Vaughan (1994) and Ixer (1990).

Native Elements

Both metals and non-metals exist in nature in the native form, where essentially only one element exists in the structure. Metals occurring in the native form include copper, silver, gold, and platinum which are all characterized by cubic close packing of atoms, high densities, and are malleable and soft. The carbon atoms in diamond are linked in tetrahedral groups forming well cleaved, very hard, translucent crystals. Sulfur also occurs as rings of eight atoms and forms bipyramids or is amorphous.

Metals

Gold – Au
Silver – Ag
Platinum – Pt
Palladium – Pd
Copper – Cu

Non-metals

Sulfur – S
Diamond – C
Graphite – C

Halides

The halide mineral group comprises compounds made up by ionic bonding. Minerals such as halite and sylvite are cubic, have simple chemical formulae, and are highly soluble in water. Halides sometimes form as ore minerals, such as chlorargyrite and atacamite.

Halite – NaCl
Sylvite – KCl
Chlorargyrite – AgCl
Fluorite – CaF_2
Atacamite – $Cu_2Cl(OH)_3$

Sulfides and Sulfosalts

This is a large group of minerals in which bonding is both ionic and covalent in character. The sulfide group has the general formula $A_M X_P$, where X is typically S but can be As, Sb, Te, Bi, or Se, and A is one or more of the metals. The sulfosalts, which are less common than sulfides, have the general formula $A_M B_N X_P$, where A is usually Ag, Cu, or Pb, B is commonly As, Sb, or Bi, and X is S. The sulfide and sulfosalt minerals are generally opaque, dense, and have a metallic to sub-metallic luster.

Sulfides

Chalcocite – Cu_2S
Bornite – Cu_5FeS_4
Galena – PbS
Sphalerite – ZnS
Chalcopyrite – $CuFeS_2$
Pyrrhotite – $Fe_{1-x}S$
Pentlandite – $(Fe,Ni)_9S_8$
Millerite – NiS
Covellite – CuS
Cinnabar – HgS
Skutterudite – $(Co,Ni)As_3$
Sperrylite – $PtAs_2$
Braggite/cooperite – $(Pt,Pd,Ni)S$
Moncheite – $(Pt,Pd)(Te,Bi)_2$
Laurite – RuS_2
Cobaltite – $CoAsS$
Gersdorffite – $NiAsS$
Loellingite – $FeAs_2$
Arsenopyrite – $FeAsS$
Molybdenite – MoS_2
Realgar – AsS
Orpiment – As_2S_3
Stibnite – Sb_2S_3
Bismuthinite – Bi_2S_3
Argentite – Ag_2S
Calaverite – $AuTe_2$
Pyrite – FeS_2

Sulfosalts

Tetrahedrite – $(Cu,Ag)_{12}Sb_4S_{13}$
Tennantite – $(Cu,Ag)_{12}As_4S_{13}$
Enargite – Cu_3AsS_4

Oxides and Hydroxides

This group of minerals is variable in its properties but is characterized by one or more metals in combination with oxygen or a hydroxyl group. The oxides and hydroxides typically exhibit ionic bonding. The oxide minerals can be hard, dense, and refractory in nature (magnetite, cassiterite) but can also be softer and less dense, forming as products of hydrothermal alteration and weathering (hematite, anatase, pyrolusite). Hydroxides, such as goethite and gibbsite, are typically the products of extreme weathering and alteration.

Oxides

Cuprite – Cu_2O
Hematite – Fe_2O_3
Ilmenite – $FeTiO_3$
Hercynite – $FeAl_2O_4$
Gahnite – $ZnAl_2O_4$
Magnetite – Fe_3O_4
Chromite – $FeCr_2O_4$
Rutile – TiO_2
Anatase – TiO_2
Pyrolusite – MnO_2
Cassiterite – SnO_2
Uraninite – UO_2
Thorianite – ThO_2
Columbite-tantalite – $(Fe,Mn)(Nb,Ta)_2O_6$

Hydroxides (or Oxyhydroxides)

Goethite – $FeO(OH)$
Gibbsite – $Al(OH)_3$
Boehmite – $AlO(OH)$
Manganite – $MnO(OH)$

Oxysalts

The carbonate group of minerals form when anionic carbonate groups (CO_3^{2-}) are linked by intermediate cations such as Ca, Mg, and Fe. Hydroxyl bearing and hydrated carbonates can also form, usually as a result of weathering and alteration. The other oxysalts, such as the tungstates, sulfates, phosphates, and vanadates, are analogous to the carbonates, but are built around an anionic group in the form XO_4^{n-}.

Carbonates

Calcite – $CaCO_3$
Dolomite – $CaMg(CO_3)_2$
Ankerite – $CaFe(CO_3)_2$
Siderite – $FeCO_3$
Rhodochrosite – $MnCO_3$
Smithsonite – $ZnCO_3$
Cerussite – $PbCO_3$
Azurite – $Cu_3(OH)_2(CO_3)_2$
Malachite – $Cu_2(OH)_2CO_3$

Tungstates

Scheelite – $CaWO_4$
Wolframite – $(Fe,Mn)WO_4$

Sulfates

Baryte(s) – $BaSO_4$
Anhydrite – $CaSO_4$
Alunite – $KAl_3(OH)_6(SO_4)_2$
Gypsum – $CaSO_4 \cdot 2H_2O$
Epsomite – $MgSO_4 \cdot 7H_2O$

Phosphates

Xenotime – YPO_4
Monazite – $(Ce,La,Th)PO_4$
Apatite – $Ca_5(PO_4)_3(F,Cl,OH)$

Vanadates

Carnotite – $K_2(UO_2)(VO_4)_2 \cdot 3H_2O$

Silicates

The bulk of the Earth's crust and mantle is made up of silicate minerals that can be subdivided into several mineral series based on the structure and coordination of the tetrahedral SiO_4^{4-} anionic group. Silicate minerals are generally hard, refractory, and translucent. Most of them cannot be regarded as ore minerals in that they do not represent the extractable part of an ore body, and the list provided below shows only some of the silicates more commonly associated with mineral occurrences as gangue or alteration products. Some silicate minerals, such as zircon and spodumene, are ore minerals and represent important sources of metals such as zirconium and lithium, respectively. Others, such as kaolinite, are mined for their intrinsic properties (i.e. as a clay for the ceramics industry).

Tekto (framework)

Quartz – SiO_2
Orthoclase – $(K,Na)AlSi_3O_8$
Albite – $(Na,Ca)AlSi_3O_8$
Scapolite – $(Na,Ca)_4(Al,Si)_4O_8)_3 (Cl, CO_3)$
Zeolite (analcime) – $NaAlSi_2O_6 \cdot H_2O$

Neso (ortho)

Zircon – $Zr(SiO_4)$
Garnet (almandine) – $Fe_3Al_2(SiO_4)_3$
Garnet (grossular) – $Ca_3Al_2(SiO_4)_3$
Sillimanite – Al_2SiO_5
Topaz – $Al_2SiO_4(F,OH)_2$
Chloritoid – $(Fe,Mg,Mn)_2(Al,Fe)Al_3O_2(SiO_4)_2$
  $(OH)_4$

Cyclo (ring)

Beryl – $Be_3Al_2Si_6O_{18}$
Tourmaline – $(Na,Ca)(Mg,Fe,Mn,Al)_3(Al,Mg,$
  $Fe)_6Si_6O_{18}(BO_3)_3(OH,F)_4$

Soro (di)

Lawsonite – $CaAl_2Si_2O_7(OH)_2 \cdot H_2O$
Epidote – $Ca_2(Al,Fe)_3Si_3O_{12}(OH)$

Phyllo (sheet)

Kaolinite – $Al_4Si_4O_{10}(OH)_8$
Montmorillonite – $(Na,Ca)_{0.3}(Al,Mg)_2Si_4O_{10}$
  $(OH)_2 \cdot nH_2O$
Illite – $KAl_2(Si,Al)_4O_{10}(H_2O)(OH)_2$
Pyrophyllite – $Al_2Si_4O_{10}(OH)_2$
Talc – $Mg_3Si_4O_{10}(OH)_2$
Muscovite – $KAl_2(AlSi_3O_{10})(OH)_2$
Biotite – $K(Fe,Mg)_3(Al,Fe)Si_3O_{10}(OH,F)_2$
Lepidolite – $K(Li,Al)_3(Si,Al)_4O_{10}(OH,F)_2$
Chlorite – $(Fe,Mg,Al)_{5-6}(Si,Al)_4O_{10}(OH)_8$

Ino (chain)

Tremolite-actinolite – $Ca_2(Fe,Mg)_5Si_8O_{22}(OH)_2$

Spodumene – $LiAlSi_2O_6$

Wollastonite – $CaSiO_3$

Unknown Structure

Chrysocolla – $(Cu,Al)_2H_2Si_2O_5(OH)_4 \cdot nH_2O$

Geological Time Scale

The development of a geological time scale has been the subject of a considerable amount of thought and research over the past few decades and continues to occupy the minds and activities of stratigraphers and geochronologists around the world. The definition of a framework within which to describe the secular evolution of rocks, and hence the Earth, has been, and continues to be, a contentious exercise. The International Commission on Stratigraphy (ICS is a working group of the International Union of Geological Sciences: IUGS) has assumed the official role of developing the geological time scale, a task that is continuously being modified and improved upon. The work of the ICS is periodically published as a book, such as Harland's (1989) seminal *A Geologic Time Scale* – which has now been superseded a number of times by works such as Gradstein et al. (2004 and 2012) and Ogg et al. (2016). In these books reference is made to the timing of various events and processes and the provision of a time scale to which readers can refer. Figure 5 is a time scale based on the 2018 version of the International Stratigraphic Chart, published and sanctioned by the ICS and IUGS (http://www.stratigraphy.org). In this diagram global chronostratigraphic terms are presented in terms of eons, eras, periods, and epochs, and defined by absolute ages in millions of years before present (Ma). Also shown are the approximate positions on the time scale of many of the ore deposits and metallogenic provinces referred to in the text.

Natural Resources, Sustainability, and the Environment

One of the major issues that characterized social and economic development toward the end of the twentieth century revolved around the widespread acceptance that the Earth's natural resources are finite, and that their exploitation should be carried out in a manner that will not detrimentally affect future generations. The concept of "sustainable development" in terms of the exploitation of mineral occurrences implies that current social and economic practice should endeavor not to deplete natural resources to the point where the needs of the future cannot be met. This would seem to be an impossible goal given the unprecedented population growth over the past century and the fact that many commodities may become depleted within the next 100 years. The challenge for commodity supply over the next century is a multifaceted one and will require a better understanding of the Earth system, improved incentives to promote more efficient recycling of existing resources, and the means to find alternative sources for commodities that are in danger of depletion.

There has been a dramatic rise in global population over the past 150 years. The number of humans on Earth has risen from one billion in 1830 to over seven billion at the start of the twenty-first century. Most predictions suggest that the populations of most countries will start to level off over the next 30 years and that global numbers will stabilize at around eleven billion people by the end of the twenty-first century. Societies in the next 100 years are, nevertheless, facing a scenario in which the demand for, and utilization of, natural resources continues to increase, and certain commodities might well become depleted in this interval. Production trends for commodities such as oil, bauxite, copper, and gold (Figure 6) confirm that demand

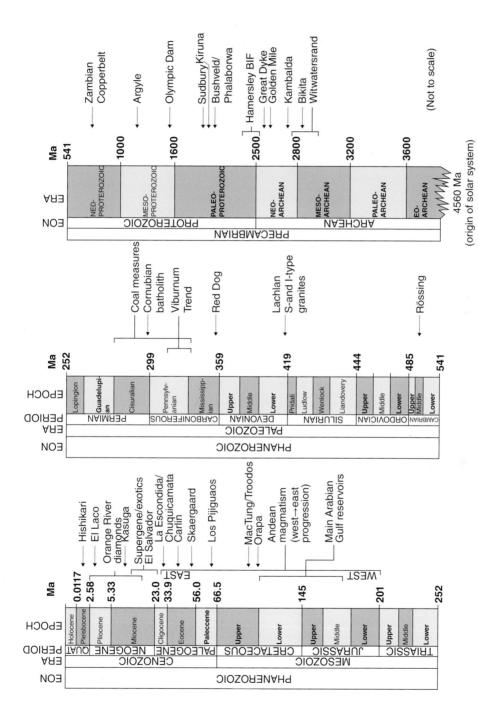

Figure 5 Geological time scale after the International Commission on Stratigraphy (http://www.stratigraphy.org/index.php/ics-chart-timescale). Also shown are the ages of the various deposits and metallogenic provinces mentioned in the book.

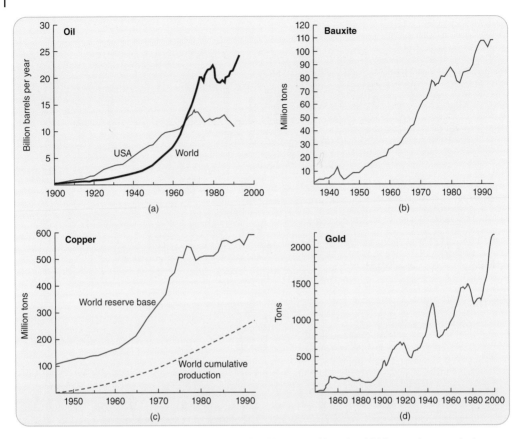

Figure 6 Global production trends for oil (a), bauxite (b), copper (c), and gold (d) over the twentieth century. Source: After compilations in Craig et al. (1996).

for resources reflects population growth and is likely to continue to do so over the next few decades. World oil production increased precipitously until the late 1970s, but since then a variety of political and economic factors have contributed to moderating demand (Figure 6a), thereby ensuring a longer-term reserve base. A similar leveling of production is evident for bauxite (Figure 6b) but such a trend is not yet evident for the precious metals such as gold or platinum. For some commodities, such as copper (Figure 6c), the world reserve base is also leveling off, a feature that in part also reflects fewer new and large discoveries. Critical shortages of most natural commodities are not likely to present a problem during the

early part of the twenty-first century (Einaudi 2000), but this situation will deteriorate unless strategies for sustainability are put into place immediately. Another area of concern is related to those strategically important metals for which the security of supply has become an issue. Many metals, including the rare earth elements, tungsten, the platinum group elements, tantalum and niobium, have been cited as "critical metals" because their supply has been affected, not by naturally diminishing resource bases, but by socio-economic and political factors.

The depletion of commodities in the Earth's crust is particularly serious for those metals that are already scarce in terms of crustal abundances and for which high degrees of enrichment are required in order to make viable ore deposits. Figure 2 illustrates the point by referring to the production of iron as a baseline measure against which extraction of other metals can be compared (Skinner 1976; Einaudi 2000). Those elements which fall above the Fe production line (notably Au, Ag, Bi, Sb, Sn, Cu, Pb, and Zn) are being extracted or depleted at faster rates, relative to their crustal abundances, than Fe. It is these metals that are in most danger of depletion in the next 50 years or so unless production is ameliorated or the reserve base is replaced. Conversely, those metals that plot beneath the Fe production line (such as Ti, Mg, and Al) are being extracted at slower rates than Fe and are in less danger of serious depletion during this century.

One of the ways in which metallogeny can assist in the creation of a sustainable pattern of resource utilization is to better understand the processes by which ores are concentrated in the Earth's crust. The replacement of the global commodity reserve base is dependent on exploration success and the ability to find new ore deposits that can replace those that are being depleted. It is, of course, increasingly difficult to find new and large deposits of conventional ores, since most of the accessible parts of the globe have been extensively surveyed and assessed for their mineral potential. The search for deeper deposits is an option but this is dependent to a large extent on the availability of technologies that will enable mining to take place safely and profitably at depths in excess of 4000 m (currently the deepest level of mining in South African gold mines). Another option is to extract material from inaccessible parts of the globe, such as the ocean floor, a proposal that has received serious consideration with respect to metals such as Mn and Cu. Again, there are technological barriers to such processes at present, but these can be overcome, as demonstrated by the now widespread exploration for, and extraction of, oil and gas from the sea floor. Environmental

barriers to sea floor exploitation are more serious and difficult to overcome, as evident from catastrophic oil spillages in many parts of the globe. A third option to improve the sustainability of resource exploitation is to extract useful commodities from rocks that traditionally have not been thought of as viable ores. Such a development can only be achieved if the so-called "mineralogical barrier" (Skinner 1976) is overcome. This concept can be described in terms of the amount of energy (or cost) required to extract a commodity from its ore. It is, for example, considerably cheaper to extract Fe from a banded iron-formation than it is from olivine or orthopyroxene in an igneous rock, even though both rock types might contain significant amounts of the metal. The economics of mining and the widespread availability of banded iron-formations dictate that extraction of Fe from silicate minerals is essentially not feasible. The same is not true of nickel. Although it is cheaper and easier to extract Ni from sulfide ore minerals (such as pentlandite) there is now widespread extraction of the metal from nickeliferous silicate minerals (garnierite) that form during the lateritic weathering of ultramafic rocks. Even though Ni is more difficult and expensive to extract from laterite than from sulfide ores, the high tonnages and grades, as well as the widespread development and ease of access of the former, mean that they represent viable mining propositions despite the extractive difficulties. Ultimately, it may also become desirable to consider mining iron laterites, but this would only happen if conventional banded iron-formation hosted deposits were depleted, or if the economics of the whole operation favored laterites over iron-formations. This is not likely to happen in the short term, but, if planned for, the scenario does offer hope for sustainability in the long term. In short, sustainable production of mineral resources requires a thorough understanding of ore-forming processes and the means to apply these to the discovery of new mineral occurrences. It also requires the timely development of technologies, both in the earth sciences and in related fields of mining and extractive metallurgy, that will enable alternative supplies of mineral resources to be economically exploited in the future.

Mining and Environmental Responsibility

A global population of possibly eleven billion people by the end of the century presents a major challenge in terms of the supply of most of the world's natural resources. What is even more serious, though, is the enormous strain it will place on the Earth's fragile environment arising from the justifiable expectation that future societies will provide an adequate standard of living, in terms of food, water, housing, technology, recreation, and material benefits, to all their people. In addition to commodity supply problems, the twenty-first century will also be characterized by unprecedented depletion of even more critical resources in the form of soil, water, and clean air (Fyfe 2000). Legislation that is aimed at dealing with issues such as atmospheric pollution and greenhouse gas emissions, erosion, factory waste and acid drainage, de-forestation, the protection of endangered species, overgrazing, and crop fertilization, is highly desirable but far from globally achievable because it is perceived as a luxury that only the developed world can afford.

The study of ore-forming processes is occasionally viewed as an undesirable topic that ultimately contributes to the exploitation of the world's precious natural resources. Nothing could be further from the truth. An understanding of the processes by which metals are concentrated in the Earth's crust is essential knowledge for anyone concerned with the preservation and remediation of the environment. The principles that underpin the natural concentration of ores in the crust are the same as those that can be utilized to tackle issues such as the control of acid mine drainage, and soil and erosion management. Mining operations around the world are required to assume responsibility for reclamation of the landscape once the resource has been depleted. The industry now encompasses a range of activities extending from geological exploration and evaluation, through mining and beneficiation, and eventually to remediation and environmental reclamation. This is the mining cycle and its effective management in the future will be a multidisciplinary exercise carried out by highly skilled scientists and engineers. Earth systems science, and in particular the geological processes that gave rise to the formation of mineral deposits, will be central to the future custodianship of the Earth's natural resources.

Summary

The discipline of "economic geology" and in particular the field of metallogeny (the study of the genesis of ore deposits) remains critical to the teaching of earth systems science. A holistic approach involving the integration of knowledge relevant to the atmosphere, biosphere, and lithosphere is now regarded as essential to understanding the complexities of the Earth system. The development of environmentally responsible policies for the sustainable production of all natural resources will demand a thorough knowledge of the nature and workings of the Earth system. Central to this is an understanding of metallogeny and the nature and origin of the entire spectrum of mineral resources, including the fossil fuels. The classification and description of ore forming processes can most effectively be achieved in terms of host rock associations, namely igneous, hydrothermal, and sedimentary. This breakdown forms the basis for the layout of this book.

Further Reading

Blunden, J. (1983). *Mineral Resources and their Management*. Harlow: Longman, 302 pp.

Craig, J.R., Vaughan, D.J., and Skinner, B.J. (1996). *Resources of the Earth – Origin, Use and Environmental Impact*. Englewood Cliffs, NJ: Prentice Hall, 472 pp.

Ernst, W.G. (2000). *Earth Systems – Processes and Issues*. Cambridge: Cambridge University Press, 559 pp.

Kesler, S.E. and Simon, A.C. (2015). *Mineral Resources, Economics and the Environment*. Cambridge University Press, 434 pp.

Part I

Igneous Processes

1

Igneous Ore-Forming Processes

TOPICS

Metallogeny of oceanic and continental crust
Fundamental magma types and their metal endowment
The relative fertility of magmas and the "inheritance factor"
  "late-veneer" hypothesis
  diamonds and kimberlite/lamproite
  metal concentrations in metasomatized mantle
  S- and I-type granites
Partial melting and crystal fractionation as ore-forming processes
Trace element distribution during partial melting
Trace element distribution during fractional crystallization
Monomineralic chromitite layers
Liquid immiscibility as an ore-forming process
Special emphasis on mineralization processes in layered mafic intrusions
  sulfide solubility
  sulfide–silicate partition coefficients
  the R factor
  PGE clusters and hiatus models

CASE STUDIES

Box 1.1 Diamondiferous Kimberlites and Lamproites: The Orapa Diamond Mine, Botswana and the Argyle Diamond Mine, Western Australia
Box 1.2 Carbonatites and Alkaline Intrusions as Sources of Critical Metals
Box 1.3 Partial Melting and Concentration of Incompatible Elements: The Rössing Uranium Deposit
Box 1.4 Boundary Layer Differentiation in Granites and Incompatible Element Concentration: The Zaaiplaats Tin Deposit, Bushveld Complex
Box 1.5 Crystal Fractionation and Formation of Monomineralic Chromitite Layers: The UG1 Chromitite Seam, Bushveld Complex
Box 1.6 Addition of External Sulfur and Sulfide Immiscibility: The Komatiite-Hosted Ni–Cu Deposits at Kambalda, Western Australia
Box 1.7 New Magma Injection and Magma Mixing: The Merensky Reef, Bushveld Complex
Box 1.8 Magma Contamination and Sulfide Immiscibility: The Sudbury Ni–Cu Deposits

---

Introduction to Ore-Forming Processes, Second Edition. Laurence Robb.
© 2021 John Wiley & Sons Ltd. Published 2021 by John Wiley & Sons Ltd.

1.1 Introduction

Igneous rocks host a large number of different ore deposit types. Both mafic and felsic rocks are linked to mineral deposits, examples of which range from the chromite ores resulting from crystal fractionation of mafic magmas to tin deposits associated with certain types of granites. The processes described in this chapter relate to properties that are intrinsic to the magma itself and can be linked genetically to its cooling and solidification. Discussion of related processes, whereby an aqueous fluid phase separates or "exsolves" from the magma as it crystallizes, is placed in Chapter 2. The topics discussed under the banners of igneous and magmatic–hydrothermal ore-forming processes are intimately linked and form Part I of this book.

A measure of the economic importance of ore deposits hosted in igneous rocks can be obtained from a compilation of mineral production data as a function of host rock type. A country like South Africa, for example, is underlain dominantly by sedimentary rocks and these undoubtedly host many of the valuable mineral resources (especially if the fossil fuels are taken into consideration). Nevertheless, the value of ores hosted in igneous rocks *per unit area of outcrop* can be comparable with that for sedimentary rocks, as indicated in Table 1.1. Although South Africa is characterized by a rather special endowment of mineral wealth related to the huge Bushveld

Complex, the importance of igneous-hosted ore deposits is nevertheless apparent.

1.2 Magmas and Metallogeny

It is well known that magmatic ore deposits vary as a function of the composition of the igneous host rock, and that the variable metal endowments in magmas are inherited from the source rocks that were melted to form them. It is widely recognized that many of the chalcophile and siderophile elements (such as Ni, Co, Pt, Pd, and Au), for example, are more likely to be associated with mafic rock types, whereas concentrations of lithophile elements (such as Li, Sn, Zr, U, and W) are typically associated with felsic or alkaline rock types. This has implications for understanding ore genesis and, consequently, some of the factors related to these differences are discussed below.

1.2.1 Crustal Architecture and Mineral Wealth

Although the highest concentrations of siderophile and chalcophile elements almost certainly reside in the mantle and core of the Earth, these are generally inaccessible due to their very great depths. In fact, most of the world's economically exploitable mineral wealth effectively lies on the surface or just below the surface of the Earth. The world's deepest mines, in the Witwatersrand Basin

Table 1.1 A comparison of the value of mineral production from igneous and sedimentary rocks in South Africa.

Mineralization hosted in	Area (km^2)	Value of sales, 1971 (10^6 US$)	% of total area	% of total value	Unit value (US$ km^{-2})
Granites	163 100	1 973	13.3	3.4	12 000
Mafic layered complexes	36 400	7 288	3.0	12.5	200 200
Total (igneous)	199 500	9 261	16.3	15.9	46 400
Sedimentary rocks	1 023 900	49 137	83.7	84.1	47 900

Source: After Pretorius (1976).

of South Africa, extend to over 4000 m deep and this places an effective limit on ore body exploitation, at least in terms of safety and economic viability. Nevertheless, many mineral commodities are formed much deeper in the crust than 4 km, with some even being derived from the mantle. Diamonds, for example, are hosted in kimberlite magmas that have been brought to exploitable depths by a variety of igneous or tectonic mechanisms. Understanding ore genesis processes, therefore, requires a knowledge of lithospheric (i.e. crust and upper mantle) architecture, and also of the origin and nature of the igneous rocks in this section of the Earth.

The oceanic crust, which covers some two-thirds of the Earth's surface, is thin (less than 10 km) and, compared to the continents, has a composition and structure that is relatively simple and consistent over its entire extent. The upper layer, on average only 0.4 km thick (Kearey and Vine 1996), comprises a combination of terrigenous and pelagic sediments that are distributed mainly by ocean floor turbidity currents and are often highly reduced and metal charged. This is underlain by a layer, typically 1–2.5 km thick, that is both extrusive and intrusive in character and dominantly basaltic in composition. The basalts are,

in turn, underlain by the main body of oceanic crust that is plutonic in character and formed by crystallization and fractionation of basaltic magma. This cumulate assemblage comprises mainly gabbro, pyroxenite, and peridotite. Sections of tectonized and metamorphosed oceanic lithosphere can be observed in ophiolite complexes which represent segments of the ocean crust (usually back-arc basins) that have been thrust or obducted onto continental margins during continent–ocean collision.

The types of ore deposits that one might expect to find associated with oceanic crust and on the sea floor are shown in Figure 1.1. They include the category of podiform chromite deposits that are related to crystal fractionation of mid-ocean ridge basalt (MORB), and also have potential for Ni and Pt group element (PGE) mineralization. Accumulations of manganese in nodules on the sea floor, metal-rich concentrations in pelagic muds, and exhalative volcanogenic massive sulfide (VMS) Cu–Zn deposits also occur in this tectonic setting, but are not directly related to igneous processes and are discussed elsewhere (Chapters 3 and 5).

The continental crust differs markedly from its oceanic counterpart. It is typically 35–40 km thick, but thins to around 20 km

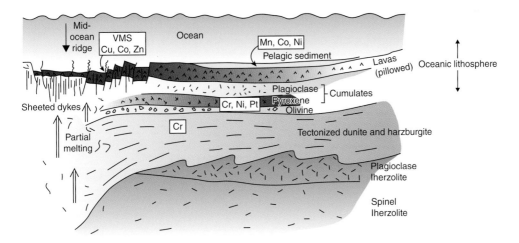

Figure 1.1 Oceanic crustal architecture showing the main types of ore deposits characteristic of this environment. Only chromite and related deposits (Cr–Ni–Pt) are related to igneous ore-forming processes; VMS (Cu, Co, Zn) and sediment-hosted deposits (Mn, Co, Ni) are discussed in Chapters 3 and 5 respectively.

under rift zones and thickens to 80 km or more beneath young mountain belts. Historically, the continental crust was thought to comprise an upper zone made up largely of granite (and its sedimentary derivatives) and a lower, more mafic zone, with the two layers separated by the Conrad discontinuity (which marks a change in seismic velocities, and, therefore crustal density). More recent geophysical and geological studies clearly indicate that crustal architecture is more complex and reflects a long-lived tectonic and magmatic history, extending back in some cases over 3800 million years (Myr) (Figure 1.2). The continental crust is currently thought of as comprising three layers, an upper, middle and lower crust (Rudnick and Gao 2014). The upper crust has a granodioritic bulk composition and includes sedimentary material derived from igneous and metamorphic precursors. In typical crustal sections the middle crust comprises rocks metamorphosed to amphibolite and lower granulite facies whereas the lower crust comprises granulite facies rocks. Lithologically the deeper crust is variable, although the middle crust typically comprises relatively evolved compositions in the granodiorite to tonalite range, whereas the lower crust is denser, has a mafic composition, and is depleted in heat-producing elements (K, U, Th).

The continents have been progressively constructed throughout geological time by a variety of magmatic, sedimentary, and orogenic processes taking place along active plate margins and, to a lesser extent, within the continents themselves. In addition, continental land masses have repeatedly broken apart and amalgamated throughout geological history. These processes, known as *Wilson cycles* (where continents rift apart – with the development of intervening oceanic crust – and then reassemble) have rearranged the configuration of continental fragments several times in the geological past. Occasionally a majority of continental fragments coalesce to form a supercontinent, and then disperse with a periodicity and range that is global in extent. In the Paleoproterozoic Era, for example, it is conceivable that segments of southern Africa and western Australia might have formed part of the same continent which might explain why the geology of these two regions is so similar. The pattern of crustal evolution with time, and the significance of both Wilson and supercontinent cycles, to global metallogeny is discussed in more detail in Chapter 6.

The upper crust, which in some continental sections is defined as extending to the Conrad discontinuity at some 6 km depth, comprises rocks of felsic to intermediate compositions (granite to diorite) together with the sedimentary detritus derived from the weathering and erosion of this material. Archean continental fragments (greater than 2500 Myr old) also

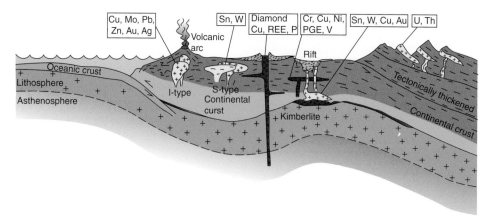

Figure 1.2 Continental crustal architecture showing the main types of igneous-related ore deposits characteristic of this environment.

contain a significant component of greenstone belt material, representing preserved fragments of ancient oceanic crust. The lower crust, between the Conrad and Mohorovicic discontinuities, is variable in composition but is typically made up of hotter, and usually more dense, material. This is because temperatures and pressures in the crust increase with depth at rates that are typically around $25\,°C\,km^{-1}$ and $30\,MPa\,km^{-1}$ respectively (Kearey and Vine 1996). The lower crust is not necessarily compositionally different from the upper crust, but exists at higher metamorphic grades. It is also likely to be more anhydrous and residual, in the sense that magma now present at higher levels may have been extracted from the lower crust, leaving a residue of modified material. Some of the lower crust may be more mafic in composition, comprising material such as amphibolite, gabbro, and anorthosite.

Most of the world's known ore deposits are, of course, hosted in rocks of the continental crust, some of which are shown in Figure 1.2. Some of the more important igneous rock-related deposit types shown include diamondiferous kimberlites, anorthosite-hosted Ti deposits, the Cr–V–Pt–Cu–Ni assemblage of ores in continental layered mafic suites, and the Sn–W–F–Li–Nb–REE–P–U family of lithophile ores related to granites and alkaline intrusions.

1.2.2 Magma Types and Metal Contents

Although their rheological properties are different, the outer two layers of the Earth comprising the more rigid lithosphere and the ductile asthenosphere, are largely solid. Zones within these layers that are anomalous in terms of pressure or temperature do, however, form and can cause localized melting of the rocks present. The nature of the rock undergoing melting and the degree to which it is melted are the main factors that control the composition of the magma that is formed. The magma composition, in turn, dictates the nature of metal concentrations that are likely to form in the rocks that solidify from that magma.

Although it is theoretically possible to form an almost infinite range of magma compositions (from ultramafic to highly alkaline), for ease of discussion this section is subdivided into four parts, each representing what is considered to be a fundamental magma type – these are basalt, andesite, rhyolite, and alkaline magmas, the latter including kimberlite.

1.2.2.1 Basalt

Basalts form in almost every tectonic environment, but the majority of basaltic magma production takes place along the mid-ocean ridges, and in response to hot-spot related plumes, to form oceanic crust. In addition, basalts are formed together with a variety of more felsic magmas, along island arcs and orogenically active continental margins. Basaltic magma may also intrude or extrude continental crust, either along well defined fractures or rifts (such as continental flood basalt provinces, or the Great Dyke of Zimbabwe) or in response to intra-plate hot-spot activity (which might have been responsible for the formation of the Bushveld Complex of South Africa).

Basalt forms by partial melting of mantle material, much of which can generally be described as peridotitic in composition. Certain mantle rocks, such as lherzolite (a peridotite which contains clinopyroxene and either garnet or spinel), have been shown experimentally to produce basaltic liquids on melting, whereas others, like alpine-type peridotite (comprising mainly olivine and orthopyroxene), are too refractory to yield basaltic liquids and may indeed represent the residues left behind after basaltic magma has been extracted from the mantle. Likewise, oceanic crust made up of hydrated (serpentinized) basalt that is drawn down into a subduction zone, is also a potential source rock for island arc and continental margin type magmatism. Komatiites, which are

Table 1.2 Average abundances of selected elements in the major magma types.

	Basalt	Andesite	Rhyolite	Alkaline magma	Kimberlite	Clarke[a]
Li	10	12	50	–	–	20
Be	0.7	1.5	4.1	4–24	–	2.8
F	380	210	480	640	–	625
P	3200	2800	1200	1800	0.6–0.9%	1050
V	266	148	72	235	–	135
Cr	307	55	4	–	–	10
Co	48	24	4.4	–	–	25
Ni	134	18	6	–	1050	75
Cu	65	60	6	–	103	55
Zn	94	87	38	108	–	70
Zr	87	205	136	1800	2200	165
Mo	0.9–2.7	0.8–1.2	1	15	–	1.5
Sn	0.9	1.5	3.6	–	–	2
Nb	5	4–11	28	140	240	20
Sb	0.1–1.4	0.2	0.1–0.6	–	–	0.2
Ta	0.9	–	2.3	10	–	2
W	1.2	1.1	2.4	16	–	1.5
Pb	6.4	5	21	15	–	13
Bi	0.02	0.12	0.12	–	–	0.17
U	0.1–0.6	0.8	5	10	–	2.7
Th	0.2	1.9	26	35	–	7.2
Ag[b]	100	80	37	–	–	70
Au[b]	3.6	–	1.5	–	–	4
Pt[b]	17–30	–	3–12	–	19	10
S	782	423	284	598	2100	260
Ge	1.1	1.2	1.0–1.3	1.3–2.1	0.5	1.5
As	0.8	1.8	3.5	–	–	1.8
Cd	0.02	0.02	0.2–0.5	0.04	–	0.2

If no average is available, a range of values is provided.
a)   Clarke is a term that refers to the average crustal abundance.
b)   Values as ppb, all other values as ppm.
Source: Taylor (1964), Wedepohl (1969), Krauskopf and Bird (1995).

ultramafic basalt magmas (with >18% MgO) mainly restricted to Archean greenstone belts, have a controversial origin but are generally believed to represent high degrees of partial melting of mantle during the high heat-flow conditions that prevailed in the early stages of crust formation prior to 2500 Ma.

Ore deposits associated with mafic igneous rocks typically comprise a distinctive (mainly siderophile and chalcophile) metal assemblage of, among others, Ni, Co, Cr, V, Cu, Pt, and Au. Examination of Table 1.2 shows that this list corresponds to those elements that are intrinsically enriched in basaltic magmas. Figure 1.3 illustrates the relative abundances of these metals in three fundamental magma types and the significantly higher concentrations in basalt compared to andesite and rhyolite.

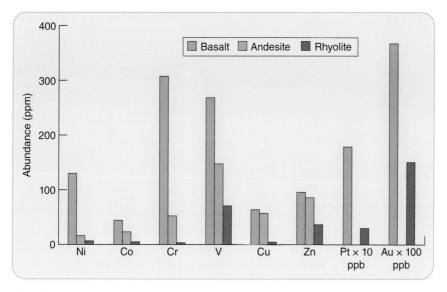

Figure 1.3 Relative abundances of selected metals in basalt, andesite, and rhyolite. Source: Data from Table 1.2.

The enhanced concentration of these metals in each case is related to the fact that the source materials from which the basalt formed must likewise have been enriched in siderophile and chalcophile metals. In addition, enhanced abundances also reflect the chemical affinity that these metals have for the major elements that characterize a basaltic magma (Mg and Fe) and dictate its mineral composition (olivine and the pyroxenes).

The chemical affinity that one element has for another is related to their atomic properties as reflected by their relative positions in the periodic table (see Figure 4, Introduction). The alkali earth elements (i.e. K, Na, Rb, Cs, etc.), for example, are all very similar to one another, but have properties that are quite different to the transition metals (such as Fe, Co, Ni, Pt, Pd). In addition, minor or trace elements, that occur in such low abundances in magmas that they cannot form a discrete mineral phase, are present in rocks by virtue of their ability either to substitute for another chemically similar element in a mineral lattice, or to occupy a defect site in a crystal lattice. This behavior is referred to as diadochy, or substitution, and explains much, but not all, of the trace element behavior in rocks. Substitution of a

trace element for a major element in a crystal takes place if their ionic radii and charges are similar. Typically, radii should be within 15% of one another and charges should differ by no more than one unit provided the charge difference can be compensated by another substitution. Bond strength and type also effects diadochy and it preferentially occurs in crystals where ionic bonding dominates.

Good examples of diadochic behavior are the substitution of Ni^{2+} for Mg^{2+} in olivine and V^{3+} for Fe^{3+} in magnetite. Analytical data for the Ni content of basalts shows an excellent correlation between Ni and MgO contents (Figure 1.4), confirming the notion that the minor metal substitutes readily for Mg. The higher intrinsic Ni content of ultramafic basalts and komatiites would suggest that the latter rocks are perhaps better suited to hosting viable magmatic nickel deposits, an observation borne out by the presence of world class nickel deposits hosted in the Archean komatiites of the Kambalda mining district in Western Australia (see Box 1.5) and elsewhere in the world.

1.2.2.2 Andesite

Andesites are rocks that crystallize from magmas of composition intermediate between

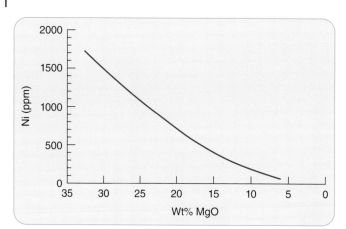

Figure 1.4 The relationship between Ni and MgO contents of basalts within which base metal mineralization does not occur. Source: Data from Naldrett (1989a).

basalt and rhyolite (typically with SiO_2 contents between 53 and 63 wt%). Their petrogenesis remains contentious, although it is well known that they tend to occur dominantly in orogenic zones, either along island arcs or on continental margins beneath which subduction of oceanic crust is taking place (Hall 1996). Discussion about the origin of andesite revolves around whether it represents a primary magma composition derived directly by an appropriate degree of melting of a suitable source rock, or an evolved melt formed by differentiation of a more mafic magma such as basalt.

Geological observations support the notion that andesite can be formed both as a primary magma composition and by in situ fractionation. The observation that andesitic volcanoes occur directly above aseismic sections of a Benioff zone (i.e. the subducted slab that produces earthquakes due to movement and fracturing of rock) suggests that melt production (and damping of seismic waves) has occurred in these areas. This would support the notion that andesitic magma is produced by direct melting of hydrous oceanic crust or, more likely, the mantle wedge overlying the subduction zone as it is permeated by fluids expelled from the subjacent oceanic crust. Alternatively, andesitic magma can be produced by fractionation of phases such as hornblende and magnetite from relatively water-rich parental magmas (Osborn 1979), or

by contamination of an originally more mafic melt by felsic material or melt.

Irrespective of the mode of formation of andesite it is apparent that as a magma type it does not exhibit a primary association with any particular suite of metals or ore deposits. It appears instead that ore deposits tend to be associated with magmas representing the ends of the compositional spectrum, and that intermediate melt compositions are simply characterized by intermediate trace element abundances. Examination of Table 1.2 shows that andesites appear to have little or no metal specificity and are characterized by trace element abundances that are intermediate between those of basalt on the one hand and either granite or alkaline rocks on the other.

1.2.2.3 Rhyolite

Felsic magmas can also form in a variety of geological environments. They crystallize at depth in the crust to form a spectrum of rock compositions ranging from Na-rich tonalite to K-rich alkali granite, or they extrude on surface to form dacitic to rhyolitic volcanic rocks. Very little granite magma forms in oceanic crust or along island arcs that have formed between two oceanic plates. Where oceanic granite does occur, it is typically the result of differentiation of a more mafic magma type originally formed by mantle melting. Along the mid-Atlantic ridge in Iceland, for example, eruptions of the volcano Hekla are initiated by a pulse of felsic ash production which is rapidly followed by eruption of more typical basaltic andesite. This suggests that the intervening period between eruptions was characterized by differentiation of the magma and that the accompanying build-up of volatiles may have been responsible for the subsequent eruption (Baldridge et al. 1973). These observations, among many others, clearly indicate that granitic melts can be the products of differentiation of more mafic magmas in oceanic settings.

Most felsic magmas, however, are derived from the partial melting of predominantly crustal material along ocean–continent island arcs and orogenic continental margins. Although Andean-type subduction zones might facilitate partial melting of the downgoing slab itself, the much higher proportion of felsic magma formed in this environment compared to oceanic settings points to a significant role for continental crust as a source. There is now general agreement that Andean-type subduction-related magmatism receives melt contributions from both the mantle lithosphere and the continental crust, with the wide-ranging compositions of so-called "calc–alkaline" igneous suites being attributed to a combination of both magma mingling and fractional crystallization (Best 2003).

Significant quantities of felsic magma are produced in the latter stages of continent–continent collision and also in anorogenic continental settings where rifting and crustal thinning has taken place. Himalayan-type continent collision, for example, is usually accompanied by crustal thickening associated with intense thrusting, tectonic duplication and reverse metamorphic gradients. These processes cause dewatering of crustal material, which, in turn, promotes partial melting to form high-level leucogranite magmas derived from source rocks that often contain significant proportions of sedimentary material (Le Fort 1975). Anorogenic continental magmatism, on the other hand, is usually related to crustal thinning (accompanying plume or hot-spot activity?) and is typified by the production of

magmas with bimodal compositions (i.e. basalt plus rhyolite). A good example is the 2056 Myr old Bushveld Complex in South Africa, where early mafic magmas intruded to form the world's largest layered igneous complex, followed by emplacement of a voluminous suite of granites. Other examples include the alkaline ring complexes of Niger and Nigeria – these intrusions, which young in age progressively from Ordovician to late Jurassic in a southerly direction, suggesting a plume track origin, have produced significant Sn and Nb mineralization especially in the Jos Plateau region of central Nigeria (Kinnaird 1987).

Ore deposits associated with felsic igneous rocks often comprise concentrations of the lithophile elements such as Li, Be, F, Sn, W, U, and Th. Table 1.2 shows that this list corresponds to those elements that are intrinsically enriched in rhyolitic magmas and Figure 1.5 illustrates, in bar graph form, the relative abundances of these elements and, in particular, the higher abundances in rhyolite by comparison with andesite and basalt.

The relative enrichment of certain lithophile elements in rhyolitic magmas is partially related to their geochemical incompatibility. An incompatible element is one whose ionic charge and radius make it difficult to substitute for any of the stoichiometric elements in rock-forming minerals. Thus, incompatible elements tend to be excluded from the products of crystallization and concentrated into residual or differentiated magmas (such as the granitic magmas that might form by crystal fractionation of mafic magmas in oceanic settings). Alternatively, incompatible elements also tend to be concentrated in crustal melts derived from low degrees of partial melting of source rocks that may themselves have been endowed in the lithophile elements. These concepts are discussed in more detail in Section 1.4 below.

A well known and interesting feature of ore deposits that are genetically associated with granite intrusions is that the origin and composition of the magma generally controls

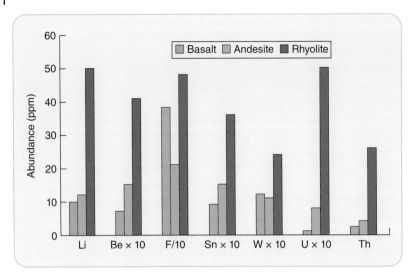

Figure 1.5 Relative abundances of selected "granitophile" elements in basalt, andesite, and rhyolite. Source: Data from Table 1.2.

the nature of the metal assemblage in the deposit (Chappell and White 1974; Ishihara 1978, 1981). This control is almost certainly related in part to the metal endowment inherited by the magma from the rocks that were melted to produce it. Where a felsic magma is derived from melting of a sedimentary or supracrustal protolith (termed *S-type granites*), associated ore deposits are characterized by concentrations of metals such as Sn, W, U, and Th. Where it is derived from melting of predominantly igneous protoliths in the crust (*I-type granite*) the ore association is typified by metals such as Cu, Mo, Pb, Zn, and Au. This association is metallogenically significant and is discussed in more detail in Section 1.3.4 below and again in Chapter 2.

1.2.2.4 Alkaline Magmas, Carbonatite and Kimberlite

Although most magma compositions can be represented by the basalt–andesite–rhyolite spectrum, some deviate from this trend and are compositionally unusual. For example, magmas that are depleted in SiO_2 but highly enriched in the alkali elements (Na, K, and Ca) are relatively rare, but may be economically important as they frequently contain impressive concentrations of a wide range of ore-forming metals (such as Cu, Fe, P, Zr, Nb, REE [rare earth elements], F, U, and Th). In addition, kimberlitic and related magma types

(such as lamproites) are the main primary source of diamonds (see Section 1.3.2 and Box 1.1).

The most common alkaline mafic magma is nephelinite, which crystallizes to give a range of rock types (the ijolite suite; Hall 1996) comprising rather unusual minerals, such as feldspathoid, calcic-pyroxene, and carbonate assemblages. Nephelinite lavas are observed in oceanic settings such as the Cape Verde and the Hawaiian Islands, but are best seen in young (Paleocene to recent) continental volcanic settings such as the East African rift valley, central Europe, and southeast Australia. Old alkaline igneous complexes are rare, one of the best preserved being the c. 2050 Myr old Phalaborwa Complex in South Africa, which is mined for copper and phosphate as well as a host of minor by-products.

Nephelinite, as well as the associated, but rare, carbonatite melts (i.e. magmas comprising essentially $CaCO_3$ and lesser Na_2CO_3; Section 1.3.4), are undoubtedly primary magma types derived from the mantle by very low degrees of partial melting under conditions of high P_{total} and P_{CO2} (Hall 1996). The relationship between nephelinitic and carbonatitic magmas is generally attributed to liquid immiscibility, whereby an original alkali-rich silicate magma rich in a carbonate component exsolves into two liquid fractions, one a silicate and the other a carbonate (Ferguson and Currie 1971; Le Bas 1987). Low degrees (2%) of partial melting of a garnet lherzolitic source in the mantle will typically yield olivine nephelinite compositions and these magmas may be spatially and temporally associated with basaltic volcanism (Le Bas 1987). Nephelinite magma associated with carbonatite, on the other hand, is only considered possible if the source material also contained a carbonate phase (such as dolomite) and a soda-amphibole. This type of mantle source rock is likely to be the result of extensive metasomatism, a process that involves fluid ingress and enrichment of volatile and other incompatible elements. Melting of a fertile mantle source rock is probably the main reason why alkaline magmas are so enriched in the variety of ore constituents mentioned above (see Box 1.2). The extent of metal enrichment relative to average basalt is illustrated in Figure 1.6.

Kimberlitic and related ultramafic magmas crystallize to form rare and unusual rocks, containing, among other minerals, both mica and olivine. Kimberlites are rich in potassium (K_2O typically 1–3 wt%) and, although derived from deep in the mantle, are also hydrated and carbonated. They usually occur as small (<1 km diameter) pipe-like bodies, or dykes and sills, and commonly extrude in highly explosive, gas-charged eruptions. The deep-seated origin of kimberlite is evident from the fact that it commonly transports garnet lherzolitic and eclogitic xenoliths to the surface, rock types made up of high pressure mineral assemblages that could only have come from the mantle. In addition, a small proportion of kimberlites also contain diamond xenocrysts. Diamond is the stable carbon polymorph under reducing conditions, and at depths in excess of about 100 km and temperatures greater than 900 °C (Haggerty 1999).

The origin of kimberlitic magmas is not too different from that of the alkaline rocks described above, and high pressure partial melting of a garnet peridotite source rock containing additional phlogopite or K-amphibole (richterite), as well as a carbonate phase, is viewed as a likely scenario (Hall 1996). The enrichment of incompatible constituents (such as K, Rb, H_2O, and CO_2) in kimberlite, as with alkaline magmas in general, again suggests that metasomatism of the mantle has played an important role in the provision of a deep-seated environment capable of producing highly enriched, or fertile, magmas. These aspects are discussed in more detail in Section 1.3 below.

1.3 Why Are Some Magmas More Fertile than Others? The "Inheritance Factor"

Geochemical inheritance is clearly an important factor in understanding the nature of

Figure 1.6 Relative abundances of selected metals in alkaline magmas (and kimberlite in the case of Cu and P) relative to average basalt. Source: Data from Table 1.2.

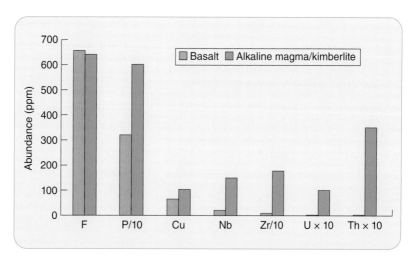

ore-forming processes in igneous rocks. Magmas may be endowed with ore-forming trace elements because the source materials from which they were derived were themselves enriched in these components. Further concentration of incompatible trace elements into residual magma, or of compatible trace elements into crystallizing phases, will take place during cooling and solidification of the magma, and these processes are discussed in more detail in Section 1.4 below.

A pertinent question that relates to the issue of geochemical inheritance is why certain portions of the Earth's crust appear to be so much better endowed in mineral deposits than others. The spectacular concentrations of, for example, gold and platinum deposits in South Africa, within the Witwatersrand Basin and Bushveld Complex respectively, perhaps point to some form of (mantle-related?) enrichment of these metals that is specific to this region. Why is it that Sudbury is so rich in Ni and the Andes are so well endowed with large Cu deposits, and is there an explanation for the fact that the Mesozoic granites of Southeast Asia are so well mineralized with Sn and W deposits in a belt that extends for over 2000 km in length? These questions point to the fact that metals are heterogeneously distributed in the lithosphere such that similar processes taking place in different parts of the world may result in variable metal endowments. Understanding the reasons for this variation, however, remains a major challenge.

1.3.1 The "Late Veneer" Hypothesis of Siderophile Metal Concentration – An Extraterrestrial Origin for Au and Pt?

In the very early stages of Earth's evolution, prevailing theory suggests that an originally homogeneous, molten planet differentiated into a metallic core, comprising essentially Fe and FeO with lesser Ni, and a mantle with a silicate composition. As this differentiation took place the siderophile metals (i.e. those with a strong affinity for Fe, such as Au and

the platinum group elements, or PGE) were comprehensively partitioned into the core. Experiments by Holzheid et al. (2000) indicate that the average concentrations of elements such as Au, Pt, and Pd in the Earth's mantle should be at least 10^{-4} times less than average chondritic abundances. Such concentrations are, in fact, so low that they virtually preclude the possibility of forming ore deposits in rocks extracted from the mantle (i.e. the crust). Yet the actual concentration of these precious metals in the mantle, although depleted, is only about 150 times lower than average chondritic abundances, a depletion that might account for the fact that there have been numerous ore deposits formed over geological time that have extracted these metals from the mantle. Another explanation is that the efficiency with which siderophile metals are partitioned between metal core and silicate mantle decreases with increasing pressure (depth) and that this could explain why the mantle is not as depleted as theory predicts. Although the latter notion probably applies to nickel, recent experiments suggest that it is not applicable to the precious metals and that some other explanation must be sought for the higher than expected concentrations of siderophilic metals such as Au and Pt in the mantle.

A clue as to why the mantle might be relatively enriched in siderophile metals lies in the fact that their abundance ratios (i.e. the abundance of one element relative to another, such as Au/Pt or Pt/Pd) are generally similar to chondritic abundance ratios as determined from analyses of meteorites that have fallen to Earth. One way to explain this is by having a substantial proportion of the precious metals in the mantle derived from meteorites that impacted the proto-crust during the early stages of Earth evolution, but *after* the differentiation of core and mantle (Figure 1.7). This idea, known as the "late-veneer" hypothesis (Kimura et al. 1974), suggests that much, if not all, the Au and Pt that is mined from ore deposits on the Earth's surface today ultimately had an extraterrestrial origin and that the

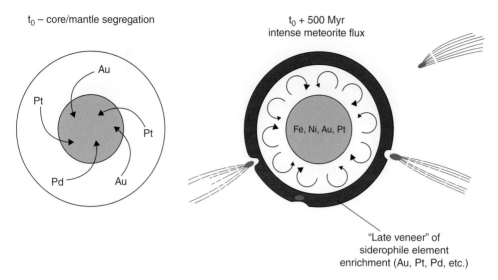

t_0 – core/mantle segregation

Au

Pt

Pt

Pd   Au

$t_0 + 500$ Myr
intense meteorite flux

Fe, Ni, Au, Pt

"Late veneer" of
siderophile element
enrichment (Au, Pd, Pt, etc.)

Figure 1.7 Schematic representation of the "late veneer" hypothesis for the siderophile (precious) metal enrichment of the Earth's lithosphere. During initial segregation of the Earth (t_0) siderophile metals were comprehensively partitioned into the Fe–Ni core. Approximately 500 Myr later ($t_0 + 500$ Myr) intense meteorite bombardment of the Earth added to the siderophile metal budget of the Earth's lithosphere.

planet's own inventory of these metals effectively resides in the core. Since both the meteorite flux itself and the subsequent distribution of this material through the mantle are likely to have been irregular, this hypothesis is also consistent with the heterogeneous distribution of precious metals over the Earth's surface.

As a footnote it is intriguing to note that precious metals may not have been the only valuable commodity introduced to Earth by meteorites. The enigmatic "carbonado" diamonds, found only in 1500 Myr old metasediments of Bahia State in Brazil and the Central African Republic, have mineralogical and isotopic characteristics unlike any diamond of terrestrial origin. Haggerty (1999) has suggested that carbonado diamonds are derived from the fall-back of a fragmented carbon-type asteroid that impacted the Earth's crust at a time when the relevant parts of Brazil and Africa formed a single continental entity. Although contentious, the notion of an extraterrestrial origin for certain constituents of the Earth's surface (such as water and even life itself?) is one that is likely to continue attracting attention in the future.

1.3.2 Diamonds and the Story They Tell

The Earth's mantle, between about 35 and 2900 km depth, is the ultimate source of material that, over geological time, has contributed, either directly or by recycling, to formation of the crust. Given that the mantle is essentially inaccessible, it has nevertheless been the subject of numerous studies by both geologists and geophysicists, such that its structure and composition are now reasonably well known. One reason for studying the mantle is to understand the origin of diamond. This remarkable mineral, together with the magmas that bring it to the surface, has provided a great deal of information about the deep Earth, much of which is very relevant to understanding those properties of the mantle that also relate to the source of metals in other igneous ore deposits.

Most diamonds are brought to the Earth's surface by kimberlitic magmas (see Section 1.2) or a compositionally similar melt known as lamproite. Most kimberlites and lamproites are barren, and diamondiferous magmas only intrude into ancient, stable continental crust that is typically older than 2500 Myr, but sometimes as young as 1500 Myr. The kimberlite magmas that transport diamonds to the surface, however, are typically much younger than the rocks they intrude, forming in discrete episodes in the Mesozoic and Cenozoic eras. Older intrusive episodes have also been observed in the Devonian, as well as at around 500 Ma and again at 1000 Ma (Haggerty 1999). Diamondiferous kimberlites must also have been emplaced during the Archean, at least on the Kaapvaal Craton in southern Africa, since the Witwatersrand conglomerates in South Africa are known to contain green detrital diamonds. To further complicate the story, the diamonds themselves tend to be much older than their kimberlitic host rocks and range in age from 1500 to 3000 Ma, indicating that they

resided in the mantle for considerable periods of time prior to their eruption onto the Earth's surface. Diamonds did not, therefore, crystallize from the kimberlite but were introduced to the Earth's surface as xenocrysts within the magma (Richardson et al. 1984; Eldridge et al. 1995). Diamond xenocrysts occur either as isolated single crystals in the kimberlitic matrix, or as minerals within discrete xenoliths of either peridotite (P-type diamonds – the more common) or eclogite (E-type diamonds). The high pressure phase relations that characterize mineral assemblages in these mantle xenoliths indicate that diamonds are derived from zones of thickened, sub-cratonic lithosphere, at least 200 km thick, that extend beneath stable Archean and Proterozoic shield areas (Figure 1.8). These lithospheric keels comprise

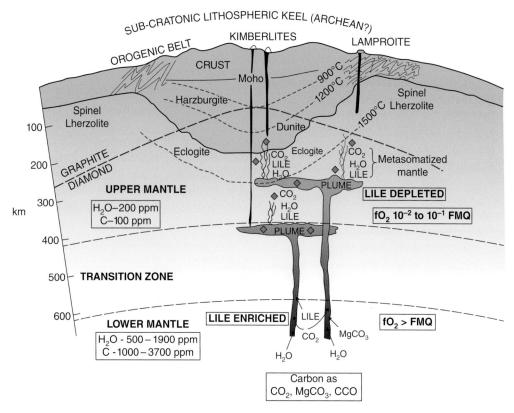

Figure 1.8 Schematic diagram illustrating features pertinent to the formation of diamond and the fertilization of the Earth's mantle by plume-related magmas and their associated aqueo-carbonic fluids. LILE refers to the large ion lithophile elements; FMQ refers to the fayalite–magnetite–quartz oxygen buffer. Source: After Haggerty (1999).

old, depleted, peridotite (i.e. from which mantle melts have already been extracted) as well as primitive, but younger, eclogite that has generally not had a melt fraction extracted and is, therefore, more fertile with respect to crust-forming elements (Haggerty 1999).

It is now generally accepted that diamonds were generated from deep in the mantle, in the layer known as the Transition Zone between the lower and upper mantle at around 400–600 km depth. Because the upper mantle is relatively depleted in carbon (100 ppm compared with 1000–3700 ppm in the lower mantle; Wood et al. 1996) it is unlikely to be a viable source for the primordial carbon that makes up diamond. The more fertile lower mantle is more likely to be the source of the carbon, and this is supported by the presence of very high pressure minerals occurring as tiny inclusions in many diamonds. However, the upper mantle is more reduced than the lower mantle, the latter, in addition to its higher carbon content, also containing substantially more water (500–1900 ppm compared to only 200 ppm) than the upper mantle. The upper mantle is, therefore, more likely to preserve diamond because the mineral's long-term stability depends on the existence of a reducing environment. Carbon in the relatively oxidized, fluid-rich lower mantle would, despite the higher pressures, not occur as diamond at all, but as CO_2, CO, or $MgCO_3$ (Wood et al. 1996). The model for diamond formation (Figure 1.8), therefore, suggests that plumes transfer melt and volatiles from the lower mantle, and precipitate diamond at higher levels, either in the reduced environment represented by the Transition Zone, or in the keels extending below thick, cratonic lithosphere. Thus, the more common P-type diamonds form when the relatively oxidized carbonic fluids dissolved in ascending plumes interact with reduced mantle at higher levels and precipitate elemental carbon. This mass transfer process is referred to as *metasomatism* and involves the movement of fluids and volatiles from deep in the Earth's mantle to higher levels. This process is turning out to be very relevant to the concepts of mantle fertilization and geochemical inheritance. The more rare E-type diamonds, by contrast, are considered to have crystallized directly from a magma intruded into or ponded below the lithospheric keel (Haggerty 1999).

Formation of the kimberlitic magma that transports diamond to the Earth's surface has also been attributed to plume activity and the metasomatic transfer of volatile constituents from a fertile lower mantle into depleted upper mantle. Evidence for this process comes from the observation that many of the major episodes of kimberlite intrusion mentioned above correlate with "superchron" events that are defined as geologically long time periods of unidirectional polarity in the Earth's magnetic field. Superchrons are caused by core–mantle boundary disruptions which increase the rate of liquid core convection, causing a damping of the geomagnetic field intensity but promoting plume activity and mantle metasomatism. Intrusion of diamondiferous kimberlites has also been linked in time to major geological events, such as continental breakup and flood basaltic magmatism (Haggerty 1999). England and Houseman (1984) suggested that enhanced kimberlite intrusion could be related to periods of low plate velocity when uninterrupted mantle convection gave rise to partial melting and volatile production in the lithosphere, and the associated development of plume related magmatism. Accompanying epeirogenic uplift created the fractures that allowed kimberlite magma to intrude rapidly upwards and, in many cases, to extrude violently onto the Earth's surface. This explanation is consistent with the geodynamic setting of kimberlites, such as in cratonic parts of Africa where many intrusions were emplaced in Mesozoic times (see Chapter 6). The relative rarity of kimberlite formation and penetration to the surface is also consistent with this model, since magma formation requires that a number of coincidentally optimal conditions apply.

Box 1.1 Diamondiferous Kimberlites and Lamproites: The Orapa Diamond Mine, Botswana and the Argyle Diamond Mine, Western Australia

---

Diamond mining around the world was worth some US$7 billion in 2001, a substantial proportion of which was derived from exploitation of primary kimberlitic and lamproitic deposits. The biggest single deposit is at Argyle in Western Australia, which is hosted in lamproite and produces some 26 million carats per year. Most of the diamonds produced here, however, are of low value. The Orapa and Jwaneng deposits of Botswana, by contrast, produce less than half the number of carats per year, but their stones are much more valuable. Orapa and Jwaneng together are amongst the richest diamond deposits in the world.

Kimberlites are by far the most important primary source of diamonds (see Section 1.3.2) and there are over 5000 occurrences known worldwide (Nixon 1995). By contrast there are only some 24 known occurrences of lamproite. Both kimberlites and lamproites are emplaced into the Earth's crust as "diatreme–maar" volcanoes which are the product of highly overpressured, volatile-rich magma. Kimberlitic or lamproitic magma is injected into the crust, along zones of structural weakness, to within 2–3 km of the surface. At this point volatiles (H_2O and CO_2) either exsolve from the magma itself (see Chapter 2), or the magma interacts with groundwater, with the resultant appearance of a vapor phase that causes violent phreatomagmatic eruption of magma, disruption of country rock, and pyroclastic eruptions (Gernon et al. 2009). Figure 1.1.1 shows the anatomy of a diatreme–maar system that has applicability to the nature and geometry of both kimberlites and lamproites.

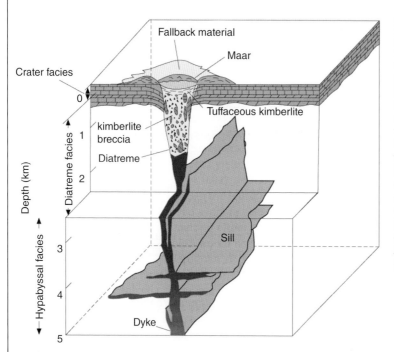

Figure 1.1.1 Idealized geometry of a diatreme–maar type volcano, showing the nature of the hypabyssal facies of magma emplacement along crustal weaknesses. Kimberlites around the world are seen as hypabyssal, diatreme, or crater facies, depending on the extent of preservation (or depth of erosion) of the bodies. Source: After Smith et al. (1979).

Kimberlites are richer in CO_2 than lamproites and since CO_2 has a lower solubility than water in silicate melts, kimberlite magmas will usually exsolve a volatile CO_2 fraction at lower

Box 1.1 (Continued)

depths than lamproites (Nixon 1995). Lamproite venting is quite often a function of magma interaction with groundwater, the availability and depth of which dictates the geometry of the crater. These factors account for the carrot-like shape of kimberlites compared to a broader, champagne glass shape for many lamproites.

The distribution of diamonds in any one kimberlite may be highly erratic and there seems to be little or no relationship between grade and depth (Nixon 1995). Some hypabyssal dykes are very rich in diamonds, such as the Marsfontein mine in South Africa which had an average grade of 200–300 carats per 100 tons of ore mined. Diatreme facies kimberlites are often characterized by multiple injections of magma, some of which are barren and others economically viable. Kimberlitic pyroclastic sediments in crater facies are also often richer than associated diatreme rocks, possibly due to enrichment of heavy minerals, including diamond, by wind (during eruption) or water (in the crater lake).

(a)

(b)

Figure 1.1.2 (a) View of the Orapa diamond mine, Botswana. Pyroclastic rocks of a kimberlitic maar-diatreme are seen overlain by a thin veneer of Karoo cover. Box shows the location of the close-up in (b). (b) Closer view of the southeast face of the pit showing the contact between a lower agglomeratic breccia and overlying lapilli tuff. Source: Photos are courtesy of Thomas Gernon.

(Continued)

Box 1.1 (Continued)

Argyle

The Argyle diamond mine occurs in a 1200 Myr old lamproite diatreme that has intruded older Proterozoic sediments in the Kimberley region of Western Australia (Boxer et al. 1989). It forms an elongate 2 km long body that varies from 100 to 500 m in width and is the result of at least two coalesced vents along a fault line. It is mined in the southern section where it has grades in excess of 5 carats per ton. Most of the body is made up of pyroclastic or tuffaceous material, with marginal breccia and occasional lamproitic dykes. The diatreme was formed when lamproite magma encountered groundwater in largely unconsolidated sediments, resulting in multiple phreatomagmatic eruptions and venting of lamproite at the surface.

Orapa

The Orapa diamond mine occurs in a mid-Cretaceous kimberlite in the north of Botswana and is well known for the excellent preservation of its crater facies (Figure 1.1.2). The kimberlite was emplaced in two pulses that merge at about 200 m depth into a single maar. The diatremes, comprising tuffisitic kimberlite breccia, grade progressively into crater facies made up of both epiclastic and pyroclastic kimberlite debris (Gernon et al. 2009). All these phases are diamondiferous. At Orapa the northern diatreme is believed to have been emplaced first, followed by residual volatile build-up and explosive volcanic activity. This was shortly followed by a similar sequence of events to form the southern diatreme, with the subsequent merging of its crater facies into a single maar (Field et al. 1997).

1.3.3 Metal Concentrations in Metasomatized Mantle and Their Transfer into the Crust

Although it has been evident for some time that there is an association between mantle metasomatism and diamond formation, the link between mantle fertilization and concentration of base and precious metals, has only been recognized more recently. A comparison of Re–Os isotopic ratios of epithermal Cu–Au ores from the Ladolam mine on Lihir Island, near Papua New Guinea, with peridotite xenoliths transported to the ocean surface by volcanic activity from the underlying mantle wedge, shows that the ore constituents are derived from the mantle. Although this is not unexpected, closer study of the peridotitic material reveals that some of this material has been extensively metasomatized to form a high temperature hydrothermal mineral assemblage comprising olivine, pyroxene, phlogopite, magnetite,

and Fe–Ni sulfides (McInnes et al. 1999). Metasomatism is considered to be the result of dehydration of the oceanic slab as it moves down the subduction zone, yielding fluids that migrate upwards into the overlying mantle wedge. Metasomatized peridotite contains precious and base metal concentrations that are up to two orders of magnitude enriched relative to unaltered mantle (Figure 1.9).

Subduction of oceanic crust beneath an island arc, such as that of which Lihir Island is part, has resulted in the formation of alkaline basalt which builds up the arc and ultimately forms the host rocks to the Ladolam Cu–Au deposit. The deposit itself occurs at a high level in the crust and formed by the circulation of metal-charged hydrothermal fluids, a process that is not relevant to the present discussion except for the fact that these fluids have dissolved the ore constituents that they carry from the immediate country rocks (i.e.

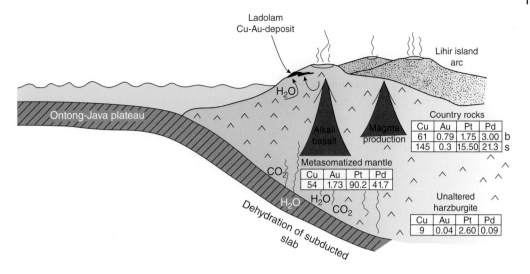

Figure 1.9 Schematic diagram illustrating the concept of mantle metasomatism and metal enrichment associated with subduction, and the subsequent inheritance of an enhanced metal budget by magmas derived from melting of metasomatized mantle. Metal abundances of the relevant rock types (in ppm) are from McInnes et al. (1999); the two analyses showing metal abundances for the magmatic products of subduction refer to basalt (b) and syenite (s).

the alkaline basalts). An interesting feature of the study by McInnes et al. (1999) is that basaltic and syenitic country rocks to the deposit (refer to b and s in Figure 1.9) were analyzed well away from the mineralization itself and also exhibit significant enrichments in their base and precious metal contents. The inference is, therefore, that although the ore metals resided originally in the mantle, they were redistributed, and their concentrations significantly upgraded, by the high temperature metasomatic processes observed in the mantle peridotite. Because it is the metasomatized peridotite that is likely to have been preferentially melted above the subduction zone (by virtue of its hydrous nature and lower solidus temperature), the resulting alkali basalt magma will have inherited a significant metal endowment, thus enhancing the chances of creating a substantial ore deposit during subsequent (hydrothermal) stages of ore formation. The Ladolam study illustrates three important features: first, that metasomatism and fluid flow are important processes in redistributing and concentrating incompatible elements in the mantle; second, that partial melting is the main process by which matter is transferred from the mantle to the crust; and, third, that inheritance is critical to the nature and formation of igneous hosted ore deposits, as well as to whether subsequent fluid circulation is likely to form viable hydrothermal deposits.

1.3.4 Metal Enrichment in Carbonatitic and Peralkaline Magmas

Carbonatites and associated igneous rocks such as nephelinites and kimberlites represent exotic and intriguing examples of magma formation and metal enrichment – their origin has been the subject of intense study and debate for several decades and continues to perplex workers to this day. Carbonatites are igneous rocks defined as comprising >50% modal abundance of carbonate minerals such as calcite or dolomite. They typically intrude into mature continental crust, although temporally they appear to be associated with orogenic events on the margins of cratons and have increasingly been found in more active tectonic environments and even in oceanic settings (Jones et al. 2013). Carbonatites are perhaps best known for the fact that they contain abnormal concentrations of a wide variety of metals such as P, F, Nb, Ti, Cu, Zr, Sr, Ba, U, Th, and the REE. Ample evidence exists for a genetic link between carbonatites and kimberlites with the latter also known for their anomalous concentrations of metals (Figure 1.6) in addition to bearing diamonds (see Section 1.3.2).

The question of how carbonatite magmas form and why they are so enriched in a wide range of metals remains poorly understood and contentious. Consensus that the magmas are mantle derived is now widespread (Harmer and Gittins 1998) and the notion that they were derived by melting of a pre-existing carbonate sediment has long been dismissed. Carbonatites would, nevertheless, appear to be polygenetic and could have formed from any one of three possible processes (Bell et al. 1999; Jones et al. 2013):

(i) Fractionation of an originally carbonated peralkaline melt such as nephelinite or melilitite;

(ii) Direct partial melting of a carbonated ultramafic rock such as peridotite; and

(iii) Formation of an immiscible liquid fraction from a CO_2-saturated silicate melt.

Fractionation (i.e. the progressive crystallization of solid phases and the concentration of incompatible trace elements into the residual magma) is a process, more fully described in Section 1.4.2, that could yield carbonatitic magmas that are enriched in a variety of metals such as those listed above. Likewise, small degrees of partial melting of an ultramafic source rock (Section 1.4.1) could also result in the required magma composition and metal enrichments. Experimental studies related to immiscibility between carbonatitic and silicate melt fractions (Martin et al. 2013) suggest that under anhydrous conditions certain metals, such as the alkalis, P, Mo, Cu, and V, exhibit weak to moderate partitioning into a carbonatitic melt fraction. Under hydrous conditions, however, a broader range of metals, including the REE, Fe, Co, Cu, Zn, and Th, are even more strongly partitioned into the carbonatite melt. Clearly, carbonatitic magmas are particularly well endowed with a wide range of metals and their concentration is a complex and multi-faceted process that probably also depends on the fertility of the source rocks being melted (see Section 1.3.3) and the appearance of an aqueous hydrothermal fluid in the magma (Chapter 2).

Although only approximately 500 carbonatite intrusions are known throughout the world (Woolley and Kjarsgaard 2008) it is clear that they are metallogenically important, especially as sources of "critical metals" – see Box 1.2. Over 90% of the world's niobium comes from just two carbonatite-hosted mines in Brazil – Barreiro/Araxá and Catalão (Linnen et al. 2014). The majority of global REE supply since the 1980s has come from the Bayan Obo deposits in China (Box 1.2; Xu et al. 2008; Wall 2014), the latter hosted in a sedimentary dolomite unit that has been intruded by carbonatite and subsequently metasomatized by fluids possibly derived from subduction between the North China craton and Central Asian Orogenic Belt (Ling et al. 2013). The Phalaborwa carbonatite-hosted deposit in South Africa has been mined for several decades and is a substantial producer of refined Cu, phosphate (from apatite), magnetite, and vermiculite, as well as Ni, Se, Te, and precious metals as by-products (see Box 1.2). Many other carbonatite deposits in different parts of the world have the potential to be mined for a variety of metals in the future.

Box 1.2 Carbonatites and Alkaline Intrusions as Sources of Critical Metals

Critical Metals

The technological advances of the twenty-first Century have placed significant demands on a number of metals whose overall supply is at risk for reasons that include difficulties of extraction, rarity, limited geographic distribution, and political/socio-economic constraints. Such metals, often referred to as "critical metals," include the rare earth elements (REE), niobium (Nb), tantalum (Ta), tungsten (W), fluorine (F), germanium (Ge), gallium (Ga), indium (In), and the platinum group elements (PGE), amongst others. Carbonatites and alkaline intrusions frequently contain significant concentrations of many of the critical metals and are increasingly the object of exploration ventures that target these strategically important commodities.

Carbonatites, and to a lesser extent alkaline intrusions, are mined in many parts of the world for phosphates (PO_4), Nb, F, and the REE, but in addition they may also be markedly enriched in Cu, Fe, Ba, Sr, Zr, V, Ta, Th, U, Au, Ag, and the PGE, many of which are produced as by-products of mining activity (Mariano 1989). The best known alkaline intrusion and carbonatite-hosted ore deposits that have been exploited around the world include Mountain Pass in California (REE), Barreiro/Araxá in Brazil (Nb, PO_4, and REE), Bayan Obo in China (REE), Okorusu in Namibia (F), Khibiny (PO_4), and Lovozero (REE, Ta, Nb) in Russia and the Phalaborwa Complex in South Africa (PO_4, Cu, and vermiculite). This case study focuses on the Barreiro carbonatite in Brazil, the world's largest producer of Nb, and the Bayan Obo deposit in China, the world's largest producer of REE. The case study also presents isotopic evidence from the Phalaborwa carbonatite in South Africa suggesting that the source of these magmas was fertile (or metasomatized) mantle (see Section 1.3.3).

Barreiro (also known as Araxá), Minas Gerais, Brazil

The world's largest Nb deposit is hosted by the Cretaceous Barreiro carbonatite–alkaline igneous complex, near Araxá, Minas Gerais, Brazil. The carbonatite intrudes metasediments of the Meso- to Neoproterozoic Brasilia fold belt on the margins of the São Francisco craton. It comprises a circular, 4.5 km diameter, intrusion (Figure 1.2.1), the central part of which is mainly occupied by a carbonatite that is predominantly beforsitic in composition (Issa Filho et al. 2014; Traversa et al. 2001).

The carbonatite comprises predominantly dolomite with lesser calcite and siderite. These rocks are intimately associated with phoscorite, an unusual rock type made up of magnetite, ilmenite, apatite, olivine, and phlogopite, in addition to carbonate minerals. Pyrochlore is the main Nb ore mineral that occurs as a common accessory phase throughout the carbonatite but exhibits the highest concentrations in the phoscorite. The carbonatites are also associated with ultramafic plutonic rocks dominated by pyroxenite and peridotite, as well as "glimmerite," the latter being an alteration product of the ultramafic rocks dominated by phlogopite.

The Barreiro carbonatite is intensely weathered, and a 250 m thick lateritic residue marks the surface expression of the intrusion. The laterite overburden comprises mainly clay minerals (kaolinite and gibbsite) and Fe oxyhydroxides (goethite). The primary, unweathered ore material has a mean grade of 1.5% Nb_2O_5, whereas the laterite typically runs at 2.5% Nb_2O_5. The pyrochlore itself is essentially resistant to weathering whereas the host rock is chemically

(Continued)

Box 1.2 (Continued)

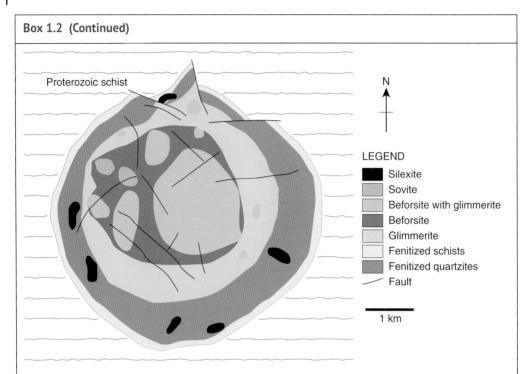

Proterozoic schist

N

LEGEND
- Silexite
- Sovite
- Beforsite with glimmerite
- Beforsite
- Glimmerite
- Fenitized schists
- Fenitized quartzites
- Fault

1 km

Figure 1.2.1 Simplified geological map of the Barreiro Complex. Source: After Traversa et al. (2001).

modified and highly altered. Volume reduction during lateritization results in a residual enrichment of pyrochlore and an upgrade of the primary ore-grade. Only the laterite is mined from a large open-pit, not only because is it higher-grade, but also because the material is soft and can be extracted without drilling and blasting. One disadvantage of the weathering process, however, is that pyrochlore is partially altered to a barium-rich pyrochlore (Figure 1.2.2) by the replacement of Ca and Na by Ba, the latter introduced by meteoric fluids.

The Nb resource in the laterite alone is estimated at more than 450 Mt. at a grade of 2.5% Nb_2O_5 (with significant REE and PO_4 resources as well).

Bayan Obo, China

The Bayan Obo REE–Nb–Fe deposits occur in north-central China, on the northern margin of the North China Craton. The geology of the Bayan Obo region is complex and polygenetic (Smith and Henderson 2000; Xu et al. 2008, 2010; Ling et al. 2013). Carbonatitic dykes are emplaced into a Mesoproterozoic metasedimentary sequence (the Bayan Obo Group) comprising dominantly meta-arkose/sandstone and slate, and a thick intercalated dolomitic marble that is the main ore-bearing horizon (Figure 1.2.3). The metasediments are intruded by numerous carbonatitic dykes that appear to have been emplaced mainly around 1400–1200 Ma and possibly again at 450 Ma. Fluid inclusion and geochronological studies point to a long-lived sequence of events that commenced with deposition in the Mesoproterozoic of a predominantly clastic sedimentary sequence onto an older Archaean gneissic basement, followed by deformation and episodic carbonatite emplacement, with further deformation terminated by an event of alkali granite emplacement in the Permian. REE mineralization appears to have been epigenetic, with U–Pb monazite ages in carbonatites ranging between 330 and 760 Ma,

Box 1.2 (Continued)

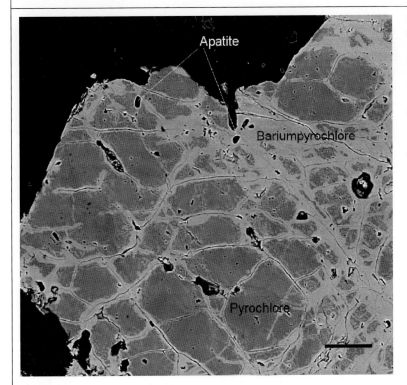

Figure 1.2.2 Back-scattered electron microscope image of pyrochlore being replaced by a barium-rich pyrochlore in a sample from the lateritized ore at Barreiro. Source: After Issa Filho et al. (2014).

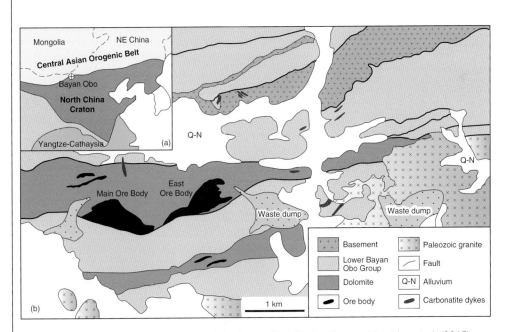

Figure 1.2.3 Simplified geological map of the Bayan Obo district. Source: After Ling et al. (2013).

(Continued)

Box 1.2 (Continued)

substantially younger than the circa 1300 Ma age of the carbonatites themselves – a process attributed to metasomatism by fluids derived by subduction occurring during the Central Asian Orogenic Belt (Ling et al. 2013).

The bulk of the mineralization at Bayan Obo is described as occurring in replacement bodies within the dolomitic marble unit. The REE mineralization occurs as bastnaesite and monazite, whereas the Nb occurs mainly as the REE–Nb oxide mineral aeschynite. Substantial iron ore, as magnetite and hematite, is also extracted from the dolomitic marble unit as part of the overall mining operation. Fluorite, apatite, and aegerine are the principal gangue minerals in the ore zones. The mineralization paragenesis is complex – early monazite, formed along fractures and grain boundaries in relatively unaltered dolomite, has been dated at 550 Ma (Wang et al. 1994). The main stage of REE mineralization overprints the earlier event and is dated at c. 430 Ma. This event is also associated with the development of magnetite and hematite and is followed by deformation and further fluid flow, precipitation of fluorite and barite, and remobilization of earlier formed ores (Smith and Henderson 2000).

China currently produces some 97% of the world's REE supply (Wall 2014), most of which comes from Bayan Obo. Actual resource figures for the REE deposits at Bayan Obo are not known, but estimates of between 48 and 100 Mt at an average grade of around 6% REE_2O_3 (Drew et al. 1990; Xu et al. 2008) have been presented. Substantial Fe ore resources also exist at Bayan Obo (up to 1.5 billion tones at 35% Fe), as well as a smaller Nb resource (1 Mt at 0.13% Nb).

Why are Alkaline and Carbonatitic Magmas So Fertile?

The diversely enriched nature of carbonatitic and alkaline magmas may be related to the fertile character of specific zones in the mantle that were melted to produce them (see Section 1.3.3) This concept is examined with reference to the Phalaborwa carbonatite in northeastern South Africa, a body that has been mined for several decades to produce a variety of metals, of which Cu and P are the most important, with Au–Ag–PGE–U–Th–Zr–Te–Se–Ni as important by-products.

Phalaborwa is a 2047 Myr old alkaline–carbonatite complex comprising several intrusive events (Eriksson 1989); an initial magma injection crystallized as a large, 6 × 3 km, pipe of clinopyroxene-apatite-phlogopite rock which was followed by intrusion of at least three pulses of ultramafic pegmatoidal rock of similar modal mineralogy. The was followed by the emplacement of phoscorite which was in turn intruded by two pulses of carbonatite. Apatite, as a source of phosphate for fertilizers, was produced from the central phoscorite intrusion, but most of the mining activity was concentrated on the central carbonatite intrusions which contain large tonnages of low-grade copper ore as chalcopyrite and bornite.

A clue to the highly fertile nature of the mantle source rocks from which the Phalaborwa Complex is likely to have been derived is provided by analysis of the Sr and Nd isotopic compositions of the host rocks. Eriksson (1989) has shown that the complex is characterized by high initial $^{87}Sr/^{86}Sr$ ratios (0.704–0.714) and low initial $^{143}Nd/^{144}Nd$ ratios (0.5096–0.5098), which point to a mantle source that was enriched in both large ion lithophile (K, Rb, Ba, and Sr etc.) and high field strength elements (i.e. the REE, Zr, Nb, U, Th, etc.). The εNd versus εSr diagram in Figure 1.2.4 confirms the enriched nature of the mantle source rocks from which the Phalaborwa alkaline and carbonatitic magmas were likely derived.

Box 1.2 (Continued)

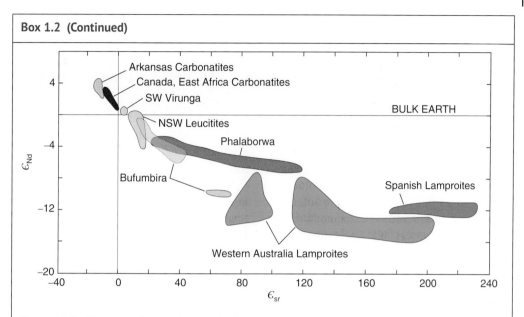

Figure 1.2.4 εNd versus εSr plot showing the field of Phalaborwa data relative to other alkaline igneous rocks. The ε notation refers to the measured $^{143}Nd/^{144}Nd$ or $^{87}Sr/^{86}Sr$ ratios normalized against the chondritic uniform reservoir (or bulk earth). Source: After Eriksson (1989).

The existence of zones of fertile mantle that can give rise to melts that are enriched in both large ion lithophile and high field strength elements, as well as providing the exotic, $H_2O–CO_2$ rich, magma compositions observed in the alkaline–carbonatite suite, is attributed to metasomatized layers called metasomes (Haggerty 1989). Metasomes are enriched in volatiles, specifically H_2O, CO_2, and S, and form from upwardly migrating fluids which have dissolved within them appreciable quantities of large ion lithophile and high field strength elements. Alkali–carbonatitic melts are derived from the shallowest metasome (50–60 km depth) which is well developed beneath thinned (rifted) or marginal continental lithosphere – this explains the preferential occurrence of these rock types in extensional tectonic settings (Haggerty 1989). Kimberlites, by contrast are derived from deeper metasomes in the more reduced asthenospheric keels that exist below old, stable cratons (see Figure 1.8).

1.3.5 I- and S-Type Granite Magmas and Metal Specificity

It is now well known that the different types of granite, and more specifically the origins of felsic magma, can be linked to distinct metal associations. Of the many classification schemes that exist for granitic rocks one of the most relevant, with respect to studies of ore deposits, is the I- and S-type scheme, originally devised for the Lachlan Fold Belt in southeast Australia (Figure 1.10a) by Chappell and White (1974). In its simplest form the scheme implies that orogenic granites can be subdivided on the basis of whether their parental magmas were derived by partial melting of predominantly igneous (I-type) or sedimentary (S-type) source rock. In general, I-type granites tend to be metaluminous and typified by tonalitic (or quartz-dioritic) and granodioritic compositions, whereas S-types are often peraluminous and have adamellitic (or quartz-monzonitic) and granitic compositions. Also important from a metallogenic viewpoint is the fact that I-type granites tend to be more oxidized (i.e. they have a higher

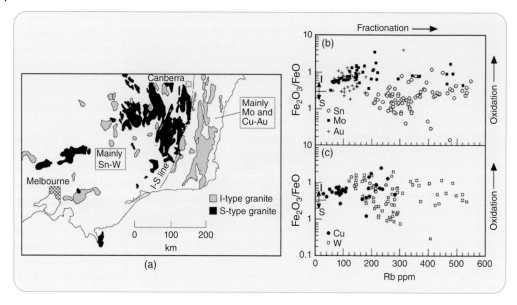

Figure 1.10 (a) Simplified map showing the distribution of S- and I-type granites, and associated metallogenic trends, in the Lachlan Fold Belt of southeastern Australia. Source: After Chappell and White (1974). (b and c) Plots of Fe_2O_3/FeO versus Rb for granites of the Lachlan Fold Belt that are mineralized with respect to Sn–W and Cu–Mo–Au. Source: After Blevin and Chappell (1992).

magmatic fO_2) than S-type granites, whose magmas were originally fairly reduced because of the presence of material such as carbon (graphite) in their source rocks. An approximate indication of the oxidation state of granitic magmas can be obtained from their whole rock Fe_2O_3/FeO ratio (which effectively records the ferric/ferrous ratio). Blevin and Chappell (1992) have shown that a Fe_2O_3/FeO ratio of about 0.3 provides a useful discriminant between I- (with Fe_2O_3/FeO > 0.3) and S-type (with Fe_2O_3/FeO < 0.3) granites, at least for the Australian case (Figure 1.10b and c). A classification of granites according to oxidation state was, in fact, made initially by Ishihara (1977), who distinguished between reduced granite magmas (forming ilmenite-series granitoids) and more oxidized equivalents (forming magnetite-series granitoids). The metallogenic significance of this type of granite classification was also recognized by Ishihara (1981), who indicated that Sn–W deposits were preferentially associated with reduced ilmenite-series granitoids, whereas Cu–Mo–Au ores could be linked genetically

to oxidized magnetite-series granitoids. Magnetite-series granitoids are equivalent to most I-types, whereas ilmenite-series granitoids encompass all S-types as well as the more reduced I-types.

Although now regarded as somewhat oversimplified, at least with respect to more recent ideas regarding granite petrogenesis, the S- and I-type classification scheme is nevertheless appealing because it has tectonic implications and can be used to infer positioning relative to subduction along Andean-type continental margins (see Figure 1.2). The scheme also has metallogenic significance because of the empirical observation that porphyry Cu–Mo mineralization (with associated Pb–Zn–Au–Ag ores) is typically associated with I-type granites, whereas Sn–W mineralization (together with concentrations of U and Th) is more generally hosted by S-type granites. Although this relationship is broadly applicable, it too, is oversimplified. Some granites, notably those that are post-tectonic or anorogenic, do not accord with the scheme, such as the alkaline granites of the Bushveld Complex, which are polymetallic and contain both Sn–W and

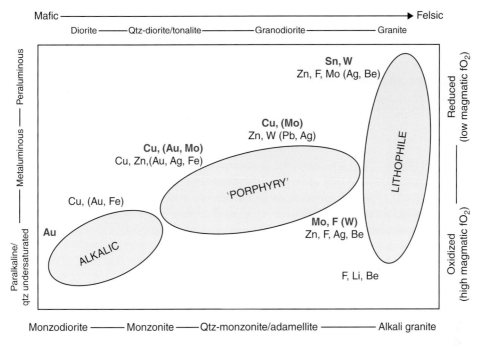

Figure 1.11 Generalized scheme that links granite compositions and magmatic oxidation state to metal associations and intrusion-related ore deposit types. Ore deposit types referred to are "alkalic," "porphyry," and "lithophile" and are discussed in more detail in Chapter 2. Metals shown in bold reflect the more important associations. Source: Modified after Barton (1996).

base metal mineralization (Robb et al. 2000). A more accurate appraisal of the relationship between magma composition (including oxidation state) and metallogenic association is given by Barton (1996). Figure 1.11 illustrates Barton's scheme, which regards granites as a continuum of compositional types and shows their metal associations in terms of different intrusion-related ore deposit types.

Adjacent to Andean type subduction zones a clearly defined spatial pattern exists with respect to the distribution of I- and S-type granite intrusions, as well as associated metallogenic zonation (Sillitoe 1976; Clark et al. 1990; Blevin et al. 1996). The leading edge (i.e. the oceanic side) of the subduction zone tends to correlate with the production of I-type granite magmas and is associated with the formation of porphyry Cu styles of mineralization. By contrast, the continental side of the subduction zone contains more differentiated granite types that are often S-type in character and with which Sn–W styles of mineralization

are associated. Other examples of this type of regional zonation are seen in the paired granitic belts of Myanmar and Malaysia (Gardiner et al. 2016), in the Cordilleran granites of the western United States and in the Lachlan Fold Belt of southeast Australia (Figure 1.10a). Again, although exceptions and complications do exist, the recognition and delineation of these patterns is clearly important with respect to understanding the spatial distribution of different types of ore deposits hosted in granitoid rocks.

The Lachlan Fold Belt provides an excellent example of the relationships between magma type and metallogenic association and, as predicted, granite related mineralization is dominated by Sn–W in the mainly S-type granites to the west of the I–S line, whereas largely Mo with lesser Cu–Au ores are found in the I-type terrane to its east (Figure 1.10a). Another important feature of the metal content of granites is, however, also apparent in the Lachlan Fold Belt. When the Fe_2O_3/FeO ratio (an indication of magmatic oxidation state) is plotted against Rb content (an indicator of degree of fractionation) for granites that are mineralized in terms of either Sn–W or Cu–Mo–Au (Figure 1.10b and c) it is clear that metal content is not simply a function of magma type alone. It is apparent that Cu–Mo–Au related intrusions typically have higher Fe_2O_3/FeO ratios than those associated with Sn–W mineralization, and are, therefore, preferentially associated with I-type granites. What is perhaps more apparent, however, is that Sn and W mineralization is associated with intrusions that are *more highly fractionated* than those containing Cu–Mo–Au (Blevin and Chappell 1992). Metal contents are, therefore, also a function of processes that happen as the magma cools and fractionates and these processes are discussed in more detail in Section 1.4 below. In addition, mineralization in granites also involves hydrothermal processes that are quite distinct from either geochemical inheritance or fractionation, and these are examined in Chapter 2.

1.4 Partial Melting and Crystal Fractionation as Ore-Forming Processes

The previous section discussed various magma forming processes and some of the reasons why they are variably endowed with respect to their trace element and mineral contents. This section examines how trace elements and minerals behave in the magma, both as it is forming and then subsequently during cooling and solidification.

Trace element abundances can be useful indicators of petrogenetic processes, during both partial melting and crystal fractionation. Since many magmatic ore deposits arise out of concentrations of metals that were originally present in very small abundances, trace element behavior during igneous processes is also very useful in understanding ore formation.

A trace element is defined as an element that is present in a rock at concentrations lower than 0.1 wt% (or 1000 ppm), although this limit places a rather artificial constraint on the definition. In general, trace elements substitute for major elements in the rock forming minerals, but in certain cases they can and do form the stoichiometric components of accessory mineral phases (Rollinson 1993). Many of the ores associated with igneous rocks are formed from elements (such as Cu, Ni, Cr, Ti, P, Sn, W, U, etc.) that originally existed at trace concentrations in a magma or rock and were subsequently enriched to ore grades by processes discussed in this and later sections.

When rocks undergo partial melting, trace elements partition themselves between the melt phase and solid residue. Those that prefer the solid are referred to as compatible (i.e. they have an affinity with elements making up the crystal lattice of an existing mineral), whereas those whose preference is the melt are termed incompatible. Likewise, during cooling and solidification of magma, compatible elements are preferentially taken up in the crystalline solids, whereas incompatible elements are enriched in the residual melt. Enrichment of trace elements and potential ore formation can, therefore, be linked to the concentration of incompatible elements in the early melt phase of a rock undergoing anatexis (see Box 1.3), or in the residual magma during progressive crystallization (see Box 1.4). Compatible trace elements tend to be "locked up" in early formed rock-forming minerals and are typically not concentrated efficiently enough to form viable ore grade material. An exception to this is provided, for example, during the formation of chromitite layers, as discussed in Section 1.4.3 below.

A brief description of how one can quantify trace element distribution during igneous processes is presented below, with an indication of how these processes can be applied to the understanding of selected ore-forming processes.

1.4.1 Partial Melting

Despite the fact that temperatures in the upper mantle reach 1500 °C and more, melting (or anatexis) is not as widespread as might be expected because of the positive correlation that exists between pressure and the beginning of melting of a rock (i.e. the solidus temperature). The asthenosphere, which is defined as that zone in the mantle where rocks are closest to their solidus and where deformation occurs in a ductile fashion (which explains the lower strength of the asthenosphere relative to the elastically deformable lithosphere), is the "engine room" where a considerable amount of magma is formed. Major magma-generating episodes, however, do not occur randomly and without cause, but are catalyzed by processes such as a decrease of pressure (caused, for example, by crustal thinning in an extensional tectonic regime) or addition of volatiles to lower the prevailing solidus temperature (such as during subduction and metasomatism). An increase in local heat supply is generally not important in the promotion of partial melting,

although it is possible that mantle plumes might be implicated in this role.

Partial melting, so called because source rocks very seldom melt to completion, invariably leaves behind a solid residue. It is a complex process that is affected by a number of variables, the most important being the nature of the mineral assemblage making up the protolith, as well as local pressure, temperature and water/volatile content. Early experimental work on the progressive fusion of peridotite with increasing temperature (Mysen and Kushiro 1976) helps to explain the process of partial melting in the asthenosphere, even though the description outlined below is probably not a particularly good indication of the very complicated processes actually involved.

In Figure 1.12 melting of peridotite starts just above 1400 °C with the formation of a small melt fraction in equilibrium with pyroxenes and olivine. Melting can continue without significant addition of heat until about 30–40% of the rock is molten and the clinopyroxene is totally consumed. Once the clinopyroxene has been consumed, further melting can only continue with addition of heat (the first inflection

Figure 1.12 The sequential melting behavior of peridotite in the mantle at 20 kbar pressure, (a) without water and (b) in the presence of some 2% water. Source: After Mysen and Kushiro (1976).

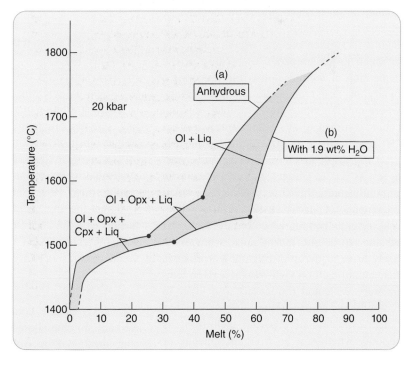

point on the curves in Figure 1.12), which, if present, would then promote the melting of orthopyroxene. Once orthopyroxene has been consumed, after some 50–60% of the rock has melted, a further input of heat is required (the second inflection point) if olivine is to be wholly included into the melt product. A small degree of partial melting of peridotite will, therefore, yield magma with a composition (alkali basalt) that is dominated by the melt products of clinopyroxene, with the residue reflecting the bulk composition of orthopyroxene + olivine. In reality, melting processes are more complex and usually involve the fusion of more than just one mineral at a time. It is also pertinent to note that the presence of even a small amount of water in the system will catalyze the melting process and also allow anatexis to occur at lower temperatures.

Anatexis in the crust is better explained by considering the partial melting of a metasedimentary protolith, to form, for example, an S-type granite. If the sediment consisted only of quartz, then the temperature would have to exceed 1170 °C (the melting temperature of SiO_2 at $P_{H_2O} = 1$ kbar) in order for it to start melting. If the sediment were an arkose, however, made up of an assemblage of quartz + orthoclase + albite (as shown in Figure 1.13), then melting would commence at much lower temperatures because of the depression of melting points that occurs for binary or ternary eutectic mixtures. Thus, where quartz + albite are in contact, melting could commence at temperatures around 790 °C and, where quartz + albite + orthoclase meet at a triple junction, melting could start as low as 720 °C. Disaggregation of the arkose protolith would, therefore, take place by small increments of partial melt forming along selected grain boundaries within the rock. The residue left behind during such a process is likely to be made up of fragmented minerals that cannot melt on their own at a given temperature. Melt and residue are also likely to have different compositions.

The extraction of a partial melt from its residue, whether it be from an igneous or

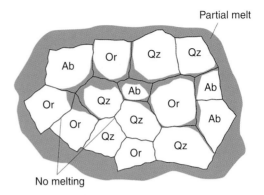

Figure 1.13 The pattern of grain boundary melting of an arkosic protolith subjected to temperatures of 700–800 °C at $P_{H2O} = 1$ kbar. Source: After Hall (1996).

sedimentary protolith, is a process which segregates chemical components and is referred to as fractionation. Partial melts can be considerably enriched in certain elements, but depleted in others, relative to the source rock. The mechanisms whereby melt is removed from its residue may differ, and this is discussed in more detail below. More detailed, quantitative descriptions of partial melt processes are provided in Cox et al. (1979), Rollinson (1993), Albarede (1996), and Best (2003).

1.4.1.1 Trace Element Distribution During Partial Melting

In theory there are two limiting extremes by which partial melting can occur. The first envisages formation of a single melt increment that remains in equilibrium with its solid residue until physical removal and emplacement as a magmatic body. This process may be applicable to the formation of high viscosity granitic melts (Rollinson 1993) and is known as "batch melting." It is quantified by the following equation, whose derivation (together with others discussed below) is provided in Wood and Fraser (1976):

$$C_{liq}/C_o = 1/[D_{res} + F(1 - D_{res})] \quad (1.1)$$

where: C_{liq} is the concentration of a trace element in the liquid (melt); C_o is the concentration of trace element in the protolith

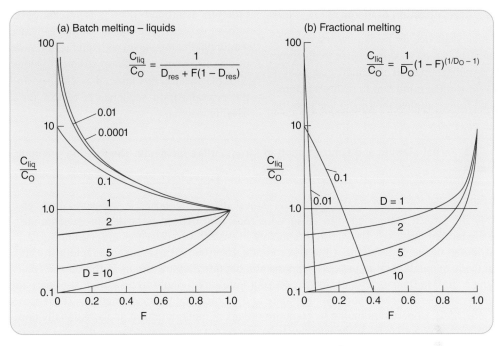

Figure 1.14 The enrichment/depletion of a trace element in a partial melt relative to its concentration in the source rock (C_{liq}/C_o) as a function of increasing degrees of melting (F). (a) Batch melting using Eq. (1.1), and (b) fractional melting using Eq. (1.2). Source: After Rollinson (1993).

(unmelted parental material); D_{res} is the bulk partition coefficient of the residual solid (after the melt is extracted); F is the weight fraction of melt produced.

A diagrammatic illustration of the extent of enrichment (or depletion) of an incompatible (or compatible) element in a batch partial melt is presented in Figure 1.14a, where the ratio C_{liq}/C_o is plotted as a function of the degree of melt produced (F) for a variety of values of D_{res}. Marked enrichments of highly incompatible elements (i.e. those with very small values of D_{res}) can occur in small melt fractions, with the maximum enrichment factor being $1/D_{res}$ as F approaches 0.

The second partial melt process is referred to as "fractional melting" and is the process whereby small increments of melt are instantaneously removed from their solid residue, aggregating elsewhere to form a magma body. This process may be more applicable to low viscosity basaltic magmas where small melt fractions can be removed from their source

regions. The distribution of trace elements during fractional melting is quantified in terms of the following equation:

$$C_{liq}/C_o = 1/D_o(1 - F)^{(1/D_o - 1)} \qquad (1.2)$$

where: D_o is the bulk partition coefficient of the original protolith (prior to melting); and the other symbols are the same as for Eq. (1.1).

The extent of trace element enrichment and depletion in a fractional partial melt is shown in Figure 1.14b. For very small degrees of melting the changes in trace element concentrations relative to the source material are extreme and vary from a maximum value of enrichment ($1/D_o$ as F approaches 0) to depletions in the magma as melting progresses. Unlike the batch melt situation, compatible element enrichment can also occur during fractional melting, but this is only likely to happen in the unlikely event of more than 70% partial melting. An example of incompatible element concentration in a granitic partial melt is provided in Box 1.3.

1.4.2 Crystallization of Magma

The melt sequence for peridotite described in Figure 1.12 can be used in reverse to illustrate the crystallization of an ultramafic magma as it cools (bearing in mind that in reality a magma derived from complete melting of a peridotitic protolith at around 1800 °C is unlikely ever to have existed). The crystallization sequence would have commenced with olivine, followed by orthopyroxene and clinopyroxene. Together with plagioclase, this mineral assemblage and the crystallization sequence is typical for a basaltic magma.

Box 1.3 Partial Melting and Concentration of Incompatible Elements: The Rössing Uranium Deposit

The Rössing mine in Namibia is one of the largest uranium deposits in the world, and certainly the largest associated with an igneous rock. The deposit is hosted in leucogranite, locally termed "alaskite". The ore occurs mainly in the form of disseminated uraninite (UO_2) distributed unevenly throughout the leucogranite, together with secondary uranium silicate and oxide minerals such as beta-uranophane and betafite. Rössing occurs in the central zone of the c. 500 Ma Damara Orogen and the leucogranites are thought to be a product of partial melting of older basement comprising granite and supracrustal (metasedimentary) sequences (Figure 1.3.1). Upper amphibolite to granulite grades of metamorphism apply regionally and melts are considered to have been derived from depths not significantly greater than the actual level of emplacement, as indicated by the profusion of leucogranitic dykes in the region. The Rössing mine is located where several of the larger leucogranitic dykes have coalesced to form a significant mass of mineable material. The deposit comprises several hundred million tons of low grade uranium ore at an average grade of about 0.031% (or 310 ppm) U_3O_8. Although the deposit is low grade, a concentration of 310 ppm uranium in the leucogranite nevertheless represents a significant enrichment factor relative to the Clarke value (i.e. about 2.7 ppm for U; see Table 1.2). This degree of enrichment is consistent with what might be expected for a magma formed by a low degree of partial melting of an already reasonably enriched protolith.

This can be tested very simply in terms of the partial (batch) melt equation, given as Eq. (1.1) in Section 1.4.1 of this chapter. If the protolith contained 10 ppm U and its bulk partition coefficient relative to the melt fraction was significantly less than 1 (i.e. D_{res} between 0.01 and 0.1) then a 5% batch melt would yield a magma that is enriched by factors of between 7 (for $D_{res} = 0.1$) and 17 (for $D_{res} = 0.01$), as shown in Figure 1.3.2. This process would have resulted in uranium concentrations in the leucogranite of between 70 and 170 ppm, which is at best only one-half of the observed average concentration. In order to achieve concentrations of around 300 ppm U by batch melting either the degree of melting would have to have been very low (i.e. <5%), or the protolith would itself have to have been significantly enriched, and have contained substantially more than 10 ppm U.

Research by Nex et al. (2001) has shown that partial melting in the central zone of the Damaran orogeny was an episodic affair and that only one of several generations of leucogranite is significantly enriched in uranium. Since all are likely to have been generated by approximately the same degrees of partial melting, this would suggest that the single enriched episode was derived from a particularly fertile protolith. It is apparent that deposits representing enrichment of incompatible trace elements in rocks formed by small degrees of melting, like the Rössing leucogranites, are not very common in the Earth's crust. Another example might be the Johan Beetz uranium deposit in Quebec, Canada.

Box 1.3 (Continued)

Figure 1.3.1 Sheets of leucogranite cutting through high-grade metasediments in the vicinity of the Rössing uranium mine, Namibia. Photograph courtesy of Paul Nex.

Figure 1.3.2 Simplified batch melt model showing the degree of enrichment expected for an incompatible element such as uranium (with D_{res} of either 0.1 or 0.01) after 5% melting (further details provided in Section 1.4.1).

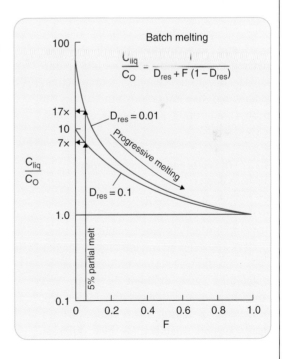

Batch melting

$$\frac{C_{liq}}{C_O} = \frac{1}{D_{res} + F\,(1 - D_{res})}$$

$D_{res} = 0.01$

Progressive melting

$D_{res} = 0.1$

5% partial melt

The sequential crystallization of a mafic magma is, of course, another way of fractionating chemical components, since minerals of one composition are often physically separated from the compositionally different magma from which they form. Once minerals are removed from the magma (by a process such as crystal settling) there is little or no further chemical communication (or equilibration) between the solid and liquid components of the chamber. This process is referred to as *fractional crystallization* and is a form of *Rayleigh fractionation*. The nature of fractional crystallization, and the geometry of cumulate rocks that form during these processes, are both very relevant to an understanding of ore deposits hosted in igneous rocks. It is, therefore, necessary to consider these processes in more detail and, in particular, their relationship to metal concentration and ore formation. It should also be emphasized at this stage that emplacement and crystallization of magma is generally accompanied by varying degrees of assimilation of country rock by the magma. Consequently, chemical trends reflecting fractionation processes are accompanied by the effects of magma contamination and this, too, has important implications for igneous ore-forming processes (see Sections 1.4.3 and 1.6).

1.4.2.1 The Form and Internal Zonation of Igneous Bodies

The emplacement mechanisms of magma into the Earth's crust and the resulting shapes of igneous intrusions are variable and these processes are reviewed in Hall (1996), Pitcher (1997), and Best (2003). Basaltic intrusions are typically quite different in shape and form from granitic batholiths. In addition, the mechanisms of crystal fractionation within a mafic magma chamber are also different to those applicable inside granite intrusions. These differences are relevant to ore formation in crystallizing magmas, as exemplified by the stratiform nature of chromitite seams in mafic intrusions such as the Bushveld Complex (see Box 1.5), compared to tin deposits in granites of the same complex which tend to be concentrated in disseminated form toward the center of such intrusions (see Box 1.4).

Mafic Intrusions The viscosity of mafic magma is relatively low, as illustrated by the ease with which basaltic magmas flow on eruption. The densities of mafic minerals are typically greater than $3\,g\,cm^{-3}$, whereas mafic magma has a density of around $2.6\,g\,cm^{-3}$. The low viscosity of mafic magma and the high densities of minerals crystallizing from it imply that minerals such as olivine and the pyroxenes will typically sink in a magma at velocities of anywhere between 40 and $1000\,m\,yr^{-1}$ depending on their composition and size (Hall 1996). By contrast, less dense minerals, such as the feldspathoids, might float in an alkaline magma as they have densities less than $2.5\,g\,cm^{-3}$. Plagioclase would float in a basaltic magma at pressures greater than about 5 kbar, but would sink in the same magma emplaced at high crustal levels (Kushiro 1980).

The layered rocks that result from gravitationally induced crystal settling are referred to as cumulates and their compositions differ from that of the starting magma. Crystal settling is, therefore, a form of fractional crystallization and this process could explain the segregation of chemical constituents and their possible concentration into either of the solid or liquid phases of the chamber. Trace elements that are readily incorporated (by substitution) into cumulus minerals are referred to as compatible elements. By contrast, trace elements that are excluded from the cumulate assemblage (because they cannot easily substitute into the crystal lattice sites of rock-forming minerals) are called incompatible elements – these will become progressively enriched in the residual magma as the cumulate assemblage is formed (Cawthorn and McCarthy 1985).

Gravitationally induced settling of minerals in mafic intrusions would seem to be the most logical way of explaining the well layered internal structure often observed in these bodies. Early models for explaining the crystallization and layering in mafic intrusions were derived from the classic work of Wager and Brown (1968) on the Skaergaard intrusion of Greenland. The latter is particularly well suited to petrogenetic study because it is considered to represent a single pulse of magma that crystallized to completion in a closed chamber without substantial addition or removal of magma during solidification. The sub-horizontal layers that form in mafic intrusions arise from the accumulation of dense, early formed minerals (such as olivine and orthopyroxene), followed, in a stratigraphic sense, by accumulations of later formed minerals such as the pyroxenes and plagioclase. Minor oxide minerals such as chromite and magnetite are also observed to accumulate among the major silicate mineral phases. The Skaergaard intrusion is made up of a chilled marginal phase and a relatively simple internal zone made up of sub-horizontal rhythmic and cryptic (i.e. where the layering is not visually obvious but is reflected only in chemical variations) layers, ranging in composition upwards from gabbro to ferrodiorite and granophyre (Figure 1.15).

Although it is tempting to attribute the relatively ordered, sub-horizontal layering that is so characteristic of intrusions such as Skaergaard to simple gravitationally induced crystal settling, solidification of magma chambers, including Skaergaard, is a much more complex process (McBirney and Noyes 1979; Irvine et al. 1998). The temperature gradients and associated variations in magma density that accompany the cooling of magmatic bodies have been shown to result in pronounced density stratification, with the formation of liquid layers through which elements diffuse in response to both chemical and temperature gradients (Huppert and Sparks 1980; Turner 1980; Irvine et al. 1983; McBirney 1985). Density variations over time in a magma chamber are also a product of crystal fractionation itself. In Figure 1.16a it is apparent that during early stages of olivine crystallization the residual magma density decreases because the olivine itself has a higher density than that of the starting liquid. The trend changes, however, with the appearance on the liquidus of a mineral such as plagioclase, which has a density lower than that of the magma. In such a situation it is possible, after an interval of crystallization, for a residual magma to become more dense than when it started solidifying.

This pattern of density variation has major implications for the behavior of crystals settling in a magma chamber, particularly when

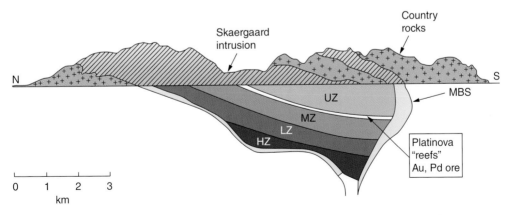

Figure 1.15 Schematic illustration showing the nature of layering in the Skaergaard intrusion, east Greenland. The relative stratigraphic position of the Au- and Pd-bearing Platinova Reefs is shown in relation to the Marginal Border Series (MBS) and the Layered Series, which is subdivided into the Hidden Zone (HZ), Lower Zone (LZ), Middle Zone (MZ), and Upper Zone (UZ). Source: After Anderson et al. (1998).

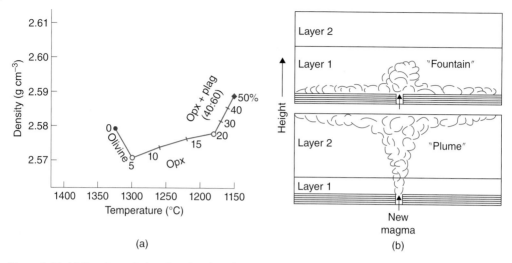

(a)

(b)

Figure 1.16 (a) Density variations in a fractionating magma similar in starting composition to that of the Bushveld Complex. (b) Contrasting behavior of a new magma injected into a density stratified magma chamber within which crystal fractionation has already occurred. Source: After Campbell et al. (1983), Naldrett and von Grüenewaldt (1989).

a new injection of magma takes place into an already evolved chamber. In Figure 1.16b two situations are considered. If a newly injected magma has a density that is greater than the liquid residue in the chamber (as might be the case if the new magma was injected fairly early after crystallization had commenced), then a rather inconspicuous fountain-like feature will form and mixing between the new and evolved liquids will be limited to a layer along the base of the chamber. By contrast, if the new magma is injected late in the crystallization sequence, it is possible that its density would be less than that of the residual liquid and a more conspicuous plume-like feature would form (see Figure 1.16b), with the new magma rising to its own density level, or to the top of the chamber. In the latter scenario turbulent mixing between the new and residual liquids is likely to be more complete. Injection of a plume that reached and interacted with the roof of the magma chamber has been suggested for the Bushveld Complex (Schoenberg et al. 1999). This plume has also been strongly implicated in the formation of both the chromite and PGE mineralization in the Bushveld (Kinnaird et al. 2002).

At a first glance, layered mafic intrusions such as Skaergaard may seem to be the products of relatively straightforward, gravitationally induced, crystal settling of minerals as they appear sequentially on the liquidus. In detail, the nature of the layers may be quite complex (see Box 1.5) and the chemistry of cumulate assemblages are likely to be affected by complex processes involving variations in thermal gradients and rates of chemical diffusion. In addition, convective currents in a magma chamber, as well as turbidity currents in which crystal-laden magma surges down a slope on the margins or floor of the chamber, perturb the cumulus layering. The composition of the minerals crystallizing at any one time will also be sensitive to perturbations in the chemistry of the magma and rapid changes caused either by injection of new magma or contamination of existing magma by wall rock. Finally, changes in the chemistry of interstitial minerals, even in relatively small magma chambers such as Skaergaard, point to subtle processes such as silicate–silicate immiscibility (Humphreys 2011) that are not easily detected or readily apparent. All of these processes influence the concentration of metals in magmatic environments, as well as the formation of mineral deposits in layered mafic intrusions.

The processes of crystal fractionation, and accompanying density variations and changes in magma composition, are all critical to the formation of ore grade concentrations of chromite, magnetite, platinum group elements, and base/precious metals in layered mafic complexes. As the processes are better understood, such intrusions are increasingly attracting the attention of explorationists worldwide and even well studied bodies, such as Skaergaard, have been shown to contain mineralized layers. Although the presence of sulfide minerals at Skaergaard was recognized early on (Wager et al. 1957), the recognition of a potentially significant Au and Pd resource only came much later (Bird et al. 1991; Anderson et al. 1998). The nature of magmatic processes and the role that they play in mineralization are discussed in more detail below, and also illustrated in several case studies (Boxes 1.4–1.8).

Felsic Intrusions Granite intrusions do not exhibit the well defined sub-horizontal layering that typifies large mafic intrusions. This is largely due to the fact that felsic magmas are several orders of magnitude more viscous than mafic ones (i.e. up to 10^6 Pa s for rhyolite compared to 10^2 Pa s for basalt at the same temperature; Hall 1996). In addition, the density contrasts that exist in crystallizing mafic magma chambers are not as marked as they are with respect to quartz and felspar forming in felsic chambers. Crystal settling is inconsequential in all but a few granite plutons, such as the relatively hot, hydrous alkaline granites whose parental magmas were emplaced at shallow depths in the crust. It is nevertheless evident that substantial fractionation does occur in felsic intrusions and there are many recorded examples of internal zonation in granite plutons (Pitcher 1997). Many plutons record a concentric zonation, with the outer zones preserving more mafic compositions (i.e. diorite, tonalite, granodiorite) and rock types

becoming progressively more fractionated toward the center. Since many granite plutons intrude at shallow crustal levels and most of their heat is lost to the sidewalls, this type of zonation is logically attributed to crystallization that commenced from the sides and roof of the magma chamber and progressed inwards. Although crystals are not being removed from the magma by settling, as applicable to mafic intrusions, they are effectively isolated from the residual melt by the crystallization front which advances in toward the center of the chamber. This process, referred to as *sidewall boundary layer differentiation*, is also regarded as a form of crystal fractionation and characterized by concentration of incompatible elements toward the center of the intrusion where the final increments of differentiated granite melt accumulate.

1.4.2.2 Trace Element Distribution During Fractional Crystallization

There are several ways by which crystallization can be modeled but the most appropriate would appear to be where crystals are removed from the site of formation with only limited equilibration between solid and liquid phases. In such a case, trace element distributions can be described in terms of Rayleigh fractionation, in which:

$$C_{liq}/C_o = F^{(D-1)} \tag{1.3}$$

for trace element concentration in the residual magma (liquid), and

$$C_{sol}/C_o = DF^{(D-1)} \tag{1.4}$$

for trace element concentration in the crystallizing assemblage (solid), where: C_o is the original concentration of a trace element in the parental liquid; D is the bulk partition coefficient of the fractionating assemblage; F is the weight fraction of melt remaining.

The extent of trace element enrichment and depletion during fractional crystallization of a magma is shown in Figure 1.17. The enrichment of an incompatible trace element in the residual melt relative to the

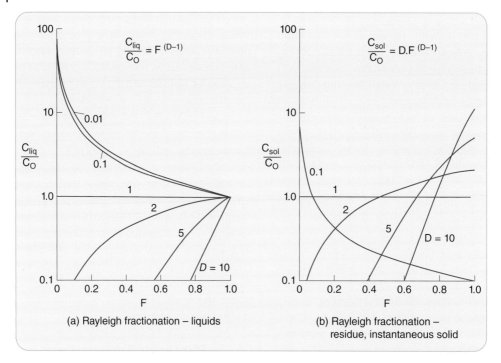

Figure 1.17 Trace element distribution during fractional crystallization. (a) The enrichment/depletion of a trace element in the residual magma relative to its concentration in the original melt (C_{liq}/C_o) as a function of the fraction of the remaining magma (F) using Eq. (1.3). (b) The enrichment/depletion of a trace element in a crystallized assemblage (immediately removed from its melt) relative to its concentration in the original melt (C_{sol}/C_o) as a function of the fraction of the remaining magma (F) using Eq. (1.4) Source: After Rollinson (1993).

original concentration in the parental magma is demonstrated in Figure 1.17a, where abundances can be seen to increase exponentially as crystallization proceeds (i.e. as the fraction of melt remaining, F, decreases). Compatible elements will, of course, continuously decrease in the residual melt as they are extracted into the solid phases. Figure 1.17b shows, however, that the relative concentration of a compatible element in the crystallizing assemblage will start off as enriched only for rocks formed in the early stages of fractionation. Relative concentrations will decrease as crystallization progresses because the magma rapidly becomes depleted in compatible elements as they are fractionated into the crystallizing assemblage.

Rayleigh fractionation equations apply reasonably well to low viscosity basaltic magmas where the effects of crystal fractionation are very evident. As mentioned previously, though,

granitic magmas do not exhibit well defined igneous stratification and they tend to solidify by inward nucleation of a sidewall boundary layer. This in situ style of crystal fractionation, where the crystallizing front is spatially distinct from the magma residue, is akin to Rayleigh fractionation, but the equations that govern trace element distribution have to be modified, as shown below, to accommodate the melt fraction that is returned to the magma chamber as the solidification zone moves progressively into the chamber (Langmuir 1989):

$$C_{liq}/C_o = (M_{liq}/M_o)^{[f(D-1)/(D[1-f]+f)]} \qquad (1.5)$$

where: M_{liq} is the mass of liquid remaining in the magma chamber; M_o is the initial mass of the magma chamber; f is the fraction of liquid in the solidification zone returned to the magma chamber; and the other symbols are as in Eq. (1.4).

The distribution trends of compatible and incompatible trace elements during this style of progressive inward solidification are essentially similar to Rayleigh fractionation (Figure 1.17) although the degrees of concentration, especially at low values of f, are not as extreme (Rollinson 1993). This accords with observations that the effects of crystal fractionation in granites tend to be less evident than in mafic intrusions.

Box 1.4 Boundary Layer Differentiation in Granites and Incompatible Element Concentration: The Zaaiplaats Tin Deposit, Bushveld Complex

The granites of the Bushveld Complex, which overlie the better known layered mafic intrusion, occur as large sill-like intrusions and are also known to be strongly fractionated and mineralized (McCarthy and Hasty 1976; Groves and McCarthy 1978). The Bushveld granites are enriched in Sn (as well as W, F, Cu, REE, Ag, Au, U, and Fe) and several deposits occur in the more highly fractionated parts of the suite. One of the better studied examples is the Zaaiplaats tin mine, which obtains the bulk of its tonnage from a zone of low-grade, disseminated cassiterite (SnO_2) mineralization that occurs within the central portion of the granite body (Figure 1.4.1).

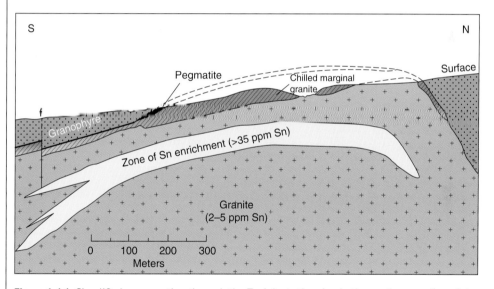

Figure 1.4.1 Simplified cross section through the Zaaiplaats tin mine in the northern portion of the Bushveld Complex, South Africa. Source: After Coetzee and Twist (1989).

A cross section through the ore body, from diamond drill core, defines the low grade zone of disseminated cassiterite toward the center of the granite sheet. This is consistent with the suggestion that solidification occurred from the margins inwards and that Sn was fractionated by processes akin to boundary layer differentiation. This type of crystal fractionation and its implications for ore-forming processes can be evaluated in terms of trace element distribution patterns. If boundary layer differentiation is modeled in terms of Rayleigh Fractionation

(Continued)

Box 1.4 (Continued)

(Eqs. (1.3) and (1.4) in Section 1.4.2), then the degree of crystallization required to concentrate Sn in the residual magma to the levels observed in the disseminated zone at Zaaiplaats can be calculated. Assuming that Sn is incompatible in the crystallizing granitic magma (i.e. D = 0.1) then some 96% crystallization is required in order to achieve an enrichment factor of 20 × (Figure 1.4.2). Coetzee and Twist (1989) have shown that unmineralized granite of the type that hosts tin mineralization in the Bushveld Complex contains between 8 and 14 ppm Sn, whereas mineralized portions of the suite, like the disseminated zone at Zaaiplaats, average around 270 ppm Sn. The disseminated mineralization at Zaaiplaats is, therefore, consistent with Sn concentration by crystal fractionation. After an advanced degree of solidification, it is suggested that the Sn content of the residual magma was sufficiently enriched to promote cassiterite crystallization. Other factors are also necessary in order to stabilize cassiterite in granites (Taylor and Wall 1992), and these include fO_2 and magma composition (specifically the Na/K ratio). These factors, together with the degree of enrichment required, account for the fact that cassiterite is seldom seen as an accessory mineral in normal granites. The highly fractionated granite at Zaaiplaats would appear somewhat exceptional, resulting in levels of tin concentration that facilitated cassiterite crystallization.

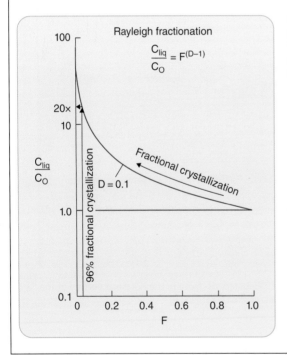

Figure 1.4.2 Rayleigh fractionation model showing the degree of enrichment expected for an incompatible element (D = 0.1) such as Sn after 96% crystallization (further details provided in Section 1.4.2).

An example of trace element concentration during inward nucleation of a sidewall boundary layer in granite is provided by the Zaaiplaats Sn deposit in the granitic phase of the Bushveld Complex in South Africa (see Box 1.4). There are many examples of trace element concentrations in layered mafic intrusions and subsequent sections discuss some of the complex processes involved in accumulation of ore grade Pt, Cu, and Ni deposits such as those of the Bushveld Complex, Kambalda, and Sudbury (see Boxes

1.5–1.8). A case where crystal fractionation alone, without the added benefit of processes such as magma replenishment or contamination, has resulted in significant enrichment of Au, Pd, and S is provided by the Platinova Reefs of the Skaergaard Complex. Figure 1.15 shows that this zone of potentially ore grade precious element enrichment occurs toward the top of the Middle Zone and formed after a substantial proportion of the magma chamber had crystallized. The incompatible nature of Au, Pd, and S with respect to the early formed crystals (dominantly plagioclase, olivine, and pyroxene with lesser magnetite and apatite) resulted in a progressive enrichment of these metals in the residual magma (Anderson et al. 1998). Once the concentration of sulfur had reached saturation (believed to be between 0.16 and 0.3 wt% in the Skaergaard magma) the metals precipitated out as a Au–Pd alloy and further concentrated as inclusions within the magmatic sulfide minerals that formed interstitially among the normal cumulus mineral assemblage at this level. The concentrations of Au and Pd that resulted from the progressive in situ crystal fractionation of the Skaergaard magma amounts to about 2 ppm for each metal and it has been estimated that some 90 million tons of potential ore grade material is available in the Platinova Reefs (Anderson et al. 1998).

1.4.3 Fractional Crystallization and the Formation of Monomineralic Chromitite Layers

Crystal settling, convective fluid flow, and diffusion-related chemical segregation across density stratified layers, are the processes which give rise to the characteristically sub-horizontal, well ordered layering evident in many layered mafic intrusions (McCarthy et al. 1985). These processes do not, however, account for the occasional development of substantial monomineralic layers of chromite or magnetite that form ore bodies in many layered mafic intrusions around the world. Nor do they explain the formation of the massive pods of chromitite that occur in the lithospheric portions of ophiolite complexes. This section provides some insight into the processes by which layers and pods of massive chromitite can form in mafic cumulate rocks.

There are many mafic intrusions that contain layers of near monomineralic chromite or (vanadium-rich) magnetite that are typically 0.5–1 m in thickness and extend laterally for tens of kilometers, representing enormous reserves of Cr and Fe–V ore (see Box 1.5). Notable among these are the Bushveld Complex in South Africa and the Great Dyke in Zimbabwe. The formation of such monomineralic layers, which might comprise up to 90% of a single oxide mineral (chromite or magnetite), would appear to require that normal crystallization of silicate minerals (dominated by olivine, pyroxene, and plagioclase) be abruptly terminated and replaced by a brief interlude where only a single oxide phase is on the liquidus. A simple, but elegant, explanation of this process was provided by Irvine (1977) with respect to the formation of chromitite layers.

1.4.3.1 The Irvine Model

The Irvine model is based on experimental studies and refers to part of the phase diagram for a basaltic system in which only the olivine–chromite–silica end-members are portrayed in a ternary plot (Figure 1.18a). The normal crystallization sequence in a basaltic magma with starting composition at A (Figure 1.18b) would commence with olivine as the only mineral on the liquidus, settling of which would result in the formation of a dunitic cumulate rock. Extraction of olivine from the magma composition at A would result in a progressive change of the magma composition away from the olivine end-member composition and toward the cotectic phase boundary at B. At B a small amount of chromite (around 1%) would also start to crystallize together with olivine, and the magma composition would then evolve along the cotectic toward C. Along the crystallization interval B–C the accumulation of cotectic proportions of olivine + chromite would result in a cumulate rock that could *not*

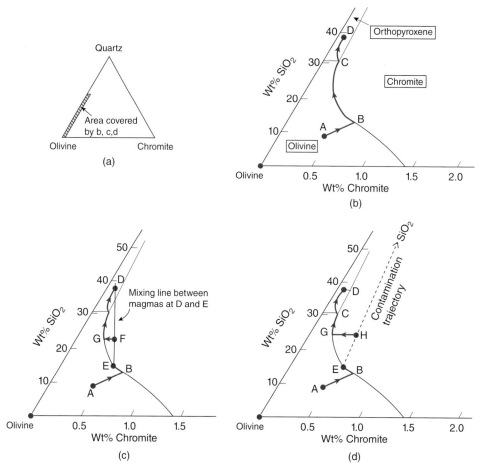

Figure 1.18 A portion of the ternary system quartz–olivine–chromite (a) showing the nature of crystallization in a mafic magma (b); scenarios in which magma mixing (c) and magma contamination (d) occur as mechanisms for promoting the transient crystallization of only chromite are also shown. Source: After Irvine (1977).

be described as a viable chromium deposit. At C the SiO₂ content of the magma has increased to a level where olivine and chromite can no longer be the stable liquidus assemblage and orthopyroxene starts to crystallize to form a bronzitite cumulate rock. From this stage magma composition evolves toward D. Continued fractional crystallization will eventually lead to the appearance of plagioclase together with orthopyroxene on the liquidus, but felspar compositions are not reflected on the simplified phase projection shown here. This crystallization sequence does not lead to the formation of a chromite seam and the

latter mineral would only occur as an accessory phase in the early formed olivine cumulates.

In order to make a monomineralic chromitite layer that might become an economically viable ore deposit, the normal crystallization sequence described above needs to be perturbed. One way of disturbing the normal crystallization sequence is to introduce, at point D for example, a new magma with a composition at E (i.e. not as primitive as the original starting liquid), that is injected into the chamber and allowed to mix with the evolved liquid (Figure 1.18c). Mingling of the two magmas represented by D and E would result in a mixture whose composition must lie somewhere along a mixing line (represented by the dashed line DE in Figure 1.18c). Exactly where along DE this mixture lies depends on the relative proportions of D and E that are mixed together. For most mixtures (at point F, for example, in Figure 1.18c) the magma composition would lie within the stability field of chromite and for a brief interval of crystallization (from F to G) only chromite would crystallize from the mixture. Being relatively dense, chromite could settle fairly efficiently and a single, monomineralic layer of chromite would form along the floor of the magma chamber. In very large magma chambers, such as the one from which the Bushveld Complex must have developed, the crystallization of such a layer could lead to the formation of an ore body with potentially vast chromium resources. Experimental confirmation of these processes has been provided by Murck and Campbell (1986).

Box 1.5 Crystal Fractionation and Formation of Monomineralic Chromitite Layers: The UG1 Chromitite Seam, Bushveld Complex

The Bushveld Complex is the world's largest layered mafic intrusion (Figure 1.5.1), covering an area of over 65 000 km². It also contains a substantial proportion (more than 75%) of the world's chromite reserves. Several major chromitite seams (at least 14 in number) occur within the Critical Zone of the Bushveld Complex and these are subdivided into three groups termed the Lower Group (LG1 to LG7), the Middle Group (MG1 to MG4), and the Upper Group (UG1 to UG3). The LG6 chromitite seam is the most important in terms of production and reserves and can be traced for over 160 km in both the western and eastern portions of the complex.

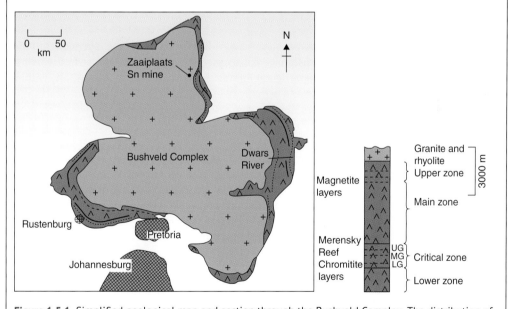

Figure 1.5.1 Simplified geological map and section through the Bushveld Complex. The distribution of chromitite layers or seams in the Critical Zone is shown in relation to the PGE-rich Merensky Reef and the vanadium-rich magnetite seams at the base of the Upper Zone. Also shown is the distribution of associated granite and rhyolite that overlies the layered mafic intrusion.

(Continued)

Box 1.5 (Continued)

Figure 1.5.2 Multiple chromitite seams exposed at Dwars River in the eastern Bushveld Complex. Source: Photo courtesy of Paul Nex.

In Section 1.4.3 of this chapter the mechanisms by which monomineralic layers of chromitite could form in a layered mafic intrusion were discussed with specific reference to the Irvine model. The UG1 chromitite layer is of particular interest because it is hosted essentially within anorthositic rocks (rather than olivine and orthopyroxene cumulates) and is also spectacularly well exposed in a gorge of the Dwars River in the eastern portion of the Bushveld Complex (Figure 1.5.2). The UG1 shows some intriguing features that require special explanation. One of the interesting aspects of the UG1 is the way in which the seam bifurcates. In some sections the seam splits into several thinner layers, whose cumulative thickness is similar to that of the single seam. Chromitite splits tend to occur in areas where anorthosite layers thicken into domal features, with the bifurcations opening out toward the core of the dome. Nex (2002) has explained this by suggesting that normal crystallization and settling of chromite (according to Irvine's model) was interrupted by liquefaction of the footwall crystalline assemblage to form a slurry of plagioclase feldspar and melt that erupted at the magma–cumulate interface. The bottom-up accumulation of the slurry is responsible for the formation of the domal features in the anorthosite and also serves to dilute the top-down settling of chromite grains. There is, therefore, a correlation between liquefaction of the footwall cumulates, doming of the anorthosites, and splitting of chromitite layers. This type of feature, together with many other intriguing textures and relationships, indicates that crystallization processes in magma chambers can be very complex.

Once the magma composition reaches point G on the cotectic (after extraction of only chromite), crystallization will again be dominated by olivine and the rocks that form in the hanging wall of the chromite seam will again contain only accessory amounts of chromite. It should also be mentioned that another way of forcing the magma composition into the chromite field is shown in Figure 1.18d, where the magma at point E (or anywhere along the cotectic for that matter) becomes contaminated with siliceous material (perhaps by assimilation of crustal material forming either the floor or the roof of the magma chamber). The contaminated magma would have a composition that lies somewhere along the mixing line joining E to the SiO_2 apex of the ternary diagram. This composition would also lie transiently in the chromite field and result in the formation of a monomineralic cumulate layer of chromite (between H and G). It is evident that there are a number of ways of perturbing the normal liquid-line-of-descent and promoting the crystallization of a single mineral phase such as chromite – similar processes, at a different stage of crystallization, might also result in the formation of monomineralic layers of magnetite.

1.4.3.2 Other Mechanisms for the Formation of Chromitite Layers or Pods

Although the Irvine model explains many of the characteristics of chromitite layers it is unlikely to apply to all situations and there are several other mechanisms that might pertain to the accumulation of monomineralic layers or pods. Two of the most likely, since they have been confirmed experimentally, include changes in oxygen fugacity (fO_2) and total pressure (P_{Tot}) of the crystallizing magma.

An increase in fO_2 will promote the stability of chromite and possibly allow the mineral to crystallize alone for a period of time (Ulmer 1969). Increasing fO_2 in the magma could be achieved by a devolatilization reaction such as Eq. (1.6) below (after Lipin 1993).

$$4FeCO_3 \leftrightarrow 2Fe_2O_3 + 4CO + O_2 \quad (1.6)$$

where CO_3^{2-} in solution breaks down to form carbon monoxide and free oxygen. However, because CO_2 and not CO is likely to be the dominant carbon species in basalt, it seems unlikely that oxidation of a magma will be easily achieved, and the process is generally not called upon as an explanation for chromite accumulation.

By contrast, small increases in P_{Tot} of the magma have now been shown to occur readily in basaltic chambers. Observations in the Kilauea volcano (Hawaii) have shown that a pressure increase (of up to 0.25 kbar) can occur in the roof of the magma chamber as a result of CO_2 exsolution and expansion of the vapor bubbles as they stream upwards in the magma chamber (Bottinga and Javoy 1990; Lipin 1993). An increase in total pressure within the magma chamber will have the effect of shifting the phase boundary between olivine and chromite such that the field of the latter phase expands. This would have the same effect as that predicted in the Irvine model, in that chromite would crystallize alone until such time as the ambient pressure is restored (by egress of magma, volatiles, or both). This mechanism has been proposed for the formation of chromitite layers in the Stillwater Complex (Montana), where it is suggested that CO_2 streaming and associated pressure increase accompanied ingress of a new magma pulse into the chamber (Lipin 1993). Exsolution of volatiles and fluids from a magma is discussed in more detail in Chapter 2.

Podiform chromitites in ophiolites, although not as large as the deposits in layered mafic intrusions, represent important resources in many parts of the world. These chromite ores are typically found as irregular, stratiform to discordant, pods within dunitic and harzburgitic host rocks which themselves are often intensely deformed. In detail the ore textures, characterized by nodular and orbicular associations of chromite and olivine, suggest that the mingling of two magmas has given rise to crystallization of the chromitite ores.

Ballhaus (1998) has suggested that the sites of chromite mineralization represent zones in the oceanic lithosphere where low viscosity, olivine-normative melt mingled with a more siliceous, higher viscosity magma. The two melt fractions remain segregated, at least for the time it takes to accumulate chromitite ore. For thermodynamic reasons chromite nucleates preferentially in the ultramafic melt globules, with crystals forming initially along the metastable liquid interface (where mixing takes place at a small scale and a situation akin to the Irvine model [Figure 1.18] pertains) and then progressively throughout the globule. Diffusion of chromium from the siliceous magma, where no chromite nucleation has occurred, across the liquid interface into the ultramafic melt globules, also takes place, with the result that the latter might ultimately be entirely replaced by chromitite. Because the siliceous magma acts as a chromium reservoir, the richest ores are considered to occur where the volume of the latter is high relative to that of the ultramafic globules within which accumulation of chromitite is occurring.

1.4.4 Filter Pressing as a Process of Crystal Fractionation

The separation of crystal phases from residual melt during the solidification of magma is generally attributed to gravitational segregation where crystals of either higher or lower density than the magma settle or float to form horizontally layered cumulate rocks. Another mechanism by which crystal melt segregation can occur is the process known as filter pressing. The residual magma that exists interstitially to accumulating crystals in a partially solidified chamber can be pressed out into regions of lower pressure such as overlying non-crystalline magma or fractures in the country rock. The process is considered to apply even in more viscous granitic magma chambers where evolved, water-saturated melts are filter pressed into adjacent fractures created during hydrofracturing.

1.4.4.1 Anorthosite Hosted Ti–Fe Deposits

Large massif-anorthosite intrusions of Mesoproterozoic age, located in the Paleohelikian and Grenvillian orogenies extending from North America into the Sveconorwegian province, are the hosts to important Ti and Fe deposits of magmatic origin (Force 1991a; Gross et al. 1997). Well-known examples of such deposits include Sanford Lake in the Adirondacks of New York State, USA, the Lac Tio deposit in the Allard Lake region of Quebec, Canada, and the Tellnes deposit in southern Norway. The deposits are typically thought to be related to large differentiated intrusive complexes made up mainly of anorthosite, gabbro, norite, and monzonite rocks emplaced in the late tectonic to extensional stages of the orogenic cycle. Although mineralogically variable, the more important category economically is the andesine anorthosite type (or Adirondack type), which contains ilmenite–hematite as the principal ore minerals. The Ti–Fe oxide ore accumulations occur as stratiform layers and disseminations within the intrusive complexes themselves, or as more massive, higher-grade, cross-cutting or dyke-like bodies.

These deposits are the product of in situ crystal fractionation. Early extraction of a plagioclase-dominated crystal assemblage results in concentration of Fe and Ti in the residual magmas, which crystallize to form ferrogabbro or ferrodiorite. Titaniferous magnetite or hemo-ilmenite (depending on the magma composition) also crystallize with disseminated layers formed by crystal settling and accumulation on the chamber floor. The more massive discordant bodies are considered to be a product of the pressing out of an Fe–Ti oxide mineral slurry – the slurry concentrates to form an intrusive body often along the margins of the largely consolidated anorthosite complex, or into fractures and breccia in the host rocks. The Tellnes ores, for example, form part of a 14 km long dyke (Gross et al. 1997). The Lac Tio orebody is an irregular, tabular intrusive mass some 1100 m long and 1000 m wide that is inconsistently layered and exhibits evidence for multiple intrusions resulting in magma mixing and the crystallization of hematite-ilmenite alone at several stages in the cooling of the complex (Charlier et al. 2010). Magma mixing is suggested to have occurred mainly in the dynamic environment represented by the magma conduit, so that concentrations of Fe–Ti are more haphazardly distributed than they are in mafic layered complexes such as the Bushveld and Great Dyke.

1.5 Liquid Immiscibility as an Ore-Forming Process

Liquid immiscibility is the segregation of two coexisting liquid fractions from an originally homogeneous magma. The two fractions may be mineralogically similar (silicate–silicate immiscibility) or very different (silicate–oxide, silicate–carbonate or silicate–sulfide immiscibility). The phenomenon of immiscibility is best observed in extrusive rocks where rapid quenching prevents the segregated products from rehomogenizing. Philpotts (1982), for example, noted two compositionally distinct glasses interstitial to cumulus minerals in a tholeiitic basalt. The two glass compositions were essentially granitic, on the one hand, and an unusual mafic assemblage, comprising pyroxene, magnetite–ilmenite, and apatite, on the other. The August 1963 eruption of the Kilauea volcano in Hawaii provided evidence of another form of liquid immiscibility where a directly observable sulfide melt separated from a cooling, basaltic magma (Skinner and Peck 1969). The Duluth Complex in Minnesota contains several small occurrences of massive Cu–Ni sulfide mineralization, as well as rare discordant bodies of ilmenite–magnetite–apatite (in oxide to apatite proportions of about 2 : 1) referred to as nelsonites. These occurrences are considered to provide evidence that both sulfide and Fe–Ti–P immiscible fractions separated from

the Duluth magma and that these represent a compositional continuum of immiscible products (Ripley et al. 1998). Although its occurrence is difficult to substantiate, immiscibility is important as an ore-forming process in mafic magmas and can lead to the formation of large and important deposits such as the PGE sulfide deposits of the Merensky Reef in the Bushveld Complex, South Africa (Box 1.7), and the Ni–Cu sulfide ores at Kambalda in Western Australia (Box 1.6) and at Sudbury in Ontario (Box 1.8). Various types of immiscibility are discussed below, with emphasis on silicate–oxide and silicate–sulfide immiscibility.

1.5.1 Silicate–Oxide Immiscibility

It is well known that unusual, discordant bodies of magnetite–apatite or ilmenite/rutile–apatite (nelsonite) are preferentially associated with some alkaline rocks, as well as with anorthosite complexes. Early experimental work showed that it is possible to create two immiscible liquids, one quenching to form a mixture of magnetite and apatite in proportions of about 2 : 1, and the other a rock that is dioritic in composition (Philpotts 1967). Further experimental studies showed that for a broad range of rock compositions under conditions of high fO_2, an immiscible FeO melt will separate from a magma of felsic composition (Naslund 1976). A significant immiscibility gap exists, for example, in the system $KAlSi_3O_8–SiO_2–FeO$ (Figure 1.19) at atmospheric fO_2 conditions, but the gap is greatly diminished in more reducing environments. The existence of immiscibility is also enhanced in magmas with high concentrations of P, Ti, and Fe, but the field diminishes with increasing Ca and Mg (Naslund 1983).

The observation that oxidized iron-rich magmas can segregate into two immiscible liquids, one which is Fe-rich and the other a more normal silicate composition, has relevance to the formation of Fe- and Ti-rich magmatic segregations in nature. In slowly

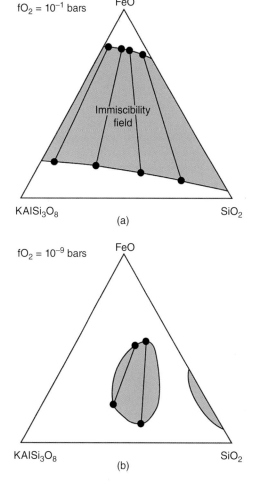

(a)

(b)

Figure 1.19 The nature of the "immiscibility gap" (shaded) that exists in felsic magmas under conditions of variable oxygen fugacity. The tie-lines indicate the coexisting liquid compositions. Source: After Naslund (1976).

cooled plutonic rocks, resorption reactions are largely responsible for the rehomogenization of segregated liquids, such that the products of immiscibility are not observed. In extrusive magmas, however, quenching can occur, or a more dense oxide liquid could separate from its silicate counterpart to be forcibly injected into a different part of the magma chamber, or into fractured country rock. Such an immiscible Fe- or Ti-rich fluid could also form a discrete magma or lava flow, a possibility that provides a theoretical basis for understanding the

important magnetite–hematite–apatite ores in, for example, the Kiruna district of northern Sweden (Freitsch 1978; Hildebrand 1986; Nyström and Henríquez 1992). Another example of what appears to be a flow of immiscible magnetite lava that separated from andesitic lavas has been documented at the El Laco volcano in northern Chile (Naslund et al. 2002). A word of caution though – in the cases of both Kiruna and El Laco there is considerable debate about the origin of the iron ores, and in the latter example in particular, a strong case has been made suggesting a hydrothermal origin for the Fe mineralization (Rhodes and Oreskes 1999; Sillitoe and Burrows 2002). The question of silicate–oxide immiscibility as a viable ore-forming process is still contentious, despite the fact that the process has been proven experimentally.

1.5.2 Silicate–Sulfide Immiscibility

By contrast with the process of silicate–oxide immiscibility, where there is some controversy over its occurrence in nature, the existence of silicate–sulfide immiscibility in mafic magmas is widely accepted as a feature of magma crystallization. Experimental data in the system SiO_2–FeO–FeS (MacLean 1969) confirms that silicate liquid can coexist with sulfide liquid over a significant range of compositions (Figure 1.20; two-liquid field). Magma at A in Figure 1.20, crystallizing Fe-rich olivine (fayalite), would evolve along A–A′ and eventually intersect the two-liquid phase boundary where the residual melt would comprise conjugate silicate and sulfide melts.

Sulfur is dissolved in most magmas as sulfide (S^{2-}), although in more oxidized melts ($fO_2 >$ FMQ) the sulfate radical (SO_4^{2-}) becomes increasingly important (Jugo 2009). In the following discussions, "solubility" is defined as the sulfur content of a magma when it is saturated in sulfide (i.e. when a sulfide liquid globule first appears in the silicate melt). Sulfur as sulfide is dissolved in magmas by displacing oxygen bonded to

Figure 1.20 Phase equilibria established experimentally in the system SiO$_2$–FeO–FeS. The field of coexisting silicate and sulfide liquids is shown by the stipple accentuated line. Oxidation of the magma would shift the phase boundary in the direction shown by the double arrow and expand the field of two-liquids. A magma represented by composition A and crystallizing fayalite would evolve along the line A–A′. Silica phases are represented by tridymite (Tr) and cristobalite (Cr) Source: After MacLean (1969).

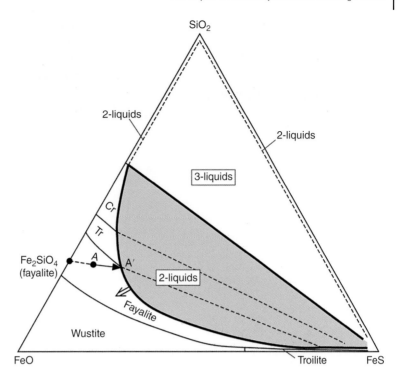

ferrous iron. Sulfide solubility is, therefore, a function of FeO activity in the magma, but is also controlled by oxygen fugacity (fO$_2$) and decreases as fO$_2$ increases (MacLean 1969). Sulfide solubility will vary as a magma progressively crystallizes and sulfide saturation can be achieved as non-sulfide bearing phases are extracted from the melt, or by a decrease in the amount of ferrous iron in the magma (such as might occur during extraction of an Fe-rich phase; see Figure 1.20). As demonstrated later, other factors, such as addition of externally derived sulfur, or ingress of new magma, may also promote saturation and the formation of an immiscible sulfide phase.

The immiscible sulfide melt that segregated from basaltic magma in the Kilauea volcano contained approximately 61% Fe, 31% S, 4% Cu, and 4% O. It subsequently solidified to form minerals such as pyrrhotite, chalcopyrite, and magnetite (Skinner and Peck 1969). Segregated sulfide melts forming in this fashion clearly have enormous potential to concentrate metals with both chalcophilic and siderophilic tendencies, such as base (Cu, Ni, Co) and precious (Au, Pt) metal ores, because these metals will partition into the sulfide fraction (Li and Naldrett 1994). There are many large and important ore deposits associated with the development of an immiscible sulfide fraction in mafic and ultramafic magmas (see Boxes 1.5–1.7). Central to the formation of all these deposits are three fundamental steps:

- the appearance of a substantial fraction of immiscible sulfide melt;
- the creation of conditions whereby the sulfide fraction can effectively equilibrate with a large volume of silicate magma; and
- the efficient accumulation of sulfide globules into a single cohesive layer or entity.

The processes whereby ore deposits are created during silicate–sulfide immiscibility are, therefore, complex and multifaceted, and are discussed in more detail in the following section.

1.6 A More Detailed Consideration of Mineralization Processes in Mafic Magmas

1.6.1 A Closer Look at Sulfide Solubility

The question of whether an immiscible fraction will develop in a magma or not is related to the amount of sulfur required to achieve sulfide saturation, or *sulfide solubility*. As mentioned previously, sulfide solubility decreases with increasing oxygen content in a magma because solution of sulfur appears to be controlled by the following equilibrium reaction (after Naldrett 1989a):

$$FeO \text{ (melt)} + \frac{1}{2}S_2 \Leftrightarrow FeS \text{ (melt)} + \frac{1}{2}O_2$$
$$(1.7)$$

Sulfide solubility increases as a function of increasing temperature and FeO content of the magma but decreases with increasing pressure and SiO_2 content. The data for sulfide solubility (reviewed in Naldrett 1989a) have been used to construct a generalized solubility curve for a fractionating Bushveld type mafic magma, as shown in Figure 1.21. The sulfur content at sulfide saturation is seen to decrease fairly rapidly from a maximum of about 0.4 wt% sulfide at the start of crystallization to below 0.1 wt% as olivine and orthopyroxene crystallize, but then levels off and may even increase again after plagioclase forms. Sulfide solubility, therefore, varies as a function of crystallization and fractionation.

A magma with an original sulfide content of 0.3 wt% would initially be undersaturated and its position would plot below the curve in Figure 1.21. Extraction of (non-sulfide bearing) olivine and pyroxene as cumulus minerals, however, would cause the sulfide content to increase in the residual magma until the saturation limit was attained (i.e. after about 10% crystallization). At this point

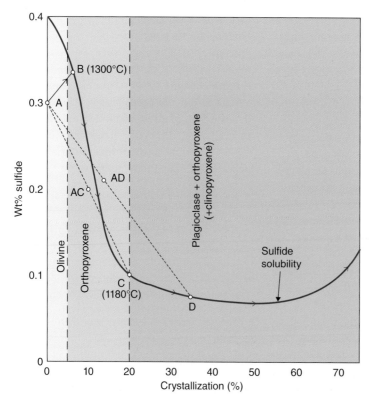

Figure 1.21 Variation in sulfide solubility as a function of progressive crystallization in a mafic magma such as that from which the Bushveld Complex formed. Source: After Naldrett and Von Grünewaldt (1989).

a small mass (in the appropriate cotectic proportion) of dispersed immiscible globules of homogeneous sulfide melt would form. If the sulfide globules are immediately extracted from the magma (by, for example, settling to the floor of the chamber) the amount of sulfide remaining in the magma would decrease, as dictated by the solubility curve. The distribution of trace elements between sulfide and silicate liquids can now be modeled in terms of crystal fractionation mechanisms, as described in Section 1.4.2. After solidification the cumulate rocks that had formed during this stage of the crystallization sequence would contain minor disseminated sulfide minerals (such as pyrrhotite, FeS, and chalcopyrite, $CuFeS_2$) trapped among the olivine, pyroxene, and plagioclase crystals. Alternatively, since the sulfide globules have a high density, they might also accumulate as a discrete sulfide layer toward the base of the magma chamber. The decrease in sulfide solubility with progressive crystallization of a mafic magma, as well as the concave-upwards shape of the solubility curve shown in Figure 1.21, both have important implications for the understanding of ore-forming processes in these rocks and this diagram is referred to again below.

1.6.2 Sulfide–Silicate Partition Coefficients

The ability of an immiscible sulfide fraction to concentrate base and precious metals depends on the extent to which metals partition themselves between sulfide and silicate melts (i.e. the magnitudes of the relevant sulfide–silicate partition coefficients). All metals that have chalcophilic tendencies are likely to partition strongly into the sulfide phase rather than remain in the silicate melt, and the data presented in Table 1.3 confirms that both base and precious metals are markedly compatible with respect to sulfide minerals. The partition coefficients indicate that although Cu, Ni, and Co partition strongly into the sulfide phase,

Table 1.3 Estimates of sulfide–silicate partition coefficients for base and precious metals in ultramafic and mafic magmas.

	Ni	Cu	Co	Pt	Pd
Komatiite					
27% MgO	100	250–3000	40	10^4–10^5	10^4–10^5
19% MgO	175	250–2000	60	10^4–10^5	10^4–10^5
Basalt	275	250–2000	80	10^4–10^5	10^4–10^5

Source: From Naldrett (1989a), Barnes and Francis (1995), Tredoux et al. (1995).

scavenging of platinum group elements (PGE)[1] by sulfide melt is even more efficient. Values of partition coefficients up to 100 000 for the PGE (Table 1.3) indicate that the presence of an immiscible sulfide fraction is potentially a powerful concentrating mechanism for these metals.

In reality, however, even though the sulfide–silicate partition coefficients for the PGE are so high, one seldom finds economically viable concentrations of these elements in layered mafic complexes, even when they do contain sulfide minerals. One reason for this is that the original concentration of PGE in magmatic reservoirs is typically very low. Another reason is that, even in magmas where immiscible globules of sulfide do form, the concentration mechanism may be diminished because sulfides are not able to interact (chemically) with the entire magma reservoir that contains the metals. In the case of the Skaergaard intrusion, for example, sulfide disseminations occur mainly in the upper portions of the layered body and sulfide immiscibility is believed to have occurred relatively late in the crystallization sequence (Anderson et al. 1998). The sulfide minerals

---

1 Six elements make up the platinum group – they are platinum (Pt), palladium (Pd), and rhodium (Rh), collectively referred to as the Pd subgroup, and osmium (Os), iridium (Ir), and ruthenium (Ru), referred to as the iridium subgroup. Both the Pd and Ir subgroups have strong affinities with Fe–Cu–Ni sulfides, whereas the iridium subgroup is also often associated with chromite and olivine cumulates.

at Skaergaard are depleted in Ni because a significant proportion of the latter metal was extracted by early formed olivine and orthopyroxene and was no longer available to be scavenged by the sulfide phase. On the other hand, the Skaergaard sulfide zone is cupriferous, and is also enriched in Au and Pd, because these metals are all incompatible and were concentrated into the residual magma during crystal fractionation. The Skaergaard sulfides appear, therefore, to have equilibrated only with the more differentiated magma fraction in the upper parts of the chamber.

The effects of early removal of trace elements from a magma are demonstrated in Figure 1.17, where it is seen that partitioning of a highly compatible trace element (for which D = 10 in Figure 1.17b) leads to enrichments in the solid cumulates only for the first 20–30% of crystallization (i.e. for values of F from 1.0 to about 0.7) and, thereafter, the solids are increasingly depleted in that element. This is because a high-D trace element will comprehensively enter the early formed solids, leaving the residual magma strongly depleted. Later solids will interact with a depleted reservoir and will likewise be depleted, irrespective of the magnitude of the partition coefficient. An immiscible sulfide melt within a silicate magma can also be regarded as a cumulus mineral phase and the same situation would, therefore, apply. In fact, the problem is particularly acute for the PGE with respect to an immiscible sulfide fraction because their partition coefficients are so high, and the original abundances in the parental magma so low (generally only a few ppb), that all but the earliest sulfide fractions will be depleted in these metals. One does, therefore, have to ask the question as to how large and economically viable concentrations of low abundance metals actually form. Or, put another way, how do the enormous concentrations of PGE-bearing sulfide ores such as those of the Bushveld Complex form when the very high partition coefficients for PGEs into sulfide and the very low abundances of metals in the original magma appear to limit the

extent to which viable concentrations of ore can be achieved? This question is addressed in the following section.

1.6.3 The R Factor and Concentration of Low Abundance Trace Elements

If a globule of sulfide melt equilibrates with an *infinite* reservoir of magma then the concentration of a trace metal in that, or any other similar, globule is given simply by the product of the partition coefficient and the initial concentration of the metal in the coexisting silicate melt. In reality, however, the sulfide globule is only likely to interact with a small, restricted mass of silicate melt, which is why the problem alluded to at the end of the previous section is very relevant. Clearly, the greater the proportion of silicate relative to sulfide melt, the higher and more persistent the concentration of compatible trace metals in the sulfide is likely to be. Campbell and Naldrett (1979) quantified the problem by proposing that the concentrations of low-abundance trace elements, such as the PGE, into an immiscible sulfide phase should be calculated in terms, not only of the original abundance in the magma and the relevant partition coefficients, but also of the silicate/sulfide liquid mass ratio termed the "R" factor. An expression which incorporates the R factor into the normal distribution law is provided in Eq. (1.8) (after Campbell and Naldrett 1979):

$$C_{sul} = C_o D(R + 1)/(R + D) \qquad (1.8)$$

where: C_{sul} is the concentration of a trace element in the sulfide fraction; C_o is the original trace element concentration in the host magma; D is the sulfide–silicate partition coefficient; and R is the mass ratio of silicate magma to sulfide melt.

Although the R factor is defined simply as a mass fraction, it is a parameter that effectively records the extent to which the immiscible sulfide fraction interacts with the silicate magma from which it is derived (and which also contains the metals that need to be concentrated if an ore deposit is to be formed). If a sulfide globule sinks through a lengthy column of magma (or is caught up in a turbulent plume) it is effectively interacting with a large volume of silicate liquid, and this can be equated to a high R factor (Campbell et al. 1983; Barnes and Maier 2002). Even though a sulfide globule may rapidly deplete the surrounding magma at any one instant of time, it may be provided with enhanced opportunities to scavenge compatible elements by continuously moving through undepleted magma. A low R factor, therefore, is analogous to a situation where a sulfide globule is static, or is removed early on from the magma, so that it is not able to efficiently scavenge compatible elements even though the partition coefficients may be very high.

The way in which the concentrations of compatible trace metals in sulfide melts vary as a function both of partition coefficient (D) and R factor is shown in Figure 1.22a (after Barnes and Francis 1995). For cases where D is much larger then R, the enrichment factor (C_{sul}/C_o) approximates the value of the R factor (i.e. the sloped portions of the curves in Figure 1.22a). Conversely, when the R factor is much larger than D, then the enrichment factor is approximately the value of D (Figure 1.22a – the flat parts of the curves). Another way of emphasizing the importance of the R factor is to examine its influence on the concentrations in a sulfide melt of two compatible trace elements, one with a moderately high partition coefficient (for example, Ni with D = 275) and the other with a very high partition coefficient (Pt with D = 100 000). Figure 1.22b shows that where the R factor is low (say 10^3), Ni concentrations in sulfide fractions are high (i.e. typical of most Ni-sulfide ores) because of the combination of high D and high initial Ni content (350 ppm) of the parental magma. By contrast, the Pt concentrations in the same sulfide fraction will be low because, despite the very large D, they have not had the opportunity of scavenging a substantial mass of the element from a magma that initially had a very low Pt abundance (5 ppb). By contrast, in situations where the R factor is high (say 10^6) the Ni concentrations of the sulfide fraction will not be significantly higher than the lower R factor case, but the Pt contents will have increased substantially because of more extensive interaction between the immiscible sulfide fraction and the silicate magma.

Variations in the R factor have important implications for the grades of sulfide phases

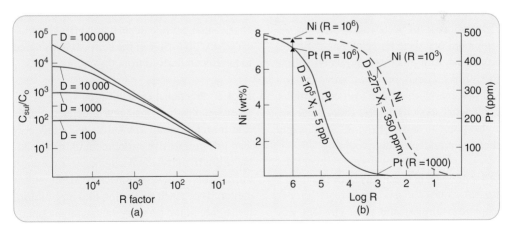

Figure 1.22 (a) Plot showing the relationships between partition coefficients (D), the R factor (i.e. the mass ratio of silicate melt to sulfide melt), and the degree of enrichment (C_{sul}/C_o) of compatible trace elements in the sulfide phase. Source: After Barnes and Francis (1995). (b) Diagram showing the effect of variations in the R factor on the concentration of Ni and Pt in an immiscible sulfide fraction that is in equilibrium with a basaltic magma. Source: After Naldrett (1989b).

within individual mafic intrusions. In the Muskox intrusion of northern Canada, for example, disseminated Cu–Ni sulfides on the margins of the body contain only moderate concentrations of PGE, whereas the more voluminous interior of the intrusion contains sulfides with higher PGE grades (Barnes and Francis 1995). This is attributed to a lower mass ratio of silicate to sulfide melt, and therefore a lower R factor, on the margins of the intrusion. Considerations such as these are likely to be relevant to exploration strategies for this type of deposit.

1.6.4 Factors that Promote Sulfide Saturation

It is conceivable that situations in which sulfide saturation are attained relatively late in the crystallization history of a magma might be disadvantageous to the formation of Ni–Cu or PGE-sulfide ores if the relevant metals had already been extracted from the residual magma. It is also conceivable that late sulfide saturation might equate with a low R factor simply because the volume of residual silicate magma has decreased because of ongoing crystallization. Consequently, scenarios whereby magmas become sulfide saturated relatively early in their crystallization history are generally considered to be advantageous for the development of a wide range of magmatic sulfide deposits. The processes by which sulfide saturation is attained, however, are many and varied – some of them are also contentious. The following sections discuss these processes and also point to some of the difficulties and challenges in understanding them.

1.6.4.1 Addition of Externally Derived Sulfur

Perhaps the simplest and most efficient way of promoting saturation in a magma whose composition is initially sulfur undersaturated is to increase the global amount of sulfur in the melt by addition from an external source (Ripley and Li 2013). Certain komatiite-hosted Ni–Cu deposits, such as Kambalda in Western Australia (Box 1.6) and several examples in Zimbabwe and Canada, are hosted in lava flows that have extruded onto sulfide-bearing footwall sediments such as chert, banded iron-formation, or shale. The komatiitic host lava commonly cuts down into its footwall, a feature attributed to a combination of thermal erosion and structural dislocation. Geological and geochemical evidence exists to suggest that the komatiitic magma arrived on the surface undersaturated with respect to sulfur, and that saturation was achieved by a combination of crystal fractionation and assimilation of crustally derived sulfur from footwall sediments (Lesher 1989). This concept is consistent with factors such as the localization of Ni–Cu ore within the footwall embayments created by extruded komatiitic lavas and is discussed in more detail in Box 1.6.

It has also been argued, for magmatic deposits that contain a large mass of actual sulfide ore relative to the size of the magma chamber, that deriving all of the ore sulfur from the magma itself is unlikely (Ripley and Li 2013), In such cases additional sulfur from an external crustal contaminant is considered essential, a view that receives support from S-isotope studies indicating crustally-derived sulfur in many magmatic ore bodies – an example is the large Ni–Cu–PGE sulfide deposit at Noril'sk in Russia where up to 50% of the sulfur is considered to have been derived from the country rocks.

1.6.4.2 Fractional Crystallization

The crystallization of silicate or oxide minerals that do not contain any sulfur will inevitably lead to a progressive enrichment of incompatible sulfur in the residual magma. At some point during this crystallization interval the sulfur content will reach saturation and an immiscible sulfide liquid (that crystallizes at magmatic temperatures to a homogeneous phase termed monosulfide solid solution or mss) will develop in the magma. This is analogous to the crystallization of chromite after an interval of olivine accumulation and described in the

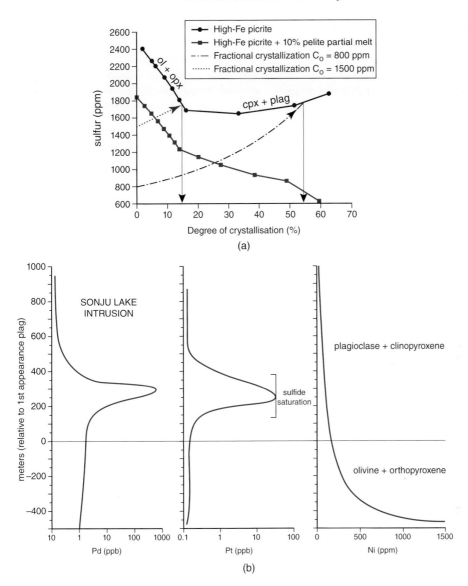

Figure 1.23 (a) Calculated models showing predicted sulfur content as a function of degree of crystallization (as a percentage). Solid curves model sulfur content at sulfide liquid saturation in an anhydrous Fe-rich picritic basalt; dashed curves model sulfur contents in the residual magma during fractional crystallization of the same basalt; (b) Pd, Pt, and Ni contents in rocks of the Sonju Lake intrusion (part of the Duluth Complex in Minnesota) plotted against vertical height relative to the first appearance of plagioclase in the intrusion. Source: Both plots modified after Ripley and Li (2013).

Irvine Model in Section 1.4.3. Predicted sulfur contents during fractional crystallization of olivine + clinopyroxene + plagioclase from a high-Fe picritic basalt composition are shown in Figure 1.23a (from Ripley and Li 2013). This plot shows the modeled sulfur content at

sulfide liquid saturation over the first 70% of crystallization and points to a trend of decreasing sulfur solubility as olivine + clinopyroxene dominate the cumulus assemblage, followed by a slight increase in solubility as the FeO content of the residual magma increases during clinopyroxene + plagioclase crystallization. Also plotted in Figure 1.23a are the calculated sulfur contents of the residual magma during progressive fractional crystallization for two model magmas that initially contained 800 and 1500 ppm sulfur respectively. These curves demonstrate that the low-S magma would attain sulfur saturation relatively late, at around 55% crystallization, whereas the high-S magma would become S-saturated much earlier at around 12% crystallization. It should be emphasized that the amount of immiscible sulfide that forms in a mafic magma on sulfide fluid saturation by fractional crystallization is small and is dictated by the cotectic proportion applicable to the system and the amount of sulfur originally contained in the magma.

It is pertinent to note that not all metals necessarily behave in an incompatible fashion and become enriched in the residual magma. Nickel in particular will be sequestered by olivine (see Section 1.2.2) and may undergo rapid depletion in the residual magma in the early stages of a mafic magma's crystallization history. This trend is well illustrated in Figure 1.23b which shows the Pd, Pt, and Ni contents of rocks from the Sonju Lake intrusion in Minnesota (after Ripley and Li 2013). Early crystallization, represented by dunitic and troctolitic cumulates at the base of the intrusion, was dominated by olivine + plagioclase fractionation resulting in a strong depletion of Ni in the residual magma (and in the rocks that formed from it). Over the same interval Pd and Pt contents remained constant or increased slightly at low levels of concentration (Figure 1.23b). At a later stage in the crystallization history of the intrusion, after the first appearance of cumulus plagioclase and clinopyroxene and the accumulation of gabbroic rocks, the Sonju Lake magma became saturated in sulfur – the presence of sulfide minerals in the sequence is marked by a significant increase in the Pd and Pt contents of rocks formed during this interval. The Ni contents over the same interval, however, show little or no corresponding

enrichment because the residual magma is now substantially depleted in this element (see Section 1.4.2 and Figure 1.17b). Any magmatic sulfide ore body that was located at this level in the Sonju Lake intrusion would almost certainly be PGE enriched, but Ni depleted.

Another example where sulfide saturation in a mafic magma has been attained by fractional crystallization, and where this process has actually resulted in the formation of a viable ore deposit, is found at Skaergaard in SE Greenland (Figure 1.15). The Platinova Reef at Skaergaard comprises a package of mineralized layers over a vertical interval of 120 m. This interval comprises sulfide mineral-enriched gabbroic rocks that accumulated at a late stage in the intrusion's crystallization history (70% solidified), when the magma had attained sulfur saturation (Holwell and Keays 2014). The mineralized zone is Cu–Au–Pd enriched and Ni–Pt depleted, consistent with the view that sulfide saturation was achieved late in the crystallization history, or from a magma that was itself already depleted in the latter two metals. The small volume of sulfide present in the Platinova Reef is consistent with a closed-system cotectic proportion, although the very small grain sizes and high Pd and Au contents of the actual sulfide minerals themselves has been attributed to partial dissolution of some of the sulfide liquid prior to solidification (Holwell and Keays 2014).

1.6.4.3 Injection of a New Magma and Magma Mixing

As mentioned in Section 1.3.2 above, large magma chambers are seldom the result of a single injection of melt and most witness one or more replenishments during the crystallization sequence. The Bushveld Complex and the Stillwater Complex in Montana contain important chromite deposits as well as PGE-enriched base metal ores associated with a sulfide-rich layer (the Merensky reef in the Bushveld and the J-M reef at Stillwater). These two layered mafic intrusions are both characterized by crystallization sequences that were interrupted by periodic injections of new magma. The magma replenishment events broadly coincide, in both cases, with the development of PGE-enriched sulfide horizons. In the Bushveld Complex numerous small injections of magma occurred early on during the formation of the Lower and Critical Zones, followed by a major magma replenishment around the level of the Merensky Reef and another toward the top of the Main Zone (Kruger 1994; see Box 1.7). At the Merensky Reef level detailed Rb–Sr and Re–Os isotopic measurements indicate that the new magma was more radiogenic (perhaps due to crustal contamination) but less differentiated than the magma remaining in the chamber at the time of injection (Kruger and Marsh 1982; Lee and Butcher 1990; Schoenberg et al. 1999). Further discussion on magma replenishment and its role in mineralization in other mafic intrusions is presented in reviews by Naldrett (1989a,b, 1997, 1999).

The sulfide solubility diagram in Figure 1.21 can be used to demonstrate how the mixing of new and residual magmas might promote sulfide saturation. Although the proportions of cumulus minerals being extracted from the magma, as well as the sulfur solubility curve itself, may not be universally applicable, the principles can be modified for application elsewhere. An initial magma composition represented by A in Figure 1.21, containing 0.3 wt% sulfide, would be undersaturated relative to sulfide (i.e. sulfide solubility at this stage is 0.4%). Extraction of olivine from the magma (A–B in Figure 1.21) would result in sulfide saturation and formation of a small, dispersed immiscible sulfide fraction. With continued crystallization and segregation of sulfides the magma would remain saturated, following the solubility curve as shown in Figure 1.21. If after some 20% crystallization (at point C just prior to the appearance of plagioclase as a cumulus phase) the chamber is replenished with the injection of a new magma (similar to A) and mixing occurs, then the composition of the mixture would lie somewhere along a mixing

line between A and C – the relative proportions of the two magmas would determine where on the mixing line the hybrid composition occurred. If the mixture is at AC, the hybrid composition would be undersaturated (i.e. plot below the sulfide saturation curve) and any sulfide globules present would be resorbed back into the magma. If, however, the chamber is replenished with the injection of a new magma at point D (after 35% crystallization, and *after* the appearance of plagioclase as a cumulus phase; Figure 1.21), then the hybrid magma composition at, for example, AD would be oversaturated with respect to sulfide. For the phase conditions and sulfide solubility applicable to this model, such mixing would result in the segregation of an immiscible sulfide fraction and a period of enhanced sulfide production. If this sulfide fraction can be efficiently collected by settling into a single layer, the sequestration of highly siderophilic metals by the liquid sulfide droplets might also explain the formation of PGE-sulfide bearing horizons such as the Merensky and J-M Reefs. A word of caution though – although the magma mixing model for the promotion of sulfide saturation in Figure 1.21 appears to provide an elegant explanation for deposits such as the Merensky Reef, it should be noted that the process is still poorly understood and somewhat contentious. In a discussion of Li et al. (2001a,b), Cawthorn (2002) has pointed out that, because the sulfur content at sulfide saturation is sensitive to melt composition, comparing the sulfide solubility of a fractionated magma with that of a hybrid magma, even though they may reflect similar degrees of fractionation, is invalid. It is also known from experimental work (Jugo et al. 2005; Jugo 2009) that sulfur contents at saturation are strongly fO_2 dependent, and correspondingly, on sulfur speciation – an exponential increase in sulfide solubility is now known to occur in more oxidized magmas where sulfate (SO_4^{2-}) species are stable alongside the more prevalent sulfide (S^{2-}) species. Accordingly, any assessment of ore forming processes based on models where sulfide solubilities are poorly constrained in terms of magma composition and oxidation state, are likely to be problematic. Nevertheless, the unquestionable link between episodes of magma replenishment and the formation of mineralized layers (both chromitites and PGE + base metal sulfides) is readily apparent in many layered mafic intrusions, despite the fact that more work is required before the detailed mechanisms of the metal concentration processes are fully understood (Barnes et al. 2001).

Box 1.6 Addition of External Sulfur and Sulfide Immiscibility: The Komatiite-Hosted Ni–Cu Deposits at Kambalda, Western Australia

An important category of magmatic Ni–Cu sulfide deposits is related to mafic and ultramafic volcanic rocks (komatiites) in Archean greenstone belts. Deposits of this type occur in Canada (the Thompson Ni belt of Manitoba) and Zimbabwe (Trojan and Shangani mines), but the largest and richest occurrences occur in the Kambalda region of Western Australia. Although a great deal is now known about these deposits (Lesher 1989) this case study focuses specifically on the source of sulfur in the immiscible globules of sulfide within the komatiitic lava flows that host the mineralization.

Komatiites are ultramafic extrusive rocks that were first described in the Barberton greenstone belt, South Africa, by Viljoen and Viljoen (1969). They comprise mainly olivine + clinopyroxene and typically contain MgO >18 wt% and low alkalis. As indicated in Section 1.2.2 of this chapter they are also characterized by high Ni contents. Komatiites are extruded onto the Earth's surface as high temperature, low viscosity lava flows characterized by a variety of volcanic forms of which pillowed lava and bladed quench textures (spinifex) are diagnostic. Most komatiites are not mineralized and their Ni is resident mainly in cumulus olivine, suggesting that on extrusion these lavas were undersaturated with respect to sulfide.

Evidence from the Kambalda region indicates that voluminous eruption of hot, low viscosity komatiitic lava formed extensive sheet flows and gradient-controlled lava rivers or channels (showing many of the features seen in parts of present day Hawaii). Hot lava rivers are believed to have thermally eroded discrete channels into the previously consolidated footwall and some of the Ni–Cu ores at Kambalda exhibit sulfide concentrations at the base of well defined, linear channelways (Figure 1.6.1). Another feature of this type of ore is that the mineralized channelways are devoid of interflow sediments even though such sediments occur laterally away from the ore zones at that level. These sediments comprise a variety of carbonaceous and sulfidic shales as well as sulfidic chert and banded iron-formation (Bavinton 1981). The nature of these sediments has led to suggestions that they represented a source of sulfur and this is supported by S isotope studies showing similarities in the isotope ratios of ores and interflow sediments (Lesher 1989). One implication of this idea is that assimilation of sediment by thermally eroding komatiitic lava channels enhanced the sulfide content within the magma and promoted local sulfide saturation and immiscibility.

Sulfur saturation and immiscibility of a sulfide fraction early in the magma crystallization history is considered to be one of the fundamental process responsible for mineralization in komatiitic lava flows. Chalcophile metals, in particular Ni and lesser Cu, were scavenged from the turbulent, flowing komatiite magma by the immiscible sulfide globules, which eventually accumulated as massive sulfide ore along the bottom of the channelways (Figure 1.6.1). The disseminated ore that overlies the massive sulfide ore reflects the static buoyancy contrast

Box 1.6 (Continued)

that existed between komatiitic crystal mush and massive sulfide. External derivation of sulfur has also been suggested for the promotion of sulfide saturation in the large Noril'sk–Talnakh Cu–Ni–PGE deposits in Russia (Grinenko 1985) and also for the Duluth Complex in Minnesota (Ripley and Al-Jassar 1987).

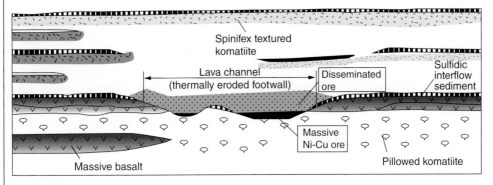

Figure 1.6.1 The characteristics of komatiite-hosted Ni–Cu deposits in the Kambalda region, Western Australia. Source: After Solomon et al. (2000).

Box 1.7 New Magma Injection and Magma Mixing: The Merensky Reef, Bushveld Complex

In addition to its huge chromite reserves, the Bushveld Complex also hosts the world's largest reserves of platinum group elements (PGE). About 80% of the world's PGE reserves are located in the complex, from three specific horizons, namely the Merensky Reef (where exploitation commenced and which itself contains about 22% of the world's total PGE reserves; Misra 2000), the UG2 chromitite layer, and the Plat Reef. A simplified outline and section of the Bushveld Complex, showing the extent of the Merensky Reef, is presented in Box 1.4. This case study focuses specifically on the Merensky Reef as it was formed at the time when a major injection of new magma occurred into the chamber, and it also marks a regional mineralogical hiatus separating the Critical Zone from the Main Zone (Kruger and Marsh 1985). It is typically represented by a <1 m thick, coarse-grained (or pegmatoidal), feldspathic pyroxenite that extends along strike for about 250 km. The origin of the Merensky Reef is a contentious issue (Naldrett 1989a) and arguments range from purely magmatic models to those involving interaction with magmatic-hydrothermal fluids (Cawthorn 1999). There is some evidence that a hydrothermal fluid may have been implicated in the formation of the Merensky Reef and its contained PGE mineralization and this topic is discussed in more detail in Chapter 2. For the purposes of this discussion, however, it is accepted that the Merensky unit (i.e. the mineralized reef and the rocks immediately overlying it, called the Bastard unit) owes its origin to the turbulence and magma mixing that accompanied a magmatic replenishment event (Schoenberg et al. 1999).

(Continued)

Box 1.7 (Continued)

Mineralization in the Merensky Reef is evident in the presence of disseminated base metal sulfides, mainly chalcopyrite and pyrrhotite–pentlandite, with which minor PGE sulfides (such as braggite, cooperite, laurite, and moncheite) and PGE metal alloys are associated. Evidence for a new magma pulse at the Merensky unit is most obvious in terms of mineral–chemical trends. The most convincing evidence for the input of a new magma, however, comes from a significant shift in the initial $^{87}Sr/^{86}Sr$ ratio (Ro) of rocks below and above the Merensky unit (Figure 1.7.1). Such a change cannot be achieved by crystal fractionation and must record the input of new magma. In this case the new magma had a higher initial $^{87}Sr/^{86}Sr$ ratio than the original magma pulse, a trait that is attributed to crustal contamination of the new magma increment. Magma mixing, sulfide segregation, and mineralization in the same interval are also, therefore, likely to be related to this event. Recognition of replenishment events should be regarded as a useful criterion for the exploration of PGE mineralization in layered mafic intrusions.

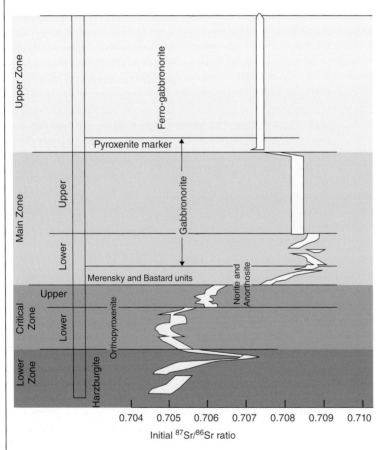

Figure 1.7.1 Sr isotope variations (in terms of the initial $^{87}Sr/^{86}Sr$ ratio or Ro) with stratigraphic height in the Bushveld Complex. Source: After Kruger (1994).

Box 1.7 (Continued)

Figure 1.7.2 Twenty meter deep pothole exposed in a quarry on the farm Maandagshoek (eastern Bushveld) showing anorthosite cutting down through melanorite at the approximate position of the UG3 and UG2 chromitite layers. Source: Photo courtesy of Wolfgang Maier.

Another intriguing feature of the Merensky Reef linked to injection of new magma is the presence of enigmatic "potholes" that are often revealed in underground mine workings, and elsewhere (Figure 1.7.2). Potholes are typically circular in plan view and range in size from 1 m to over 100 m in diameter – in the case of the Merensky Reef potholes may cut vertically down into the footwall by as much as 20–30 m. They are of considerable inconvenience to mining operations and also reflect changes in the mineralogy of the PGE, in that metal alloys tend to dominate over PGE minerals (i.e. sulfides, tellurides, etc.). Their origin is highly debated (see Buntin et al. 1985) and has been attributed to either downward erosional processes (such as convective scouring or thermal erosion) or upward fluid/melt migration (involving focused vertical migration of filter-pressed interstitial magma or a hydrothermal fluid). A different structural model, however, envisages syn-magmatic extension of the footwall cumulus assemblage just prior to injection of the new magma (Figure 1.7.3; Carr et al. 1999). The pull-apart rupture grows in a down-dip direction (perhaps due to subsidence and slumping) and is then filled by the new magma to form the Merensky unit, with the mineralized reef at its base. This model received support from studies at stratigraphic levels below the Merensky Reef where sizeable potholes have also been discovered (Maier et al. 2016) – the pothole shown in Figure 1.7.2, for example, shows an anorthositic magma slurry cutting down into a pull-apart structure that formed by extension of the footwall cumulus assemblage at the UG2 and UG3 chromitite levels.

(Continued)

Box 1.7 (Continued)

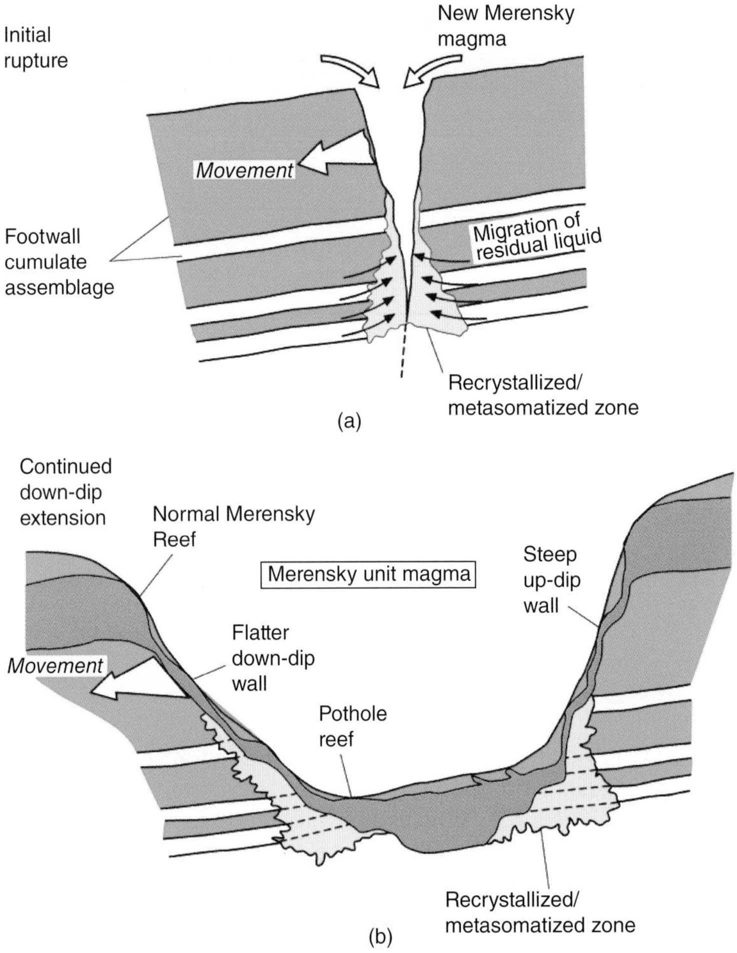

Figure 1.7.3
Schematic cross sections illustrating the progressive development of potholes in the Merensky Reef by syn-magmatic extension in the footwall cumulate rocks just prior to and during injection of a new magma. Source: After Carr et al. (1999). Reproduced with permission of Springer.

(a)

(b)

Box 1.8 Magma Contamination and Sulfide Immiscibility: The Sudbury Ni–Cu Deposits

Ni–Cu mineralization associated with the Sudbury Igneous Complex in Ontario, Canada represents a fascinating and unique ore occurrence. Although small (only 1100 km^2) in comparison to the Bushveld Complex, the Sudbury layered mafic intrusion has for a long time been the world's leading producer of nickel, and contains over 1.5 billion tons of ore at an average grade of some 1.2 wt% Ni, together with substantial Cu and PGE credits (Misra 2000). It is even more fascinating for the widely held view that it is the product of a large meteorite impact that struck Earth exactly 1849 Myr ago (Davis 2008; Krogh et al. 1984). The meteorite impacted into a composite crust containing Archean granite gneisses and the Paleoproterozoic volcano-sedimentary Huronian Supergroup.

Box 1.8 (Continued)

The main mass of the Sudbury Complex is made up of a differentiated suite of norite, quartz gabbro, and granophyre. The Ni–Cu ores, however, are found in an enigmatic mafic unit at the base of the succession termed the "sublayer" (Figure 1.8.1). The interior of the structure is underlain by a chaotic sequence of brecciated debris and volcaniclastic material interpreted as the fall-back from the meteorite impact. The granitic and sedimentary floor to the structure also contains evidence for a violent meteorite impact in the form of breccia, pseudotachylite (frictional melt rock formed by very high strain rates), and shatter cones (Lightfoot 2016 and Figure 1.8.2a). A widely accepted model suggests that Sudbury magmatism was triggered by the meteorite impact event, or more specifically by partial melting of crustal source rocks, during the rebound and pressure release that immediately followed impact (Naldrett 1989a).

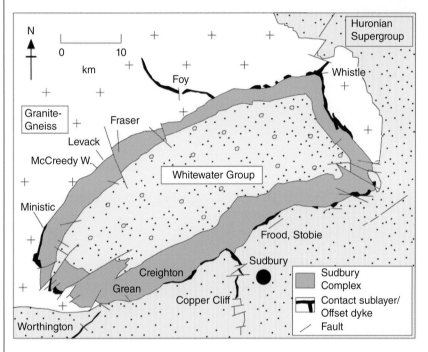

Figure 1.8.1 Simplified geological map of the Sudbury Complex showing the relationship between the main mass of the layered mafic intrusion, the Ni–Cu mineralized "sublayer" and offset dykes. Source: After Prevec et al. (2000). Reproduced with permission of Elsevier.

Ni–Cu sulfides (mainly chalcopyrite and pyrrhotite–pentlandite) are found along the basal contacts of the sublayer as massive ore which grades upwards into more disseminated mineralization. Significant sulfide mineralization also occurs in the brecciated floor of the structure (Lightfoot 2016 and Figure 1.8.2b) and also where sulfide melt percolated downwards into embayments and fractures. The offset dykes also contain steeply plunging pods of sulfide mineralization. An enigmatic feature of the Sudbury ores, however, is their high Ni contents relative to Cu. High Ni magmatic sulfide ores are generally associated with ultramafic rocks (see, for example, Box 1.5 for a description of the Kambalda deposits), with more differentiated rocks

(Continued)

Box 1.8 (Continued)

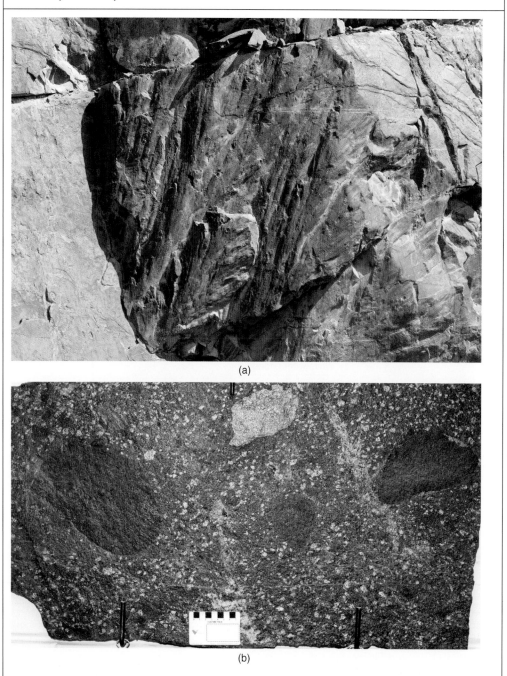

(a)

(b)

Figure 1.8.2 (a) Shatter cone in country rock surrounding the Sudbury Igneous Complex – evidence for cataclysmic deformation during a meteorite impact event; (b) slab of Ni–Cu sulfide-bearing diorite from the Frood-Stobie Mine, showing inclusions of granite and gabbro. Source: Photos are courtesy of Peter Lightfoot.

Box 1.8 (Continued)

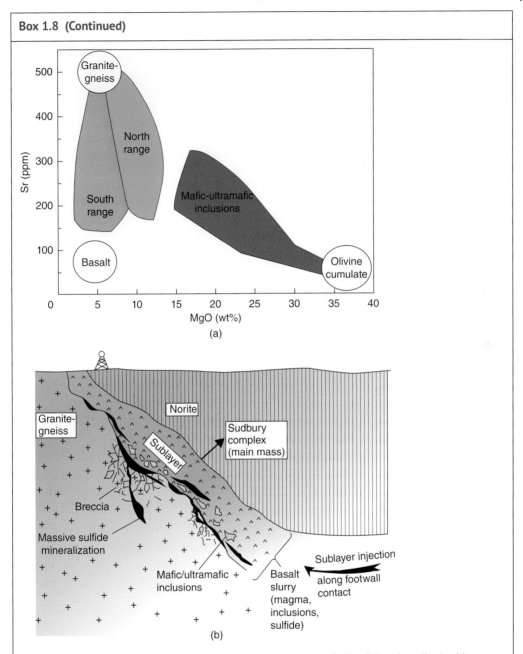

Figure 1.8.3 (a) Plot of Sr versus MgO showing the compositional fields of the mineralized sublayer from the north and south ranges of the Sudbury structure, and that of the mafic–ultramafic inclusions within the sublayer. The sublayer is a mixture of melts derived by near total melting of granite–gneiss and tholeiitic basalt of the Huronian Supergroup. The inclusions formed by olivine and pyroxene fractionation of a high-Mg melt that differentiated toward lower MgO contents. (b) Schematic and simplified cross section through a sublayer-hosted Ni–Cu sulfide orebody in the Sudbury Complex showing some of the features of the Prevec et al. (2000) model for the origin of the sublayer. Source: After Prevec et al. (2000). Reproduced with permission of Elsevier.

(Continued)

Box 1.8 (Continued)

being characterized by higher Cu/Ni ratios. An important, and relevant, feature of the Sudbury ores in the sublayer is their intimate association with exotic ultramafic inclusions. These inclusions are now known to be the same age as the main Sudbury Complex itself and display no mantle isotopic or geochemical signatures (Prevec et al. 2000). Their presence has important implications for the source of metals in the Sudbury deposits (Lightfoot 2016).

An intriguing aspect of the relationship between meteorite impact and Ni–Cu sulfide mineralization at Sudbury is the notion that the catastrophic impact was responsible for the extensive and widespread contamination of magma as it intruded into the dust and debris of the impact scar (Naldrett et al. 1986). The rocks of the Sudbury Complex are highly contaminated by crustal material and this is readily apparent in their high silica and potassium contents relative to normal continental basalts. One of the features of the Sudbury ores is that this type of contamination might have been responsible for the promotion of sulfide immiscibility and mineralization, as discussed in Section 1.6.4 of this chapter. In detail, however, the picture is more complex and the sublayer may have had an origin that is different to the remainder of the intrusive complex (Prevec et al. 2000; Lightfoot et al. 2001). The main mass of intrusive magma is thought to have been derived from the widespread melting of granite–gneiss and volcano-sedimentary target rocks immediately after meteorite impact and elastic rebound (Figure 1.8.3a). These crustally contaminated melts crystallized a dominantly plagioclase + pyroxene assemblage to form the bulk of the Sudbury complex, from which the offset dykes were also tapped. The sublayer, however, is considered to have been derived by melting of a much higher proportion of Huronian basaltic target material such that its parental magma was more mafic than that of the main mass. It crystallized an assemblage dominated initially by olivine, forming ultramafic cumulate rocks at depth in the structure (Prevec et al. 2000). This particular magma was also sulfur saturated and segregated significant volumes of immiscible Ni–Cu sulfide melt, as well as accumulating sulfides from above. It was then emplaced along the brecciated footwall contact as a contaminated basaltic slurry, comprising magma, ultramafic inclusions, country rock fragments, and sulfide melt, at a relatively late stage in the evolution of the complex (Figure 1.8.3b). Although convoluted, a meteorite impact model would appear to best explain the complex geology of the Sudbury Complex and to accommodate the inter-related processes of anatexis, crustal contamination, and sulfide segregation required to explain its contained ores.

1.6.4.4 Magma Contamination

In the discussion on the formation of monomineralic chromitite seams (Section 1.4.3 and Box 1.4) it was mentioned that one way of forcing a mafic magma composition off its cotectic crystallization path into the stability field of chromite was to contaminate the magma with siliceous crustal material. The same process will also result in a lowering of the sulfur content required for sulfide saturation, thereby promoting the formation of an immiscible sulfide fraction. This is demonstrated in Figure 1.23a where sulfide solubility calculated for a high-Fe picritic magma contaminated with 10% pelite is seen to be substantially lower than for uncontaminated picrite. Figure 1.24 shows a 1200 °C isothermal section of the SiO_2–FeO–FeS system in which the silicate–sulfide immiscibility field is identified. The composition of a homogeneous magma undersaturated in sulfide at point A can be forced into the field of two liquids by the addition of SiO_2 (toward B). At B the two liquids in equilibrium will have compositions at Y (silicate rich) and X (sulfide rich). This process could have applied to the

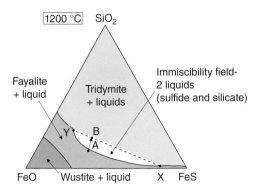

Figure 1.24 The ternary system FeO–SiO$_2$–FeS at 1200 °C showing how the addition of silica to a homogeneous, sulfide-undersaturated magma will force its composition below the solvus into the field of two liquids, one at Y (silicate-rich) and the other at X (sulfide-rich) Source: After Naldrett and MacDonald (1980).

formation of the Ni–Cu ores at Sudbury (see Box 1.8) where contamination of the host magma may have facilitated the formation of an immiscible sulfide fraction which scavenged Ni and Cu from the magma to form the ore deposits.

1.6.5 Other Models for Mineralization in Layered Mafic Intrusions

The preceding sections described processes such as magma replenishment and sulfide immiscibility that have come to be regarded as good "first-order" explanations for mineralization in layered mafic intrusions. In detail, however, it is evident that additional mechanisms are required to explain the many differences that exist from one deposit type to the next, and also the enigmatic patterns of metal distribution evident from detailed study and more rigorous exploration (Cawthorn 1999; Barnes and Maier 2002). Suggestions have been made, for example, that even in rocks as well mineralized as the Critical Zone of the Bushveld Complex, sulfide saturation might not have been achieved and the appearance of a sulfide melt was not the most important feature of the mineralization process. Instead, it is argued that, under conditions of low sulfur

fugacity (fS$_2$), PGE might crystallize directly from the silicate magma in the form of various platinum group minerals (PGM, such as braggite, laurite, malanite, moncheite, cooperite) or alloys. Although it is difficult to conceive how this might happen given the very low abundances of the PGE in the magma, a novel mechanism involving clustering of PGE ions has lent credibility to this as an alternative process to sulfide segregation (Tredoux et al. 1995; Ballhaus and Sylvester 2000).

1.6.5.1 PGE Clusters

Detailed studies of PGE mineralization have shown that these elements are heterogeneously distributed in rocks and minerals, even at a sub-microscopic level. Recent developments in the recognition of nanoparticles, as well as in the field of *cluster chemistry*, have indicated that the heavy transition metals in a magma will tend to coalesce by metal–metal bonding into clusters of 10–100 atoms (Schmid 1994). Experimental work has confirmed that clustering of siderophilic metals such as Pt and Pd to form nanoclusters well in advance of saturation and normal nucleation/crystallization will take place in silicate melts at magmatic temperatures (Helmy et al. 2013). Tredoux et al. (1995) and Wirth et al. (2013) have suggested that this mechanism might also apply to mineralization in natural PGE ores such as the Merensky Reef of the Bushveld Complex and the J-M Reef at Stillwater. Theory predicts that heavy transition metals form more stable clusters than light ones and that the PGE will therefore tend to cluster preferentially relative to Cu and Ni. Likewise, the heavy PGE (i.e. higher atomic numbers – Os, Ir, and Pt) will form more stable clusters than the light PGE (Ru, Rh, and Pd). Clustering may, therefore, provide a mechanism for fractionating metals in a magma chamber, and one that is related more to surface properties and charges (see Section 3.5.2) than to sulfide–silicate partitioning

Clustering of atoms and molecules in magmas is controlled by variables such as

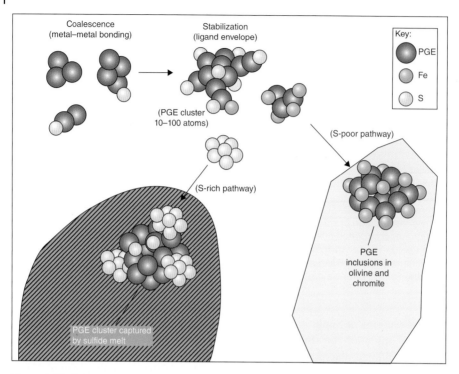

Figure 1.25 Schematic diagram illustrating the formation of PGE clusters in a magmatic system and their eventual inclusion either in an immiscible sulfide fraction (S-rich pathway) or in a silicate or oxide cumulus phase such as olivine or chromite (S-poor pathway) Source: After Tredoux et al. (1995), Ballhaus and Sylvester (2000).

temperature and composition. Figure 1.25 schematically illustrates the nature of PGE clustering in magma. Initial coalescence of metals occurs by metal–metal bonding, but the clusters are then stabilized by an envelope of ligands (see Chapter 3) such as sulfur or aluminosilicate species. Circumstantial evidence for PGE clusters comes from features such as light/heavy PGE fractionation and microbeam detection of minute PGE metal alloys in natural materials (Mathez 1999; Wirth et al. 2013). It has also been noted that cumulus olivine- and chromite-rich rocks, such as the harzburgites and chilled margins of the Bushveld Complex which show no association with sulfide segregations, are notably enriched in PGE relative to less mafic rocks (Davies and Tredoux 1985; Ballhaus and Sylvester 2000). Such an enrichment could be attributed to inclusion of PGE clusters into

early formed cumulus phases such as olivine and chromite. Stability of PGE clusters and their precipitation from the magma as discrete PGM or metal alloys is likely to be promoted by a low fS_2 environment.

The existence of PGE clusters in mafic magmas has important implications for understanding the mineralization process. If PGE clusters do form then crystal-chemical considerations and partitioning behavior might be less important than mechanical concentration mechanisms involving trapping of clusters as micro-inclusions in any suitable cumulus phase, be it silicate, oxide, or sulfide. Figure 1.25 illustrates two options, one where PGE clusters are incorporated into sulfide globules and the other where they are trapped as micro-inclusions in either olivine or chromite (Tredoux et al. 1995; Ballhaus and Sylvester 2000).

Detailed analyses of PGE in minerals of the Merensky Reef indicate that cluster theory might well explain some of the metal distribution patterns not previously explained by conventional sulfide partitioning behavior. Micro-inclusions of Os–Ir–Pt alloy are found in sulfide minerals such as pyrrhotite and pentlandite, whereas in olivine and chromite, micro-inclusions are dominated by a Pt–Pd–Au assemblage (Ballhaus and Sylvester 2000). The latter category is particularly significant as it suggests that the Bushveld magma contained abundant Pt– (Pd–Au) clusters early on in its crystallization history and that concentration of these metals (by trapping them as inclusions in early cumulus phases) can be achieved without sulfide saturation having occurred at all.

1.6.5.2 The Role of Chromite in PGE Concentration

The significant concentrations of PGE commonly associated with chromitite suggest that mineralization processes other than sulfide scavenging must have played a role (Scoon and Teigler 1994; Zaccerini et al. 2002). The UG2 chromitite layer also contains a different assemblage of the PGE than the sulfite-bearing Merensky Reef, for example, and is relatively enriched in Ru, Os, and Ir, but with similar Pt contents (von Grünewaldt et al. 1986). At a first glance this would seem to suggest that heavy PGE clusters had been preferentially concentrated as metal-alloy inclusions in cumulus chromite grains (similar to the process shown in Figure 1.25), a model first suggested by Hiemstra (1979). More detailed work, however, reveals the following characteristics of PGE mineralization in chromitite layers:

1) The chromitite layers do, in fact, contain minor sulfide minerals, both interstitially to the chromite grains and as inclusions (of which laurite RuS_2 is predominant; Merkle 1992) within them.

2) Pyrrhotite is virtually absent from the chromitites probably because of reaction between sulfides and chromite which partitions additional Fe into chromite and liberates sulfur (as shown in Eq. (1.9), after Naldrett and Lehmann 1988):

$$4Fe_2O_3 \text{ (chr)} + FeS \rightarrow 3Fe_3O_4 \text{ (chr)} + 0.5S_2$$
$$(1.9)$$

3) The assemblage Pt–(Pd–Au) occasionally occurs as discrete metal-alloy inclusions in chromite (Ballhaus and Sylvester 2000), indicating that the other PGE are associated with sulfide minerals, either as inclusions within, or interstitially to, chromite.

These characteristics led Barnes and Maier (2002) to suggest that an association between PGE concentration and chromitite layers could be related to both sulfide accumulation and metal clustering. In their model sulfide segregation under high R factor conditions leads to the accumulation of PGE-enriched sulfide cumulates together with the normal silicate minerals and chromite. The sulfide phase reacts with chromite consuming Fe and liberating S, resulting in a Cu–Ni rich residual sulfide assemblage that crystallizes minor chalcopyrite and pentlandite. The localized lowering of fS_2 that results from sulfide–chromite interaction promotes the direct crystallization of PGM or metal alloys from clusters in the magma so that PGE concentration in the immiscible sulfide fraction is dictated by clustering principles rather than strict partitioning behavior. The unusual PGE content of chromite-rich cumulate assemblages arises from the fact that Ru and, to a lesser extent, Rh partition strongly into the oxide phase whereas other PGE (Pd and Pt) are specifically excluded from the chromite lattice and remain in solution. Since Ru and Rh are fractionated into the chromite, the remainder of the PGE cluster (i.e. Os, Ir, and Pt) destabilizes and is likely to precipitate as PGM or metal alloys. The latter are scavenged by any interstitial sulfides present in and around the chromitite layer.

Application of composite models, as illustrated above, seems to provide an adequate explanation for several of the intriguing characteristics of PGE mineralization in both sulfide-rich and sulfide-poor environments. Detailed observations, such as the location of PGE in different mineralogical sites, clearly provide the sort of constraints that are needed to improve our understanding of the relevant ore-forming processes.

1.6.5.3 Hiatus Models

One of the tendencies of traditional models for the formation of magmatic deposits is to consider the processes in terms of accumulation of cumulus phases, be they silicate or oxide minerals, or sulfide melt. An intriguing hypothesis by Cawthorn (1999) has suggested that the opposite, namely the lack of accumulation of cumulus minerals, might be the essential ingredient in the formation of PGE and base metal sulfide mineralization styles in certain types of deposits. The Merensky Reef, for example, represents a layer in the Bushveld magma chamber where a major injection of new magma occurred and where there is a pronounced mineralogical hiatus in the magmatic "stratigraphy" (see Box 1.7). Magma replenishment could create a zone of mixing where the liquidus temperature is substantially above that relevant to local crystallization, resulting in a transient stage when no silicate minerals formed. This would also result in a hiatus in the solidification process, equating to a period, at least in terms of simplified gravity settling, with little or no accumulation of minerals on the chamber floor. Sulfide accumulation and/or metal clustering could, nevertheless, have continued during this interval so that enrichment and mineralization was assisted by the absence of phases that might normally contribute to dilution of the PGE. The view of Ballhaus and Sylvester (2000) with respect to Merensky Reef formation is somewhat analogous. They suggest that PGE concentration occurred by metal clustering in the magma to form an in situ stratiform PGE anomaly. The anomaly was effectively "frozen in" at the hiatus created by magma replenishment and local sulfide saturation, as the sulfide globules scavenged PGE clusters from the magma and then settled into the underlying cumulate mush.

1.6.5.4 Fluid-Related Infiltration of PGE

There is abundant mineralogical evidence to suggest that many layered mafic intrusions, as well as their contained mineralization, have been affected by alteration related to hydrothermal fluids derived either from the evolving magma itself or from an external source. There is little doubt that such fluids could have played a role in the remobilization of existing magmatic sulfide mineralization. A more pertinent question, however, is whether such magmatic-hydrothermal processes may not in fact have been dominant in ore formation (Schiffries and Rye 1990; Boudreau and Meurer 1999). Discussion of this topic falls outside the confines of this chapter but it is revisited in Chapter 2 (Section 2.12).

1.7 A Model for Mineralization in Layered Mafic Intrusions

The Bushveld Complex in South Africa, together with the Great Dyke of Zimbabwe and the Stillwater Complex of Montana, represent prime examples of chromite and PGE-sulfide ore formation in layered mafic intrusions. These examples have many features in common and the work of researchers such as Irvine (1977), Naldrett (1989a,b), Naldrett and von Grünewaldt (1986), Naldrett and Wilson (1991), and Campbell et al. (1983), as well as many others, facilitates the compilation of a relatively simple model to explain the occurrence and formation of these important deposits. The nature and origin of other Cu–Ni sulfide deposits such as Sudbury (Ontario), Noril'sk-Talnakh (Siberia), and Duluth (Minnesota) are somewhat different, but their formation can nevertheless be explained by some of the processes described below. The model described in this section represents a useful summary, but for ease of explanation is simplified, both conceptually and with respect to the actual nature of the ore deposits that it represents. Readers should be cautious in applying this model too rigorously to the real situation. Additional detail can be obtained from the original reference articles, in particular Naldrett (1989b, 1997).

Previous sections of this chapter have emphasized the following features as being important with respect to the formation of ores in igneous intrusions:

1) Crystal fractionation and gravity-induced crystal settling.
2) Density stratification of magma chambers and the ability of magmas to undergo density changes as crystallization proceeds.
3) Repeated recharge of chambers by injection of new magma.
4) The ability of new magma to find its own density level in the chamber and to turbulently mix with the residual magma.
5) The existence of transient periods of crystallization when only a single phase, such as chromite, is on the liquidus.
6) The formation of immiscible globules of sulfide liquid once the magma becomes saturated with respect to sulfide.
7) The very high partition coefficients of siderophilic elements for the immiscible sulfide fraction.

These features all contribute to the model presented in Figure 1.26, which shows a schematized layered mafic intrusion containing elements of mineralization representing well known deposits such as the PGE-sulfide bearing Merensky, J-M (Stillwater), and LSZ (Lower Sulfide Zone, Great Dyke) reefs, and the chromitiferous UG and LG (upper and lower groups respectively) seams of the Bushveld Complex. The UG2 chromitite is particularly important because it contains large reserves of Cr, as well as PGE, occurring together in the same, laterally extensive, horizon.

One empirical observation which has not been previously mentioned, but which turns out to be of considerable importance to understanding both the position and tenor of these ores, is the "stratigraphic" level in the layered intrusion where plagioclase first appears as a cumulus phase. This seemingly innocuous point in the evolution of an igneous intrusion (clearly marked in Figure 1.26) divides ore horizons which tend to be less PGE-enriched below it, from those above it which tend to be both more enriched and more substantial (at least in the Bushveld and Stillwater). In Figure 1.16a it is clear that an evolving magma could become progressively more dense once significant plagioclase is extracted. This has important implications for the behavior of newly injected melt into the magma chamber and affects the formation of fountains and plumes (Figure 1.16b) as well as the magnitude of the R factor, which in turn controls metal concentration into the immiscible sulfide fraction (Figure 1.22). Thus, with reference to Figure 1.26, any new magma which is introduced into the chamber early in its crystallization sequence (i.e. *prior to the appearance of plagioclase on the liquidus*) will tend to behave as a fountain. The limited degree of mixing that does occur (point AC in the "Irvine model," Figure 1.26) will promote the crystallization of chromite alone, and give rise to a chromitite layer associated with ultramafic cumulates such as those, for example, of the Great Dyke. The magma composition at this stage will be undersaturated with respect to sulfide (point AC in the "Naldrett and von Grünewaldt model," Figure 1.26) and no sulfide ores will, therefore, form. As crystallization continues, however (A to B in the "Naldrett and von Grünewaldt model," Figure 1.26), saturation will be achieved and immiscible sulfides will segregate to form disseminations of sulfide ore in an orthopyroxene cumulate rock. The very high partition coefficients of Cu, Ni, and the PGE into the immiscible fraction will mean that even with limited buoyancy and mixing the earliest sulfides will be enriched in these elements.

As further sulfides exsolve, however, they will rapidly become depleted in low abundance metals because the residual magma in

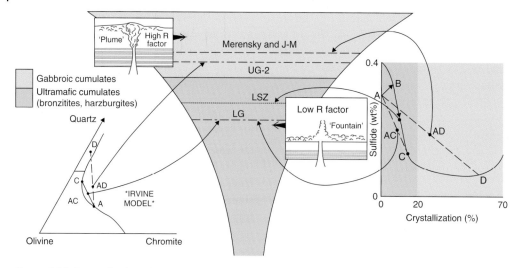

Figure 1.26 Generalized model showing the nature of igneous processes that give rise to the chromite and PGE–base metal sulfide deposits associated with layered mafic intrusions. LG and UG refer to the Lower and Upper Group (UG2 specifically) chromite seams of the Bushveld Complex; LSZ is the Lower Sulfide Zone of PGE mineralization in the Great Dyke; Merensky refers to the Merensky Reef of the Bushveld Complex and J-M is the J-M Reef of the Stillwater Complex, both of which contain PGE-sulfide mineralization. The difference between ultramafic and gabbroic cumulates in this model is marked by the first appearance of cumulus plagioclase in the latter. Source: Adapted from Naldrett (1997).

equilibrium with these sulfides, which itself originally had very low abundances of the PGE, becomes depleted (refer to Figure 1.22 for a quantification of this effect). This feature characterizes the Pt–Pd ores of the LSZ of the Great Dyke, for example, where despite the existence of Cu–F sulfides over a 2–3 m interval, the ores contain viable grades of PGE over only a few centimeters at the base of the mineralized zone.

By contrast, if a new injection of magma takes place *after the first appearance of plagioclase* then it will likely have a lower density than the residual magma and emplace itself in a turbulent, buoyant fashion at a relatively high level in the chamber. The more efficient mixing that occurs will again promote chromite precipitation (point AD in the "Irvine model," Figure 1.26). These chromite seams will be associated with either gabbroic or ultramafic cumulate rocks, depending on the relative proportions of liquids A and D

that are mixed together. Mixing of primitive magma A with an evolved liquid at D will also, in this case, result in a mixture that is sulfide oversaturated (point AD in the "Naldrett and von Grünewaldt model," Figure 1.26), resulting in the formation of a sulfide immiscible fraction. The plume effect that is created by the injection of the lower density primitive magma will result in buoyancy and a high degree of convective circulation of these sulfides, which translates into a high R factor. PGE are again strongly partitioned into the sulfide fraction but the grades will be higher and more evenly distributed because the sulfides have the opportunity of equilibrating with a much larger mass of the magma. This situation is applicable to the formation of the Merensky and J-M reefs, both of which are formed well above the first appearance of plagioclase cumulate rocks. It also applies to the formation of the UG2 seam where PGE-sulfides are intimately associated with a major chromitite layer.

1.8   Summary

There are many different types of ore deposits associated with igneous rocks; there is also a wide range of igneous rock compositions with which ore deposits are linked. Magmas tend to inherit their metal endowment from the source area from which they are partially melted. Fertile source areas, such as metasomatized mantle or sedimentary rock, are usually themselves a product of some sort of metal concentration process. Felsic magmas crystallize to form granites, or their extrusive equivalents, and are associated with concentrations of elements such as Sn, W, U, Th, Li, Be, and Cs, as well as Cu, Mo, Pb, Zn, and Au. Incompatible elements in felsic magmas are concentrated into the products of very small degrees of partial melting or into the residual magma at an advanced stage of crystallization. Such processes, however, do not very often result in economically viable ore deposits. Crystal fractionation in mafic magmas, on the other hand, results in important concentrations of elements such as Cr, Ti, Fe, and V, while associated sulfide immiscibility in these rocks results in accumulations of PGE, Cu, Ni, and Au. Layered mafic intrusions are important exploration targets for this suite of metals worldwide. Primary diamond deposits represent the very unusual situation where deep-seated mafic magmas vent to the surface as explosive diatreme-maar type volcanoes, bringing with them older, xenocrystic diamond from fertilized mantle. In both mafic and felsic magmas the latter stages of crystallization are accompanied by the exsolution of a dominantly aqueous and carbonic fluid phase that ultimately plays a very important role in ore formation. It is this process that is the subject of Chapter 2.

Further Reading

For those readers wishing to delve further into magmatic ore-forming processes, the following references to books and journal special issues will help:

Best, M.G. (2003). *Igneous and Metamorphic Petrology*, 2e, 729. Blackwell Publishing.

Special Issue on the Bushveld Complex (1985). *Economic Geology* 80 (4): 803–1211.

Chakhmouradian, A.R. and Wall, F. (2012). Rare earth elements. *Elements* 8 (5): 333–340.

Gunn, G. (ed.) (2014). *Critical Metals Handbook*, 439. Wiley and AGU.

Kirkham, R.V., Sinclair, W.D., Thorpe, R.I., and Duke, J.M. (eds.) (1993) *Mineral Deposit Modeling*. Geological Association of Canada, Special Paper 40.

Lightfoot, P.C. (2016). *Nickel Sulphide Ores and Impact Melts: Origin of the Sudbury Igneous Complex*, 680. Elsevier.

Misra, K.C. (2000). *Understanding Mineral Deposits*, 845. Dordrecht: Kluwer Academic Publishers.

Naldrett, A.J. (1989). *Magmatic Sulfide Deposits*. Oxford Monographs on Geology and Geophysics, 186. Oxford: Clarendon Press.

Naldrett, A.J. (1999). World-class Ni–Cu–PGE deposits: key factors in their genesis. *Mineralium Deposita* 34: 227–240.

Taylor, R.P. and Strong, D.F. (eds.) (1988). Recent advances in the geology of granite-related mineral deposits. *Canadian Institute of Mining and Metallurgy*, special volume 39: 445.

Whitney, J.A. and Naldrett, A.J. (eds.) (1989). *Ore deposition associated with magmas*, Reviews in Economic Geology, vol. 4, 250. El Paso, TX: Society of Economic Geologists.

2

Magmatic-Hydrothermal Ore-Forming Processes

TOPICS

Some physical and chemical properties of water
Magmatic-hydrothermal fluids
   water solubility in magmas
   first boiling, second boiling
   granite-related magmatic-hydrothermal ore deposits
Composition and characteristics of magmatic-hydrothermal fluids
Pegmatites and their significance
Metal transport in magmatic-hydrothermal fluids
   fluid-melt partitioning of trace elements
Water content and depth of emplacement of granitic magmas
Origin of porphyry Cu, Mo, and W deposits
Epithermal Au–Ag (Cu) deposits
Polymetallic skarn deposits
Fluid flow in and around granite intrusions
Role of fluids in mineralized mafic rocks

CASE STUDIES
Box 2.1 Magmatic-Hydrothermal Fluids Associated with Granite Intrusions (1) – La Escondida Porphyry Copper Deposit, Chile
Box 2.2 Magmatic-Hydrothermal Processes in Volcanic Environments – the Kasuga and Hishikari Epithermal Au–Ag Deposits, Kyushu, Japan
Box 2.3 Magmatic-Hydrothermal Fluids Associated with Granite Intrusions (2) – The MacTung Tungsten Skarn Deposit, Yukon, Canada
Box 2.4 Fluid Flow In and Around Granite Intrusions – Polymetallic Mineralization Associated with the Granites of Cornwall, Southwest England

2.1 Introduction

In Chapter 1 emphasis was placed on the concentration of metals during the igneous processes associated specifically with magma formation and its subsequent cooling and crystallization. Little mention was made in the previous chapter of aqueous solutions (or hydrothermal fluids), despite the fact that such fluids have often played a role in either the formation or modification of various magmatic deposits. The association between igneous and

Introduction to Ore-Forming Processes, Second Edition. Laurence Robb.
© 2021 John Wiley & Sons Ltd. Published 2021 by John Wiley & Sons Ltd.

hydrothermal processes is important for the formation of a wide variety of ore deposit types, especially in near surface environments where magmas and fluids tend to coexist as ore-forming entities. This chapter introduces the concept of a "magmatic-hydrothermal" fluid, focusing on the properties of fluids derived from within the body of magma itself. Particular emphasis is given to the fluids derived from *granitic magmas* crystallizing in the Earth's crust.

There is little doubt that the majority of ore deposits around the world either are a direct product of concentration processes arising from the transport of metals dissolved in hot, aqueous solutions circulating through the Earth's crust, or have been significantly modified by such fluids. A wide variety of ore-forming processes are associated with hydrothermal fluids and these can be applied to both igneous and sedimentary environments, and at pressures and temperatures that range from shallow crustal levels to those deep in the lithosphere. Many different types of fluids are involved in hydrothermal ore-forming processes. The most primitive or "juvenile" of these are the magmatic-hydrothermal fluids that originate from magmas as they cool and crystallize at various levels in the Earth's crust, and a number of important ore deposit types are related to the concentration of metals that arise from circulation of such solutions.

This chapter has particular relevance to ore deposits such as the important category of porphyry Cu and Mo deposits, as well as the "high-sulfidation" epithermal Au–Ag deposits that often represent the surface or volcanic manifestations of porphyry deposits. In addition, the formation of greisen-related Sn–W ores, polymetallic skarn mineralization, and pegmatite-related deposits are also explained in this chapter. A discussion of processes implicated in the formation of hydrothermal ores that are not directly linked to magmatic activity follows in Part II of the book (Chapter 3).

2.2 Some Physical and Chemical Properties of Water

Water (H_2O) is a liquid at standard ambient temperature and pressure (SATP – 25 °C and 100 kPa). By contrast, other light molecules (CH_4, NH_3, H_2S, etc.) at ambient conditions exist as vapor and form gaseous hydrides. H_2O boils at a much higher temperature than the hydrides formed by other light elements. The main reason for this is because of the net dipole moment that characterizes water molecules and the creation of temporary hydrogen bonds formed between neighboring water molecules (Figure 2.1). Water is, therefore, a liquid polymer at SATP, and this is responsible for its many anomalous properties.

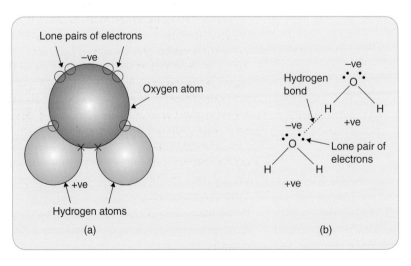

Figure 2.1 Idealized illustration of the water molecule and the nature of hydrogen bonding.

The physical properties of water that make it so important for chemical, biological, and geological processes include:

1) A high heat capacity, which means that it can conduct heat more readily than other liquids.
2) High surface tension implying that it can easily "wet" mineral surfaces.
3) A density maximum at temperatures just above the freezing point, which means that solid-H_2O (ice) will float on liquid-H_2O.
4) A high dielectric constant (also referred to as relative permittivity) and, hence, an ability to dissolve more ionic substances, and in greater quantities, than any other natural liquid.

The latter property is particularly relevant to the formation of ore deposits, since water is largely responsible for the dissolution, transport, and, hence, concentration of a wide range of elements and compounds, including metals, in the Earth's crust.

The chemical properties of water are dominated by its net dipole moment, a feature related to the fact that oxygen effectively attracts electrons more strongly than hydrogen, so that when they covalently bond, a net negative charge is associated with the oxygen compared to a net positive charge on the adjacent hydrogen. Consequently, the water molecule is not symmetrical but comprises two hydrogen atoms that are not directly opposite to one another but offset relative to the diametric axis of the single, much larger, oxygen atom (Figure 2.1). The H_2O molecule is, therefore, an electric dipole in which the center of positive charge does not coincide with the center of negative charge. As a result, water molecules can align themselves in an electrical field or orientate themselves around other charged ionic species. The clustering of water molecules around a dissolved ion is referred to as hydration, and this promotes the stability of ions in aqueous solution. The dielectric constant is a measure of chemical polarity and a fluid's ability to separate ions or other dipoles – the high dielectric constant of water has the effect of enhancing dissolution and increasing solubility.

In nature water exists as three different phases, solid (or ice), vapor (or steam), and liquid (Figure 2.2a). H_2O-ice typically exists below about 0 °C and forms when water molecules are arranged in a hexagonal crystalline structure similar to many rock forming silicate minerals. The density of H_2O-ice decreases markedly on freezing (to a value of about 0.92 g cm^{-3}) because of a reduction in the coordination number and expansion of the crystalline state. As H_2O-ice melts only a small proportion of the hydrogen bonds are broken and H_2O-liquid retains a large measure of tetrahedral coordination. The density maximum for water occurs at temperatures just above the freezing point (i.e. 4 °C), which explains why the liquid form of water is denser than its solid, and ice floats. Above the freezing point the density of H_2O-liquid decreases noticeably as a function of temperature, but increases slightly as a function of pressure, as illustrated in Figure 2.2b,c. Density defines the difference between the liquid phase (typically just below 1 g cm^{-3} at ambient temperature) and its coexisting vapor (where densities are one to two orders of magnitude lower). Lines of equal density (or volume) in P–T space are referred to as isochores (Figure 2.2c). The phase boundary along which liquid and vapor are in equilibrium (i.e. between the triple point and the critical point in Figure 2.2a) defines the equilibrium or saturation vapor pressure, also known simply as the boiling point curve. On the Earth's surface, water effectively "boils" when equilibrium vapor pressure exceeds the prevailing atmospheric pressure and vapor can bubble off from the liquid.

As temperature rises and the density of H_2O-liquid decreases, the latter must coexist in equilibrium with a vapor whose partial vapor pressure (and density) is increasing. The boiling point temperature of pure water increases progressively with pressure until a maximum limit is reached, called the critical

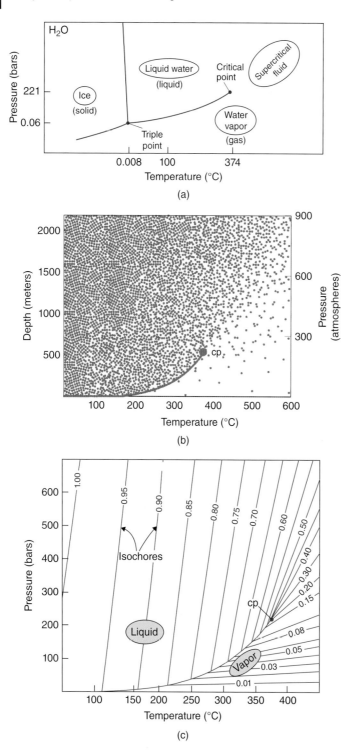

Figure 2.2 (a) Pressure–temperature phase diagram (not to scale) for pure H_2O showing the occurrences of the three phases (solid/ice, liquid water, water vapor/gas). The Triple Point, where solid, liquid, and vapor coexist, is at 0.008 °C and 0.06 bar. The Critical Point, beyond which there is no longer a physical distinction between liquid and vapor, is at 374 °C and 221 bar. The critical density is 0.322 g cm^{-3}. (b) Schematic visual representation of H_2O fluid densities in relation to the boiling point curve and critical point, cp. Source: After Helgeson (1964). (c) Part of the H_2O phase diagram showing actual variations of liquid and vapor equilibrium densities in pressure–temperature space (isochoral densities in g cm^{-3}).

point, at 374 °C and 221 bar (Figure 2.2a). The critical point is where it is no longer possible to increase the boiling point by increase of pressure and is effectively defined as the stage where there is no longer a physical distinction (i.e. a difference in density) between liquid and vapor. The densities of liquid and vapor at the critical point have merged to a value of around $0.3\,\mathrm{g\,cm^{-3}}$ (Figure 2.2c). Since the terms "liquid" and "vapor" no longer have any meaning at the critical point, the term "*supercritical fluid*" (see Section 2.3.3 below) is used to describe the homogeneous single phase that exists at pressures and temperatures above the critical point (Figure 2.2a). It should be noted that although the precise nature of a supercritical fluid is unclear and still the subject of debate among physical chemists (Woodcock 2014), it is the fact that supercritical fluids have both vapor- and liquid-like properties that is important in the present context.

For an ideal gas which obeys the Gas Law, the relation between pressure (P), temperature (T), and volume (V), or the reciprocal of density, is expressed by the well-known equation:

$$PV = RT \qquad (2.1)$$

where R is the specific gas constant.

Water only behaves as an ideal gas at high temperatures and low pressures and Eq. (2.1) cannot, therefore, be used to predict aqueous phase relations under natural conditions in the Earth's crust. Consequently, many attempts have been made to introduce correction factors and modifications to the ideal gas equation such that it will more accurately reflect non-ideal behavior. The modified forms of the equation, of which there are many that have been derived by both empirical and theoretical means, are referred to as equations of state. The simplest, general form of an equation of state, which incorporates two corrective terms, is an expression known as van der Waals' equation:

$$P = [RT/(V - b)] - a(V) \qquad (2.2)$$

where a is the corrective term that accounts for the attractive potential between molecules, and b is a term that accounts for the volume occupied by the molecules.

A widely utilized equation of state for water at higher pressures and temperatures is the modified Redlich–Kwong equation, a more detailed discussion of which, together with its applicability to the chemical and thermodynamic properties of magmatic fluids, can be found in Holloway (1987). A more detailed account of the properties of water as a solvent and their applicability to hydrothermal ore-forming environments can be obtained in Seward et al. (2014).

2.3 Formation of a Magmatic Aqueous Phase

2.3.1 Magmatic Water – Where Does It Come from?

In the very early stages of the Earth's development there was probably very little water either in the atmosphere or on the surface of the planet. It seems likely that the incipient oceans developed as soon as the early crust had stabilized and rain water ponded on its surface. Substantial bodies of water (oceans or seas) existed by at least 3800 Myr ago as evident from the preservation of early Archean sediments and subaqueously deposited volcanic rocks. A significant proportion of this early water is thought to have been derived from the degassing of volcanic magmas as they extruded onto the early crust and is referred to as juvenile water. Although still debated, a proportion of the Earth's water budget may also have been derived extraterrestrially, from comets and asteroids that impacted during the Late Heavy Bombardment. As plate tectonic processes progressively dominated Earth processes, water has been subjected to extensive recycling and it is likely that much of the fluid introduced by magmatic activity at the surface in more recent geological times is no longer juvenile, although it is still referred to as magmatic.

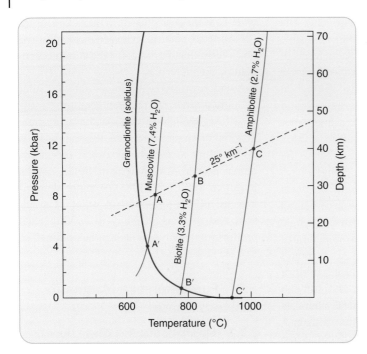

Figure 2.3 Pressure–temperature plot showing the approximate conditions under which dehydration melting of muscovite-, biotite-, and hornblende-bearing assemblages would take place in relation to a 25 °C km^{-1} geothermal gradient. Melts formed at A, B, and C respectively are likely to contain different initial water contents. The solidus curve for granite is for water-saturated conditions. Source: Redrawn after Burnham (1997).

Subduction of oceanic crust that has been highly altered and hydrated by percolating sea water is a scenario that can be used to explain the H_2O contents of most arc-related andesitic and basaltic magmas. Arc-related magmas probably, therefore, contain dissolved water derived from mixing of primitive, mantle-derived fluids and sea water. A minor component from meteoric fluid (i.e. derived from the hydrosphere) is also possible. It is interesting to note that hydrous basaltic magmas that pond at the base of the crust, and whose heat is responsible for initiating anatexis of the lower crust to form granitic melts, may contribute their water to these melts. Whitney (1989) has suggested that underplating of a felsic magma chamber with a denser mafic magma will result in diffusive transfer of elements and volatile species across the boundary layer.

Most of the water present in granitic magmas is, however, derived from the dehydration of minerals in the crust that were themselves melted to form the magma. The concept of "fluid-absent" or dehydration melting (i.e. where melting occurs without the presence of free water in the rock) is regarded as a realistic model for the generation of granites in the Earth's crust. The process can best be explained conceptually by considering the phase equilibria of essentially three minerals, namely muscovite, biotite, and hornblende. Although the detailed reactions are complex (see Clemens and Vielzeuf 1987; Whitney 1989; Burnham 1997), the approximate conditions for dehydration melting of muscovite, biotite, and hornblende in relation to an average geothermal gradient are shown in Figure 2.3. The amount of H_2O contained within these three minerals decreases from around 8–10% in muscovite, to 3–5% in biotite and 2–3% in hornblende. Accordingly, the water activity in melts formed at breakdown of these hydrous minerals will vary considerably and a magma derived from anatexis of a muscovite-bearing precursor is likely to contain more dissolved water than one derived by an equivalent melt proportion of amphibolite. Source material comprising dominantly muscovite and progressively buried along the 25 °C km^{-1} geotherm in Figure 2.3 will start to melt at point A. The water content of that first-formed magma will be 7.4 wt% (Burnham 1997). If the source material contained mainly biotite as the hydrous phase, melting would begin at a significantly higher temperature and pressure, at point B, and in this case the first-formed melt would contain only 3.3 wt% H_2O. Likewise, dehydration melting of an amphibolitic source rock would only produce a melt at even higher P–T (point C, Figure 2.3), and this would contain 2.7 wt% H_2O. It is apparent, therefore, that granites of variable composition derived from different levels in the crust might *initially* contain very different H_2O contents.

It is evident that substantial volumes of magmatic water will be added to the Earth's crust as granite magmas progressively build the continents. It is relevant to point out that melting of a muscovite- or a muscovite + biotite-bearing source rock (represented in nature by rocks such as metasediments) is likely to yield peraluminous, S-type granite compositions (see Section 1.3.5). The types of granites with which most Sn–W–U ore assemblages are typically associated will, therefore, have been derived from relatively hydrous magmas containing high initial H_2O contents. S-type granites generally correlate with the relatively reduced "ilmenite-series" granites of Ishihara (1977, 1981), which inherit their low fO_2 character by melting of graphite-bearing metasedimentary material. By contrast, melting of biotite- or biotite + hornblende-bearing protoliths (represented by meta-igneous rocks) will yield metaluminous, I-type granites. The porphyry Cu–Mo suite of ore deposits are associated globally with I-type granites which are initially relatively anhydrous compared to S-type granites. Ishihara's "magnetite-series" granites are often, but not always, correlatable with I-type granites and these are characterized by higher fO_2 magmas (Ishihara 1977, 1981).

2.3.2 H_2O Solubility in Silicate Magmas

It is evident from the above discussion that magmas derived in different tectonic settings are likely to have had variable initial H_2O contents, and that this is partly a function of the amount of water supplied by the source material during melting. There is, however, a limit to the amount of H_2O that a magma can dissolve, and this is defined by the solubility of water in any given magma (Lowenstern 1994). The solubility of H_2O in silicate magmas is determined mainly by pressure and to a lesser extent temperature. Figure 2.4a shows the results of experimental determinations of water content in melts of basaltic, andesitic, and granitic compositions. The experiments suggest that water content is strongly dependent on pressure, with magmas at the base of the crust (c. 10 kbar) able to dissolve up to 10–15% H_2O. It would also appear from Figure 2.4a that for any given pressure, felsic melts are able to dissolve more water than mafic ones.

When water dissolves in a magma it exists essentially as structurally bound hydroxyl (OH) groups and as discrete molecular water (H_2O) (Stolper 1982; Mysen 2014), although the relative proportions of the two species vary as a function of temperature (Figure 2.4b) and total water content (Figure 2.4c). Experiments demonstrate that in melts of granitic composition the abundance of OH groups increases with temperature and at the expense of molecular H_2O (Figure 2.4b). The relative proportions of OH and H_2O also vary, however, as a function of the total water concentration in a melt as, shown in Figure 2.4c.

The solubility of water in silicate magmas is thought to be governed by the following equilibrium reaction:

$$H_2O_{(molecular)} + O^o \leftrightarrow 2OH \qquad (2.3)$$

where O^o refers to the oxygen that bridges or polymerizes the silicate structure of the magma. Low viscosity basaltic magmas (with lower SiO_2 contents) comprise a smaller proportion of bridging O^o than more highly polymerized granitic melts. Basaltic melts may, therefore, accommodate fewer OH groups in O^o substituted sites, which could explain their inability to dissolve as much water as granite.

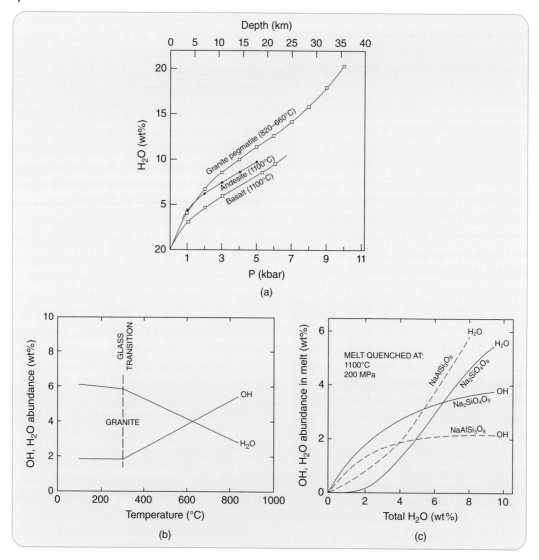

Figure 2.4 (a) Experimentally determined solubilities (in wt%) of H_2O in silicate melts as a function of pressure. 1, basalt (at 1100 °C); 2, andesite (at 1100 °C); 3, granitic pegmatite (at 660–820 °C). Source: After Burnham (1979). (b) Abundance of structurally bound OH groups and molecular H_2O in glass and melt of granitic composition as a function of temperature. (c) Abundance of OH groups and molecular H_2O as a function of total water concentration in quenched melts of compositions shown – solid lines $Na_2Si_4O_9$; dashed lines $NaAlSi_3O_8$. Source: (b) and (c) Redrawn after Mysen (2014). https://link.springer.com/article/10.1186/2197-4284-1-4#copyrightInformation. Licensed under CC BY 2.0.

The inter-relationships between water solubility and magma composition, temperature, pressure, and viscosity are, nevertheless, complex (Mysen 2014). In general, melts become progressively depolymerised as water contents rise so that water saturated magmas are less viscous then their anhydrous equivalents and, more importantly, water solubilities in silicate magmas decrease significantly with decreasing pressure. These relationships are fundamentally important to many geological processes as they control magma properties and also the appearance of an aqueous phase in magmas (discussed in more detail in the following section). Aqueous phase separation plays a critical role in the nature of volcanic eruptions and also underpins the understanding of magmatic-hydrothermal ore-forming systems, as discussed in the remainder of this chapter.

2.3.3 The Burnham Model

The importance of processes whereby zones of H_2O-saturated magma are formed and localized toward the roof of a granite intrusion, and their significance with respect to granitoid related ore deposits, was emphasized by the work of C.W. Burnham (1967, 1979, 1997). The concept has stimulated much fruitful research and continues to receive experimental and theoretical refinement through the work of Whitney (1975, 1989), Candela (1989a,b, 1991, 1992, 1997), Shinohara (1994), Williams-Jones and Heinrich (2005) and many others.

When a granitic magma crystallizes, the liquidus assemblage is dominated by anhydrous minerals and the concentration of dissolved incompatible constituents, including H_2O and other volatile species, increases by processes akin to Rayleigh fractionation (see Chapter 1). At some stage, either early or late in the crystallization sequence, granitic magma will become water-saturated, resulting in the appearance, or "*exsolution*," of an aqueous fluid to form a chemically distinct phase in the silicate melt. This process is called H_2O-saturation, but it is also often referred to as either "*boiling*" or "*vapor-saturation*." These terms often lead to semantic confusion and the footnote[1] should

provide some clarity on the issue. Because the aqueous fluid has a density that is considerably lower (usually less than or around $1\,g\,cm^{-3}$, see Figure 2.2c) than that of the granitic magma (which is typically around $2.5\,g\,cm^{-3}$) it will tend to rise and concentrate in the roof, or carapace, of the magma chamber. Although some of the original OH^- in the magma may be utilized to form hydrous rock-forming minerals (such as biotite and hornblende), the amount of magmatic-hydrothermal water formed in this way can be very substantial.

The concept of the formation of a zone of H_2O-saturated magma in a high-level granite intrusion (2 km depth) which initially contained some 2.7 wt% H_2O is schematically illustrated in Figure 2.5. At these shallow depths, H_2O-saturation is achieved after only about 10% crystallization, when the water content of the residual magma reaches 3.3 wt%. At low pressures such as those pertaining in Figure 2.5, the fluid does in fact boil, since the equilibrium vapor pressure equals that of the load pressure on the magmatic system and bubbles of gas (i.e. water vapor plus other volatiles such as CO_2) vesiculate. The process whereby vapor saturation is achieved by virtue of decreasing pressure (i.e. because of upward emplacement of magma or mechanical failure of the chamber) is called "*first boiling*" and is particularly applicable to high crustal levels. As mentioned earlier, it is also possible to achieve saturation with respect to an aqueous fluid by progressive crystallization of dominantly anhydrous minerals under isobaric conditions, and this process is referred to specifically as "*second boiling*." Second boiling generally occurs in more deep-seated magmatic systems and occurs only after a relatively advanced stage of crystallization. As shown later, the differences between first and second boiling, and more specifically the timing of H_2O fluid saturation relative to the progress of solidification of the magma, is very important

---

1 Unless the pressure, temperature, and composition of a magmatic aqueous solution are specified, one cannot say whether it exists as liquid or vapor, or a homogeneous supercritical phase (see Figure 2.2). In such a case it should be referred to by the generic term "*H_2O fluid*." Since the magmatic aqueous phase is so much less dense than the silicate melt from which it was derived, and because it may contain other low solubility volatile species such as CO_2 or SO_2, it is often referred to as the "vapor" or "volatile" phase. In addition, a homogeneous supercritical fluid is one that would effectively fill its container, and in this sense should be regarded as a gas or vapor, even though its density might be much higher than a gas as we might envision one at the Earth's surface. Accordingly, in this

book the terms H_2O-saturation, boiling, degassing, and vapor-saturation are used interchangeably. The concepts are discussed again in Section 2.4.4.

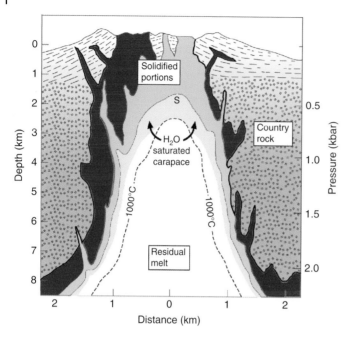

Figure 2.5 Schematic section through a high-level granodioritic intrusion undergoing progressive crystallization and showing the hypothetical position in space of the H_2O-saturated granite solidus (S), as well as a zone (in blue) where aqueous fluids might concentrate in the residual magma. Source: After Burnham (1979).

in understanding how different granite related ore deposits form (see Sections 2.6 and 2.7).

In addition to the strong dependence of water solubility on pressure, it is obvious that fluid saturation will also be a function of the original water content of the initial melt. Melts that are initially more enriched in water and volatiles will achieve saturation earlier (relative to the progression of crystallization) than those that are depleted in these components. Experimental studies (Whitney 1989) have confirmed the effects of pressure and initial water content on fluid saturation. Figure 2.6 compares the attainment of H_2O-saturation in a typical granite melt at high and low pressures, and in terms of crystallization sequences and initial H_2O content. In a situation deep in the Earth's crust (i.e. at 8 kbar), where the original granite melt contained 2 wt% H_2O, crystallization would have commenced with nucleation of plagioclase at temperatures around 1100 °C, followed by the appearance of K-feldspar and quartz on the liquidus at lower temperatures (path A–A′–A″, Figure 2.6a). H_2O-saturation is achieved at temperatures that are just a few degrees above the solidus (at A′) and only after over 80% of the melt had crystallized.

The solidus is intersected at 650 °C, at which point the magma has totally solidified. In the unlikely event that this granite was initially H_2O saturated at 8 kbar it would have contained at least 12 wt% H_2O and, as illustrated by the crystallization path B–B′ (Figure 2.6a), would not have started to crystallize plagioclase until the melt temperature had cooled to around 750 °C. In this case solidification would have progressed quite rapidly between 750 and 650 °C and entirely in the presence of H_2O fluid.

At shallower levels in the crust (2 kbar) the situation is quite different (Figure 2.6b). The same granite magma composition would be saturated in water if it originally contained only 6–7 wt% H_2O and in this situation (path D–D′) crystallization in the presence of H_2O fluid would take place over a wider temperature interval than at greater depth. The same crystallization path at 8 kbar would have existed over much of its temperature range in the H_2O undersaturated field. A granitic melt at 2 kbar with low initial water content (2 wt%; path C–C′–C″ in Figure 2.6b) would also crystallize over a significant temperature interval in the undersaturated field, as with deeper

Figure 2.6 Plots of temperature versus H_2O content showing the crystallization sequences for granitic melts cooling and solidifying at (a) deep crustal levels (8 kbar) and (b) shallower crustal levels (2 kbar). The bold lines in both cases refer to the H_2O-saturation curve, and also the liquidus and solidus. Pl – plagioclase; Q – quartz; Af – alkali feldspar; L – melt; V – H_2O fluid. Crystallization paths refer to hypothetical situations where the granite was either initially oversaturated in H_2O (B–B′ and D–D′) or markedly undersaturated in H_2O (A–A′–A″ and C–C′–C″). Source: After Whitney (1989).

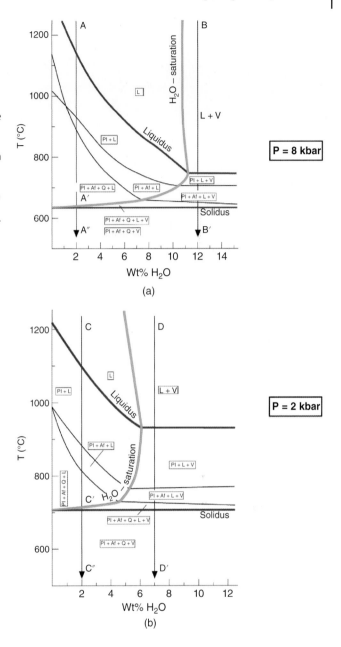

in the crust, but in this case H_2O-saturation would be achieved at a higher temperature (around 700 °C at C′) and after only some 60–70% crystallization.

These experimental data reinforce the concept that an aqueous fluid will exsolve from a granitic melt as a normal consequence of its crystallization. The Burnham model has great relevance to the formation of a wide range of ore deposit types. The porphyry Cu–Mo suite of deposits, epithermal precious metal ores, and polymetallic skarn type deposits are all examples of deposits whose origins are related to the processes conceptualized in this model. More detailed discussion of these deposit types is presented in later sections.

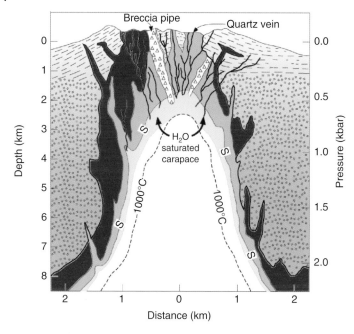

Figure 2.7 Schematic section through a high-level granodioritic intrusion (as in Figure 2.5) showing the nature of veining and breccia pipe formation that could form as a result of hydrofracturing around the apical portion of a granite body. Source: After Burnham (1979).

2.3.3.1 A Note on the Mechanical Effects of Boiling

An aqueous fluid constrained to the roof zone of a granite magma chamber will have limited effect on the distribution and concentration of metals unless it has the opportunity to circulate effectively in and around the intrusive complex from which it is derived. The appearance of an exsolved H_2O fluid within a magma is, however, also accompanied by the release of mechanical energy, since the volume per unit mass of silicate melt plus low density H_2O fluid is greater than the equivalent mass of water-undersaturated magma (Burnham 1979). This is because the densities of aqueous and volatile species are significantly lower than that of the silicate magma from which they exsolve. At shallow levels in the crust the volume change accompanying H_2O fluid production may be as much as 30% (at P_{total} of 1 kbar). This results in overpressuring of the chamber interior and can cause brittle failure of the surrounding rocks. The *hydrofracturing* that results from this type of failure usually forms fractures with a steep dip, because expansion of the rock mass takes place in the direction of least principal stress, which is usually in the horizontal plane. Hydrofractures tend to emanate from zones of H_2O fluid production in the apical portions of the granite body and may propagate into the country rock and even reach the surface. This concept is schematized in Figure 2.7.

Experimental work has confirmed that high level granite emplacement enhances the likelihood of brittle failure, both in the intrusion itself and in the surrounding country rocks (Dingwell et al. 1997), thereby providing excellent ground preparation for the efficient circulation of ore-bearing fluids. The factors that help to promote brittle failure in high level granite-related ore-forming systems include volatile saturation, which increases magma viscosity because of bubble vesiculation, and rapid cooling.

2.4 The Composition and Characteristics of Magmatic-Hydrothermal Solutions

2.4.1 Quartz Veins – What Do They Tell Us About Fluid Compositions?

Quartz veins are ubiquitous indicators of hydrothermal activity and mineralization, and represent the products of silica precipitation from hot aqueous solutions percolating through fractures in the Earth's crust. As pressure and temperature increases, water becomes an increasingly effective solvent and can dissolve significant quantities of most rock-forming minerals. The solubility of quartz in water increases to about 8 wt% at temperatures of 900 °C and pressures up to 7 kbar (Anderson and Burnham 1965). When dissolved in water, including ocean waters, silica exists in the form H_4SiO_4 (silicic acid). This is an abundant constituent of hydrothermal solutions and explains the common occurrence of quartz in veins. The variation of quartz solubility in aqueous solution is, however, quite complex (Rimstidt 1997) and other parameters in addition to P and T, such as pH and salinity, play an important role. The combined effects of P, T, and pH on quartz solubility in aqueous solutions is shown in Figure 2.8. The salinity of an aqueous solution (represented by its NaCl content) plays an important role in its ability to dissolve silica (Newton and Manning 2000). At P < 4 kbar quartz solubility increases slightly with increasing salinity, but at higher pressures there is a marked change in solution behavior and silica solubility decreases significantly, even at high temperatures. Silica solubilities in mixed H_2O–CO_2 solutions are known to be much lower than in aqueous brines indicating that it is the latter fluid type that is more likely to transport silica in the Earth's crust (Newton and Manning 2000).

2.4.2 Major Elements in Magmatic Aqueous Solutions

In addition to silica, water can dissolve significant amounts of other major elements, such as the alkali metals. In a classic experiment, Burnham (1967) reacted granite with pure water under a variety of conditions and showed that at high pressures and temperatures (10 kbar and 650 °C) the total solute content of the solution was about 9 wt% and comprised Si, Na, and K *in proportions that are approximately the same as the granite eutectic (or minimum melt) composition* (Figure 2.9).

This indicated that material precipitating from an aqueous solution at high P–T could have the same composition and mineralogy as granite that crystallized from a silicate melt (i.e. quartz + plagioclase + K-feldspar in approximately equal proportions). At progressively lower pressures and temperatures, however, the total solute content of the solution decreased (to a minimum of about 0.7 wt% at 2 kbar), with the alkali metals (i.e. Na + K) also decreasing relative to silica (Figure 2.9). Close to the surface, therefore, the products precipitating from an aqueous hydrothermal solution comprise mainly silica.

In addition to their ability to dissolve cationic species (i.e. electron acceptors) such as Na^+, K^+, and Si^{4+}, magmatic aqueous solutions can also transport significant amounts of Ca^{2+}, Mg^{2+}, and Fe^{2+}, as well as a variety of anionic substances, in particular Cl^-. The anions are referred to as ligands (electron donors) and there are several others which are also commonly found in aqueous solutions, including HS^-, HCO_3^-, and SO_4^{2-}. These additional components are important in ore-forming processes and are discussed in more detail below. Further discussion of solution chemistry, and of the ability of hydrothermal fluids to transport different metals, is presented in Chapter 3 and is also summarized in Seward et al. (2014).

2.4.3 Other Important Components of Magmatic Aqueous Solutions

The compositions of rocks can be represented essentially by 10 major element oxides (SiO_2, Al_2O_3, TiO_2, Fe_2O_3, MgO, MnO, CaO, K_2O, Na_2O, and P_2O_5) that are used to form the more abundant rock-forming minerals. Any given rock sample also comprises most other known elements, albeit for many of them in vanishingly small quantities. Their detection, for the most part, depends on the barriers imposed by analytical technology. To a certain extent the same is true of the solute content of aqueous solutions in the Earth's crust, which, although

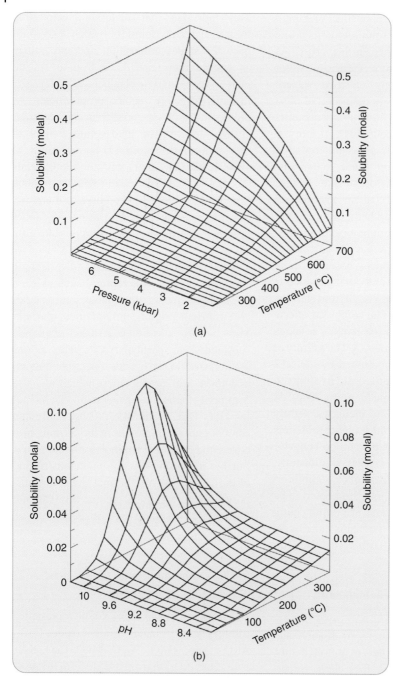

Figure 2.8 The variation of quartz solubility in aqueous solution as a function of (a) pressure and temperature and (b) temperature and pH. Source: After Rimstidt (1979).

dominated by a few highly soluble species, contain a wide range of other constituents, some of which are easily detectable, and the remainder occurring in only trace quantities. The trace constituents of an aqueous solution cannot simply be dismissed, especially in an

ore-forming context, as it is these ingredients that distinguish an ore-forming fluid from one that is barren.

The typical composition of a magmatic aqueous solution in the Earth's crust is probably best obtained by direct analysis of waters

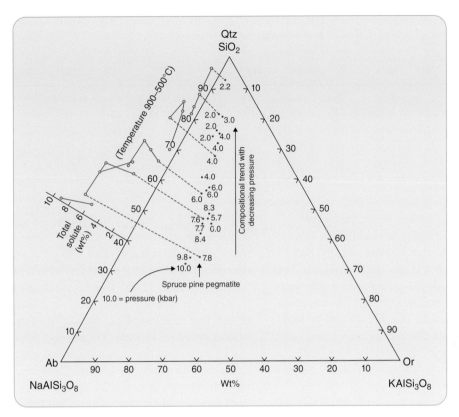

Figure 2.9 Normalized compositions of aqueous fluid in equilibrium with a granitic pegmatite at pressures varying from 2 to 10 kbar (pressure shown next to each point on the diagram) and temperatures between 600 and 900 °C. The points plotted along the left axis indicate the corresponding variation in total solute content of the aqueous solutions as a function of pressure. Source: After Burnham (1967).

produced adjacent to active volcanoes or geothermal springs. Although it is now known that such fluids are not entirely magmatic in origin (many have significant contributions from meteoric waters), their composition gives a reasonable indication of the types and quantities of dissolved components in aqueous solutions. Table 2.1 shows the chemical compositions of volcanic and geothermal fluids from a variety of places around the world. The data show that these natural fluids are variable in composition, reflecting the different rocks through which they have been circulating, as well as, possibly, contamination by other types of fluid. Typically, the solute content of magmatic fluids is dominated by the alkali and alkali-earth metal cations and by chlorine as the major anion, although exceptions do occur (e.g. Rotorua; Table 2.1).

A few additional comments are relevant. There is often a significant amount of carbon dioxide associated with magmatic fluids and this is discussed in more detail below. The amount of sulfur in magmatic fluids is generally low, but this may reflect the fact that at high crustal levels SO_2 partitions into the vapor phase on boiling. The oxidized and reduced forms of sulfur are essentially mutually exclusive and exist either as the SO_4^{2-} complex (with S^{6+}), or as the reduced HS^- complex (with S^{2-}). Relatively oxidized I-type magmatic fluids tend to comprise sulfur as SO_4^{2-} which fractionates into the aqueous liquid phase. Porphyry Cu and Mo type deposits, therefore, are associated with sulfide minerals in the form of pyrite and chalcopyrite. S-type magmas, which are more reducing because of equilibration

Table 2.1 Typical solute concentrations (in mg kg^{-1} or ppm) of the main components that comprise natural aqueous solutions associated with geothermal and volcanic environments.

Location	pH	Na$^+$	K$^+$	Ca^{2+}	Mg^{2+}	Fe^{2+}	Cl$^-$	HCO$_3^-$	SO$_4^{2-}$	SiO$_2$
White Island	1.1	8630	960	2010	3200	6 100	7 300	<1	6 600	—
Mahagnao	5.8	20 340	4840	2900	95	<0.1	46 235	20	138	—
Rotorua	6.8	147	20	8	1	<0.1	13	560	<5	—
Guanacaste	8.0	1720	167	181	18	—	3 070	201	59	110
Sea water	—	10-560	380	400	1270	—	18-900	140	2 710	—

Sea water is shown for comparison.
Source: Adapted from Giggenbach (1997) and Giggenbach and Soto (1992).

with graphite-bearing metasediments, exsolve aqueous fluids that contain mainly H$_2$S or HS$^-$ and have lower total sulfur contents. The presence of reduced sulfur species promotes the stability of sulfide minerals down to lower temperatures, such that the Sn and W oxide minerals (cassiterite and scheelite) associated with S-type granites tend to form early in the mineralizing sequence of these systems (Burnham 1997).

It is important to point out that Table 2.1 illustrates the *major* dissolved components of natural aqueous solutions. The concentrations of ore metals, which should be regarded as trace constituents in hydrothermal fluids, are usually considerably lower than those shown in the table. Recent analytical developments have allowed measurements of magmatic ore-forming fluids *directly* from the tiny volumes of fluid trapped in individual fluid inclusions and this has provided some important new insights into the compositions of both the aqueous liquid and vapor phases in granite ore-forming systems.

2.4.3.1 Magmatic Fluid Compositions from Fluid Inclusion Analysis

Ore-forming fluids can be directly studied by examining the tiny fluid inclusions (typically 5–30 μm in diameter) that exist in quartz, as well as in many other rock- and ore-forming minerals. Studies of high-level porphyry Cu–Mo systems have shown that the early generation of fluid trapped in primary fluid

inclusions is characterized by high temperatures and high salinities, and also provides evidence of liquid–vapor phase separation, or boiling (Roedder 1984). The high salinities of these ore-forming fluids are confirmed by the fact that the fluid inclusions often contain daughter crystals (i.e. minerals that precipitated from the fluid on cooling after it was trapped within the inclusion; see inset microphotograph in Figure 2.10a). The majority of these daughter crystals are identified as halite (NaCl) or, less commonly sylvite (KCl), and point to the fact that magmatic fluids contain significant amounts of dissolved K, Na, and Cl. A plot of fluid compositions derived from microthermometric homogenization experiments on individual fluid inclusions for a wide range of porphyry Cu deposits is shown in Figure 2.10a. A more detailed description of these data and their significance can be found in Roedder (1984).

A much more accurate and complete assessment of porphyry Cu-related ore fluid composition, however, is obtained from the quantitative analysis of individual fluid inclusions by laser-ablation inductively coupled plasma mass-spectrometry (LA-ICP-MS). This remarkable technique allows recognition and analysis of a wide range of cationic species in the tiny volume of inclusion fluid, and down to levels of about 1 ppm. Figure 2.10b shows the results of LA-ICP-MS analysis of two coexisting fluid inclusions in quartz from the Bajo de la Alumbrera porphyry Cu–Au deposit

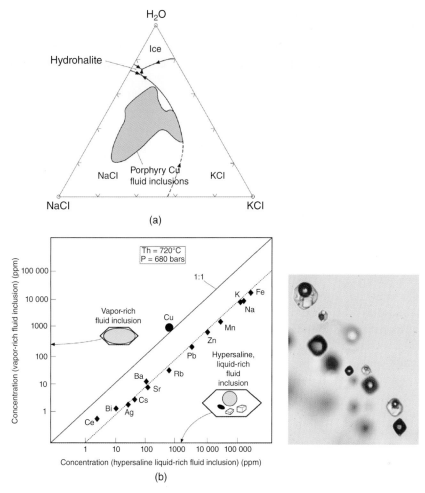

Figure 2.10 (a) Plot of the system H_2O–NaCl–KCl showing the compositional range of fluids in fluid inclusions from a variety of different porphyry Cu deposits. Source: After Roedder (1984). (b) Plot of metal concentrations (obtained by laser-ablation ICP-MS analysis) in vapor-rich fluid inclusions versus metal concentrations in coexisting hypersaline liquid-rich fluid inclusions. The plot illustrates the abundances and distribution of metals in coexisting liquid and vapor phases in a system that may be undergoing boiling. The data pertain to the Bajo de la Alumbrera porphyry Cu–Au deposit in Argentina and the plot is redrawn after Ulrich et al. (2001). Inset is a microphotograph showing coexisting hypersaline fluid inclusions (containing a vapor bubble and daughter crystals) and vapor-rich fluid inclusions (dark and appearing empty) from the tin-mineralized Mole Granite in Australia Source: (Audétat et al. 1998) – the photo is courtesy of Christopher Heinrich.

(Ulrich et al. 2001). The two coexisting fluid inclusions, one a vapor-rich inclusion and the other a hypersaline liquid-rich inclusion containing several daughter crystals, indicate that the ore fluid was boiling as it was trapped. Their analysis reflects the extent to which elements were partitioned between liquid and vapor, as well as the metal abundances in each phase. The plot shows that the majority of elements analyzed partition preferentially (by a factor of about 10–20times) into the liquid phase rather than the vapor (Figure 2.10b). As expected, the Na and K contents of the ore fluid are high, but so too is the Fe content. The liquid also contains significant quantities of Mn, Cu, Pb, and Zn in solution. Interestingly,

Cu behaves differently to the other metals in that it is preferentially fractionated into the vapor phase, a feature also previously noted by Lowenstern et al. (1991) and Heinrich et al. (1999). This feature indicates that the vapor phase may also be implicated in metal transport, despite the fact that previous studies have suggested that metal solubilities in gases are typically low. Vapor transport of metals is discussed in more detail in Section 2.6.2.

The direct analysis of fluid inclusions, and consequently of the fluids implicated in the transport and concentration of metals during the formation of hydrothermal ore deposits, facilitates the distinction of an "ore" fluid, where metal concentrations may be very high, from the much more common barren fluids that typically circulate through the Earth's crust. Table 2.2 shows the metal contents of a variety of magmatic-hydrothermal, and other, ore fluids – it is evident that aqueous ore fluids can dissolve and transport significant quantities of a wide range of metals, which explains the global preponderance and diversity of hydrothermal mineral deposits.

2.4.4 Carbon Dioxide in Magmatic Fluids

After water, CO_2 is the most common gas emanating from volcanic eruptions and its presence is important in a variety of ore-forming processes, albeit in a more obscure way than for H_2O. Many different studies have confirmed that the solubility of CO_2 in magmas is typically an order of magnitude lower than that of H_2O (Mysen 1976; Brooker et al. 2001; Behrens and Gaillard 2006) and, although it exhibits complex dissolution behavior, solubility increases with pressure but decreases with increasing melt SiO_2 content. It dissolves as molecular CO_2 in normal felsic or mafic melts, but in alkaline magmas it also exists as carbonate ionic complexes in solution (Lowenstern 2001). As an indication, rhyolite and basalt melts at 3 kbar will normally contain between 1000 and 2000 ppm dissolved CO_2 compared to values at least four to five times higher in a leucitite melt. CO_2 solubilities, therefore, increase as a function of pressure and magma alkalinity.

Because CO_2 solubilities are much lower than those of H_2O, it follows that CO_2 will exsolve early such that vapors generated at an early stage of solidification will tend to be more CO_2-rich than those forming later. Carbon dioxide will partition preferentially into the vapor phase and it is probable that many normal magmas, both felsic and mafic, will be vapor-saturated at significant crustal depths (Lowenstern 2001). Other low solubility gases are also likely to be associated with the vapor phase created by early CO_2 saturation or effervescence. Volatiles that are generated in systems that have undergone protracted differentiation, or reach shallow crustal levels, will tend to be H_2O-dominated since the relatively insoluble gases like CO_2, as well as N_2 and others, will already have exsolved. It is, therefore, possible for magmas to become

Table 2.2 Selected metal contents (in ppm) of fluid inclusions from a variety of mineral deposit types.

Location	Fe	Cu	Zn	Pb	Sn	W	Ag	As	Sb	Ce
Mole (Sn granite)	73 000	2 300	3 600	3 400	390	56	290	120	110	2
Santa Rita (Cu porphyry)	74 700	2 300	3 400	800	—	57	7	8	—	6
El Mochito (Zn–Ag skarn)	—	—	5 930	4 350	—	—	50	248	365	—
Capitan (REE granite)	43 700	230	2 500	470	63	30	3	25	—	300
Tri-State (Pb–Zn MVT)	—	6	13	180	—	—	—	—	—	—

Source: Audétat et al. (2000), Audétat et al. (2008), Samson et al. (2008), Banks et al. (1994), Stoffell et al. (2004), and Seward et al. (2014).

volatile-saturated before, and independently of, H_2O fluid saturation. The presence of CO_2 in an evolving aqueous fluid within a crystallizing granite will promote immiscibility between vapor and saline liquid phases of the solution. Such processes can be very important during ore-forming processes since they promote phase changes and hence the precipitation of metals from solution (Lowenstern 2001). Effervescence of CO_2 from the fluid will also promote certain types of alteration in the host rocks and increase pH in the remaining fluid, further influencing ore-forming processes (see Chapter 3 for more discussion of these and related topics). Thus, although CO_2 does not appear to play a major role in the transport and concentration of metals in hydrothermal solutions, it nevertheless has a role to play in the distribution and precipitation of metals, especially via the volatile or vapor phase.

2.4.5 Other Important Features of Magmatic Fluids

An important characteristic of magmatic aqueous solutions at low pressures is that they tend to segregate into two phases of differing densities and composition. This feature of hydrothermal solutions is often overlooked, and it also further complicates the description of these fluids. It is evident that the so-called boiling-point curve (Figure 2.2) is actually part of a solvus which separates lower density H_2O from higher density H_2O. As emphasized above, this distinction becomes meaningless in P–T space above the critical point (or the solvus) where densities of liquid and vapor have merged and only a single "fluid" phase exists. The critical point of pure water, at 374 °C, 220 bar, and 0.322 g cm^{-3}, migrates toward higher temperatures and pressures as dissolved salts (usually in the form of alkali chloride complexes) are added to water. A representation of the critical points for hydrothermal solutions of differing salinities and the effect on fluid densities (isochores) is shown in Figure 2.11a. The locus of critical points, as a function of increasing salinity, represents a solvus boundary which governs the existence or otherwise of immiscible behavior in aqueous solutions.

The phase relationships of magmatic aqueous solutions (brines) are also, therefore, linked to this solvus which, depending on the pressure, temperature, and composition of the fluid, dictates whether a single homogeneous phase exists above the critical point, or two immiscible phases below it. Figure 2.11b shows a projection of the H_2O–NaCl system onto the T–X plane in which the solvus curves at various pressures are shown. The phase relationships show that, for a granite at 600 bar in which an exsolved H_2O fluid has formed with a composition represented by point A, the fluid exists below the relevant solvus. Such a fluid must segregate into two phases, a small proportion of brine containing 78 wt% NaCl (A″) and a more voluminous low density fluid (A′) containing only 1 wt% NaCl (Burnham 1997). The same granite at 1000 bar (with an exsolved H_2O fluid represented by point B, Figure 2.11b) would also exsolve two aqueous solutions, but with different compositions. The brine would in this case contain 53 wt% NaCl (B″), whereas the remaining fluid would have 2 wt% NaCl (B′). The granite at 1500 bar (with an exsolved H_2O fluid composition at point C), however, would exist well above the relevant solvus and would exsolve a *single*, homogeneous fluid phase containing about 12 wt% NaCl. It should be noted that the presence of CO_2 in the fluid will promote immiscibility to much higher pressures, this being another important role played by CO_2 in magmatic-hydrothermal fluid systems. The processes whereby an H_2O fluid exsolves from a granite magma are important to the understanding of ore-formation because metals partition themselves differently between the high solute content brine and the more pure H_2O phase. This topic is discussed in more detail in Section 2.6 below.

The discussion of magmatic-hydrothermal fluids and ore-forming processes in the

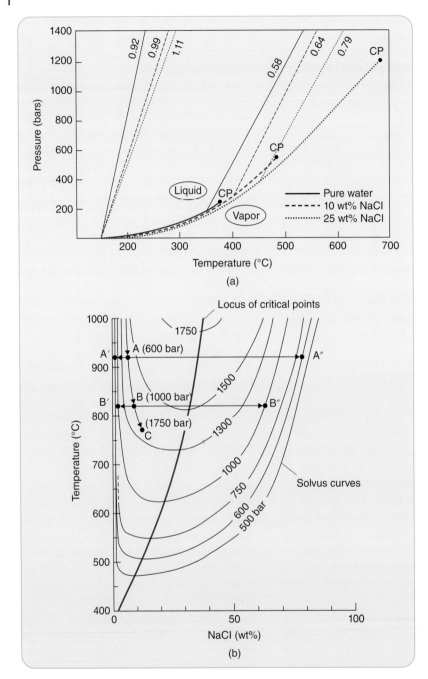

Figure 2.11 (a) Boiling point (liquid–vapor) curves for pure water and for two aqueous solutions containing 10 and 25 wt% NaCl, showing the shift of critical point to higher temperatures and pressures with increasing salinity. Selected isochores, with relevant densities (in g cm^{-3}), are also shown for each of the three cases. Source: The diagram is redrawn after Roedder and Bodnar (1980). (b) A projection of the system H$_2$O–NaCl showing the solvus curves, for a variety of pressures between 500 and 1750 bar, that define whether a homogeneous aqueous solution will segregate into two immiscible phases (such as at A [600 bar] and B [1000 bar], which are below their relevant solvi) or remain as a single homogeneous phase (such as at C which is above the solvus for 1500 bar). Source: The diagram is redrawn after Bodnar et al. (1985).

preceding sections has given rise to some confusing terminology. As a summary, and before we proceed further, the following definitions are presented in order to clarify some of the concepts referred to thus far.

- **H$_2$O-saturation**. At some stage, either early or late in the crystallization sequence, granitic magma will become water saturated, resulting in the exsolution of a chemically distinct, aqueous phase from the silicate melt. The magmatic aqueous phase can exist as a liquid, vapor, or homogeneous supercritical fluid. In the last case the fluid should be regarded as gas-like since it is a substance that would fill its container. For this reason the magmatic aqueous fluid is often referred to as the "*vapor phase*." The process of H$_2$O-saturation can be achieved in two ways, either by decreasing the pressure of the system (first boiling) or by progressive crystallization of magma (second boiling):
- **Vapor-saturation**, also referred to as boiling, occurs when the equilibrium vapor pressure of the magma equals that of the load pressure on the system and bubbles of gas (i.e. steam or water vapor, CO$_2$, N$_2$, SO$_2$, H$_2$S, etc.) nucleate in the magma. Vapor-saturation of low solubility volatiles such as CO$_2$ may be unrelated to, and can precede, H$_2$O-saturation.
  - **First boiling** refers to the case where vapor-saturation is achieved as a result of decreasing pressure (i.e. because of upward emplacement of magma or mechanical failure of the chamber) and is particularly applicable to high level systems. It reflects the fact that solubilities of volatile phases in a melt are enhanced as a function of increasing pressure, and vice versa.
  - **Second boiling** refers to the achievement of H$_2$O fluid saturation by progressive crystallization of dominantly anhydrous minerals under isobaric conditions. It pertains to more deep-seated magmatic systems and occurs only after a relatively advanced stage of crystallization.
- **Immiscibility** refers to the tendency for brine solutions at low pressures, or in the presence of CO$_2$, to segregate into two phases, one a dense, more saline brine and the other a lower density, low salinity aqueous solution.

2.5 A Note on Pegmatites and Their Significance to Granite-Related Ore-Forming Processes

Pegmatites are commonly regarded as rocks derived from magma that *may* have crystallized in the presence of a magmatic aqueous fluid. They are defined as very coarse-grained rocks, typically associated with granites and comprising the major granite rock-forming minerals. The large crystals of quartz, feldspar, and muscovite that make up the bulk of most pegmatites are often extracted for industrial purposes. In addition, they can comprise a wide variety of minor minerals of more exotic and semi-precious character, such as tourmaline, topaz, and beryl. Pegmatites also contain concentrations of the large ion lithophile and high field strength elements, such as Sn, W, U, Th, Li, Be, B, Ta, Nb, Cs, Ce, and Zr (Thomas and Webster 2000). The origin of pegmatites is a topic that provides a fruitful area of research because of the clues they provide for ore-forming processes and igneous petrogenesis in general (Cerny and Meintzer 1988).

Although there are several classification schemes for pegmatites, the model by Cerny (1991) is particularly convenient because it separates them into characteristic metal assemblages that have an implied genetic connotation. Two families of pegmatites are recognized, the Nb–Y–F suite associated with sub-alkaline to metaluminous (largely I-type) granites and the Li–Cs–Ta suite, also enriched in boron, and typically associated with peraluminous (dominantly S-type) granites. Some pegmatitic magmas arise from small degrees of partial melting and form minor dykes and segregations in high grade metamorphic terranes. Other pegmatites are spatially associated with the cupola zones of large granite intrusions and may be genetically linked to the most highly differentiated, water-saturated portions of such bodies. The origin of pegmatites is one of the more fascinating problems in igneous petrology and, even though the processes are now well understood, they remain complex and somewhat contentious. Both the early and later developments in the understanding of these rocks are instructive to ore-forming processes and are discussed below.

2.5.1 Early Models of Pegmatite Genesis

Prior to the classic paper by Jahns and Burnham (1969), pegmatites were viewed as the products of extreme crystal fractionation of mainly granitic magmas. Experimental work, however, suggested that pegmatites were distinguished from "normally textured" granite by crystallization in the presence of an exsolved H_2O fluid phase. More specifically, Jahns and Burnham (1969) suggested that the transition from granite to pegmatite marked the point at which H_2O fluid saturation occurred in the crystallization sequence and that pegmatites, therefore, formed in the presence of a discrete H_2O + volatile fraction. The existence of H_2O fluid in the magma was used to explain the large crystal size of pegmatites, the latter growing because of extended crystallization due to either volatile-related depression of the granite solidus, or more efficient diffusion of major elements into low viscosity H_2O fluid-saturated magma. Mineral zonation in pegmatites was attributed largely to the separation of relatively dense silicate melt from the low density H_2O fluid phase, and the attendant segregation of alkalis such that Na remained in the silicate melt and K partitioned in the aqueous phase. In addition, it was envisaged that precipitation of dissolved constituents from the H_2O fluid phase could explain the pockets of exotic minerals (including gemstone-quality crystals formed in an open space or vug from fluids enriched in Li, B, Cs, Be, etc.) so often characteristic of pegmatites. The Jahns and Burnham model was particularly attractive because it was supported by experimental evidence demonstrating that an aqueous fluid at high temperatures and pressures is capable of dissolving very significant proportions of solute (see Figure 2.9).

2.5.2 More Recent Ideas on the Origin of Pegmatites

The Jahns–Burnham model represented a major advance in the understanding of pegmatite genesis, as well as the crystallization and mineralization of granites generally. More recent work, however, has shown that there are other factors, in addition to H_2O-saturation, which need to be considered in order to fully understand the formation of these rocks and their associated ore deposits. The research of D. London (1990, 1992, 1996, 2005) in particular has shown that it is possible to generate pegmatites from H_2O undersaturated granitic melts by *undercooling* the magma below its normal liquidus temperature. It was demonstrated experimentally that kinetic delays in the initiation of crystallization in felsic magmas that are rapidly undercooled means that melts persist in a metastable condition and result in non-equilibrium crystal growth. Many of the features of pegmatites, including mineral zonation, large grain size, variable textures, and highly fractionated chemistry, can be replicated experimentally in terms of the delayed or metastable crystallization response to undercooling of granitic melts. This work has enabled pegmatites to be explained by magmatic crystal fractionation process (as had been suggested prior to the Jahns–Burnham model) rather than to silicate melt–H_2O fluid exsolution and alkali element segregation. The London model has removed a fundamental difficulty faced by the Jahns–Burnham model, namely, how to explain the paragenesis and symmetric mineral zonation pattern typical of many pegmatites. Progressive inward crystallization of granitic melts (such as sidewall boundary layer differentiation; Section 1.4.2 and Figure 2.5) is more likely to result in symmetric mineral and chemical zonation than is H_2O fluid exsolution which is intrinsically asymmetric because aqueous fluids tend to coalesce in the upper portions of the melt chamber. It also provides a rather neat explanation for why pegmatitic textures are uncommon in mafic rocks. Mafic melts are less viscous and nucleate crystals much more readily than their felsic equivalents, resulting

in a tendency toward equilibrium crystallization conditions and the formation of more
uniform textures.

It should be emphasized, however, that the
more recent models of pegmatite genesis have
not, by any means, precluded the important
role of water and other volatiles in the formation of these rocks. A fluid inclusion study of
the Tanco Li–Cs–Ta pegmatite in Manitoba,
Canada, showed that it crystallized from a
mixture of alumina–silicate melt plus H_2O
fluid (with minor CO_2) over a temperature
interval that extended from about 700 °C to
below 300 °C (Thomas et al. 1988). These
observations were interpreted as support for
the Jahns–Burnham model. The processes
are, however, further complicated when one
takes into account the major influence of other
volatile species, such as B, F, and P, on pegmatite formation (London 1996). These three
elements, individually and collectively, lower
the granite solidus temperature to below 500 °C
and increase the range of temperatures over
which magmatic crystallization occurs. They
also dramatically increase the solubility of H_2O
in the melt and promote the crystallization of
quartz. In a study of the Ehrenfriedersdorf
Sn–W–B–F–P-rich pegmatites of the Erzge-
birge in southeast Germany, it has been shown
that melt inclusions in pegmatitic quartz are
made up of coexisting silicate-rich H_2O-poor

and silicate-poor H_2O-rich melts (Thomas
et al. 2000). On progressive reheating from
500 to 700 °C at 1 kbar pressure, the latter
inclusions become poorer in H_2O, whereas
the former undergo a steady increase in water
content (Figure 2.12). At just above 700 °C the
two inclusion populations are effectively identical, with both containing about 20 wt% H_2O.
This indicates that the silicate-rich H_2O-poor
and silicate-poor H_2O-rich melts were derived
from a single, originally homogeneous, melt
above 700 °C. Even at low pressures (the
Ehrenfriedersdorf pegmatites crystallized at
about 1 kbar) these B–F–P-rich pegmatitic
magmas exhibit complete miscibility between
silicate melt and aqueous fluid such that
water can be regarded as infinitely soluble in
this melt. On cooling below 700 °C, however,
immiscibility occurs to form two coexisting
melts. One of these is a silicate melt with low
H_2O content, while the other is a silicate melt
with high H_2O content (Figure 2.12). The
latter, however, has a density, viscosity, and
diffusivity that is more akin to an aqueous
solution than to a silicate melt. Although its
properties are water-like, it is by definition
different from the exsolved H_2O fluid envisaged to have formed in the Jahns–Burnham
model.

It is apparent, therefore, that a magmatic
aqueous phase can be represented by a fluid

Figure 2.12 Plot of H_2O content
versus temperature for
rehomogenized (at 1 kbar)
silicate-rich H_2O-poor (A) and
silicate-poor H_2O-rich (B) melt
inclusions from the
Ehrenfriedersdorf pegmatite. The
curve defines a solvus with a critical
point (above which there is no
longer a distinction between the
two melts and they are miscible) at
about 700 °C and 20 wt% H_2O
content. Source: After Thomas et al.
(2000).

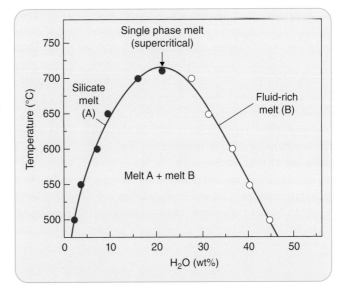

that is more complex than simply exsolved H_2O. At high pressures and temperatures water can contain significant quantities of dissolved silica and alkalis which, on cooling, will yield precipitates with bulk compositions not dissimilar to a granite. At lower pressures H_2O fluid will exsolve from a normal granitic magma, such that crystallization will occur in the presence of an aqueous phase. At low pressures and in the presence of significant concentrations of elements such as B, F, and P, however, granite melt solidus temperatures are significantly depressed and H_2O solubility increases to such an extent that H_2O-saturation might not occur at all. In this case two immiscible melts form, one of which is H_2O-rich and has physical properties akin to an aqueous solution with high solute content. Pegmatites could conceivably form in all these situations.

2.6 Fluid–Melt Trace Element Partitioning

The rise of magma to shallow levels of the Earth's crust inevitably leads to saturation with respect to an aqueous fluid or vapor phase. Even though the diffusivity of H_2O is relatively low in most felsic magmas, the buoyant, low density H_2O-fluid, together with entrained melt and crystals, will move to the apical portions of the magma chamber (Candela 1991). In an equilibrium situation, trace components of the magma must then partition themselves between melt, crystals, and H_2O-fluid. The extent to which trace components, such as metals, distribute themselves between these phases is a process that is potentially quantifiable if the partition coefficients are known. It is possible to derive equations that express the partitioning of trace constituents between melt and H_2O-fluid in exactly the same way as was discussed for the partitioning of trace elements between melt and crystals in Section 1.3.4 of Chapter 1. However, considerations of melt–fluid partitioning are complicated by the fact that it is necessary to consider partitioning

behavior prior to water saturation, and also because the sequestration of incompatible constituents by the H_2O-fluid depends on solubilities of both the fluid and the magma from which it is derived. A detailed account of the quantification of melt–fluid partitioning behavior is provided by the work of Candela and co-workers, summarized in Candela and Holland (1984, 1986), Candela (1989a,b, 1992), and Candela and Piccoli (1995). Before the results of this work are discussed in more detail, brief mention will be made of some of the classic experimental work that provided initial insights into why hydrothermal fluids are such good solvents and, therefore, able to so effectively scavenge and transport metals. Again, more details on this topic are provided in Chapter 3.

2.6.1 Early Experiments on Metal Solubilities in Aqueous Solution

Early experimental studies by workers such as Burnham (1967) and Holland (1972) showed that the solubilities of many metals in a magmatic H_2O-fluid are strongly dependent on its Cl^- concentration, in addition to other parameters such as temperature and pH. When experimentally reacting an aqueous solution with a granitic melt it was observed that metals such as Zn, Mn, Fe, and Pb are strongly partitioned into the H_2O-fluid and that the ratio of metal concentration between fluid and melt is exponentially proportional to the Cl^- concentration of the fluid (Figure 2.13a,b). These experiments demonstrated quite clearly that certain metals will only readily dissolve in an aqueous solution if the latter contains appreciable amounts of the chloride anion (Cl^-). Furthermore, metals will exhibit different magnitudes of partitioning into H_2O-fluid according to their variable abilities to bond, or form complexes, with Cl^-.

Figure 2.13b also shows that the total Fe content of the H_2O-fluid varies as a function of temperature, and is higher for any value of Cl^- at 750 °C than it is at 650 °C. This confirms

Figure 2.13 (a) The ratio of the concentration of Zn in H_2O-fluid $[Zn]^f$ to that in an associated granitic melt $[Zn]^m$ as a function of the Cl^- concentration of the fluid; experiments carried out at various temperatures between 770 and 880 °C and pressures between 1.4 and 2.4 kbar. Source: After Holland (1972). (b) The total Fe content (left ordinate) and the total Fe to Al ratio (right ordinate) in H_2O-fluid that equilibrated with a granitic melt as a function of the Cl^- concentration; experiments carried out at temperatures of 650 and 750 °C and a pressure of 2.0 kbar. Source: After Burnham (1967).

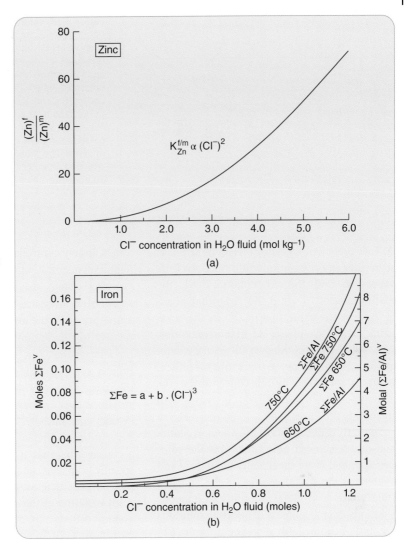

the intuitive notion that solubility normally increases as a function of temperature. The diagram also shows that the Al content of the fluid remains constant (because the total Fe content changes in the same way as does the total Fe/Al ratio) and that it is not affected by the Cl^- concentration of the aqueous fluid. This indicates that metal solubilities are not always controlled by Cl^- concentrations in the fluid (although they may still be affected by the presence of other ligands) and that some metals may not even choose to partition into the aqueous phase at all.

The dependence of metal solubility on Cl^- concentrations presupposes that the ligand itself will partition effectively into the H_2O-fluid phase. Kilinc and Burnham (1972), Shinohara et al. (1989) and Webster and Holloway (1990) have confirmed experimentally that chloride partitions strongly into the magmatic aqueous phase, according to the reaction:

$$Cl^-(melt) + OH^-(melt) \rightarrow HCl(fluid) + O^{2-}(melt)$$

$$(2.4)$$

Reaction between granitic melt and a chloride-bearing aqueous fluid showed that the Cl^- ion partitions strongly into the fluid, but that the value of the partition coefficient varies in a complex fashion (Figure 2.14).

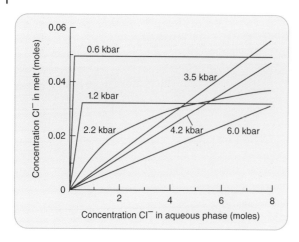

Figure 2.14 Variations in the concentration of Cl^- in a silicate melt with that in an exsolved H_2O-fluid for a variety of different pressures. Cl^- is partitioned strongly into the aqueous phase relative to the melt. At low pressures, where liquid–vapor immiscibility occurs (0.6 and 1.2 kbar), the concentration of Cl^- is fixed by the saturation of Cl^- in the coexisting vapor-rich phase; at higher pressures, for any given concentration of Cl^- in the melt, a homogeneous H_2O-fluid phase will contain more Cl^- at higher pressures. Source: After Shinohara et al. (1989).

At low pressures where the H_2O-fluid phase segregates into two immiscible fractions (one vapor-rich and the other liquid-rich), the partitioning of Cl^- into the H_2O-fluid phase is fixed by the saturation of Cl^- in the coexisting vapor-rich phase. At higher pressures, however, the partitioning of Cl^- into a homogeneous H_2O-fluid phase (i.e. above the critical point) is controlled by the concentration (or activity) of Cl^- in the melt, but increases with increasing pressure.

The early experiments of Burnham, Holland, and others showed how important the composition, and more specifically the nature and content of the ligand species, of the aqueous fluid phase is to metal solubility in aqueous solutions. Metal contents of aqueous solutions are not, of course, only dependent on the Cl^- concentration – many other factors affect the ability of hydrothermal fluids to dissolve and transport metals. A more complete discussion of this topic is presented in Section 3.4 of Chapter 3. In the section below, a more detailed consideration is given to the partitioning of metals between silicate melt and an exsolved magmatic H_2O-fluid in the light of recent theoretical and experimental work.

2.6.2 A More Detailed Look at Fluid–Melt Partitioning of Metals

The partitioning of metals between a silicate melt and an exsolved H_2O-fluid is a complex matter. The extent to which metals are partitioned between melt and fluid is variable and is largely controlled by the Cl^- (and other ligand) concentration of the fluid which, in an evolving system, itself changes continuously. In addition, however, factors such as temperature, pressure, the amount of water exsolved relative to the amount of water remaining in the silicate melt (i.e. when H_2O-fluid saturation is achieved during crystallization), and the oxygen fugacity (fO_2) of the silicate–fluid system, also influence metal partitioning (Wood and Blundy 2002). This means that a partition coefficient cannot be regarded as a constant because it pertains only to a very specific, transient, set of conditions. Melt–fluid partition coefficients, therefore, exhibit significant variation as the magmatic-hydrothermal system evolves. In addition, the contrasted partitioning of metals between associated vapor and liquid phases in fluids undergoing boiling is now known to be relevant to ore forming processes and is also discussed briefly below.

Candela (1989a) pointed out that the metal content of an exsolved H_2O-fluid will depend to a large extent on when in the crystallization sequence water saturation is achieved. This is because crystal–melt partitioning can be as efficient a means of distributing and concentrating trace metals in a magma as can H_2O fluid–melt partitioning. It is particularly important to distinguish between "first boiling" (when vapor-saturation is caused by pressure decrease) and "second boiling" (when crystallization of a dominantly anhydrous assemblage causes the H_2O content of the residual melt to rise to the saturation water content at a given pressure or level of emplacement). First boiling processes are simpler to model mathematically and pertain to ore-forming processes associated with high level granite intrusions such as those associated with porphyry Cu and Mo deposits. Second boiling requires a more complicated mathematical model since the metals have to partition among melt, crystal, and H_2O fluid phases.

2.6.2.1 Fluid–Melt Partitioning During "First Boiling"

Candela (1989b) has provided equations that quantify the distribution of a trace component between H_2O fluid and silicate melt for a first boiling situation. These equations are applicable to those components whose partitioning does not change as a function of the ligand concentration of the fluid. They are somewhat similar in form to the exponential equations applicable to fractional melting and crystallization presented in Chapter 1, and are expressed as:

$$C_{fluid}^i = D_{fluid/melt}^i C_O^i (1 - [(1-F)C_O^{water}])^{D_{fluid/melt}^i - 1}$$
(2.5)

where: C_{fluid}^i is the concentration of a component (i) in the H_2O fluid at any instant in the evolution of the aqueous phase; $D_{fluid/melt}^i$ is the H_2O fluid/silicate melt partition coefficient for the component (i); C_O^i is the initial concentration of the component (i) in the silicate melt (defined at the instant of water saturation); C_O^{water} is the initial concentration of water in the silicate melt (defined at the instant of water saturation); and F is the ratio of the mass of water in the silicate melt (at any given instant after water saturation) to the initial mass of water in the silicate melt.

The concentration of the component (i) in the associated silicate melt (C_{melt}^i) is simply given by:

$$C_{melt}^i = C_{fluid}^i / D_{fluid/melt}^i$$
(2.6)

where C_{fluid}^i is the value obtained from Eq. (2.5).

If the magnitude of the partition coefficient ($D_{fluid/melt}^i$) is known, the concentrations of a halogen such as chloride in an evolving aqueous phase, as well as in the associated silicate melt, can be calculated. Diagrammatic representations of Cl concentration trends in an evolving silicate melt–H_2O fluid system are shown in Figure 2.15. The appearance of a magmatic aqueous fluid will typically result in depletion of Cl in the melt (Figure 2.15b) and this will be accompanied by a concomitant decrease of Cl in the H_2O-fluid (Figure 2.15a), as the system evolves toward complete solidification. The efficiency with which the aqueous phase extracts components, such as chlorine or other chloride-complexing metals, from the melt is effectively measured by the ratio of the total amount of the component in the H_2O-fluid to the amount initially present in the melt at the time of water-saturation (Candela 1989b). The efficiency factor will be dependent not only on obvious criteria such as $D_{fluid/melt}^i$, but also on parameters such as F (the ratio of the mass of water in the silicate melt at any given instant after water saturation to the initial mass of water in the silicate melt). Note that not all volatile components will act in the same way, as is the case for fluorine, whose concentration in the melt is predicted to remain relatively constant during first boiling (Figure 2.15b).

2.6.2.2 Fluid–Melt Partitioning During "Second Boiling"

Calculation of fluid–melt partitioning for metals such as Cu or Zn is more difficult because their behavior varies, not only as a function of pressure, but is also strongly controlled by the Cl^- concentration of the aqueous fluid phase. Likewise, the partitioning of metals into the aqueous fluid phase in a situation where "second boiling" has taken place also involves many additional assumptions and a more complex mathematical derivation. The detailed discussion of both of these situations is beyond the scope of this book, and the interested reader is referred to Candela (1989a,b) for more detail.

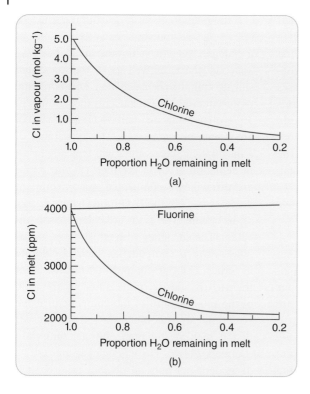

Figure 2.15 Calculated concentrations of chlorine in (a) an aqueous fluid exsolved from a granitic magma, and (b) chlorine and fluorine in the residual magma itself – in both cases as a function of the proportion of water remaining in the melt. The calculations were carried out for a situation where water saturation is achieved in a "first boiling" scenario with a magma rising adiabatically upwards from 5.4 km depth (where water content at saturation is 5 wt%) to the surface (after which 1 wt% water remains in the rock). C_0^{Cl} is assumed to be 4000 ppm and $D_{fluid/melt}^{Cl}$ is calculated to vary from 39 at the point of water saturation to 1 as the magma reaches the surface. Source: After Candela (1989b).

It is pertinent to note, however, that any metal which bonds strongly with Cl⁻ in an aqueous solution (such as Cu) will tend to exhibit partitioning behavior that is similar to that shown in Figure 2.15a. In the case of metals such as copper, therefore, the maximum concentration into the aqueous fluid occurs immediately after H_2O-fluid saturation has been achieved. By contrast, metals such as molybdenum, which do not bond strongly with Cl⁻, and are not, therefore, controlled by concentrations of this ligand, partition differently and may be more efficiently concentrated in the last remaining portions of the evolving H_2O-fluid phase. The contrasting behavior of Cu, Mo and W in the magmatic-hydrothermal environment, suggested by these theoretical calculations, has been confirmed by direct experimental evidence and this is discussed in Section 2.6.3.

2.6.2.3 Partitioning of Metals into H_2O-Vapor

Most genetic models for hydrothermal deposits assume that metals are transported predominantly in the aqueous phase, and that metal solubilities in the vapor phase, with the exception of very volatile elements such as Hg, As, and Sb, are insignificant. This assumption is now known to be incorrect and direct measurements of aqueous vapors reveal that metal transport by gases is an important process (Lowenstern et al. 1991; Heinrich et al. 1992, 1999; Ulrich et al. 2002; Baker et al. 2004). Examination of sublimates (i.e. minerals or elements that crystallize directly from a vapor phase) associated with volcanic gases indicate that a variety of metals, including Cu, Pb, Zn, Au, Ag, Sn, W, and Mo could have been transported in the vapor state. Compilation of analyses of fumarole vapor condensates indicate that concentrations of these metals are typically in the ppb to low ppm range, with vapors from Mt. Etna being exceptional and containing up to 24 ppb Au, 120 ppm Ag, 12 ppm Pb, and 13 ppm Zn (Williams-Jones and Heinrich 2005). Microanalysis of fluid inclusions in mineralizing systems indicate that metal solubilities may be elevated in the vapor phase, and the Cu and Au contents of vapors associated with porphyry-epithermal systems (see Boxes 2.1 and 2.2), for example, may be higher than they are in the coexisting liquid phase.

It is evident that boiling, and the appearance of coexisting liquid and vapor phases derived from an originally homogenous fluid,

can result in marked fractionation of metals between the two phases (Heinrich et al. 1992, 1999). Liquid-vapor metal fractionation has been detected in different ore-forming environments (Williams-Jones and Heinrich 2005) although it is most frequently observed in magmatic-hydrothermal ore fluids that circulate at shallow levels in the crust (and hence undergo phase separation). Figure 2.16 plots

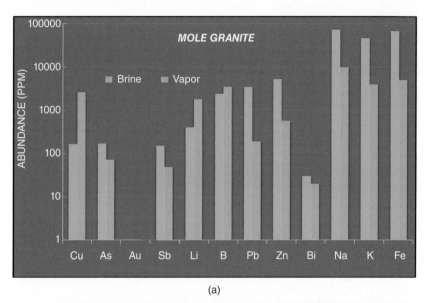

(a)

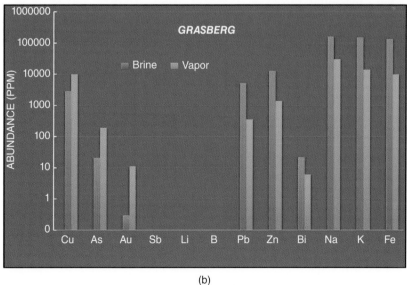

(b)

Figure 2.16 Bar graph showing the abundances of metals in selected, coexisting brine and vapor rich fluid inclusions from (a) the Sn mineralized Mole Granite in Australia; and (b) the host granites of the porphyry Cu–Au deposit at Grasberg, Indonesia Source: Data from Williams-Jones and Heinrich (2005).

the metal contents of coexisting vapor and liquid analyzed in selected fluid inclusions from the Sn-mineralized Mole Granite in Australia (Figure 2.16a) and the porphyry Cu–Au deposit at Grasberg in Indonesia (Figure 2.16b). The Grasberg porphyry, along with many other examples, exhibits preferential enrichment of Cu, Au, and As in the vapor phase, with most other metals occurring predominantly in the brine. In the Mole Granite, Cu is also enriched in the vapor phase, as are Li and B, with the remaining metals analyzed concentrated into the brine.

Experiments to determine Cu and Zn solubilities in the system H_2O–NaCl indicate that when solutions are seeded with elemental sulfur the vapor is enriched in Cu (about 3000 ppm) but less so in Zn (100 ppm), whereas sulfur-free experiments exhibited negligible Cu and Zn in the vapor phase (Nagaseki and Hayashi 2004). In the latter case both Cu and Zn partition strongly into the liquid phase as a function of the chloride ion concentration (see Figures 2.13a and 2.17). These results suggest that sulfur, in the form either of HS^- or SO_4^{2-}, promotes the fractionation of Cu (and also Au and other chalcophile metals) into the vapor phase because the sulfur species themselves partition preferentially into the vapor, and in so doing carry bonded chalcophile and siderophile metals with them. The significance of metal transport by vapors in epithermal ore-forming systems is discussed in more detail in Section 2.9.1 below and the solubility of metals in the aqueous vapor phase is covered again in Chapter 3 (Section 3.4.2).

2.6.3 Partitioning of Cu, Mo, and W Between Melt and H_2O-Fluid

In a series of experiments run at 750 °C and 1.4 kbar, Candela and Holland (1984) were able to measure the extent to which Cu and Mo partition between granitic melt and a coexisting aqueous fluid containing both Cl^- and F^-. The results of these experiments are shown in Figure 2.17 and clearly demonstrate that Cu is

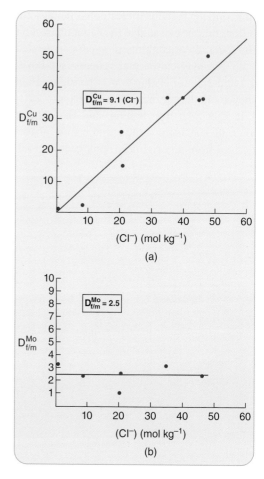

Figure 2.17 (a) The relationship between H_2O fluid/silicate melt partition coefficient for Cu ($D_{f/m}^{Cu}$) and the Cl^- concentration of the aqueous fluid. (b) The relationship between the H_2O fluid/silicate melt partition coefficient for Mo ($D_{f/m}^{Mo}$) and the Cl^- concentration of the aqueous fluid. Source: After Candela and Holland (1984).

a metal whose partitioning behavior is strongly controlled by the Cl^- concentration of the fluid phase, whereas Mo remains unaffected by the latter. The value $D_{fluid/melt}^{Cu}$ varies from about 1, for low salinity fluids, up to about 50 at high salinities (Figure 2.17a), and is determined by the relationship $D_{fluid/melt}^{Cu} = 9.1[Cl^-]$. By contrast, although Mo is also preferentially partitioned into the H_2O-fluid ($D_{fluid/melt}^{Mo} = 2.5$), its partition coefficient remains constant irrespective of the Cl^- concentration (Figure 2.17b).

It is interesting to note that the partitioning behavior of both Cu and Mo is unaffected by the presence of F^- in the aqueous solution. This is because fluorine tends to partition more strongly into the silicate melt ($D^F_{fluid/melt}$ = 0.2–0.3) than into the aqueous fluid and is, therefore, generally not present as a complexing agent for metals in the fluid phase.

The fluid–melt partitioning behavior of W is not as well constrained as that for Cu and Mo, although experiments do suggest that its properties are similar to those for Mo, but that its partition coefficient is even lower ($D^W_{fluid/melt} \approx 1$: Manning and Henderson 1984; Keppler and Wyllie 1991). However, it is also known that W behaves as an incompatible element in terms of its *crystal–melt* partitioning behavior in relatively reduced (i.e. S-type) granitic magmas solidifying at some depth in the Earth's crust. By contrast, Mo is less incompatible in the same magma types, such that crystal fractionation will tend to result in increasing W/Mo ratios (Candela 1992). The efficiency of metal extraction by an aqueous phase exsolving late in the crystallization sequence from such a granite will, therefore, favor concentration of W over Mo, despite the lower $D_{fluid/melt}$ values. The crystal–melt partitioning behavior of W also contrasts with that of copper, which tends to act as a compatible element in virtually any magma composition (Candela and Holland 1986). A value of $D^{Cu}_{crystal/melt}$ = 3 was calculated for a crystallizing basaltic magma composition and is likely to be >1 for a granite too, since Cu is able to substitute very efficiently into accessory sulfide phases and, to a lesser extent, into biotite and magnetite.

It is apparent that both fluid–melt and crystal–melt partition coefficients play an important role in the distribution of metals in and around crystallizing granite plutons. These parameters, when coupled with considerations of granite type, depth of emplacement, and the timing of water-saturation relative to the crystallization sequence, can be used to explain the nature and origin of many different magmatic-hydrothermal ore deposit types. A model that summarizes some of the features discussed in this chapter is presented in Section 2.8 as an explanation for the characteristics of granite-related Cu, Mo, and W deposits.

2.7 Water Content and Depth of Emplacement of Granites – Relationships to Ore-Forming Processes

Many of the ore deposits associated with granites, such as porphyry Cu and epithermal Au–Ag ores, are related to magma emplacement at high levels of the crust where volatile saturation and hydrofracturing often take place (Hannah and Stein 1990). These types of deposits are commonly located in the volcanic or subvolcanic environment and have formed as much from the action of surface-derived (or meteoric) water as they have from the circulation of magmatic waters. Other deposit types, such as porphyry Mo and granitoid hosted Sn–W deposits, are generally associated with magmas emplaced at slightly deeper levels in the crust. The depth of emplacement of a granite magma, together with related parameters such as magma composition and initial water content, plays an important role in determining the nature and origin of ore deposits associated with felsic igneous rocks, as discussed below.

Section 1.3.1 described how granitic magmas derived from the melting of different source materials, at different pressures and temperatures in the crust, might contain variable initial amounts of water. A melt derived by anatexis of a rock comprising mainly muscovite, for example, could contain in the region of 7–8 wt% H_2O. By contrast, dehydration melting of an amphibolitic source rock would produce a melt at higher pressure and temperature (i.e. deeper in the crust) containing lesser amounts (perhaps only 2–3 wt%) of H_2O (Figure 2.3). It was suggested on this basis that S-type granite magmas might initially contain more water than I-type granite magmas.

In terms of this model drier I-type granite magmas would be derived from the deep crust (possibly with contributions from the upper mantle), whereas S-type granites come from material melted in the mid- to lower-crust. Several workers have used these concepts to develop models that link granite emplacement depths to their metallogenic characteristics (Ishihara and Takenouchi 1980; Hyndman 1981; Strong 1981).

Figure 2.18a shows the same P–T diagram as that in Figure 2.3 but inverted to reflect the surface (i.e. low pressures) at the top of the diagram. Hypothetical zones of melting are shown for each of three cases where the water required to initiate melting is supplied by the breakdown of muscovite, biotite, and amphibole. If sufficient melt is allowed to accumulate and then to rise upwards in the crust along an adiabatic cooling path, it is apparent, at least conceptually, that each of these magmas would crystallize at different levels in the crust. Adiabatic upward movement of magma (i.e. where conductive heat loss to the wall rocks is ignored) would involve cooling at a rate of about $1.5\,°C\,kbar^{-1}$ and in P–T space would approximately follow the steep curves that define mineral phase boundaries. Conceptually, therefore, magmas would rise upwards in the crust until they intersect the water-saturated granite solidus, by which time they become completely solid and cannot intrude any further. In reality crystallization is likely to have occurred prior to this level because of heat loss to the wall rocks, and because the water-saturated solidus effectively represents the depth above which a magma is unlikely to be emplaced. These considerations suggest that S-type granite magmas would be emplaced at mid-crustal depths (4–5 kbar), as dictated by the intersection of the muscovite breakdown curve with the water-saturated granite solidus. By contrast, intersection of the biotite or amphibole breakdown curves with the water-saturated granite solidus suggests that an I-type magma could move to much shallower crustal levels (1 kbar or less) before completely solidifying (Figure 2.18a).

Figure 2.18b is a schematic crustal profile showing the relationships between depth of emplacement and the metallogenic character of various granite-related deposit types. I-type magmas generated deep in the lithosphere usually form adjacent to subduction zones and commonly receive a contribution from mantle-derived mafic melts. Forming at high temperatures (1000 °C or more) and being relatively water-poor (H_2O contents < 3–4 wt%) they will rise to shallow levels of the crust and may even extrude to form substantial volcanic structures. Such magmas will typically exsolve a magmatic vapor phase by first boiling, an event that will also promote hydrofracturing, brecciation, and the widespread circulation of hydrothermal solutions in and around the high-level sites of magmatic activity. These are the environments in which porphyry Cu, as well as epithermal Au–Ag deposit types, occur.

By contrast, S-type magmas are generated in the mid- to lower-crust by partial melting of a source rock that comprises a substantial proportion of metasedimentary material. These melts will form at relatively low temperatures (around 700 °C) and initially comprise significant amounts of H_2O dissolved in the magma. Such a magma type will crystallize in the mid-crust, not too far from its site of generation, and will typically be barren. If substantial crystal fractionation occurs, however, incompatible trace elements will become concentrated in residual melts, and in the rocks that form by the crystallization of differentiated magma (see Box 1.3, Chapter 1). H_2O-fluid saturation will also eventually occur in the residual magma by second boiling, to form pegmatites and related magmatic-hydrothermal deposits. The concentration of volatiles, as well as elements such as lithium, boron and phosphorus, will, together with exsolved water in the residual magma, depress the temperature of the water-saturated granite solidus to lower temperatures (500–600 °C) such that the magma might continue its upward migration

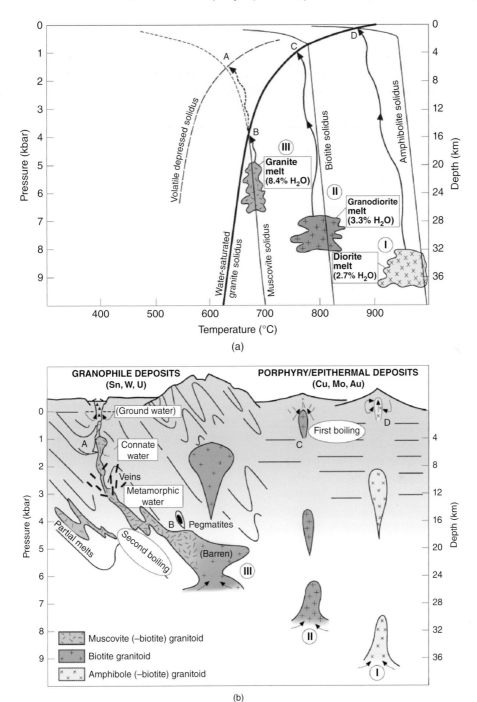

Figure 2.18 Strong's model showing the relationship between level of granite emplacement and metallogenic character. (a) Pressure–temperature plot showing the approximate conditions where fluid-absent melting of amphibolite- (I), biotite- (II), and muscovite-bearing (III) protoliths would occur, and the expected levels in the Earth's crust to which melt fractions would rise adiabatically, as a function of the water-saturated solidus. (b) Schematic diagram illustrating the emplacement style and metallogenic character of granites formed under each of the conditions portrayed in part (a). Source: After Strong (1988).

to higher levels in the crust (Figure 2.18b). Fluid pressures would probably not be sufficient to fracture rock at these depths, but other structural controls would nevertheless promote fluid circulation, and it is these environments in which Sn greisen deposits, porphyry Mo deposits, polymetallic skarn ores, and mesothermal veins might form.

2.8 Models for the Formation of Porphyry-Type Cu, Mo, and W Deposits

As a generalization, porphyry Cu–Mo deposits are associated with arc-related "calc–alkaline" or I-type magmas generated adjacent to subduction zones. Sn–W deposits are more often associated with S-type granites that are derived by partial melting of continental crust that would normally include a significant proportion of metasedimentary material. In certain settings the latter deposits are also regarded as "porphyry-type" even though they may differ significantly in their geological characteristics from the Cu–Mo types. The plutonic portions of porphyry systems grade upwards into sub-volcanic and volcanic regimes, the latter representing the setting in which epithermal Au–Ag deposits form – and if the intrusions are emplaced into carbonate sediments, this environment can host polymetallic skarn-type deposits. Arc related magmatism, therefore, represents a very important metallotect that is

implicated in the formation of numerous, often very substantial, base and precious metal ore deposits across geologic time and around the globe. Considerable progress has been made in understanding the processes that occur in a crystallizing granite intrusion in order to form genetically associated base and precious metal deposits (Burnham and Ohmoto 1980; Candela and Holland 1984, 1986; Candela 1989a,b, 1991, 1992 ; Candela and Piccoli 1995; Williams-Jones and Heinrich 2005; Audétat et al. 2008; Wilkinson 2013; Blundy et al. 2015). Some of these processes are described below.

2.8.1 The Origin of Porphyry Cu–(Mo) and Porphyry Mo–(Cu) Type Deposits

The family of deposit types known as porphyry Cu–Mo ores can be subdivided into two groups, one in which Cu is the dominant exploitable metal (together with minor Mo and occasionally Au), and the other where Mo is the dominant exploitable metal (with minor Cu and occasionally W). The two groups are referred to below as the Cu–(Mo) and the Mo–(Cu) porphyry types, respectively. Porphyry deposits are the world's most important source of both Cu and Mo and, especially in the circum-Pacific region, there are many world-class deposits that exploit these metals. A description of one of the world's great porphyry Cu–(Mo) deposits, La Escondida in Chile, is provided in Box 2.1.

Box 2.1 Magmatic-Hydrothermal Fluids Associated with Granite Intrusions (1) – La Escondida Porphyry Copper Deposit, Chile

Chile is the world's leading copper producer, with an output in 2000 of about 4.6 million tons of fine copper, or 36% of global production (Camus and Dilles 2001). Northern Chile in particular contains several giant porphyry copper deposits, including world-renowned mines such as Chuquicamata, El Teniente, and El Salvador. The porphyry deposits of this huge metallogenic province formed during five discrete magmatic episodes extending from the Cretaceous to the Pliocene. The most prolific of these is the late Eocene–Oligocene period of magmatism, during which time at least 10 major porphyry copper deposits were formed, including two of the biggest copper mines in the world, La Escondida and Chuquicamata. La Escondida was

Box 2.1 (Continued)

discovered as recently as 1981 and in 1999 produced over 800 000 tons of fine copper from a reserve base of more than 2 billion tons of ore at 1.15 wt% copper (Padilla Garza et al. 2001).

La Escondida, together with other giant porphyry systems such as Potrerillos, El Salvador, Chuquicamata, El Abra, and Collahuasi, to its north and south, lies on the Domeyko fault system (Figure 2.1.1). This faulting was dominantly transcurrent during Eocene–Oligocene magmatism. Locally developed areas of transtension and dilation are considered to have created an environment conducive to high level ascent of magma and accompanying magmatic-hydrothermal mineralization (Richards et al. 2001). The Escondida district actually comprises six separate deposits (of which La Escondida is the largest) contained within a system of left-lateral strike-slip faults. Mineralization is associated with quartz monzonite (adamellite) and granodiorite intrusions within which alteration evolved from early potassic to sericite–chlorite and quartz–sericite types. These assemblages have been overprinted by a younger advanced argillic alteration assemblage, and both hypogene sulfide and supergene sulfide + oxide ores occur (Padilla Garza et al. 2001).

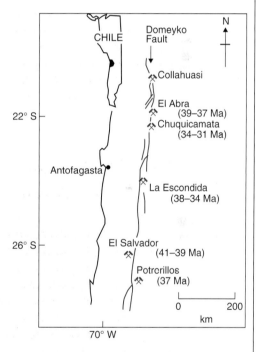

Figure 2.1.1 Simplified map showing the position, in northern Chile, of the Domeyko fault system and the six giant porphyry copper deposits located on it.

Cross sections through La Escondida (Figure 2.1.2) show that the system is related to a multiphase, 38 Myr old, intrusion, the Escondida stock, that cuts through Paleocene andesites. This was capped, 35 Myr ago, by a rhyolite dome. Three phases of sequential alteration are recorded in the deposit (Figure 2.1.3). The earliest stage involves potassic alteration (K-feldspar + biotite) with associated silicification and propylitic (chlorite–sericite) alteration, and is linked to magnetite, chalcopyrite, and bornite. This was followed by phyllic alteration (sericite–chlorite–quartz) with which chalcopyrite, pyrite, and molybdenite are associated. The final hydrothermal alteration stage involves acid-sulfate or advanced argillic

(Continued)

Box 2.1 (Continued)

alteration (quartz–pyrophyllite–alunite). Precipitation of the hypogene sulfide assemblages, mainly during the first two stages of alteration, is attributed to decreasing temperature as the magmatic system cooled and was uplifted, as well as to progressive dilution of magmatic fluids by meteoric solutions (Padilla Garza et al. 2001).

(a)

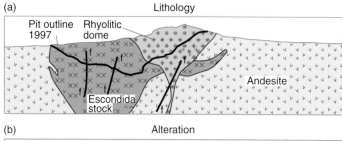

(b)

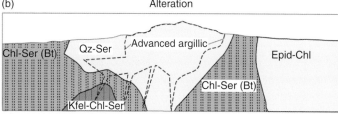

(c)

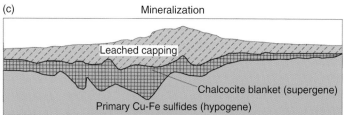

Figure 2.1.2 North–south sections through La Escondida showing the lithologies, alteration zonation, and mineralization characteristics. Source: After Padilla Garza et al. (2001).

Figure 2.1.3 View of the Escondida open pit with the inset diagram showing the distribution of alteration types in the pit. Source: After Riveros et al. (2014). Reproduced with permission of Springer. The photo is courtesy of BHP Billiton.

Box 2.1 (Continued)

An important aspect of the ore-forming system at La Escondida is the development of supergene mineralization (see Chapter 4 and Box 4.2) in areas where the three stages of hydrothermal alteration overlap (i.e. where maximum sulfides were concentrated). Supergene processes were active between 18 and 15 Ma and were responsible for concentrating 65% of the total copper resources of the mine (Figure 2.1.2). The upper portions of the supergene zone comprise a leached cap characterized by low Cu and Mo contents and limonite, hematite, and goethite. This is underlain by a supergene-enriched sulfide blanket containing mainly chalcocite, with lesser covellite and digenite. The best grades (up to 3.5 wt% Cu) occur in the thickest parts of the supergene zone, and also coincide with zones of intense vein stockwork and overlapping hydrothermal alteration. Copper "oxide" minerals such as chrysocolla, brochantite, and atacamite also occur, mainly in late fractures.

As illustrated in Figure 2.19, both types are associated with the generation of oxidized I-type granite magmas associated with melting processes adjacent to subducted oceanic crust. Porphyry Cu–(Mo) can be explained in terms of a body of magma with a relatively low initial H_2O content (perhaps inherited from the fluid absent melting of an amphibolitic protolith) rising to high levels in the crust before significant crystallization takes place. It is considered likely that some melt fractions from this high level magma chamber will be tapped off and extrude on the surface. These fractions will crystallize to form volcanic and subvolcanic (porphyry) suites of rocks whose compositions (i.e. granodioritic or rhyodacitic) will not be highly differentiated because of the low degree of fractionation that has taken place prior to extrusion. Because the magma is emplaced at low load pressures the saturation water content will be relatively low and probably not significantly different from the initial water content. Vapor-saturation and volatile exsolution will, therefore, occur *early* in the crystallization sequence, essentially due to "first boiling." Even though Cu is a compatible element in a crystallizing granitic melt (sequestration of Cu into accessory minerals such as sulfides, magnetite and biotite can result in $D_{crystal/melt}^{Cu} > 1$), the low degree of crystallization means that very little copper will have been removed from the melt by

the time water-saturation occurs. The vapor phase, by contrast, is characterized by high Cl^- concentrations and it will, therefore, efficiently scavenge Cu from the silicate melt.

In this setting, however, the situation with respect to Mo is different. Mo is an incompatible element in a crystallizing granitic melt ($D_{crystal/melt}^{Mo} < 1$) but, because of the low degree of crystallization, its concentration in the residual magma will not have increased significantly prior to water saturation. On saturation, Mo will partition into the H_2O-fluid phase but, because its partition coefficient is relatively low and unaffected by the Cl^- concentration ($D_{fluid/melt}^{Mo} = 2.5$) it will never attain very significant concentrations in the fluid phase. In this setting, therefore, a high level granodioritic I-type magma will exsolve an aqueous fluid phase that is significantly enriched in Cu, but only moderately so in Mo, and form a typical porphyry Cu–(Mo) deposit.

A different scenario can be envisaged for the formation of Mo–(Cu) porphyry deposits, where the parental I-type magma might originally have contained a slightly higher initial water content (perhaps due to the fluid absent melting of a biotite-bearing protolith) than the Cu-dominant situation. As shown in Figure 2.19, this magma would not normally rise to the same shallow crustal levels as its drier equivalent. The saturation water content of the magma is also significantly

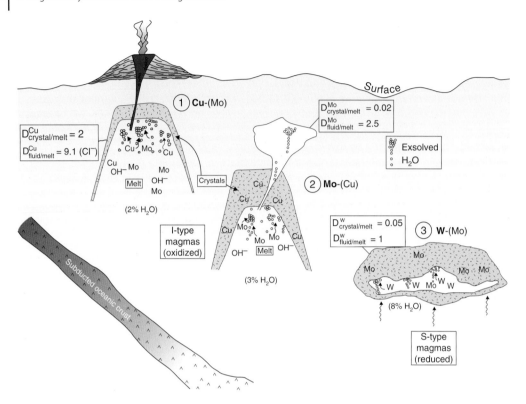

Figure 2.19 Schematic model for the origin and formation of porphyry-type Cu, Mo, and W deposits. Detailed descriptions are provided in the text Source: Modified after the models of Candela and Holland (1984, 1986), Strong (1988), Candela (1992).

higher in this situation and, consequently, a greater degree of crystallization needs to take place before water-saturation is achieved (in this case second boiling will be more important in the achievement of vapor-saturation). As the magma crystallizes, copper will be extracted from the melt and end up distributed somewhat evenly in early formed crystals throughout the rocks representing the marginal zones of the pluton. This Cu will not, of course, be available for partitioning into the aqueous fluid once saturation is achieved. Mo, on the other hand, is incompatible and its concentration will continue to increase in the residual melt. When saturation does occur, Mo will be concentrated even further into the H_2O-fluid phase because of its favorable partition coefficient. Despite the fact that Cu will also be strongly partitioned into the highly saline H_2O-fluid, its abundance in the

melt is now significantly depleted due to earlier crystal–melt sequestration and the vapor phase will not be significantly concentrated in Cu. The extraction of a melt fraction from a more deep-seated magma chamber will, therefore, give rise to a pluton that is less likely to reach the surface and whose composition is more highly differentiated due to the greater degree of crystal fractionation prior to extraction. Porphyry-style mineralization associated with the circulation of a magmatic hydrothermal fluid in and around such an intrusion may well have Mo concentrations in excess of Cu, giving rise to Mo–(Cu) porphyry-type mineralization.

2.8.2 The Origin of Porphyry W ($\pm$Sn) Type Deposits

The above model can also be used to explain some of the characteristics of tungsten and tin dominated porphyry-type ore deposits that are associated with many magmatic arc settings. Porphyry style Sn-W deposits are well known from the Andean Eastern Cordillera that runs through southern Peru and Bolivia and contains giant deposits such as Llallagua. Candela (1992) has emphasized the relationship that exists in particular between W-rich porphyry-type deposits and the more reduced, ilmenite-bearing, S-type granites that crystallize at relatively deep levels of the Earth's crust and whose setting is illustrated under (3) in Figure 2.19. In oxidized environments, such as those applicable to I-type, magnetite-bearing granites, the efficiency of extracting W from a magmatic-hydrothermal system into an ore body is relatively low and this metal is not a normal component of porphyry Cu–Mo deposits. Under reducing conditions, however, W behaves as an incompatible element in terms of its crystal–melt partitioning behavior and its concentration will increase in the melt during crystal fractionation. When a magma forms from the partial melting of a metasedimentary precursor, the melt formed will be relatively hydrous and will tend to crystallize deep in the crust. Such a melt may also be peraluminous and relatively reduced since it might have equilibrated with meta-sedimentary material that contained organic carbon. The saturation water content under these conditions will be high and a significant degree of crystallization will occur before saturation is attained (by second boiling). Because W is incompatible under these conditions, its concentration will rise in the residual melt, whereas Mo concentrations will tend to decrease because it behaves in a more compatible way under reducing conditions. When the H_2O-fluid phase does exsolve it will interact with a highly differentiated melt that is significantly enriched in W. The fluid phase will scavenge some W but the relatively low partition coefficient ($D^W_{fluid/melt} \approx 1$) will ensure that most of the concentration is achieved prior to water saturation. The deposits that form in this setting will, therefore, be associated with deep-level, highly differentiated S-type granites with metal concentrations dominated by W and only minor Mo. Any magmas forming in this environment that are endowed with Sn – which is also incompatible and likely to fractionate strongly into the aqueous fluid – will produce tin mineralization alongside tungsten.

2.8.3 The Role of Sulfur in the Formation of Porphyry Copper Deposits

Metal deposits associated with high-level arc-related magmatism and the exsolution of hydrothermal fluids from these magmas (i.e. the porphyry, epithermal and skarn family of deposit types) represent the most important global source of Cu, as well as providing significant Mo, W, Sn, Zn, Pb, Au, and Ag. The recognition in the field of these deposit types is facilitated by the fact that they are (i) hosted in rocks that have been subjected to intense acid hydrolysis and alteration (see Chapter 3, Section 3.6); and (ii) they are delineated by large and prominent sulfur anomalies (Blundy et al. 2015) that manifest mainly as disseminated pyrite and chalcopyrite. Sulfur in particular plays a key role, in a number of ways, during the formation of these deposits, as discussed below.

2.8.3.1 The Role of Sulfur in Concentrating Metals in Porphyry Systems

As discussed in Chapter 1, the presence of an immiscible sulfide melt in a silicate magma serves to promote the sequestration of siderophile and chalcophile metals by the sulfide fraction, a process that promotes the concentration of these metals into phases that can accumulate and crystallize to form ore bodies (for example Box 1.6). Wilkinson (2013) has suggested that an analogous process may be an important trigger for the formation of large porphyry copper deposits. It is argued that sulfide-saturation in arc-related magmas could be responsible for the accumulation of metals such as Fe, Cu, and Au into sulfurous parts of the magmatic system that on crystallization represent zones of metal preconcentration. Although it can be argued that this process would deplete the more fractionated magmatic phases in both sulfur and metals, interaction of a subsequently exsolved, saline, aqueous fluid phase with such a protore could result in very efficient partitioning of the ore ingredients into the fluid phase. This fluid would then migrate upwards in the normal way, carrying with it the products of a previous cycle of ore concentration – this feature is consistent with the fact that many ore fluids in this environment are extraordinarily enriched in metals (Figure 2.10b), perhaps because they have interacted directly with a sulfide protore, whereas others have not done so and as a consequence are less fertile.

2.8.3.2 The Role of Sulfur in Precipitating Ore Minerals in Porphyry Systems

One of the enigmas of porphyry copper deposits is that they are associated with relatively oxidized (I-type) magmas from which high-temperature, saline aqueous solutions are exsolved carrying abundant Cu and other metals, but the ores are marked by a sulfide mineral assemblage (mainly pyrite + chalcopyrite) for which the source of sulfur is not readily evident. Blundy et al. (2015) have suggested that the sulfur associated with porphyry copper deposits comes from the degassing (as H_2S and SO_2) of mafic magmas residing at a lower crustal level than the more felsic magmas from which the Cu-bearing aqueous fluids are derived. The importance of this source of reduced sulfur is that it is suggested to have played a role as a trigger for the precipitation of Fe- and Cu-sulfide minerals from the fluids carrying these metals. They propose that the interaction of reduced sulfur bearing gases from an underlying mafic magma chamber with more oxidized Cu-bearing brines/vapors in overlying, felsic intrusions, is a viable mechanism for promoting the precipitation of chalcopyrite for example, as suggested by the following reaction:

$$\underset{\text{(gas)}}{11H_2S + SO_2} + \underset{\text{(brine)}}{6CuCl^-_2 + 6FeCl_2} \rightarrow$$

$$\underset{\text{(chalcopyrite)}}{6CuFeS_2} + 6Cl^- + 18HCl + 2H_2O$$

$$(2.7)$$

Blundy et al. (2015) point out that this process results in the production of significant volumes of acid (as HCl – see reaction (2.7)) and is consistent with the occurrence of intense alteration and acid hydrolysis associated with porphyry ore forming process. More detailed discussions of the precipitation mechanisms and alteration processes applicable to hydrothermal ores is provided in Chapter 3.

2.9 Near-Surface Magmatic-Hydrothermal Processes – The "Epithermal" Family of Au–Ag–(Cu) Deposits

Exploration for gold, especially in the circum-Pacific region, has led to the discovery of a large number of world-class gold deposits associated with either active or geologically recent volcanic environments (Mitchell and Leach 1991). These deposits are now regarded as an important and very prospective category of gold deposit type, termed epithermal deposits (Hedenquist et al. 1996). The term "epithermal" is derived from Lindgren's (1933) classification of ore deposits and refers to those that formed at shallow crustal levels (i.e. the epizone). Many studies of this ore-forming environment, and in particular comparisons with active, modern analogues such as the Taupo volcanic zone on the north island of New Zealand, have shown that epithermal deposits typically form at temperatures between 160 and 270 °C and pressures equivalent to depths of between 50 and 1500 m (Cooke and Simmons 2000; Hedenquist et al. 2000; Simmons et al. 2005; Taylor 2007; Saunders et al. 2014).

There are two contrasting styles of mineralization that are now recognized in epithermal deposits, and these are referred to as *high-sulfidation* and *low-sulfidation* types. These terms refer specifically to the oxidation state of sulfur in the ore fluid, the chemistry and pH of which also relates to the nature of alteration associated with each type. Unfortunately, the geological literature contains a confusing plethora of synonyms for these two types. Table 2.3 attempts to summarize these terms and also illustrates the main characteristics of each. Note that the two terms do not relate to the abundance of sulfur, as this is highly variable in each deposit type, but in some publications "high sulfur" and "low sulfur" have been equated with high-sulfidation and low-sulfidation respectively. High- and low-sulfidation epithermal deposits can be viewed as end-members of processes related to fluid evolution and circulation in and around volcanoes. High-sulfidation deposits occur in proximal settings and are commonly found within or close to the volcanic vent itself (Hedenquist et al. 1993). The fluids involved with mineralization are derived directly from the magma as a product of fluid-saturation and are usually boiling in the ore-forming environment. The fluids are very acidic (pH of 1–3) and oxidized, carrying the oxidized S^{4+} or S^{6+} species as SO_2, SO_4^{2-}, or HSO_4^- in solution. As this fluid boils and SO_2 and CO_2 are partitioned into the vapor phase, the remaining liquid carries a surplus of H^+ which makes it very acidic (pH = 1; Hedenquist et al. 2000). This acidic fluid is also capable of leaching most of the major elements from the host volcanic or volcano-sedimentary rocks through which it circulates, resulting in vuggy textures and an advanced argillic style of alteration (see Table 2.2 and also the more detailed account of hydrothermal alteration in Chapter 3). By contrast, low-sulfidation deposits are associated with fluids that are similar to those involved with hot springs and other geothermal manifestations in areas of enhanced heat flow. These fluids have equilibrated with their host rocks and generally comprise a dominantly meteoric component, although it is likely that this will have been mixed with an evolved magmatic fluid if active volcanism is located nearby. Consequently, low-sulfidation deposits may form within the volcanic edifice, especially during the waning stages of magmatic activity when draw-down of meteoric fluids is perhaps more likely. More typically they form at locations that are somewhat removed from the forms of volcanism. The fluid involved is near-neutral and has low salinities, but as with high-sulfidation environments, it is also likely to have boiled in and around the zone of ore formation.

Table 2.3 Characteristics of high- and low-sulfidation epithermal deposits.

High-sulfidation	Low-sulfidation
Oxidized sulfur species (SO_2, SO_4^{2-}, HSO_4^-) in ore fluid/vapor	Reduced sulfur species (HS^-, H_2S) in ore fluid/vapor
Also referred to as	
Gold–alunite, acid–sulfate, alunite–kaolinite	Adularia–sericite, hot spring-related
Fluids	
Acidic pH, probably saline initially, dominantly magmatic	Near-neutral pH, low salinity, gas-rich (CO_2, H_2S), dominantly meteoric
Alteration assemblage	
Advanced argillic (zonation: quartz–alunite–kaolinite–illite–montmorillonite–chlorite)	Adularia–sericite (zonation: quartz/chalcedony–calcite–adularia–sericite–chlorite)
Metal associations	
Au–Cu (lesser Ag, Bi, Te)	Au–Ag (lesser As, Sb, Se, Hg)

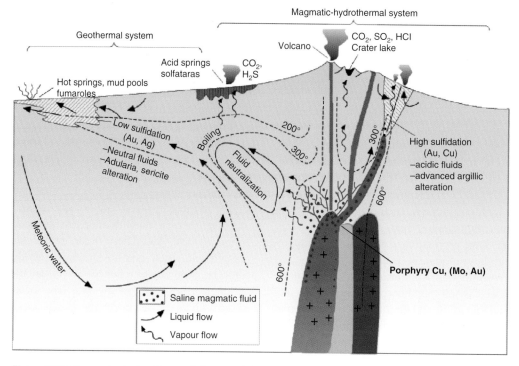

Figure 2.20 The geological setting and characteristics of high-sulfidation and low-sulfidation epithermal deposits. A genetic link between high-sulfidation epithermal Au–Cu and sub-volcanic porphyry type Cu–Au deposits is also suggested. Source: After Hedenquist et al. (2000).

The characteristics of high- and low-sulfidation deposits are shown schematically in Figure 2.20. This diagram suggests that a spatial and genetic link exists between the two types, but it should be emphasized that this may not be the case. Many gold districts contain either low-sulfidation epithermal deposits (such as the major deposits in Nevada, USA, including Round Mountain, Comstock Lode, Midas, and Sleeper) or high-sulfidation deposits (such as many of the Andean deposits, including Yanacocha, Pierina, and El Indio–Tambo). There is, however, an increasing number of cases where spatial and genetic links are evident. One area where both high- and low-sulfidation epithermal deposits occur in close proximity to one another is in Kyushu, the southernmost island of Japan (Box 2.4). It is evident in Figure 2.20 that there should also be a genetic link between porphyry Cu–(Mo–Au) deposits

formed in the sub-volcanic environment and high-sulfidation Au–Cu deposits at surface. This link can be demonstrated in the case of the Lepanto epithermal Au–Cu and the Far Southeast porphyry Cu–Au deposits on Luzon in the Philippines. These two deposits are adjacent to one another in late Pliocene rocks, with the mineralization in both having formed over a brief 300 000 year time period at around 1.3 Ma (Arribas et al. 1995). It is also pertinent to note that low-sulfidation Au–Ag mineralization is located within a few kilometers of the Lepanto–Far Southeast system, indicating that spatial and genetic links between high- and low-sulfidation epithermal and porphyry styles of mineralization are likely (Shinohara and Hedenquist 1997). The fact that their coexistence is only rarely documented is, at least partially, therefore, a function of erosion and lack of preservation.

2.9.1 Gold Precipitation Mechanisms in Epithermal Deposits

The detailed mechanisms of gold precipitation in high- and low-sulfidation deposits are complex and vary depending on the geological setting and nature of fluids involved (Cooke and Simmons 2000). The following section briefly considers gold precipitation mechanisms as applicable to epithermal deposits, although the principles discussed here also apply to other gold deposit types. The reader is encouraged to read the relevant sections on metal solubility and precipitation mechanisms in Chapter 3 (Sections 3.4 and 3.5) before continuing with this section.

Anionic species, or ligands, of chloride (Cl^-) and sulfide (specifically HS^-) are regarded as being essential in order to solubilize gold in aqueous solutions (Seward and Barnes 1997; Seward et al. 2014). In reduced, near-neutral pH, aqueous solutions Au is likely to be transported as the $Au(HS)_2^-$ complex and this is also likely to be the preferred medium for the movement of gold in low-sulfidation environments. By contrast, at higher temperatures ($>300\,°C$) and for solutions that are both more acidic and saline, gold is preferentially transported as the $Au(Cl)_2^-$ complex, and this mode of transport probably applies to high-sulfidation environments. Since the mechanism whereby gold is transported could be fundamentally different in the two epithermal deposit types, it follows that the chemical and physical controls that precipitate the gold from the hydrothermal fluids are also likely to differ.

Box 2.2 Magmatic-Hydrothermal Processes in Volcanic Environments – the Kasuga and Hishikari Epithermal Au–Ag Deposits, Kyushu, Japan

Kyushu, the southernmost major island of the Japanese arc, is the country's principal gold producing region. Late Cenozoic volcanic activity in this region has given rise to several epithermal Au–Ag deposits, all related to either extinct or waning magmatic-hydrothermal systems arising from this volcanism. Both styles of epithermal gold mineralization are preserved on Kyushu, namely high-sulfidation (or acid-sulfate) mineralization represented by the Nansatsu type ores of which Kasuga is an example, and low-sulfidation (or adularia–sericite) mineralization represented by the very rich Hishikari deposit.

Kasuga

The Nansatsu district in the southern portion of Kyushu has been the site of calc–alkaline volcanism for the past 10 Myr. Several deposits occur in the district, including Akeshi, Iwato, and Kasuga. Kasuga is a small deposit that produced about 120 000 tons of ore annually at an average Au grade of about $3\,g\,t^{-1}$. The ore body is associated with a residual high-silica zone in andesites of the Nansatsu Group, which overlies basement metasediments of the Cretaceous Shimanto Supergroup (Hedenquist et al. 1994). Alteration in the open pit is characterized by a central quartz-rich zone in which the volcanic host rock has been almost entirely leached of all elements except Si (Figure 2.2.1a). The quartz body is surrounded by a zone of advanced argillic alteration (comprising alunite, dickite, and kaolinite), which in turn grades out into a propylitic zone (chlorite + illite). Ore minerals consist mainly of pyrite and enargite with paragenetically-later native sulfur, covellite, and goethite. The Au–Ag (electrum) ore is largely contained within the quartz body in the center of the alteration halo, although considerable remobilization into late oxidized phases also occurs. Mineralization is linked to the exsolution of a metal charged fluid phase from the host andesite, with subsequent segregation into vapor

(Continued)

Box 2.2 (Continued)

and liquid phases. Highly acidic vapors are implicated in the leaching of volcanic rocks to form the porous quartz body and may also have transported much of the Au and Cu. The metal charged aqueous phases subsequently percolated through the porous rock and precipitated Cu, Ag, and Au. Progressive mixing of this fluid with meteoric waters was probably the main precipitation mechanism and was also responsible for the zonation in the alteration halo, as well as the late oxidation of pre-existing assemblages (Hedenquist et al. 1994).

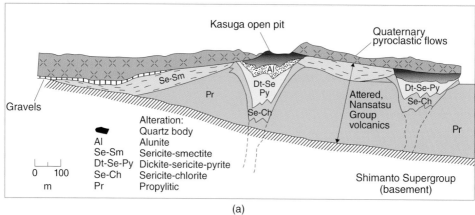

(a)

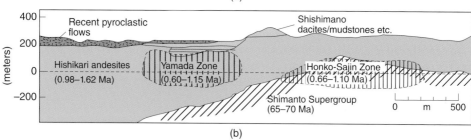

(b)

Figure 2.2.1 (a) Cross section through the Kasuga deposit showing the nature and geometry of the alteration zoning. Source: After Hedenquist et al. (1994). (b) Cross section through the Hishikari deposits illustrating the concentration of mineralization at one particular elevation. Source: After Izawa et al. (2001).

Hishikari

The Hishikari deposit, discovered in 1981, is a very rich low-sulfidation epithermal deposit with an initial resource of about 260 tons of mineable gold (Izawa et al. 2001). It is located some 20 km to the northwest of the recently active Kirishima volcano in south-central Kyushu. The deposit occurs as a series of vertical, en echelon veins largely within the Cretaceous Shimanto Supergroup basement, but also extending short distances into overlying Pleistocene andesites. The veins consist mainly of quartz and adularia, with lesser smectite clay. Vein formation and mineralization is believed to be linked to volcanism that started at about 1.6 Ma and terminated at about 0.7 Ma. Vein related adularia, however, has been dated at between 1.15 and 0.60 Ma, suggesting that the geothermal system and mineralization commenced some 0.5 Myr after volcanism (Izawa et al. 2001). The spectacularly rich Hosen vein (which records grades

---

Box 2.2 (Continued)

---

of over 3000 g t⁻¹ Au and 2000 g t⁻¹ Ag; Figure 2.2.2) records evidence of three phases of vein deposition. Each phase consists mainly of quartz, adularia, and electrum. Minor pyrite and chalcopyrite also occur but only during the early phases of deposition. The high grade ores at Hishikari are restricted to a specific elevation, extending over about 200 m of vertical extent (Figure 2.2.1b), and veins either terminate or become sub-economic above and below this range. This, together with fluid inclusion evidence, suggests that boiling of a dominantly meteoric fluid was responsible for the precipitation of gold and silver from solution.

Figure 2.2.2 Gold and silver (electrum) rich Hosen vein as seen underground at the Hishikari mine, Kyushu. Source: The photo is courtesy of Noel White.

For low-sulfidation deposits gold precipitation is relatively straightforward and is linked to one, or both, of the two processes that characterize fluid evolution in this environment, namely boiling and fluid mixing. Boiling of an ore fluid in this case will result in loss of sulfur to the vapor phase, which causes destabilization of the Au(HS)$_2^-$ complex and precipitation of Au, as described in Eq. (2.8) (after Cooke and Simmons 2000):

$$Au(HS)_2^- + H^+ + 0.5H_2 \leftrightarrow Au + 2H_2S \tag{2.8}$$

Mixing of an oxidized meteoric water with the same ore fluid would also have the effect of precipitating gold, as shown in Eq. (2.9):

$$Au(HS)_2^- + 8H_2O \leftrightarrow Au + 2SO_4^{2-} + 3H^+ + 7.5H_2 \tag{2.9}$$

Evidence for boiling-induced Au precipitation is provided from modern geothermal systems that exploit steam to drive electricity-generating turbines, such as at Broadlands in New Zealand (Cooke and Simmons 2000). It is well known in these power stations that the siliceous scale that plates the inside of pipes, and accompanies the flashing of water to steam, is often enriched in both Au and Ag. In actual low-sulfidation deposits the narrow vertical interval over which vein-hosted mineralization occurs (such as at Hishikari; Box 2.2) is another indication that boiling acts as a fundamental control on ore precipitation. Although it is clear that fluid mixing also plays a role in deposits such as Creede in Colorado, the evidence for a widespread role for this mechanism is less clear (Cooke and Simmons 2000).

For high-sulfidation deposits, the Au deposition mechanisms are less well understood and more complex, since it is feasible to transport gold, not only as a Au(Cl)$_2^-$ complex, as suggested above, but also as a bisulfide complex, Au(HS) in fluids with a high sulfur activity.

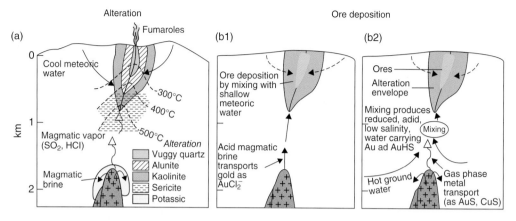

Figure 2.21 Two stage model for the formation of high-sulfidation epithermal deposits. (a) Initial stage where a dominantly magmatic vapor phase is responsible for leaching of the country rock and development of an advanced argillic alteration halo around the main fumarolic conduit. (b1) Ore deposition stage, in this case where gold is transported as a chloride complex; and (b2) ore deposition stage where gold is transported as a bisulfide complex. Source: After Arribas et al. (1995).

It is also considered possible that Au is transported together with Cu in the vapor phase (see Section 2.4.3 above). If gold is transported as a bisulfide complex in high-sulfidation environments then the precipitation mechanisms are also likely to be related to boiling and fluid mixing, as for low-sulfidation systems. If, on the other hand, gold is transported as a chloride complex, then boiling and extraction of oxidized sulfur species (SO_2 or SO_2^-) into the vapor phase will have little effect on its stability. Fluid mixing, between a hot, acidic, and saline $Au(Cl)_2^-$ bearing ore fluid and a cooler, neutral meteoric solution, could, however, be an important precipitation mechanism, since it will have the effect of increasing pH and decreasing salinity (by dilution) of the ore fluid. Studies at Lepanto, for example, indicate that fluid mixing did occur as indicated by variations in the ratio of magmatic to meteoric fluids from 9:1 to 1:1 over the ore deposit (Hedenquist et al. 1998).

Arribas et al. (1995) have provided a model for the formation of high-sulfidation epithermal systems that reflects the complexities related to variations of gold speciation in this environment. The model envisages two stages of ore formation. The first involves degassing of hot, acid rich magmatic vapors that are responsible for intense leaching of the host rock to form the porous vuggy quartz zone in the fumarolic conduit, as well as the advanced argillic alteration halo around it (Figure 2.21a). The vapor phase could also mix with meteoric waters to form an acid sulfate fluid that has a low Au solubility but is still implicated in the alteration process. The secondary porosity and permeability created during this alteration process is considered to be a necessary preparatory stage for the influx of later metal-bearing fluids. Subsequent ore deposition can occur in one of two ways. A hot, acidic, and saline ore fluid, carrying gold as a $Au(Cl)_2^-$ complex, could be derived directly from the subjacent magma and move directly up into the alteration zone. Ore precipitation would occur as a result of mixing and dilution of this fluid by cooler meteoric waters (Figure 2.21b). Alternatively, it is suggested that Cu and Au are initially removed from the magma in the vapor phase (see Section 2.6.2) and that these metal charged gases mix with heated ground waters circulating around the intrusion to form a low salinity fluid in which gold is transported as Au(HS). This fluid would then precipitate metals by boiling in the near surface environment, or mixing with meteoric waters, or both (Figure 2.21b).

2.10 Skarn Deposits

"Skarn" is a Swedish term that originally referred to the very hard rocks composed dominantly of calc–silicate minerals (i.e. Ca-rich garnet, clinopyroxene, amphibole, and epidote) that characterize the alteration assemblages associated with magnetite and chalcopyrite deposits in that country. It is now widely used to refer to the metasomatic replacement of carbonate rocks (limestone and dolomite) by calc–silicate mineral assemblages during either contact or regional metamorphic processes. Mineral deposits associated with skarn assemblages are referred to as skarn deposits and are typically the product of contact metamorphism and metasomatism associated with intrusion of granite into carbonate rocks. A wide variety of deposit types and metal associations are grouped into the category of skarn deposits, and these include W, Sn, Mo, Cu, Fe, Pb–Zn, and Au ores. The different metals found in skarn deposits are a product of the differing compositions, oxidation states, and metallogenic affinities of the igneous intrusion, as described in Chapter 1 (Einaudi et al. 1981; Misra 2000). A simple diagram relating granitoid composition to skarn deposit type is shown in Figure 2.22.

As a general rule Fe and Au skarn deposits tend to be associated with intrusions of more mafic to intermediate compositions. Cu, Pb, Zn, and W are linked to calc–alkaline, magnetite-bearing, oxidized (I-type) granitic intrusions, and Mo and Sn with more differentiated granites that might be reduced (S-type) and ilmenite-bearing. It should be noted that there are exceptions to this general trend.

Skarn deposits can be classified into calcic or magnesian types, depending largely on whether the host rock is limestone or dolomite. They are also described as either endo- or exo-skarns, depending on whether the metasomatic assemblage is internal or external to the intruding pluton. Most of the large, economically viable skarn deposits are associated with calcic exoskarns. Tungsten skarns produce a significant proportion of the world's W production and are typically associated with intrusion of calc–alkaline intrusions, emplaced relatively deep in the crust. Examples include the King Island mine of Tasmania and the MacTung deposit in the Yukon territory of Canada (see Box 2.3). Copper skarns, by contrast, are often associated with high level porphyry-style intrusions and many porphyry copper systems that intrude carbonate host rocks have copper skarns associated with

Figure 2.22 Plot of $[FeO + Fe_2O_3 + CaO + Na_2O]/K_2O$ versus SiO_2 showing the relationship between the composition of igneous intrusions and the dominant metal in various skarn deposits. Source: After Meinert (1992).

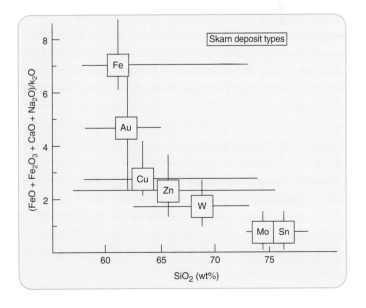

them. A classic example is the Bingham district of Utah, USA, which contains not only a huge porphyry Cu deposit, but also the world's largest Cu skarn deposit. In addition, the skarn ores at Bingham contain economically viable Pb–Zn–Ag ores in limestones that are distal to the copper mineralization (Einaudi 1982). It is interesting to note that one of the world's largest gold mines, at Grasberg in the Ertsberg district of West Papua, also exhibits a porphyry–skarn association and actually produces gold as a by-product of the copper mining process (Meinert 2000). Gold skarns associated with porphyry Cu mineralization are associated with emplacement of high level,

oxidized, magnetite-bearing granitoids. Other gold-specific skarn deposits, where Au occurs in association with a Bi–Te–As metal assemblage, are linked to more reducing, ilmenite bearing granitoid intrusions (Meinert 2000). Fe skarns, which occasionally form large, economically viable magnetite deposits, such as at Sverdlovsk and Sarbai in Russia, are associated with more mafic gabbroic to granodioritic intrusions, and are typified by endoskarn alteration and sodium metasomatism (Einaudi et al. 1981). Tin skarns are generally associated with highly differentiated S-type (or ilmenite-bearing) granitoids, a good example of which is the Renison Bell mine in Tasmania.

Box 2.3 Magmatic-Hydrothermal Fluids Associated with Granite Intrusions (2) – The MacTung Tungsten Skarn Deposit, Yukon, Canada

The northern Cordillera of Canada, extending from British Columbia into the Yukon Territory, is a highly prospective zone that contains several major porphyry, SEDEX, and VMS base metal mines, as well as gold deposits. In addition, several large W–Cu–(Zn–Mo) skarn deposits occur in the region, including one of the world's largest known W deposits in the Macmillan Pass area. The deposit, known as MacTung, has yet to be mined but contains ore reserves of some 63 million tons at 0.95 wt% WO_3, with minor copper (Misra 2000).

The MacTung deposit occurs in Cambrian–Ordovician clastic and carbonate sediments deposited on a continental margin in a platformal shelf setting (Atkinson and Baker 1986). The sequence is characterized by alternating siltstones, carbonaceous shales, and limestones. It is intruded by the 90 Myr old Cirque Lake stock, a biotite- and garnet-bearing quartz monzonite of probable S-type affinity (Figure 2.3.1). This granite is implicated in the mineralization process, although the actual source of the magmatic-hydrothermal fluids associated with alteration and metal deposition is believed to be a hidden intrusion at depth to the south of the deposit (Figure 2.3.1). Two distinct zones of skarn mineralization are evident at MacTung. The lower zone is hosted within a folded limestone breccia, whereas the upper zone straddles three separate lithological units, namely a lower limestone breccia, an intermediate pelitic unit, and an upper unit comprising alternating shale and limestone (Figure 2.3.1). Prograde hydrothermal alteration varies as a function of the host sediment composition and reactivity with fluids. Non-reactive and less porous shales are characterized mainly by quartz veining with narrow bleached halos and low concentrations of scheelite. Highly reactive, porous limestones are the main hosts to skarn mineralization, and are pervasively altered. A distinct zonation is evident and alteration progresses from marginal limestone cut by occasional garnet–pyroxene veins, to a pervasively altered intermediate zone comprising extensive limestone replacement by garnet–pyroxene skarn, and finally a core of pyroxene and pyroxene–pyrrhotite skarn. Ore grades improve progressively from the margins to the core, where WO_3 and Cu concentrations can exceed 1.5 and 0.2 wt% respectively (Atkinson and Baker 1986). Retrograde alteration is minimal at MacTung and is characterized by late quartz

Box 2.3 (Continued)

and calcite veins with coarse grained scheelite, and minor amphibole formation representing hydration of the pyroxene skarn.

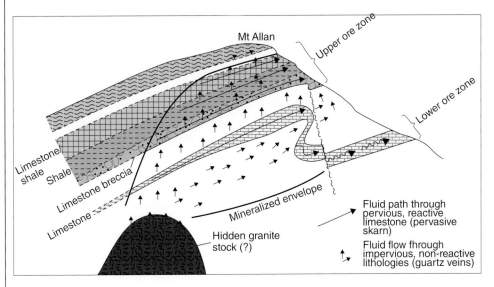

Figure 2.3.1 Generalized geology of the MacTung deposit and a cross section of the ore zone showing mineralized lithologies and the possible location of a hidden intrusion suggested to be the source of the ore-bearing magmatic-hydrothermal fluids. Source: After Atkinson and Baker (1986).

Figure 2.3.2 View of the north face of Mount Allan showing the exposed tungsten skarn mineralization of the MacTung deposit. The dashed line marks the contact between the intrusive Cirque Lake granite stock (lower reaches) and Cambrian–Ordovician clastic and carbonate sediments that build the upper reaches of Mount Allan. The sediments are altered to hornfels and skarn and host the tungsten mineralization, mainly in the lighter colored bands. Source: The photo is from Fischer et al. (2018), with permission of the Northwest Territories Geological Survey.

(Continued)

Box 2.3 (Continued)

The pattern of alteration zonation, which is discordant to the Cirque Lake stock contact, as well as the distribution of quartz veins, suggest that fluids migrated up-dip and toward the stock. For these reasons the source of magmatic-hydrothermal fluids involved in skarn formation is believed to be a hidden intrusion at depth to the south of the mineralization, as suggested in Figure 2.3.2. It is not known whether such an intrusion is merely part of the Cirque Lake stock or an unrelated event.

Even though there are many different metal associations in skarn deposits, the processes by which they form are similar, namely granitoid emplacement and magmatic-hydrothermal activity, albeit at different levels in the crust. An association with granite intrusion cannot always be demonstrated but is usually inferred. Skarn deposits typically form as a result of three sequential processes (Einaudi et al. 1981; Meinert 1992). These are isochemical contact metamorphism during the early stages of pluton emplacement and crystallization, followed by open system metasomatism and alteration during magmatic fluid saturation, and, finally, draw-down and mixing with meteoric fluids (for a definition see Chapter 3) during cooling of the pluton.

2.10.1 Prograde – Isochemical Contact Metamorphism

As the granite pluton intrudes the country rock, sediments are subjected to contact metamorphism and the formation of a variety of hornfelsic textures. The mineral assemblages that form at this stage reflect the composition of the lithotypes within which they form. Contact metamorphism is largely a thermal effect although fluids are likely to circulate during this process and are a product of prograde metamorphic reactions (Figure 2.23a), comprising mainly H_2O and CO_2 (see Chapter 3 for a description of metamorphic fluids). In dolomitic rocks metamorphic mineral zonation approximates the sequence garnet–clinopyroxene–tremolite–talc/phlogopite, reflecting increased distance and progressively more hydrous assemblages away from the intrusion. In limestones the mineral zonation is garnet–vesuvianite – wollastonite–marble. There is no mineralization associated with this stage, although the process of dehydration close to the pluton margins may be important for increasing porosity of the source rocks and facilitating fluid flow during later episodes of mineralization.

2.10.2 Prograde – Metasomatism and Replacement

The second stage in the formation of skarn deposits involves H_2O-fluid and vapor-saturation of the intruding magma (as a function of either first or second boiling or both) and the egress of the fluid phase into the surrounding contact metamorphic halo (Figure 2.23a). At deeper crustal levels fluid flow is likely to be concentrated along discrete structural or bedding parallel conduits, whereas at higher levels fluid flow will be more pervasively distributed, perhaps due to hydrofracturing, in a broad halo in and around the granite cupola (see Section 2.11). Metasomatic mineral assemblages will be similar to those formed during contact metamorphism, but alteration will be more pervasive and coarser-grained and will replace earlier formed assemblages. Si, Al, and Fe, as well as other components, are introduced into the calcareous sediments by the aqueous magmatic fluid, while Ca, Mg, and CO_2 are locally derived and also introduced into the metasomatic system. Sulfide mineralization does not form at this stage, although magnetite and scheelite (in W skarns) do precipitate in the waning stages of prograde metasomatism.

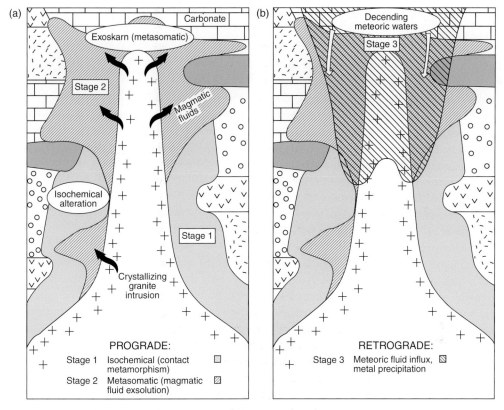

Figure 2.23 The evolution of intrusion-related skarn deposits showing the three sequential stages of formation. (a) Prograde stages, and (b) retrograde stage Source: Adapted from Corbett and Leach (1998).

2.10.3 Retrograde – Meteoric Fluid Influx and Main Metal Precipitation

All magmatic-hydrothermal systems undergo progressive cooling and decay of the high temperature magmatic fluid system. As fluids become progressively dominated by shallow meteoric waters, a series of retrograde reactions takes place, as well as the precipitation of the main stages of base and precious metal, sulfide-related mineralization (Einaudi et al. 1981). The retrograde alteration assemblages are superimposed onto earlier metamorphic and metasomatic minerals (Figure 2.23b) and, typically, this process is recognized by paragenetically late formation of epidote, biotite, chlorite, plagioclase, calcite, quartz (all after various garnet types), tremolite–actinolite, and talc (after pyroxenes) and serpentine (after olivine). Sulfide ore minerals, as well as magnetite and hematite, occur as disseminations, or veins, that cut across prograde assemblages. Assemblages such as pyrite–chalcopyrite–magnetite characterize proximal settings, whereas bornite and sphalerite–galena are typically more distal in occurrence. The paragenetically late precipitation of most skarn-related ores suggests that metal precipitation is related to decreasing temperature of the ore fluids (and a resulting drop in solubilities), fluid mixing, or neutralization of the ore fluid by reaction with carbonate lithologies. Mixing of the magmatic ore fluid with a late meteoric component, and related redox reactions in the fluid, may be additional controls on the ore formation process. A more detailed description of precipitation mechanisms in hydrothermal solutions is provided in Chapter 3.

2.11 Fluid Flow in and Around Granite Plutons

Discussion has focused thus far on the mechanisms by which a magmatic aqueous fluid phase is formed and its role in the transport of metals derived essentially from the melt. Of equal importance with respect to the formation of magmatic-hydrothermal ore deposits are the flow patterns in and around the intrusion from which the fluids are derived, and the length of time that magmatic-hydrothermal fluids can remain active subsequent to magma emplacement. Fossilized fluid flow pathways are revealed by recognizing the effects of alteration and mineralization that a hydrothermal solution imposes on the rocks through which it circulates (see Chapter 3). The prediction of where such fluids have circulated, however, is best achieved by modeling the thermal effects of an intrusion and its exsolved fluid phase in terms of both conductive and convective heat loss in and around the pluton. There have been many attempts to model fluid flow around granite plutons and these have proven useful in predicting where optimal fluid flux, and hence mineralization, should be located in relation to the intrusion.

Cathles (1981) used a thermal modeling approach to demonstrate that intrusions cool very rapidly, at least in terms of the geological time scale. In Figure 2.24a model curves are plotted which relate the time for an intrusion to cool to 25% of its initial temperature, and the size and geometry of the intrusion. The models can be erected for cases where heat is lost either by conduction alone, or by a combination of conduction and convective fluid flow. For the case where a small (1–2 km wide) intrusion cools by conduction of heat to the wall-rocks, solidification will be complete and magmatic-hydrothermal fluid circulation terminated in about 10^5–10^6 years subsequent to emplacement. If cooling is accompanied by convective heat loss caused by the circulation of a hydrothermal fluid through a permeable network of fractures (such as would be the case

if hydrofracturing had occurred; see Section 2.3.3 and Figure 2.7) then such a pluton would cool to 25% of its initial temperature within as little as 10^4 years. Thus, in a highly permeable system where fluid flow is maximized by high permeabilities, heat loss will be very rapid, and a single intrusion will not be able to sustain a geologically long-lived period of hydrothermal circulation. This feature might work against the development of a viable ore body in situations where only magmatic fluids are responsible for ore formation. It should, however, be noted that mineralization in and around magmatic intrusions can also be the result of externally derived fluids and that sustained fluid flow might arise from multiple, long-lived magmatic episodes. It should also be noted that some evidence exists for the very rapid development of hydrothermal deposits – for example, the giant, epithermal Ladolam Au deposit on Lihir Island, Papua New Guinea, containing over 1300 t of contained gold, is thought to have formed in around 55 000 years (Simmons and Brown 2006), a geologically very short-lived episode.

Norton and Cathles (1979) have been able to model the evolution of hydrothermal fluid flow with time in and around a granite intrusion. Figure 2.24b,c illustrate the isotherms and fluid flow streamlines that are likely to exist around a granite intrusion at two discrete periods (i.e. 2×10^4 and 10^5 years after magma emplacement and fluid saturation) for situations where heat is lost by both conduction and convection. Soon after intrusion of even a small granite body (Figure 2.24b) a substantial thermal anomaly is created for up to 2 km around the intrusion and an active system of circulatory fluid flow is set up from both the sides and top of the body. Aqueous solutions establish a convective cell where flow is upwards from the intrusion and circulates back downwards at some distance away. These solutions could incorporate waters from the surrounding country rocks as well as the magmatic fluid derived from the cooling pluton itself. An indication of the fluid flux around

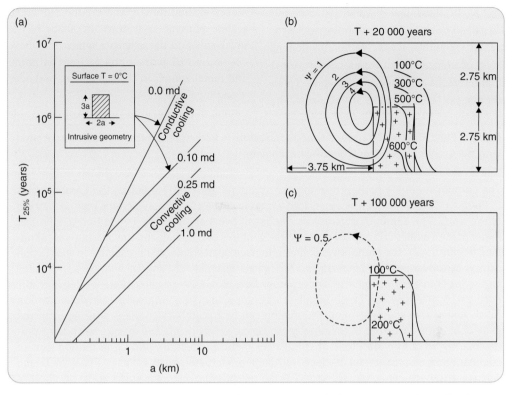

Figure 2.24 Models illustrating the thermal and fluid flow characteristics in and around a cooling igneous intrusive body. (a) Plot of the time taken for the temperature of an intrusion to fall to 25% of its initial temperature ($T_{25\%}$) as a function of the intrusion dimensions (a in km and defined in the inset diagram). Two situations, one where heat is lost by conduction alone (steep slopes) and another where a combination of conduction and convective heat loss occurs (shallow slopes), are shown. Permeabilities (in millidarcy, md) are shown for the two cases. Source: After Cathles (1981). (b) Thermal model showing fluid flow and temperatures in and around an intrusion some 20 000 years after emplacement. For clarity, streamlines (quantified in terms of the stream function ψ) are shown on the left and isotherms on the right hand side, although the model is symmetric about the intrusion. The intrusion was initially at 750 °C with a permeability of 0.07 md and the country rocks at 0 °C with a permeability of 0.13 md. (c) Similar model for the situation 100 000 years after emplacement. Parts b and c are after Norton and Cathles (1979).

the intrusion is obtained in the diagrams from the gradients of the stream function which defines the streamlines (i.e. the closer together the streamlines, the greater the fluid flux). The demise of the hydrothermal fluid cell is demonstrated in Figure 2.24c, where it is clear that the thermal anomaly providing the energy for fluid circulation, as well as the fluid convection itself, has diminished to insignificant proportions within 10^5 years of magma emplacement and fluid saturation. This again reinforces the view that ore-forming events are likely to be short-lived in and around small granite intrusions representing just a single pulse of magma. Although, intuitively, one

might expect that sizeable ore bodies will form only in association with larger, multi-episodic intrusions, the Ladolam example indicates that even short-lived hydrothermal events can give rise to significant metal concentrations under optimal conditions.

A good example of where the nature of fluid flow in and around a granite intrusion contributes greatly to understanding the distribution of magmatic-hydrothermal mineralization is provided by the polymetallic (Sn–W–Cu–Pb–Zn) deposits of the Cornubian batholith in Cornwall and Devon, southwest England. In this classic mining district mineralized veins occur along the margins of individual granite plutons, but also extend out into the surrounding metasedimentary country rocks. Mineralization is also characterized by a pronounced regional zonation in the distribution of metals. These patterns can be explained in terms of the nature of fluid flow in and around individual granite plutons, as well as the shape of the intrusions and the extent to which they have been exhumed. A description of mineralization and metal zoning in the Cornubian batholith is provided in Box 2.4.

Box 2.4 Fluid Flow in and Around Granite Intrusions – Polymetallic Mineralization Associated with the Granites of Cornwall, Southwest England

The Cornubian granite batholith, emplaced at around 280–290 Ma during the Carboniferous to Permian Variscan Orogen, is associated with one of the classic polymetallic mining districts of the world. The batholith outcrops as a series of five major plutons, in addition to a host of minor granite bodies and related dykes, in the counties of Cornwall and Devon, southwest England (Figure 2.4.1). The batholith is believed to have been emplaced in response to subduction caused by convergence of Laurasia and Africa (the assembly of Pangea), with similar granites and styles of mineralization also evident elsewhere along the orogenic belt, notably in Spain, Portugal, France, and the Czech Republic in Europe, and New Brunswick–Newfoundland in North America. In southwest England mining can be traced back to the Bronze Age, but the peak of activities occurred in the nineteenth century. Historically, the district has been a predominantly Sn (2.5 million tons of cumulative metal production) and Cu (2 million tons of cumulative metal production) producer, although other metals such as Fe, As, Pb, Zn, W, U, and Ag have also been extracted (Alderton 1993). Presently the district is still a major producer of kaolinite (China clay) for the ceramics industry (see Section 4.4.1 of Chapter 4).

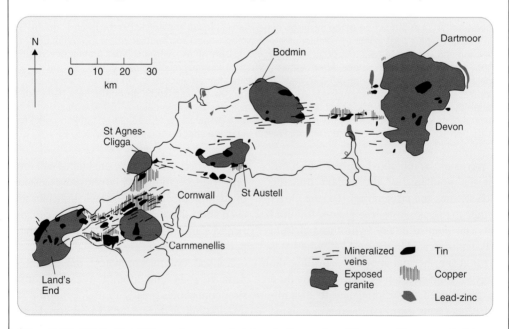

Figure 2.4.1 Distribution of the major exposed plutons in Cornwall and Devon making up part of the Cornubian granite batholith, and the distribution of Sn, Cu, and Pb–Zn mineralization in and around these plutons. Source: After Atkinson and Baker (1986).

Box 2.4 (Continued)

The Cornish granite, as with many similarly mineralized bodies elsewhere along the Variscan belt, is peraluminous in composition, enriched in Rb, Li, F, B, and Be, and can be classified as an S-type granite (Willis-Richards and Jackson 1989). Mineralization in the granite is related to the exsolution of a voluminous aqueous phase from the magma which is responsible for widespread alteration of pluton margins and much of the mineralization (Jackson et al. 1989). Alteration is typically zoned and characterized by the sequence tourmaline – potassic alteration (K-feldspar and biotite) – sericite – chlorite as one progresses away from mineralization. Magmatic-hydrothermal fluids are also responsible for the formation of mineralized quartz veins that are both endo- and exogranitic. Particularly characteristic of the Cornubian mineralization is the presence of sheeted, greisen-bordered vein sets which generally represent the sites of preferential Sn–W mineralization (Figure 2.4.2). These veins consist essentially of quartz and tourmaline with variable cassiterite and wolframite, and a greisen alteration halo comprising quartz, muscovite, Li-mica, and topaz. These vein systems often extend for several kilometers out into the surrounding country rock and record a well defined metallogenic zoning or paragenetic sequence. The ore mineral zonation progresses from an early oxide-dominated assemblage (cassiterite and wolframite together with tourmaline) to a later sulfide assemblage comprising chalcopyrite, sphalerite, and galena, together with fluorite and chlorite. The zonation pattern is evident both vertically within a single vein system and laterally away from a granite body as shown in Figures 2.4.1 and 2.4.3. Although metal zonation is complex (Willis-Richards and Jackson 1989), the progression from Sn–W to Cu to Pb–Zn reflects an ore fluid that evolved, in terms of both cooling and chemical changes, due to fluid–fluid and fluid–rock interactions.

Figure 2.4.2 Sheeted Sn–W-bearing, greisen-bordered veins (dark coloration) exposed in the cliffs at Cligga Head, Cornwall. The photo is courtesy of Helen Wilkinson.

Box 2.4 (Continued)

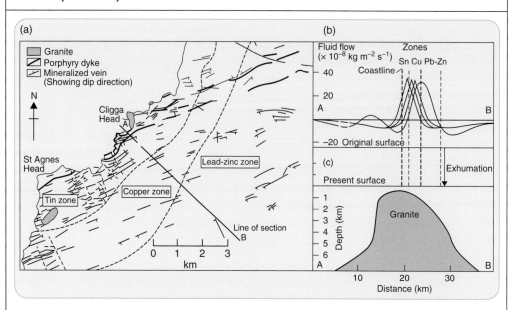

Figure 2.4.3 (a) The regional pattern of metal zonation around the southeast margin of the St Agnes–Cligga Head granite; (b and c) numerical simulation of fluid flow in and around the granite as a function of progressive unroofing of the granite with time. Source: After Sams and Thomas-Betts (1988).

Numerical simulations of the fluid flow in and around the Cornubian granites have provided useful insights into the distribution of mineralization, as well as the controls on metal zonation that is such a feature of the district. One of the characteristics of the Cornubian batholith is that mineralization is preferentially distributed along one margin of the exposed granite pluton, such as the northern margin of the Carnmenellis pluton and the southern margins of the St Austell and Bodmin plutons (Figure 2.4.1). Two of the factors that influence the pattern of magmatic-hydrothermal fluid flow in and around granite plutons are the shape of the body and its depth of emplacement (Sams and Thomas-Betts 1988). Maximum fluid flow tends to be concentrated on the more gently dipping margins of a granite body, which accounts for the asymmetry in distribution of mineralization. In addition, progressive unroofing of a granite by erosion will cause the locus of convective fluid flow to migrate outwards into the surrounding sediments. A model for the southeastern margin of the St Agnes–Cligga Head granite (Sams and Thomas-Betts 1988), much of which is under the sea, illustrates how progressive unroofing of the body might cause the locus of vertical fluid flow to move progressively away from the pluton margins. Thermal and chemical evolution of the mineralizing solutions over the time interval represented by unroofing could account for the successive increments of metal deposition and the regional zonation of metals observed.

2.12 The Role of Hydrothermal Fluids in Mineralized Mafic Rocks

Much of this chapter has focused on the relationships between felsic magmas and the fluids that exsolve from them. As mentioned at the beginning of this chapter, however, it is also feasible for a mafic melt to exsolve a magmatic fluid phase, although the resultant mass fraction of exsolved water will typically be lower than in felsic melts. In mafic rocks the role of a magmatic fluid in ore-forming processes is commonly overlooked, even though it is now apparent that such fluids may have been important in the concentration of metals such as Cu, Ni, and the platinum group elements (PGE) in layered mafic intrusions (Mathez 1989a,b).

The solubility of H_2O in a melt is higher than that of CO_2 and it follows, therefore, that the latter will exsolve before the aqueous species. The first fluids to form from a crystallizing mafic melt, therefore, are rich in CO_2 (as well as CO in more reduced environments), but they will evolve to more H_2O rich compositions as crystallization progresses. In addition, the fluid may also contain appreciable chlorine (as HCl or $FeCl_2$) and sulfur (as HS^- or SO_4^{2-} depending on fO_2). The change in fluid composition from CO_2 to H_2O-dominated occurs quite rapidly in fluids derived from mafic melts and is marked by the precipitation of C (graphite) from the fluid. These fluids are, therefore, unlikely, *at least initially*, to be able to dissolve much metal, but this might change as the proportion of water in the solutions increases.

Considerations of metal concentration in mafic rocks, such as those discussed at some length in Chapter 1, assume that distribution patterns of Cu, Ni, and PGE reflect crystal–melt or melt–melt partitioning during progressive crystallization, or silicate–sulfide immiscibility. These assumptions presuppose that the ore deposits observed have not re-equilibrated with an aqueous fluid phase and that the metal distribution patterns are genuinely magmatic and not overprinted by hydrothermal processes. This is not always the case and mineralogical observations in serpentinites from eastern Australia, for example, show that certain of the PGE (in particular Pd, Pt, and Rh) tend to be readily redistributed by hydrothermal solutions, whereas others (specifically Os, Ir, and Ru) remain relatively unaffected by such processes (Yang and Seccombe 1994). The rare cases where nuggets, overgrowths, and dendrites of PGE are found in placer deposits, and in laterite enrichments, also indicate that under certain conditions these metals are labile and can be put into solution even at low temperatures (see Chapter 4).

In the case of the PGE it is apparent, with the possible exception of palladium (Pd), that these metals tend to exhibit *low solubilities* in magmatic-hydrothermal fluids at high temperatures and under fairly reduced (low fO_2) conditions. In the case of Pd it appears that this metal is fairly soluble in saline solutions at elevated temperatures (Sassani and Shock 1990). In general, however, the presence of ligands such as Cl^- enhances PGE solubilities only slightly and significant dissolution of these metals is only feasible under highly oxidized, acidic conditions (Wood et al. 1992). Hydroxide (OH^-) complexation is also regarded as unlikely to contribute toward PGE solubility in hydrothermal solutions, although it may be important in surficial waters, and PGE–OH^- complexes may be implicated in placer and laterite deposits. Likewise, bisulfide (HS^-) complexes result in very low (ppb) quantities of Pt and Pd being transported in most geologically pertinent solutions. These observations tend to support the conventional view that, in most cases, metal concentrations in mafic rocks have not been markedly modified by hydrothermal processes. However, under conditions of very low fS_2 (where the metal and not the metal sulfide is stable) it appears that PGE solubilities in a highly saline brine may be much higher (10^2–10^3 ppb), mainly because PGE complexation with Cl^- is not detrimentally affected by the presence of sulfide ligands. Under such conditions, a fluid could play an important role in the redistribution of these metals.

2.12.1 The Effects of a Magmatic-Hydrothermal Fluid on PGE Mineralization in the Bushveld Complex

In the Bushveld Complex the grades of PGE mineralization and the relative proportions of these elements remain uniform over tens of kilometers of strike, in both the Merensky Reef and the UG2 chromitite seam (Cawthorn et al. 2002). By contrast the mineralogy of the PGE varies markedly. The normal stratiform reef is generally dominated by PGE sulfide minerals such as cooperite, braggite, and laurite. Atypical situations, represented by mineralized potholes (see Box 1.6) and unusual discordant Pt-rich dunite pipes, are characterized by PGE–Fe alloys and tellurides. The general consensus is that the sulfide-dominated mineralogy of normal reefs is the product of magmatic processes, as discussed in Chapter 1, but that the sulfur-deficient PGE mineralogy in potholes and pipes may reflect areas where a volatile-rich fluid phase has interacted with the magmatic ores, resulting in localized low fS_2 conditions.

The mafic rocks of the Bushveld Complex contain several unusual rock types such as iron-rich ultramafic pipes, as well as iron-rich pegmatites (with concentrations of Pb, As, Sb, and Bi) and plagioclase–amphibole–phlogopite veins, that have been cited as evidence for vapor- or fluid-saturation

in the late stages of crystallization of the Bushveld magmas. Schiffries (1982) regarded the Fe- and Pt-rich dunite pipes as metasomatic in origin and implicated an aqueous magmatic brine that reacted with the host rocks at around 600 °C and 3.5 kbar. More recent work has indicated that these bodies probably were discrete magmatic intrusions, but that they did act as channelways for later, lower temperature hydrothermal fluids (Cawthorn et al. 2000). The sulfide-poor, Pt-dominated mineralogy of these bodies is considered to represent re-equilibration of an original magmatic sulfide assemblage by later hydrothermal fluids.

One of the still contentious issues concerning the Bushveld Complex, however, is the extent to which the major mineralized horizons, such as the Merensky Reef and the UG2, owe their PGE and base metal mineralization to hydrothermal rather than magmatic processes. Detailed mineralogical studies have shown that in certain environments magmatic fluids must have played a role in redistribution of sulfur and recrystallization of the PGE. Where the UG2 chromitite horizon is cut by a dunite pipe, for example, it contains the same distinctive Pt–Fe alloy and Pt–arsenide (sperrylite) phases as the pipe itself. By contrast, the UG2 horizon well removed from such bodies comprises the normal, magmatic PGE–sulfide minerals such as braggite and cooperite (Peyerl 1982).

In addition, the Merensky Reef itself often exhibits the same sort of variations in PGE mineralogy as does the UG2, but in this case the occurrence of sulfide-deficient PGE minerals is usually related to the formation of "potholes" and not necessarily to cross-cutting dunite replacement pipes (Figure 2.25). The Merensky Reef potholes were described in Chapter 1 (see Box 1.6) and are attributed to syn-magmatic faulting of the footwall cumulates just prior to injection of a new magma pulse, with accompanying metasomatism of rocks within the pothole structure by hydrothermal fluid. The occurrence of desulfidized PGE mineral assemblages in the potholes is consistent both with this process

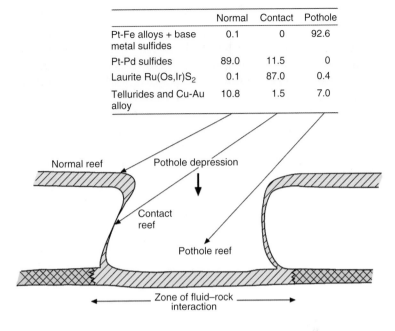

Figure 2.25 Schematic illustration showing the nature of "potholes" in the Merensky Reef of the Bushveld Complex and a table showing the distinctive mineral assemblages of these features relative to normal reef. Source: After Kinloch (1982), Mathez (1989b).

	Normal	Contact	Pothole
Pt-Fe alloys + base metal sulfides	0.1	0	92.6
Pt-Pd sulfides	89.0	11.5	0
Laurite Ru(Os,Ir)S$_2$	0.1	87.0	0.4
Tellurides and Cu-Au alloy	10.8	1.5	7.0

and with the fact that potholes were probably also the sites of fluid-saturation and enhanced circulation. The presence of saline, but relatively oxidized, fluids is considered to have been responsible for modifying the primary magmatic ore mineralogy and destabilizing the dominant base metal and PGE sulfide phases, with the resultant loss of sulfur and reaction between PGE and Fe to form alloys or telluride and arsenide minerals (Kinloch 1982). It is important to note that, during the hydrothermal overprint, PGE and base metals do not appear to be redistributed and the metal budget, therefore, remains constant (Cawthorn et al. 2002). There seems little evidence for a solely hydrothermal origin for the base and precious metal ores of the Merensky Reef, or any other major mineralized part of the Bushveld Complex, although it is clear that magmatic fluids have reacted with certain portions of this layered mafic intrusion.

In summary, it would appear that in contrast to granitoid-related systems, magmatic fluids in mafic melts have played a relatively small role in the mineralization processes that accompany the crystallization of these rock types. Vapor saturation does occur but its influence as a mineralizing agent appears to be limited to re-equilibration of existing magmatic PGE and base metal sulfides in rocks where metasomatism is clearly demonstratable, such as potholes, Fe-rich dunite pipes, pegmatites, and veins.

2.13 Summary

All magmas contain the constituents that, on crystallization, combine to exsolve discrete aqueous liquid and vapor phases. Most magmas will exsolve substantial quantities (up to several wt%) of water, as well as an order of magnitude or so less carbon dioxide, and these are the two principal magmatic-hydrothermal fluids. Water in particular has the ability to dissolve significant quantities of anionic substances, in particular Cl^-, which in turn promotes the solubility of other alkali and transition metal cations. The magmatic aqueous phase can exist as a liquid, vapor, or homogeneous supercritical fluid. The process of H_2O-saturation can be achieved in two ways, either by decreasing the pressure of the system (called first boiling) or by progressive crystallization of magma (second boiling).

H_2O-saturation is particularly relevant to ore-forming processes during the emplacement and crystallization of granitic magmas at moderate to shallow crustal levels. This environment gives rise to the formation of a wide variety of important ore deposit types including porphyry Cu and Mo deposits, polymetallic skarn ores, granite-related Sn–W deposits, and the family of volcanic-related epithermal Au–Ag–(Cu) deposit types. Many metals will partition strongly into the liquid or vapor that forms on H_2O-saturation and, in such cases, mineralization accompanies the alteration of host rocks, both within and external to the intrusion. The formation of either Cu-dominant or Mo-dominant porphyry deposits reflects a subtle interplay between the depth of intrusion of a granitic body (itself a function of the original water content), the timing of H_2O-saturation relative to the progress of crystallization, and the behavior of metals during melt–fluid partitioning. Egress of fluids and vapor from the magma and their subsequent circulation are dependent on the permeability of the surrounds and may be modified by boiling-related hydrofracturing. Polymetallic skarn deposits reflect the interaction between the exsolved magmatic fluids from different types of granite and calcareous sediments. H_2O-saturation and boiling in volcanic environments, producing significant volumes of volatile-rich vapor, is conducive to the formation of epithermal deposits. High- and low-sulfidation epithermal deposits reflect end-members in a continuum of magmatic-hydrothermal processes that progressively incorporate more non-magmatic waters as the volcanic system wanes, or as one moves away from the volcanic center. Many ore deposit types are the product of fluids that are unrelated to a magmatic source and these are the subject of Chapter 3.

Further Reading

For those readers wishing to read more about magmatic-hydrothermal ore-forming processes, the following references to books and journal special issues will be useful.

Barnes, H.L. (ed.) (1967). *Geochemistry of Hydrothermal Ore Deposits*, 1e. Holt, Rinehart and Winston Inc.

Barnes, H.L. (1979). *Geochemistry of Hydrothermal Ore Deposits*, 2e. Wiley.

Barnes, H.L. (1997). *Geochemistry of Hydrothermal Ore Deposits*, 3e. Wiley.

Berger, B.R. and Bethke, B.M. (eds.) (1985). *Geology and Geochemistry of Epithermal Systems*, Reviews in Economic Geology, vol. 2, 298. El Paso, TX: Society of Economic Geologists.

Cooke, D.R., Hollings, P., Wilkinson, J.J., and Tosdal, R.M. (2014). Geochemistry of porphyry deposits. In: *Treatise on Geochemistry*, 2e, vol. 13 (eds. H.D. Holland and K.K. Turekian), 357–381. Oxford: Elsevier.

Corbett, G.J. and Leach, T.M. (1998). *Southwest Pacific Rim Gold–Copper Systems: Structure, Alteration and Mineralization*. Special Publication 6, 237. El Paso, TX: Society of Economic Geologists.

Hagemann, S.G. and Brown, P.E. (eds.) (2000). *Gold in 2000*, Reviews in Economic Geology, vol. 13, 559. El Paso, TX: Society of Economic Geologists.

Lowell, J.D. and Guilbert, J.M. (1970). Lateral and vertical alteration-mineralization zoning in porphyry ore deposits. *Economic Geology* 56: 373–408.

Misra, K.C. (2000). *Understanding Mineral Deposits*, 845. Dordrecht: Kluwer Academic Publishers.

Pirajno, F. (1992). *Hydrothermal Mineral Deposits*, 709. New York: Springer-Verlag.

Seltmann, R., Lehmann, B., Lowenstern, J.B., and Candela, P.A. (eds.) (1997). High-level silicic magmatism and related hydrothermal systems. Special issue. *Journal of Petrology* 38 (12): 1617–1807.

Seward, T.M., Williams-Jones, A.E., and Migdisov, A.A. (2014). The chemistry of metal transport and deposition by ore-forming hydrothermal fluids. In: *Treatise on Geochemistry*, 2e, vol. 13 (eds. H.D. Holland and K.K. Turekian), 29–57. Oxford: Elsevier.

Taylor, R.P. and Strong, D.F. (eds.) (1988). Recent advances in the geology of granite-related mineral deposits. *Canadian Institute of Mining and Metallurgy*, Special Volume 39: 445.

Thompson, J.F.H. (ed.) (1995). *Magmas, Fluids and Ore Deposits*, vol. 23, 525. Mineralogical Association of Canada, Short Course Handbook.

Whitney, J.A. and Naldrett, A.J. (eds.) (1989). *Ore Deposition Associated with Magmas, Reviews in Economic Geology*, vol. 4, 250. El Paso, TX: Society of Economic Geologists.

Williams-Jones, A.E. and Heinrich, C.A. (2005). Vapor transport of metals and the formation of magmatic-hydrothermal ore deposits. *Economic Geology* 100: 1287–1312.

Part II

Hydrothermal Processes

3

Hydrothermal Ore-Forming Processes

TOPICS

Origin of fluids in the Earth's crust
Deformation, pressure gradients, and hydrothermal fluid flow
Metal solubilities in aqueous solutions
   the nature of metal–ligand complexes and Pearson's Principle
Fluid–rock interactions and alteration
Precipitation mechanisms
  physico-chemical processes
  adsorption
  biologically mediated processes
Metal zoning and paragenetic sequence
Modern analogues of hydrothermal ore-forming processes on the ocean floor
   the VMS–SEDEX continuum
Ore deposits associated with aqueo-carbonic hydrothermal fluids
   orogenic, Carlin-type, and metamorphosed quartz pebble conglomerate hosted gold deposits
Ore deposits associated with basinal fluids
   stratiform sediment-hosted copper (SSC) and Mississippi valley type (MVT) lead–zinc deposits
Ore deposits associated with meteoric fluids
   sandstone-hosted uranium deposits

CASE STUDIES

Box 3.1 Fluid Mixing and Metal Precipitation: The Olympic Dam Iron Oxide–Copper–Gold (IOCG) Deposit, South Australia
Box 3.2 Alteration and Metal Precipitation: The Golden Mile, Kalgoorlie, Western Australia – An Archean Orogenic Gold Deposit
Box 3.3 Exhalative Venting and "Black Smokers" on the Sea Floor: The VMS Deposits of the Troodos Ophiolite Complex, Cyprus
Box 3.4 Sedimentary Exhalative (SEDEX) Processes: The Red Dog Zn–Pb–Ag Deposit, Alaska
Box 3.5 Circulation of Orogeny-Driven Aqueo-Carbonic Fluids: Twin Creeks – A Carlin-Type Gold Deposit, Nevada
Box 3.6 Circulation of Sediment-Hosted Basinal Fluids: 1 The Central African Copperbelt
Box 3.7 Circulation of Sediment-Hosted Basinal Fluids: 2 The Viburnum Trend, Missouri

---

Introduction to Ore-Forming Processes, Second Edition. Laurence Robb.
© 2021 John Wiley & Sons Ltd. Published 2021 by John Wiley & Sons Ltd.

3.1 Introduction

This chapter extends the concept of hydrothermal mineralization to deposits related to fluids derived from sources *other* than magmatic solutions. Such fluids include those formed from metamorphic dehydration reactions, from the expulsion of pore fluids during compaction of sediment, and from meteoric waters. It also considers sea water as a hydrothermal fluid with specific reference to the formation of base metal deposits on the ocean floor. Unlike the previous chapter, which was mainly concerned with granite-related ore deposits, the present chapter discusses a much broader range of ore-forming processes and environments.

Hydrothermal ore-forming processes are ubiquitous and there is scarcely an ore deposit anywhere on Earth that has not been formed directly from hot aqueous solutions flowing through the crust or modified to varying degrees by such fluids. This view is supported by the example in Section 2.12 of Chapter 2, where a compelling case for hydrothermal remobilization of platinum group elements (PGE) mineralization in the potholes of the Merensky Reef, Bushveld Complex, traditionally regarded as igneous in origin, is presented. Likewise, the Au–U ores of the Witwatersrand Basin can no longer be regarded simply as paleoplacer deposits and hydrothermal processes have clearly played a significant role in their formation (see Section 3.9 below). Similarly, oil and gas deposits (see Chapter 5) have migrated to their present locations in the presence of hot water, during processes akin to those discussed below for hydrothermal solutions. Many of the giant ore deposits of the world owe their origins to the flow of hydrothermal fluids in the Earth's crust and the ability of aqueous solutions to effectively scavenge, transport, and concentrate a wide range of economically important components.

Over the past few decades in particular, a great deal of research has been directed toward better understanding the complexity of hydrothermal processes. Concepts such as the source of hydrothermal solutions, their passage through the Earth's crust, and the precipitation mechanisms involved in the formation of ore bodies, are now relatively well understood. The three editions of Barnes's *Geochemistry of Hydrothermal Ore Deposits* (1967, 1979a, 1997) provide an account of the progress of this research over several decades. There are, of course, some features of hydrothermal ores about which we need to know more, and these include the ages and duration of ore-forming processes, the detailed recognition of ancient fluid pathways, and the depths of fluid flow in the crust, as well as the relationship between global tectonics and metallogeny (Skinner 1997). The last topic is a particularly important one (see Chapter 6) that applies not only to hydrothermal ores, but to the entire range of mineral deposit types. Finally, the role of microorganisms in the formation of ore deposits is a topic that is beginning to attract attention and may turn out to be much more important than previously thought.

It will be evident that the separation of "magmatic-hydrothermal" and "hydrothermal" processes into two sections (Chapters 2 and 3 respectively) is not so much a conceptual necessity as a requirement of the organization and structure of the book. The two chapters should be seen as covering a spectrum of processes ranging from magmatic fluid flow to shallow level meteoric infiltrations. In order to emphasize the continuum, this chapter is terminated with a summary diagram that attempts to relate the sources of hydrothermal solutions to ore deposit types and applies to both Chapters 2 and 3.

3.2 Other Fluids in the Earth's Crust and Their Origins

Figure 3.1a shows that, in addition to magmatic fluids, there are four other major water types on or near the Earth's surface. Although they may all have had similar origins, each

Figure 3.1 (a) The major types of liquid water that exist at or near the Earth's surface; magmatic water was specifically discussed in Chapter 2, while the other fluids are discussed in this chapter. (b) Plot of hydrogen (δD permil) and oxygen ($\delta^{18}O$ permil) isotopic ratios for various water types. Standard Mean Ocean Water (SMOW) is defined to be zero for both δD and $\delta^{18}O$. Some ore-forming environments are clearly related to mixed fluid reservoirs; in the cases shown (see A and B) mixing of meteoric and connate or basinal fluids has taken place. Source: After Taylor (1997).

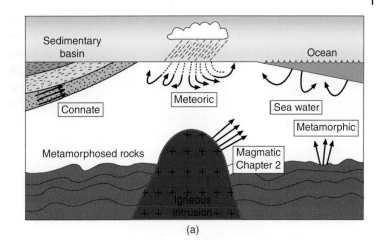

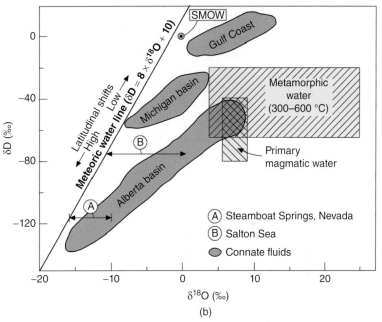

of these fluid reservoirs is different in terms of its composition and temperature and will have, therefore, played different roles in the formation of ore deposits. The major water types are defined as sea water, meteoric water, basinal (or connate) water, and metamorphic water, listed typically in order of increasing depth (and temperature) in the crust. A fifth fluid reservoir, where waters are derived from a mixing of two or more other water types, is also described below, specifically because mixed fluids can be very important in certain ore-forming environments.

In their present settings each of these fluids can be identified because the environment from which the fluid came is known. In the geological past, however, where only indirect manifestations of fluid flow are apparent, it is much more difficult to determine the origin of a particular water type. Fortunately, the hydrogen and oxygen isotope characteristics of water are reasonably diagnostic of its source and can be used to deduce the origins of an ancient fluid reservoir. Figure 3.1b is a plot of δD versus $\delta^{18}O$ that shows trends and fields which serve to fingerprint the major fluid reservoirs on or near the Earth's surface. The origins of the various water types and their stable isotopic characteristics are briefly described below.

3.2.1 Sea Water

The oceans collectively represent the largest fluid reservoir on the Earth's surface (Figure 3.2), covering some 70% of the surface and containing about 98% of its free water. As mentioned in Chapter 2, the Earth's surface has been covered by substantial volumes of water since early in Earth history. Sea water is relatively well mixed at a global scale and is weakly saline because of reaction with both continental and oceanic erosion products over time. The principal dissolved constituents in sea water are the cations Na^+, K^+, Ca^{2+}, and Mg^{2+} and anions Cl^-, HCO_3^-, and SO_4^{2-}, (Table 3.1) which typically occur at a total concentration (or salinity) of around 35 g of solids per kg of sea water (3.5 wt%).

Sea water is extensively circulated through the rocks that make up the ocean floor and is responsible for widespread alteration and metal redistribution in this portion of the Earth's crust. The drawdown of sea water into major faults associated with the mid-ocean ridges, and its subsequent emergence from exhalative vents or "black smokers," is a major oceanographic discovery that has revolutionized the understanding of volcanogenic

massive sulfide (VMS) deposits (see Section 3.8). VMS deposits occur in many different parts of the world, and in rocks of all ages, confirming the importance of sea water as a hydrothermal fluid source. Table 3.1 illustrates the concentrations of major ionic species in sea water and compares this to fluids exhaled from black smoker vents, where significant concentrations of metals have taken place due to the interaction between hot sea water and oceanic crust. Sea water is also compared to rain water and meteoric groundwaters, the latter also having higher dissolved solute contents at higher temperatures.

3.2.2 Meteoric Water

Meteoric water has its immediate origin within the hydrological cycle and has, therefore, been in contact with the atmosphere. In a geological context it refers to groundwater that has infiltrated into the upper crust, through either rainfall or seepage from standing or flowing surface water. In this sense sea water infiltrating into the crust should also be regarded as a source of meteoric groundwater. Groundwater is the second largest reservoir of liquid water (Figure 3.2) and generally exists close to the surface in the interstitial pore spaces of rocks, and in cracks and fissures. It does not refer to the water that forms part of the crystal structure of hydrous minerals, nor does it

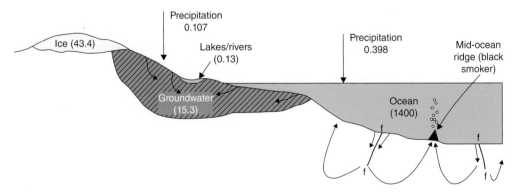

Figure 3.2 Simplified diagram illustrating the water budget on or close to the surface of the Earth. Precipitation fluxes are in 10^6 km^3 yr^{-1} and reservoir volumes in 10^6 km^3. Source: After Berner and Berner (1987).

Table 3.1 Representative concentrations (in ppm) of major ionic species in natural meteoric/hydrothermal waters.

Species	Sea water	Black smoker sea water (at 100–300 °C)	Rain water	Meteoric spring water (at 6 °C)	Meteoric geothermal water (Salton Sea)
Na^+	10 000	6 000–14 000	1.0	23	187
Cl^-	20 000	15 000–25 000	1.1	3.1	21
SO_4^{2-}	2 700	0	1.5	11	103
Mg^{2+}	1 300	0	0.2	2.4	0
Ca^{2+}	410	36	0.4	5.1	0.5
K^+	400	26	0.5	1.0	27
Si^{4+}	0.5–10	20	1.2	17	780
Cu	2.2				6.8
Pb	63.8				102
Zn	6.9				507
Ag	0.0004(ppb)				1.4

Source: After data in Krauskopf and Bird (1995), Scott (1997), and Seward et al. (2014).

refer to the micrometer-sized fluid inclusions that occur within many of the rock-forming minerals of both the crust and mantle. Meteoric fluid can nevertheless penetrate along fractures to deep levels in the crust and is, therefore, involved in widespread circulation throughout the crustal regime (Ague 2014). It typically dissolves low-levels of many metals (Table 3.1) and is responsible for the formation of a wide variety of hydrothermal ore deposits, especially those characterized by relatively low temperature transport and precipitation, such as sandstone-hosted and surficial uranium ores (see Section 3.11 and Chapter 4).

The oxygen and hydrogen isotope compositions of meteoric waters, both on the surface and as groundwater, vary systematically over the entire globe as a function of latitude and elevation. The linear relationship between δD and $\delta^{18}O$ values (the meteoric water line in Figure 3.1b) exists because deuterium and hydrogen, as well as ^{18}O and ^{16}O, are systematically fractionated between liquid water and water vapor during the evaporation–condensation processes of the hydrological cycle, and this fractionation increases as a function of temperature (Craig 1961). Thus, most meteoric waters have δD and $\delta^{18}O$ values that vary along a straight line ($\delta D = 8 \times \delta^{18}O + 10$; Figure 3.1b) and which help to broadly identify the climatic conditions or latitude from where the water came.

3.2.3 Basinal (or Connate) Water

Water that is included within the interstitial pore spaces of sediment as it is deposited is referred to as connate, formation, or basinal water. Originally this water may have been either meteoric or sea water, but it undergoes substantial modification as the sediment is buried, compacted, and lithified. The various stages of diagenesis that result in the transformation from uncompacted particles of sediment to lithified sedimentary rock produce aqueous solutions that evolve with time and depth. Such fluids invariably move through the sequence and are often involved in the formation of ore deposits.

The progressive burial of sediment to depths of around 300 m results in a rapid reduction of porosity and the initial production of a substantial volume of water. Shales are initially very porous and the early stages of burial can result in the production of significant quantities of water – over 75% of the interstitial pore fluid is likely to be expelled from a shale by the time it is buried to 300 m depth and its porosity will, accordingly, decrease rapidly at this early stage (Hanor 1979). By contrast, uncompacted sandstones are initially less porous than shales and will release significantly less water for each cubic meter of sediment deposited. Calculated average rates of water release for shales and sands are shown as a function of depth of burial in Figure 3.3a.

An important process that accompanies the burial and diagenesis of sediment is the conversion of clay minerals, such as smectite, to illite (Kharaka and Hanor 2014). An additional source of basinal water is the accompanying release of "structural" or "bound" water that occurs either as loosely bound H_2O or OH^- molecules within clay mineral particles in argillaceous sediments. Such water is liberated from the host mineral as temperatures of 50–100 °C are reached. This is shown as the episodic peak-like increases in the rate of release of water from the sediment in the depth profiles of Figure 3.3a. The nature and volume of bound water released will obviously vary from one situation to the next and depend on factors such as the local geothermal gradient and the type and proportions of clay host minerals. The two main stages of basinal fluid production (i.e. pore fluid and bound water) are also schematically illustrated in Figure 3.4, where they are compared to the production of metamorphic fluids that occurs at somewhat higher temperatures and greater burial depths.

The temperature of basinal fluids increases with depth in the sedimentary sequence, with the exact rate of increase being a function of the local geothermal gradient. The latter varies typically between 15 and 40 °C km^{-1}. Fluid pressures will also increase with depth, although the nature of pressure variation with progressive burial is likely to be complex and depend on the amount of pore fluid present in any given sedimentary unit and the interplay between hydrostatic and lithostatic pressure gradients. A more detailed description of the relationships between hydrostatic and lithostatic pressure is presented in Section 3.3.2 below, where its significance with respect to the movement of hydrothermal fluids in sedimentary sequences is discussed. Figure 3.3b illustrates the way fluid pressure varies with depth in the Gulf Coast sediments of the southern USA and points to significant deviations above the hydrostatic gradient – excess pressures (i.e. "overpressuring") occur in low permeability horizons within the sequence because they impede the expulsion of fluids so that pore water supports the weight of the overlying strata and high fluid pressures result (Hanor 1979). Zones of overpressured fluids dictate the pattern of fluid flow and mass/heat transfer, which has important implications for the migration of oil brines (see Chapter 5) and for the formation of a variety of sediment-hosted hydrothermal ore deposits.

Basinal fluids also undergo increases in density and salinity (or total dissolved solids – TDS) with depth, as illustrated in Figure 3.3c,d. The density increases are related to increases in pressure and salinity, although there is a limit to such a trend since temperature is also augmented and this has an inverse effect on density. Deep basinal waters are commonly saline to the extent that they are generally unpotable. The increase in salinity is sometimes related to interaction of connate waters with evaporitic horizons that contain easily dissolvable minerals such as halite, sylvite, gypsum, and anhydrite. Salinity increases are also, however, apparent in sedimentary sequences which do not contain soluble evaporites, a process sometimes attributed to "membrane-" or "salt" -filtration. In argillaceous layers, where clay particles are strongly compacted, the interaction of the net negative charge (caused by atomic substitutions within

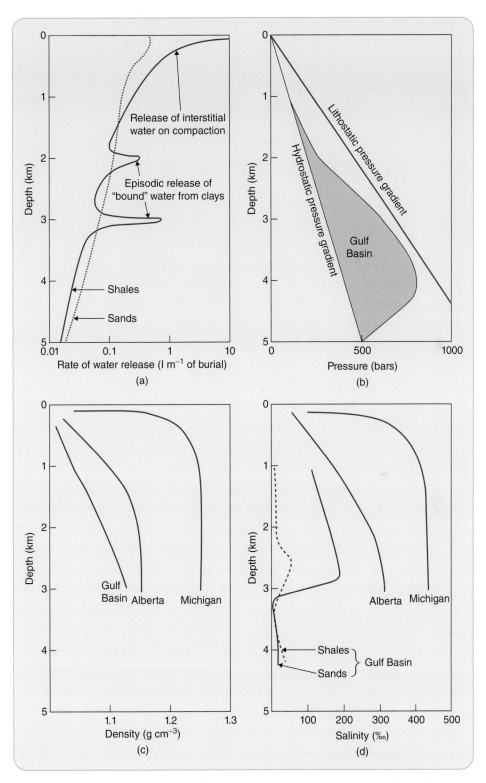

Figure 3.3 The formation and characteristics of basinal waters. (a) Depth profile illustrating the rate of water release due to initial compaction and porosity reduction followed by dehydration of clay minerals such as montmorillonite. (b–d) Depth profiles showing typical pressures, fluid densities, and fluid salinities, respectively, in a compacting sediment pile. Different profiles refer to the Michigan, Alberta, and Gulf basins in the USA. Source: After Hanor (1979).

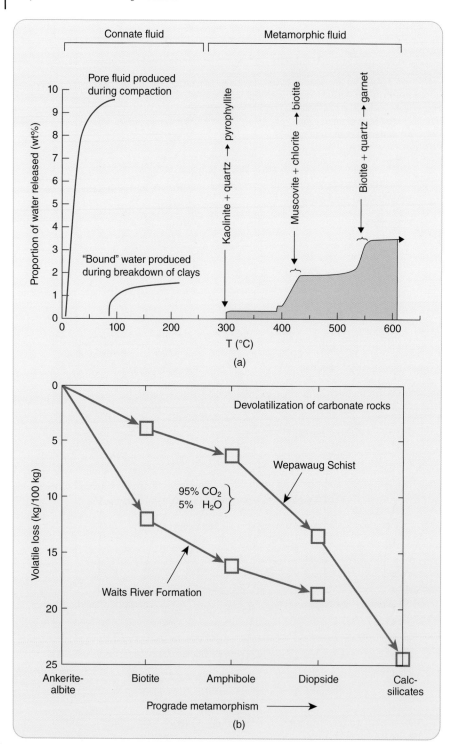

Figure 3.4 (a) The relationship between basinal fluid production during diagenesis and metamorphic fluid production due to systematic dehydration reactions involving the breakdown of one mineral assemblage to another containing less water. The conditions pertaining here were calculated specifically for the Witwatersrand Basin. Source: After Stevens et al. (1997). (b) Calculated production of volatile phases (in the proportion 95% CO_2 : 5% H_2O) during the prograde metamorphism of carbonate rocks. Source: After Ague (2014).

the crystal lattice) around each particle creates what is termed a "Gouy layer" (Berner 1971). This acts as a semipermeable membrane or filter which serves to reject anions in solutions passing through the shale layer. Since anions are excluded from the filtrate then so too are accompanying cations, in order to maintain electroneutrality. The result is that any effluent solution passing through the compacted clay will emerge with a lower salinity. Upwardly migrating brines will, therefore, become less saline, whereas the dissolved salt content of deeper residual waters will increase. Although doubts have been expressed about the efficacy of forcing solutions upwards in a sedimentary sequence across semipermeable shale layers, the process of salt filtration is generally regarded as a feasible explanation for the widespread existence of deep basinal brines. The topic is of considerable importance to ore genesis, as the availability of ligands in hydrothermal fluids, especially Cl^-, is crucial to increasing the solubility and transport efficiency of metals in solution (see Chapter 2 and Section 3.4 below).

The salinity characteristics of basinal waters will vary from one sedimentary sequence to another, as seen in Figure 3.3d. Likewise, the δD and $\delta^{18}O$ characteristics of such fluids will also vary depending on the climatic characteristics and latitude of the host basin. Since the original trapped pore fluids in most sedimentary sequences will be either meteoric or sea water, their hydrogen and oxygen isotope compositions will initially reflect either the appropriate point along the meteoric water line or Standard Mean Ocean Water (SMOW) (Figure 3.1b). As the fluids evolve, through both temperature increases and interaction with the host rocks, their stable isotope characteristics will also change and define trends that typically deviate from a position on the meteoric water line along paths of shallower slope. Fluids will tend to reflect higher $\delta^{18}O$ values as their temperatures and salinities increase in sedimentary sequences (Taylor 1997). In Figure 3.1b, for example, the low latitude Gulf

Coast basinal waters define a trend that is quite distinct from fluids circulating in the high latitude Alberta Basin of Canada.

3.2.4 Metamorphic Water

As rocks are progressively buried and temperatures exceed about 200 °C, the process of diagenesis evolves to one of metamorphism. Metamorphism is a multifaceted process but essentially involves the transformation of one mineral assemblage to another which is more in equilibrium with the prevailing conditions at higher pressure and temperature. Of importance in this discussion is the transformation of hydrous silicate and carbonate minerals to newly formed assemblages that are less volatile rich. Dehydration and decarbonation reactions during prograde metamorphism are, therefore, important processes that produce substantial volumes of metamorphic water in the mid- to lower-crust. An example from the Witwatersrand Basin in South Africa (Figure 3.4a) shows that at around 300 °C a metamorphic reaction involving the breakdown of kaolinite to pyrophyllite results in the production of metamorphic water. The latter arises from the fact that kaolinite contains more water than does pyrophyllite, with the excess water liberated by the phase transition being expelled into the surrounding sedimentary sequence. This reaction is followed at around 400 °C by the transformation of muscovite and chlorite to biotite, when more fluid is produced. Both H_2O and CO_2 (the latter specifically when the breakdown reactions involve carbonate minerals), as well as volatiles such as CH_4 and sulfur species, can be produced as metamorphic fluids during prograde reactions of this type. It has been estimated that metamorphism in the crust produces a fluid flux that is more than 10^{17} kg per million years (Ague 2014).

Fluids produced in low to medium grade metamorphic terranes are dominated compositionally by H_2O, CO_2, CH_4, and N_2 in approximately that order of abundance. High grade rocks tend to be dominated by dense CO_2

with lesser amounts of associated water and methane. Most metamorphic fluids have low salinities and low concentrations of reduced sulfur. Fluids of this nature are globally implicated in the formation of orogenic gold deposits, which represent a very important category of gold deposit types (Bodnar et al. 2014; Phillips et al. 1994; see Box 3.2). Although it has not been discussed in detail, it is clear that CO_2 may be an important component of hydrothermal solutions, not only in those originating through metamorphic devolatilization, but also in magmatic waters.

Carbon dioxide is produced by devolatilization of carbonate rocks during prograde metamorphism, as indicated in Figure 3.4b, which demonstrates the increase of mainly CO_2 fluid production with increasing metamorphic grade (Ague 2014). After water, CO_2 is the most abundant component in hydrothermal solutions and its phase equilibria relative to those of water are briefly considered below. It is a larger molecule than water and is non-polar (see Chapter 2), which accounts for its lower melting and critical points. It is of little importance as a solvent other than for non-anionic species, hydrocarbons (especially CH_4), and metal-chelate complexes.

A part of the phase diagram for pure CO_2 is shown in Figure 3.5a and when compared to a similar diagram for H_2O (Figure 2.2c) is seen to have remarkably similar densities and phase transition characteristics to water, except that it has lower melting ($-56.6\,°C$) and critical ($31.1\,°C$) points. At high temperatures H_2O and CO_2 are completely miscible and form a single fluid phase where one compound is dissolved in the other. At lower temperatures H_2O and CO_2 become immiscible (like oil and water) and exist as two separate phases (usually as H_2O liquid and CO_2 supercritical fluid). This relationship is shown in Figure 3.5b, where the solvus (i.e. the curve which defines where unmixing of the single phase mixture occurs) for pure H_2O–CO_2 at 2 kbar identifies a plateau at

around 250 °C, above which the two compounds occur as a single miscible phase. As the temperature falls the single-phase fluid progressively unmixes until at room temperature H_2O and CO_2 are essentially immiscible. This phenomenon is clearly illustrated in fluid inclusions that have trapped single phase H_2O–CO_2 mixtures above their relevant solvi. When viewed in a laboratory at room temperature, these fluid inclusions are typically characterized by a so-called "double bubble." The larger "bubble" is actually a globule of immiscible CO_2 fluid within liquid H_2O which coats the walls of the inclusion, whereas the smaller inner "bubble" comprises a mixture of H_2O and CO_2 vapor (Figure 3.5c). The widespread occurrence of H_2O- and CO_2-bearing fluid inclusions in the rock record attests to the ubiquity of aqueo-carbonic fluids especially in the deeper crust.

Metamorphic waters are characterized by relatively high $\delta^{18}O$ values (+10 to +30 permil) which are typically quite distinct compared to both meteoric and basinal waters, but which may overlap to a certain extent with magmatic waters (Figure 3.1b). The reasons for this are complex and related to the variable nature of metamorphic protoliths and the ubiquitous effects of fluid–rock interaction during regional metamorphism (Taylor 1997).

3.2.5 Waters of Mixed Origin

Although the previous discussion has emphasized the nature and characteristics of different individual fluid reservoirs in the Earth's crust, it is unreasonable to expect that in nature such waters would necessarily retain their integrity with time. In the upper crust especially, it is known that fluids undergo mixing and effectively become hybrids with more than just a single source or origin. It is apparent that the mixing of ore-bearing hydrothermal fluids may be an important consideration in the precipitation of metals from such solutions. Magmatic and meteoric

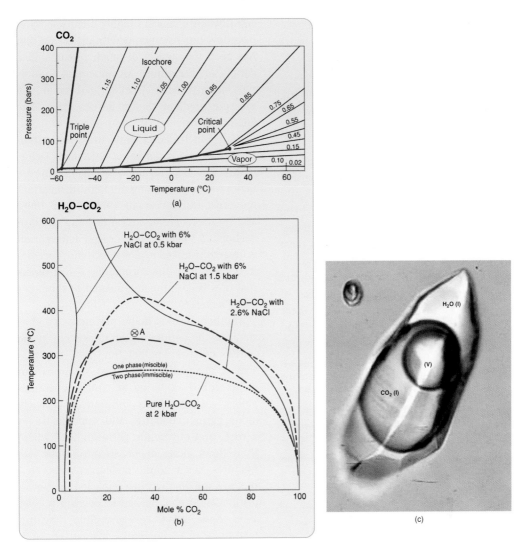

Figure 3.5 (a) Part of the phase diagram for CO_2 showing the variations in fluid density (i.e. isochores in $g\,cm^{-3}$) as a function of pressure and temperature. The diagram is similar to that for H_2O (see Figure 2.2c) except for the lower melting and critical points. (b) Various solvus curves for the system $H_2O–CO_2$; the solvus (i.e. the curve which defines where in P–T–X space H_2O and CO_2 unmix) moves to higher temperatures with decreasing pressure and increasing salinity of the mixture. Source: Compilations after Brown (1998). (c) Microphotograph showing the typical appearance of a $H_2O–CO_2$-bearing fluid inclusion. Prior to entrapment this fluid was a homogeneous mixture of mutually soluble H_2O and CO_2. On cooling it exsolved to form immiscible fractions of H_2O (outer liquid wetting the walls of the inclusion) and CO_2 (inner liquid globule). A vapor bubble (v) occurs within the inner CO_2 globule. Source: The photo is courtesy of Christopher Heinrich.

waters, for example, often undergo mixing at volcanic to subvolcanic crustal levels associated with porphyry copper and epithermal Au–Ag ore-forming environments. Mixing of basinal and meteoric waters is also a process known to characterize sedimentary basins forming in rapidly filled continental rift environments. The major effect of fluid mixing is to promote the precipitation of metals from ore-forming solutions and, consequently, this topic is discussed in more detail in Section 3.5.1 below.

3.3 The Movement of Hydrothermal Fluids in the Earth's Crust

In order to be effective as a mineralizing agent, hydrothermal fluids need to circulate through the Earth's crust. The main reason for this is that they need to interact with large volumes of rock in order to dissolve and transport the metals required to form hydrothermal ore deposits. The flow of an ore fluid should eventually become more focused so that the dissolved constituents can be concentrated into an accessible portion of the Earth's crust that has dimensions consistent with those of a potential ore body. The study of hydrothermal fluid movement has received a great deal of impetus from the oil and gas industry, where it is particularly important to understand the pathways of hydrocarbon-bearing brines during both exploration and extraction stages (see Chapter 5, Section 5.4). Exploration for metallic hydrothermal ores has also benefited over the past few years from a better understanding of fluid flow in the Earth's crust, especially with respect to Archean lode gold ores and Mississippi Valley type (MVT) Pb–Zn deposits (see Boxes 3.2 and 3.7 respectively).

In Chapter 2 the nature and duration of hydrothermal fluid flow associated with magmatic intrusions was briefly discussed. The present section looks at hydrothermal fluid flow in the Earth's crust in more general terms and also at a variety of scales. It emphasizes the role that deformation and the resulting structural features play, in terms of both movement and focusing of fluid flow in the crust. It is these features that are of particular importance to the formation and location of metallic hydrothermal ores.

3.3.1 Factors Affecting Fluid Flow at a Crustal Scale

The question of how large volumes of fluid can move around at deep levels in the Earth's crust, where rocks are highly compacted and have low intrinsic permeabilities, is one that has been the subject of much research in a variety of disciplines. Movement of fluid is typically a response to either a thermal or a pressure gradient in the Earth's crust. The latter, in particular, is often related to deformation which causes both stress and strain to vary in the rocks being deformed. Deformation is widely regarded as having played a major role in controlling fluid flow throughout the crust (Sibson et al. 1975; Oliver 1996) and its effects are discussed in more detail in Section 3.3.3 below. Since crustal deformation is invariably linked to global tectonic processes it is clear why the formation of hydrothermal ores is so intimately related to the latter (see Chapter 6).

Another question that is pertinent to understanding the nature of fluid movement in the Earth's crust is whether fluid flow is pervasive or channelized. The two scenarios can be described in terms of "intrinsic permeability," where fluids infiltrate pervasively along grain boundaries and microcracks in a rock, and "hydraulic permeability," where fluid flows along major discontinuities (i.e. fractures, shear zones, bedding planes, etc.) that are sufficiently interconnected to allow a distinct fluid channelway to develop (Oliver 1996; Ague 2014). Pervasive fluid flow can take place in porous rocks at relatively shallow levels in the crust, a situation that allows convective fluid movement to occur. Figure 3.6 illustrates a variety of tectonic scenarios at a crustal scale and the nature and distribution of major fluid flow vectors associated with each. At relatively shallow crustal levels pervasive fluid flow, along a porous aquifer for example, often occurs in response to a gravity-driven hydraulic head formed by uplifted topography (Figure 3.6a). This is believed to be the dominant type of groundwater flow in continental regions and flow rates of between 1 and $10 \, m \, yr^{-1}$ could develop depending on the permeability of the aquifer (Garven and Raffensperger 1997). A similar style of long-range

Figure 3.6 Various tectonic scenarios illustrating the mechanisms by which major fluid movements in the Earth's crust take place. (a) Gravity-driven fluid flow in response to the creation of a hydrostatic head in an area of uplift. (b) Orogeny-driven fluid flow in response to rock compaction in a fold and thrust belt. (c) Thermally driven fluid flow in the ocean crust in response to high heat flow at mid-ocean ridges. (d) Thermally driven fluid flow in a permeable unit within a basin or rift. (e) Dilatancy- or fault-driven fluid flow along major, seismically active structural features by seismic pumping and fault valve mechanisms. Source: Modified after Garven and Raffensperger (1997).

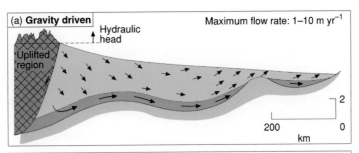

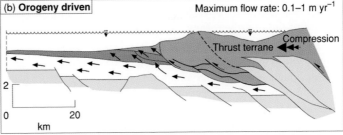

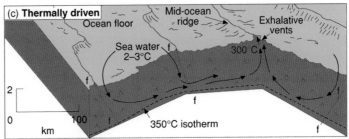

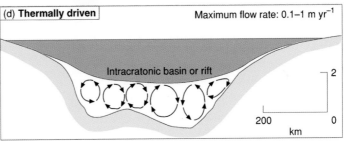

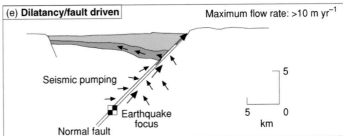

fluid flow is now known to be associated with periods of orogenic compression which effectively squeeze the fluid out of one package of rock into channelways (thrusts?) or high permeability aquifers in another (Figure 3.6b). Orogeny-driven fluids have been implicated in the formation of the major MVT Pb–Zn ore province of the southeast USA (see Box 3.7).

Fluids in the oceanic crust are believed to flow in response to thermal gradients formed because of the high heat flow that characterizes the mid-ocean ridges (Figure 3.6c). Sea water is forced down into the crust along faults and, although essentially channelized, flows largely in response to convection, reappearing close to the zones of maximum heat flow. Thermally driven convective fluid flow can occur, as mentioned in Chapter 2, in close proximity to high level magmatic intrusions. It can also occur in deep intracratonic rift basins where high heat flow or fluid density gradients result in convective flow, but only in aquifers where permeabilities are sufficiently high (Figure 3.6d).

Large-scale flow of fluid at a crustal scale can also occur in association with the dilatancy (i.e. change in volume) of a rock mass that accompanies faulting and seismic rupture (Figure 3.6e). Orogeny-driven fluids may be partially associated with this process, but more typically dilatancy accompanies faulting and shearing and this particular process is considered in more detail in Section 3.3.3 below. A detailed account of the parameters affecting crustal-scale fluid flow and associated mass transport by fluids is presented in Ague (2014).

3.3.2 A Note on Hydrostatic Versus Lithostatic Pressure Gradients

As mentioned previously, one of the main factors that controls fluid migration pathways is the existence of pressure differentials in the host rock. In a completely dry rock, the pressure at any depth is effectively provided by the weight of the mass of rock above that point and is known as *lithostatic pressure*. In Figure 3.7 the lithostatic pressure gradient is demonstrated by the line of slope 25 kPa m^{-1}, and reflects the average density (i.e. about 2.5 g cm^{-3}) of the rock sequence over that interval. The distribution of stress (i.e. force per unit area) in a rock at any point along a lithostatic depth profile will vary with orientation and will be maximized in a vertical sense and minimized horizontally. By contrast, the pressure on a rock located on the ocean floor will be given by the *hydrostatic pressure* gradient (Figure 3.7), which has a slope of about 10 kPa m^{-1}, reflecting the lower density of water relative to rock (i.e. about 1.0 g cm^{-3}). Stresses under hydrostatic pressure are distributed equally in all directions, which accounts for the relative "weightlessness" of objects immersed in water.

Uncompacted sediment on the ocean floor containing abundant pore water that is still in direct contact with the overlying ocean water column will also be pressurized along the hydrostatic gradient. As the sediment is buried and compaction proceeds, water will be driven off. If, however, the pore water remains interconnected then the applied load on that rock will still be carried by the water and pressures will be made up of partly hydrostatic and partly lithostatic components and have values intermediate to the two gradients. When drainage of pore fluid from the sediment is good then a condition known as compaction equilibrium will accompany the progressive burial and lithification of that rock. If, however, the removal of water is impeded by low permeability then compaction will likewise be retarded (and porosity maintained at a higher value) and pressures will increase to values above hydrostatic – a condition known as "*overpressuring*." Fine and coarse sediment will expel pore waters at different rates during burial and will, therefore, compact along different pressure gradients. Fluid pressures will usually be significantly higher in less permeable rock units, such as shales, relative to well drained rocks, such as sandstones. Figure 3.7 illustrates this effect with respect to a sequence of alternating sandstone and shale and plots the overpressures that occur in compacting shale horizons relative to sandstones which maintain compaction equilibrium and hydrostatic pressures. Eventually, however, with increasing depth in normal crustal profiles pore fluid pressures increase from hydrostatic to lithostatic (so that fluid pressure equals rock pressure) and this occurs typically at between 2 and 5 km depth depending on the nature of the rocks. The nature of hydrostatic and lithostatic pressure gradients

in the Earth's crust, and the deviations from these end-member scenarios, are critical to understanding the nature of fluid flow in all rock types.

3.3.3 Deformation and Hydrothermal Fluid Flow

At progressively deeper levels in the crust, rock porosity is reduced and so is the fluid mass contained within that rock. It is also more difficult for fluids to move pervasively through a rock and under these conditions fluid migration can only take place along channelways represented by structural discontinuities that form during deformation. Most of the fluid movement that occurs at deeper crustal levels, and is relevant to the formation of mineral

Figure 3.7 Plot of depth versus fluid pressure to illustrate the difference between hydrostatic and lithostatic pressure gradients. The dashed line shows the excesses in hydrostatic pressure (i.e. overpressure) that can accumulate in low permeability shale horizons (shown as SH relative to sandstone (SS) layers in the column on the left hand side) because of the inability of these rocks to efficiently drain off pore waters on compaction. Source: After Hunt (1979).

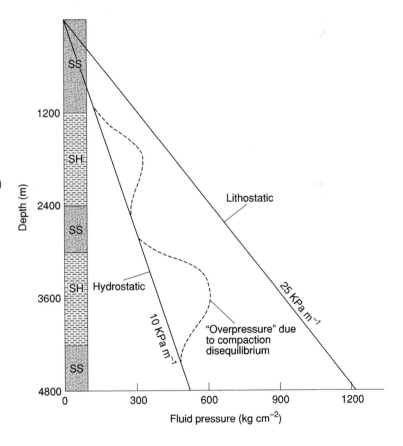

deposits, is located within faults that may, as a result, become loci for mineralization. Fault displacements are generally accomplished by increments of rapid movement that trigger earthquake events, usually in the upper 15 km (the seismogenic regime) of the crust (Sibson 1994). The relationships between earthquake events, fault propagation, and fluid flow have become an important area of study that has relevance not only to ore genesis but to hydrology and earthquake prediction. It is, for example, well known that seismic events trigger a wide range of hydrological events that can vary from the sudden drying up of wells to significant increases in the flow of rivers adjacent to faults (Muir Wood 1994). The work of Sibson and co-workers (Sibson 1986, 1987; Sibson

et al. 1975, 1988) in particular has emphasized the importance of fault-driven fluid flow (Figure 3.6e) to the formation of hydrothermal ore deposits and this is discussed in more detail below.

Sibson et al. (1975) first introduced the term *seismic pumping* to describe a theoretical model in which it was envisaged that the cyclicity of stress variations in and around a rupturing, seismically active fault system would affect local fluid pressures and promote the flow of fluids along the fault (Figure 3.8a). Prior to a seismic event friction between the opposing faces of a fault prevents rupture such that shear stress increases and the adjacent rock undergoes dilation (Figure 3.8a,b). The dilatant strain that develops in the rock around the fault causes cracks to form, into which

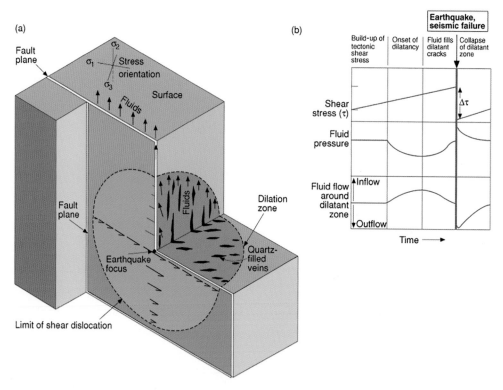

Figure 3.8 Model explaining the episodic flow of fluid along a seismically active fault zone. (a) Block diagram illustrating the geometry of a fault generated seismic pumping system. (b) Explanation of the changes in tectonic shear stress and fluid pressure with time before and after a seismic event, and the accompanying fluid flow direction around the dilatant zone. Maximum outflow of fluid from the zone of dilatancy occurs as the seismic failure event occurs and fractures close. Source: After Sibson et al. (1975).

fluids flow. Consequently, in this model fluid pressures are predicted to fall in the build up to fault failure. At the instant the fault ruptures and an earthquake occurs, however, shear stress drops significantly ($\Delta\tau$ in Figure 3.8b) but fluid pressure increases, resulting in the upward expulsion of fluids along the fault. As frictional forces are reinstated and shear stresses increase again, the entire cycle repeats itself, causing episodic fluid flow along fault-related channelways in the crust. The seismic pumping model provides an explanation and a mechanism for moving fluids through the crust related to cyclical variations in both the stress and strain state of rocks around faults.

In reality the factors controlling channelized movement of hydrothermal solutions through the Earth's crust are more complex than the mechanisms outlined in the Sibson et al. (1975) model. The model also suffers from the fact that there seems to be little or no evidence to substantiate the sort of crack propagation in rocks adjacent to faults that was originally envisaged (Sibson 1994). It is nevertheless still apparent that stress/strain cycling in rocks in and around faults is a major control of crustal fluid movement. The relevance of such processes to the formation of mesothermal or orogenic gold deposits in the mid-crust has, for example, been demonstrated by Sibson et al. (1988). These types of deposits (see Box 3.2) are typically located within high angle (i.e. subvertical) reverse faults (Figure 3.9a). Such faults are an enigma since compressive stresses in the Earth's crust generally form thrust faults that are shallowly inclined (i.e. 25°–30° to a horizontal maximum principal compression).

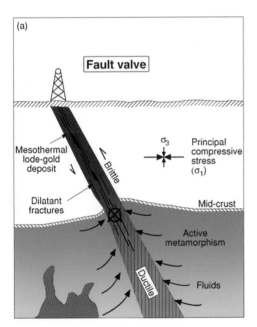

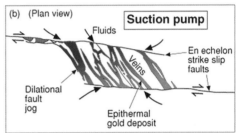

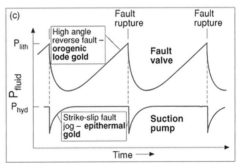

Figure 3.9 Models explaining the nature of hydrothermal fluid flow in (a) high angle reverse faults in a horizontal compressive stress regime considered to be applicable to the formation of mesothermal lode-gold deposits such as the Mother Lode in California and in many Archean greenstone belts; and (b) en echelon strike-slip fault arrays associated with extensional stresses and considered applicable to some shallow level epithermal gold deposits. (c) Plot of fluid pressure with time showing the cyclic fluctuations envisaged for the fault valve model (applicable to high angle reverse fault-related mesothermal lode-gold deposits) and the suction pump model (applicable to strike-slip related fault jogs and shallow level epithermal gold mineralization). Source: The diagrams are after Sibson (1987) and Sibson et al. (1988).

Theory suggests that reactivation of high angle faults in a horizontal compressive stress regime can only occur when the local fluid pressure reaches, or briefly exceeds, the lithostatic load. The subsequent model proposed by Sibson et al. (1988) envisages that sites of mesothermal lode-gold mineralization are located in structures where fault rupture has been caused by the attainment of lithostatic fluid pressures. Figure 3.9c illustrates how fluid pressure in a seismically active fault builds up to levels approaching the lithostatic load because the upper crust represents an impervious cap to the fluid reservoir beneath. Fault rupture occurs as fluid pressures reach the lithostatic equivalent and the seismic event is accompanied by dilatancy and the development of a fracture permeability upwards and along the fault (Sibson et al. 1988). As the fault fails and shear stresses are substantially reduced, fluids are also discharged into the open space created by the fault rupture itself. The dramatic decrease in fluid pressure (back toward hydrostatic values) that results from the creation of open space could promote the precipitation of dissolved constituents within the hydrothermal solution, causing the fault to reseal itself. Once this has happened fluid pressures are likely to increase again. This crack–seal mechanism is now known as the "*fault valve model*" and appears to explain many of the characteristics of mesothermal gold deposits (Figure 3.9a,c).

A slightly different mechanism of fluid movement is envisaged with respect to strike slip faulting associated with extensional stresses in the upper crust. Dilational fault jogs are commonly found between the end of one rupture plane and the beginning of another in extensional en echelon fault arrays (Figure 3.9b), and such sites are particularly favorable for epithermal gold mineralization at high crustal levels (Sibson 1987). Fault jogs are typically characterized by the development of extensional fractures that are preserved as a network of quartz veins. Fault jog related quartz vein arrays may also be mineralized, an example of which is the Martha Hill (or Waihi) gold deposit of the Coromandel Peninsula in New Zealand (Brathwaite and Faure 2002). The development of extensional fractures in the fault jog results from the transfer of fault slip from one en echelon fault segment to another during seismic activity along the fault trace. The opening up of space at high crustal levels and the discharge of fluid into these spaces results in a rapid fluid pressure drop (Figure 3.9c) which could, at least within 2–3 km or so of the surface, also be accompanied by boiling of the fluid. The mechanical energy released by boiling could result in further hydraulic fracturing and brecciation, with enhanced fluid circulation and mineral precipitation. This particular process is referred to as the "*suction pump model*" (Sibson 1987) and appears to have applicability to strike slip fault systems and epithermal fluid flow at shallow crustal depths (Figure 3.9b and c).

There is no doubt that deformation-controlled fluid movement along major fault systems in the Earth's crust is of major importance to the formation of hydrothermal ore deposits over a considerable range of crustal depths. This is confirmed by the fact that many hydrothermal ores are located within structural discontinuities and also that exploration geologists frequently target structural features in their quest for mineral deposits.

3.3.4 Other Factors Affecting Fluid Flow and Mineral Precipitation

Previous discussion has centered on fluid movement in large-scale structures such as faults that are tens to hundreds of kilometers in length and where mass movement is related to large-scale fluid advection within these structures or discontinuities. This section considers smaller-scale fluid migration as evident at mesoscale (as individual quartz veins) or microscale (as fluid inclusion traces in individual minerals) levels. At these scales effective flow of fluid through a rock is determined by hydraulic conductivity, or the extent to which

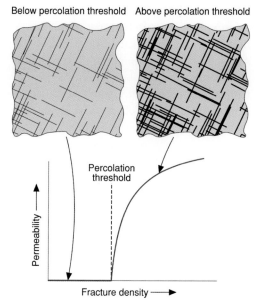

Figure 3.10 The ability of a fluid to flow through a rock at a small scale depends on the creation of fracture-related permeability. The percolation threshold, which is a function of connectivity and fracture density, is reached when one cluster of fractures is able to connect to another so that fracture connectivity is able to span the entire rock mass in question. Source: After Odling (1997).

fractures along which fluids might flow are interconnected (Odling 1997; Cox et al. 2001). In a rock containing very few fractures or cracks, and where porosity is low, permeability may be effectively zero and little or no fluid will flow through the rock (Figure 3.10). As the fracture density increases, so too does the probability of two cracks interconnecting. Eventually a point is reached, termed the *percolation threshold*, where permeability suddenly increases markedly and fluid flow across a finite volume of rock becomes possible (Figure 3.10; Odling 1997). The attainment of a percolation threshold, at any scale in a rock mass, is clearly necessary if effective fluid circulation is to take place.

3.3.4.1 How Do We Know that a Fluid Has Passed Through a Rock?

The evidence that a fluid had once flowed through a rock mass is provided either by alteration (see Section 3.6 below) or by the presence of veins that are typically filled with quartz or a carbonate phase (i.e. calcite, dolomite, siderite, etc.), together with other less common gangue and ore minerals. Veins are formed by an assemblage of minerals that precipitate from hot aqueous solutions as they pass through a fracture, effectively fossilizing and preserving the fluid conduit in the rock record. Fracture fill will develop in one of two ways (Bjørlykke 1994; Ague 2014):

1) By diffusion of solids from the surrounds and precipitation within a fracture or open space (i.e. a vug), in what is essentially a closed system. Since a mineral such as quartz has a lower solubility in aqueous solution at lower pressures and temperatures (see Section 2.4.1 in Chapter 2), it follows that it is likely to precipitate in a fracture because the open space may exist under hydrostatic pressure gradients compared to the fracture walls which carry higher pressure lithostatic loads.

2) By precipitation of minerals from fluids flowing through a fracture, in what is essentially an open system in which fluid and solute is brought in from a distant source. Minerals will precipitate in the open space as a function of many different processes (see Section 3.5.1 below), such as temperature decrease, rock alteration, solubility decrease, and boiling.

The factors controlling precipitation of minerals from a hydrothermal solution are varied and complex and this topic is discussed in more detail with respect to ore-forming constituents in Sections 3.4 and 3.5 below. Minerals also precipitate from fluids flowing pervasively through a porous rock mass and this is the mechanism responsible for diagenesis, whereby sedimentary particles are cemented together and lithified. Assuming equilibrium between pore fluid and host rock, the volume of mineral matter precipitated from a fluid circulating through a rock mass can be calculated from the relation (after Bjørlykke 1994):

$$V_m = Ft \sin \beta (dT/dZ)\alpha_T/\rho \qquad (3.1)$$

where V_m is the volume of mineral precipitated; F is the fluid flux; t is time (seconds); β is the angle between direction of flow and isotherms in the rock; dT/dZ is the geothermal gradient; α_T is a function which reflects solubility in terms of temperature, and ρ is the density of the mineral being precipitated.

A schematic illustration of mineral precipitation, and conversely of mineral dissolution, is shown in Figure 3.11, where pore fluid is considered to be flowing freely through a porous sandstone. Because the solubility of quartz is reduced along a negative temperature gradient it will precipitate along a cooling path, but dissolve along a heating path. The volumes of quartz precipitated in such a situation can be calculated from Eq. (3.1) above. Conversely, a mineral such as calcite exhibits retrograde solubility (its solubility decreases as a function of increasing temperature) and will, therefore, behave in the opposite sense to quartz, tending to dissolve at the sites of quartz precipitation and vice versa (Figure 3.11). Different types of mineral precipitates in a sandstone aquifer can, therefore, provide information about the nature and hydrology of fluid flow. In this type of environment calculations will show that the precipitation of even small volumes of quartz or calcite requires large fluid fluxes,

so that mass transfer becomes effective only when sediments have high permeabilities, or fluid fluxes are very focused (Bjørlykke 1994). Although cooling is important, precipitation of minerals from circulating fluids is not only temperature-dependent, many other factors control the formation of ore constituents from hydrothermal solutions (see Section 3.5 below).

3.4 Additional Factors Affecting Metal Solubility

Solubility is defined as the "upper limit to the amount of dissolved metal that a hydrothermal fluid can transport" (Wood and Samson 1998). The solubility of most oxide, sulfide, and silicate minerals in an aqueous hydrothermal solution increases with temperature (prograde solubility), although the reverse is true of many carbonate and sulfate minerals (retrograde solubility). Solubility behavior, however, is complex and quartz, for example, exhibits prograde solubility in pure water until around 370 °C, retrograde solubility at higher temperatures, and an increase in solubility with pressure at any temperature (Seward et al. 2014). Prograde and retrograde solubility behavior of many minerals also changes as a function of the type and concentration of ligands in the solution.

In Chapter 2 the concept of metal solubility in hydrothermal aqueous solutions was

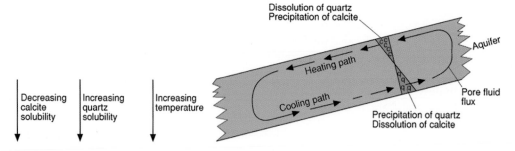

Figure 3.11 Schematic diagram showing a convecting pore fluid circulating in a porous sandstone and the contrasting behavior of mineral precipitation and dissolution. In this model it is assumed that quartz solubility decreases, but calcite solubility increases, as fluid temperatures decrease. Source: After Byørlykke (1994).

introduced and mention made of Holland's (1972) experiments in which the solubility of metals such as Pb, Zn, Mn, and Fe was found to vary as an exponential function of the chloride ion (Cl^-) concentration of the fluid. This work emphasized the fact that efficient transport of metals by hydrothermal fluids in the Earth's crust can be achieved only if aqueous solutions contain other dissolved ingredients with which the metals can bond, thereby promoting higher metal solubilities. For example, the solubility of a mineral such as galena (PbS) in pure water, even at high temperatures, is extremely small. Wood and Samson (1998) calculated that the amount of Pb that could be dissolved in an aqueous solution (with 10^{-3} m H_2S and in equilibrium with galena at 200 °C) would be a trivial 47.4 ppb. If 5.5 wt% NaCl were added to this solution, however, the solubility of Pb at the same temperature would increase significantly to 1038 ppm.

In addition to Cl^-, a variety of other ligands (electron donors or electronegative ions with lone pairs of valence electrons) will affect the solubility of metals in solution. The ligand combines with a metallic ion by the formation of a chemical bond that will typically increase the solubility of metals in aqueous ore-forming solutions.

Considerations of how metals go into solution in hydrothermal fluids can be made with respect to the concept of Lewis acids and bases. In the latter concept the simplistic definition of acids and bases (i.e. substances capable, respectively, of contributing H^+ ions to, or taking up H^+ ions from, a solution) is expanded to situations where H^+ ions are not present, a concept that has particular relevance to geological situations (Gill 1996). A Lewis acid is defined as an atom or molecule that can accept a lone pair of valence electrons, whereas a Lewis base can donate an electron pair to a bond. "Hard" Lewis acids are strongly electropositive metals (with high charges and/or small atomic radii) such as the alkali and alkaline earth metals (e.g. Na^+, K^+, Mg^{2+}, Ca^{2+}) that form ionic bonds with strongly electronegative elements like oxygen (O^{2-}). "Soft" Lewis acids have an abundance of easily accessible electrons in their outer shells and prefer to form covalent bonds with soft bases. Soft acids are typified by the chalcophile metals (i.e. those that have an affinity with sulfur, such as Cu, Pb, Zn, Ag, Bi, Cd) which tend to form covalent bonds with ligands of low electronegativity, such as S^{2-}. The principle stating that in a competitive situation hard metals (acids or electron acceptors) will tend to complex with hard ligands (bases or electron donors), and soft metals with soft ligands, is commonly referred to as *Pearson's Principle*, and underpins the understanding of solubility behavior of metals in hydrothermal solutions. A classification of metals and ligands in terms of the hard–soft breakdown and its applicability to ore-forming processes is presented in Table 3.2. The "borderline" category is added to accommodate the fact that some metals (such as Fe and Pb) can complex readily with both hard and soft ligands, forming a range of minerals such as sulfides (pyrite FeS_2, galena PbS) and carbonates (siderite $FeCO_3$, cerrusite $PbCO_3$). Likewise, Cl^- can be an effective complexing agent for both hard and soft metals, and therefore promotes the solubility of a wide range of metals in hydrothermal aqueous solutions.

The classification of aqueous ionic species into hard and soft Lewis acids/bases is useful in understanding the nature of metal–ligand complexation, and, therefore, the controls on solubility in hydrothermal solutions (Seward and Barnes 1997; Wood and Samson 1998; Seward et al. 2014). In addition, however, it is well known that temperature plays an important role in determining the degree to which metals enter solution. In most cases stabilities of metal–ligand complexes such as $PbCl^+$ and $ZnCl^+$ will increase by several orders of magnitude as temperature increases from ambient values to 300 °C (Seward and Barnes 1997). This is illustrated, together with data for other metal–chloride complexes, in Figure 3.12. By contrast, an increase in pressure will have the opposite effect to temperature and the

Table 3.2 Classification of some metals and ligands in terms of Lewis acid/base principles.

Hard metals	Borderline	Soft metals
Li^+ Na^+ K^+ Rb^+ Cs^+	Divalent transition metals	Au^+ Ag^+ Cu^+
Be^{2+} Sr^{2+} Ba^{2+} Fe^{3+}	(Zn^{2+} Pb^{2+} Fe^{2+} etc.)	Hg^{2+} Cd^{2+} Sn^{2+} Pt^{2+} Pd^{2+}
Ce^{4+} Sn^{4+} Mo^{4+} W^{4+} V^{4+} Mn^{4+}		Au^{3+} Tl^{3+}
As^{5+} Sb^{5+} U^{6+}		
$\Downarrow$	$\Downarrow$	$\Downarrow$

Hard ligands	Borderline	Soft ligands
NH_3	$\mathbf{Cl^-}$ Br^-	$\mathbf{HS^-}$ I^- CN^- H_2S $S_2O_3^{2-}$
$\mathbf{OH^-}$ F^- NO_3^- HCO_3^-		
(acetate) CH_3COO^-		
CO_3^{2-} SO_4^{2-}		
PO_4^{3-} SiO_4^{4-} WO_4^{2-}		

The most abundant ligands in natural hydrothermal solutions are shown in **bold**. Caution is required in the application of the hard–soft model since the structure of water changes at higher temperatures and metal–ligand interaction will, likewise, change (Seward and Barnes 1997).
Source: After Pearson (1963), Seward et al. (2014).

stabilities of metal–ligand complexes will tend to decrease because of bond dissociation and the formation of free metal ions. In general, however, the effect on solubilities of pressure increases are minimal and more than offset by the significant changes associated with temperature increases (Seward and Barnes 1997).

3.4.1 The Important Metal–Ligand Complexes in Hydrothermal Solutions

There are several excellent overviews of metal solubility characteristics for a wide variety of metal–ligand complexes (Wood and Samson 1998; Seward et al. 2014). These data are used below to describe metal solubilities in terms of hard, borderline, and soft metals. This summary provides an indication of the most likely metal–ligand complex that will exist in a natural fluid as a function of variables such as oxidation state, pH, temperature, and fluid composition. However, it cannot, and should not, be used to try and simply predict the nature of metal speciation in a situation where the many parameters that control solubility are imperfectly known. The concept

of hard–soft Lewis acids and bases is an idealized scenario predicated on the assumption that there is competition between metals and a variety of ligands. In natural situations metal complexation may be constrained by the limited availability of ligands such that the nature of metal–ligand complexes and their solubilities may be different to those predicted by this scheme.

3.4.1.1 Hard Metals
Tungsten (W) Tungsten tends to occur in nature as the hexavalent aqueous cation W^{6+}, although the pentavalent form W^{5+} can also occur under more reducing conditions. These ions are hard Lewis acids and are likely to complex with hard bases such as O^{2-}, OH^-, F^-, and CO_3^{2-}, with the chloride ion being relatively unimportant as a complexing agent under most geological conditions. The majority of dissolved tungsten species in hydrothermal solutions occur as tungstates and have the forms WO_4^{2-}, HWO_4^-, and H_2WO_4. The stabilities of the tungstate species are obviously dependent on hydrogen ion concentration (i.e. on pH), such that H_2WO_4 is only stable

Figure 3.12 Plot of the effective stability of a metal complex (expressed in terms of β_1, which is the equilibrium formation constant) versus temperature. The stabilities of metal–chloride complexes increase by several orders of magnitude as temperature increases, as will their solubilities in aqueous solution. Source: After Seward and Barnes (1997).

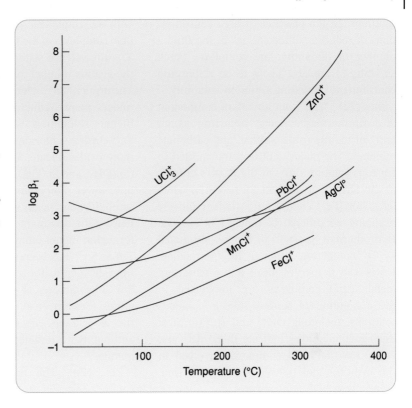

under acidic conditions, whereas WO_4^{2-} is stable at alkaline pH. Significant quantities of tungsten can be carried in solution by these tungstate complexes and they are most likely to have been implicated in the formation of most scheelite and wolframite deposits.

Molybdenum (Mo) Molybdenum tends to be more easily reduced in nature than tungsten and typically occurs in a variety of aqueous valence states as $Mo^{6+}, Mo^{5+}, Mo^{4+}$, and Mo^{3+}. It is also a hard Lewis acid and like tungsten will complex with hard bases, in particular, O^{2-} and OH^-. Chloride complexing takes place only under more acidic conditions and typically it is the oxyhydroxide complexes (such as MoO_2^+ and $Mo(OH)_3^+$) that are the most likely to occur in natural aqueous solutions.

Uranium (U) Uranium is characterized by two valence states in nature, as the quadrivalent uranous ion, U^{4+}, and the hexavalent uranyl ion, U^{6+}. The uranous ion is generally characterized by very low solubilities

in aqueous solutions and most fluid-related transport takes place as U^{6+}. As a hard acid, U^{6+} will complex with hard bases such as O^{2-}, OH^-, F^-, HCO_3^-, and CO_3^{2-}.

3.4.1.2 Borderline Metals

Arsenic (As) The trivalent cation As^{3+} dominates the valence state of arsenic in nature and in solution the neutral H_3AsO_3 complex is known to be stable over wide ranges of temperature (above 200 °C), Eh, and pH. Other less protonated arsenate complexes are also stable at higher pH values (as for tungsten) and generally chloride complexes do not play a role in fluid transport. Under conditions of high sulfide concentration and at low temperatures, As–sulfide complexes (such as $AsS_2(SH)^{2-}$) may become more important, although their exact stoichiometric proportions are unclear.

Antimony (Sb) The aqueous geochemistry of antimony is similar to arsenic, with Sb^{3+} predominating and $Sb(OH)_3$ occurring as the main stable complex in low-sulfide aqueous solutions. In the presence of high concentrations of reduced sulfide species in the fluid a number of thioantimonite complexes (such as $HSb_2S_4^-$) are also likely to be stable and contribute to increasing antimony solubility.

Iron (Fe) Iron is an abundant component of hydrothermal systems, occurring as an array of mainly oxide, sulfide, and carbonate minerals. The majority of iron transported in aqueous solution through the Earth's crust is in the ferrous or divalent form (Fe^{2+}), while ferric iron (Fe^{3+}), which forms under relatively oxidizing conditions, is much less soluble. Hydrothermal transport of iron is dominated by Cl^- complexes, although its intermediate Lewis acid properties mean that a variety of metal–ligand complexes are implicated in the dissolution of ferrous iron in aqueous solutions, including OH^-, and HCO_3^-. Most experimental studies indicate that $FeCl^+$ and $FeCl_2$ are the main complexes involved in the hydrothermal transport of ferrous iron, especially at high temperatures and salinities. The free hydrated ion itself, Fe^{2+}, is also implicated in fluid transport where hydrolysis has taken place, as are Fe–bicarbonate complexes under more alkaline conditions. Fe–bisulfide complexes are generally not involved in the transport of iron in hydrothermal solutions although they may be important in ocean floor exhalative environments (i.e. black smokers).

Manganese (Mn) Manganese is also a borderline Lewis acid and, like iron, Mn^{2+} complexation in hydrothermal solutions is likely to be dominated by the chloride ligand (i.e. $MnCl^+$ and $MnCl_2$), with hydroxide and bicarbonate complexes also contributing to Mn solubility.

Tin (Sn) Tin is a metal that exhibits both hard acid quadrivalent (as Sn^{4+}) and borderline divalent (as Sn^{2+}) traits such that it can be solubilized by complexation with a number of different ligands. Under oxidizing conditions it has been found that the Sn^{4+}–hydroxychloride complex, $Sn(OH)_2Cl_2$, is the dominant species, but that its solubility is low. Under more reducing conditions both Sn^{4+} and Sn^{2+} can complex with the simple chloride ion, forming very soluble complexes of the form $SnCl_n^{X-n}$ (i.e. where X is the valence state, either 2 or 4, and n is the ligation number for the chloride complex). The divalent Sn^{2+}–chloride complexes, formed under more reducing conditions, exhibit higher solubilities than the quadrivalent Sn^{4+}–hydroxychloride complex that exists in a more oxidized state. Sn–hydroxide complexes ($Sn(OH)_4$ and $Sn(OH)_2$) are stable under alkaline, lower temperature conditions but their solubilities are again lower than those exhibited by the dominant Sn–chloride complexes formed at higher temperatures and lower pH. Tin as Sn^{2+} has been shown to complex with fluoride ions to form stable $SnF(OH)_2^-$ and $SnFCl^0$ complexes at relatively high temperatures (Seward et al. 2014), consistent with the association between tin mineralization and highly fractionated fluorite/topaz-bearing granites.

Zinc (Zn) The dominant divalent zinc cation, Zn^{2+}, is a borderline Lewis acid that complexes with a variety of ligands, including Cl^-, HS^-, OH^-, HCO_3^-, and CO_3^{2-}. At relatively low temperatures and high pH, and in fluids with high bisulfide concentrations but low salinities, a variety of Zn^{2+}–bisulfide complexes are also stable, including $Zn(HS)_2$, $Zn(HS)_3^-$, and $Zn(HS)_4^{2-}$. In contrast, at higher temperatures and under more acidic conditions, a host of chloride complexes predominate, including $Zn(Cl)^+$, $Zn(Cl)_2$, $Zn(Cl)_3^-$, and $Zn(Cl)_4^{2-}$.

Lead (Pb) Lead is the softest of the borderline Lewis acids and forms stronger bonds with Cl^- and HS^- ligands than with OH^- or HCO_3^-. The valence state and complexing behavior of lead is very similar to that of zinc. Pb^{2+}–bisulfide complexes are more stable than chloride complexes in low temperature, neutral to alkaline, low salinity solutions. Different Pb^{2+}–bisulfide complexes form as a function of pH, and these include $Pb(HS)_2$ under acidic conditions and $Pb(HS)_3^-$ at higher pH (i.e. >6). At higher temperatures and lower pH however, a variety of Pb^{2+}–chloride complexes form and these include $Pb(Cl)^+$, $Pb(Cl)_2$, $Pb(Cl)_3^-$, and $Pb(Cl)_4^{2-}$. Pb^{2+}–carbonate or Pb^{2+}–bicarbonate complexes can form in CO_2-bearing fluids but their solubilities are typically low and they are unlikely, therefore, to have relevance in ore forming processes. There is, however, compelling evidence that lead, and to a lesser extent zinc, are transported by organo-metallic complexes (see Section 3.4.2 below), especially in relatively low temperature environments such as those related to the formation of MVT Pb–Zn deposits (see Box 3.7), where hydrothermal fluids are known to contain hydrocarbons. The acetate ligand (CH_3COO^-) is one organic molecule that has been implicated in base metal transport, although there are others, such as thiols or sulfur-bearing organic compounds, that may also be important.

3.4.1.3 Soft Metals

Copper (Cu) In nature, copper occurs in both Cu^+ and Cu^{2+} valence states, although the monovalent state is predominant in most hydrothermal fluids. It is a relatively soft acid and forms stable Cu^+–chloride and Cu^+–bisulfide complexes, as well as Cu^+–hydroxide complexes if other ligands are not available. Movement of copper in hydrothermal solutions under a wide range of conditions is most likely to take place by Cu^+–chloride complexation in the form $CuCl_2^-$. Although copper–bisulfide complexes will form in high-sulfur environments, they are not likely to play a significant role in most hydrothermal ore-forming environments, although this may not be the case in environments where Cu is transported in the vapor phase (see Section 3.4.2).

Gold (Au) Gold is the softest metal ion considered in this summary (Webster 1986; Seward 1991) and prefers complexation with soft ligands such as the bisulfide ion, rather than chloride. Gold typically occurs in a monovalent state (Au^+) in aqueous solutions and will form the complex AuHS at low pH and $Au(HS)_2^-$ under weakly acidic to basic conditions. The bisulfide complexes predominate in

most hydrothermal fluids and are stable over a wide temperature range. Under oxidizing, saline, and acidic conditions, it is possible to form both Au^+– and Au^{3+}–chloride complexes in the form of $Au(Cl)_2^-$ and $Au(Cl)_4^-$ respectively. The former complex might prevail in certain high temperature environments such as porphyry coppers, whereas the latter is only likely to be important at low temperatures in near surface settings. The solubility ranges for $Au(Cl)_2^-$ and $Au(HS)_2^-$ complexes as a function of pH and fO_2 are illustrated in Figure 3.13. Gold forms very soluble complexes with cyanide ($Au(CN)_2^-$), which is the reason why cyanidation is used so effectively at low temperatures to metallurgically recover gold from ore. The cyanide molecule is, however, unstable at higher temperatures and the complex has no relevance to hydrothermal gold transport.

Silver (Ag) Silver is also a soft metal ion, but not as soft as gold (Webster 1986). It is, therefore, more likely to form complexes with a borderline ligand such as chloride than gold is, although solution chemistry will typically be dominated by Ag^+–bisulfide bonding. The predominant complexes are AgHS and $Ag(HS)_2^-$ under acidic and neutral to alkaline conditions respectively. Several Ag^+–chloride complexes are also known to be stable over a wide range of conditions pertaining to hydrothermal fluid flow and these include AgCl, $Ag(Cl)_2^-$, and $Ag(Cl)_3^{2-}$.

Mercury (Hg) Mercury, occurring predominantly as Hg^{2+} in aqueous solutions, is a soft metal ion and will complex preferentially with bisulfide ligands. Most mercury transport under neutral to alkaline and moderately reducing conditions is likely to take place by the formation of complexes such as $Hg(HS)_2$, $HgS(HS)^-$, and HgS_2^{2-}. Harder ligands probably do not play much of a role in most hydrothermal fluids. Of interest and related to its low boiling point, however, is the fact that mercury can be transported as a vapor, or as elemental Hg in either aqueous solutions or hydrocarbon-bearing fluids.

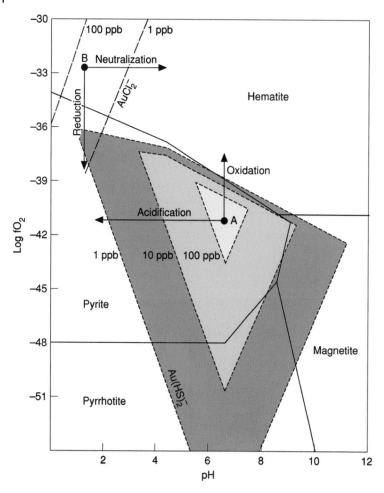

Figure 3.13 Log fO_2–pH diagram showing the stability of iron oxide and sulfide minerals in relation to gold solubility contours for $Au(HS)_2^-$ and $AuCl_2^-$ complexes in a fluid at 200 °C (also at saturated water vapor pressure, with $\Sigma S = 0.01$ and $a_{Cl^-} = 1.0$). Oxidation of fluid A across the pyrite–hematite phase boundary will result in a four order of magnitude decrease in the solubility of the $Au(HS)_2^-$ complex and likely precipitation of gold. The same result would occur if the fluid became more acidic and pH decreased by some 5 units. The converse would apply with respect to $AuCl_2^-$ in fluid B. In this case Au precipitation would occur in response to an increase in pH (neutralization) or reduction of the fluid (perhaps caused by an increase in the activity of H_2S). Source: Phase diagram and contours from Wood (1998) and S. Wood, personal communication.

In summary, it is apparent that the borderline chloride ion is the most important ligand that promotes the dissolution and transport of metals in hydrothermal fluids, followed by bisulfide. A variety of other ligands such as OH^-, HCO_3^-, CO_3^{2-}, F^-, SO_4^{2-}, NO_3^-, NH_3, and CH_3COO^- (acetate) are of lesser importance because they are either stable under abnormal conditions or exist at only low concentrations in natural fluids. In addition to its ability to complex with a wide range of both hard and soft metal cations, Cl^- is also the major anionic species in most natural fluids. This ensures that many natural fluids are capable of transporting large quantities of metal in solution and it is for this reason that the circulation of hydrothermal solutions in the Earth's crust is such an important ore-forming process. Even though metal–ligand complexes may commonly be stable in natural aqueous solutions, it is nevertheless apparent, both from direct analysis and theoretical calculation, that the actual concentrations of metals dissolved in most hydrothermal fluids are generally low. This means that the formation of a viable ore deposit requires that large volumes of fluid pass through a highly focused point in the Earth's crust and that efficient precipitation mechanisms are on hand to remove metals from solution and concentrate them in the host rock. The compilation shown in Figure 3.14 shows the minimum concentrations of precious and base metals (i.e. the vertical dashed lines) estimated to be necessary for the formation of a viable deposit, and also the ranges shown for several examples where such data are available. Hydrothermal solutions that deposit Au and Ag typically contain much lower concentrations, in the range 1 ppb to 1 ppm, than those associated with Cu, Pb, and Zn ores. In the latter case massive sulfide deposits are formed from fluids that typically carry 1–100 ppm of metal, whereas the fluids associated with MVT and porphyry deposits are relatively enriched, with concentrations in the range 100–1000 ppm.

3.4.2 More on Metal Solubilities in the Aqueous Vapor Phase

Phase separation of an aqueous fluid to form higher-density liquid and lower density vapor phases (boiling) also results in a fractionation of metals between these phases, as discussed in Chapter 2 (Section 2.6). Typically, hydrothermal liquids are dominated by solutes comprising ionic or charged species, whereas the accompanying vapor is characterized by uncharged molecular species. It was pointed out previously that partitioning of soft metals such as Cu and Au into the vapor phase was promoted by the fact that reduced, volatile sulfur species (such as HS^-) were likewise partitioned into the vapor phase, thereby creating Au-thio complexes that are

stable in the latter. Experimental work has shown, however, that Cu, Au, and Ag may have appreciable solubilities in the aqueous vapor phase even when sulfur is not involved as a complexing agent. Figure 3.15 illustrates predicted Au and Ag solubilities in an aqueous vapor phase in which solvation occurs as hydrated AuCl and AgCl molecular clusters (Migdisov and Williams-Jones 2013; Seward et al. 2014). The presence of such hydrated metal clusters enhances their solubility in the aqueous vapor phase – the metals dissolve in the gas as $CuCl:(H_2O)$, $AgCl:(H_2O)$, and $AuCl:(H_2O)$ clusters that form because of either hydrogen or van der Waals bonding in the aqueous medium (Liu et al. 1996). It is evident in Figure 3.15 that the solubility of the hydrated AgCl cluster in the aqueous vapor phase (up to 80 ppm Ag) is significantly higher than that for AuCl clusters (up to 5 ppm Au), which may contribute to the fact that high-level epithermal deposits, where boiling and vapor transport of metals may be more prevalent, are generally characterized by higher Ag/Au ratios than deeper-seated deposit types.

3.4.3 A Brief Note on Metal–Organic Complexes

A number of different hydrothermal ore types that are hosted in intracratonic and rift-related sedimentary sequences are known to have formed from fluids containing significant amounts of dissolved organic compounds. It is also well known that crude oil and bitumen invariably contains minor amounts (<1%) of inorganic material that includes dissolved metals. It is not surprising, therefore, that considerable attention has been given to the role that metal–organic complexes might have played in hydrothermal ore forming processes (Wood 1996), especially in MVT Pb–Zn and stratiform, red-bed or shale-hosted base metal ores (see Parnell 1994; Gize 1999).

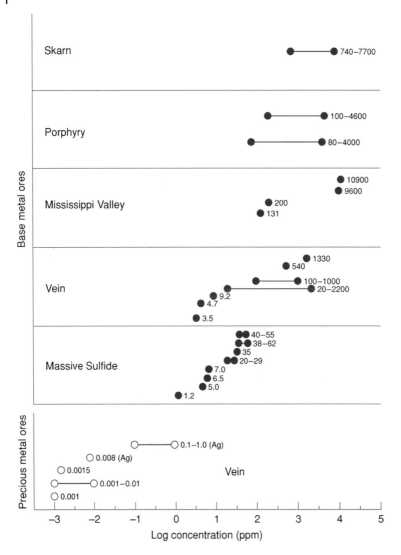

Figure 3.14 Base and precious metal concentrations in ore-forming hydrothermal solutions, in each of five major categories of deposit type. Base metals include Cu, Pb, and Zn (filled dots), and precious metals comprise Au, Ag, and Hg (open circles). The numbers next to individual points or ranges refer to average concentrations or ranges in a variety of different situations. Source: Seward and Barnes (1997).

Organic compounds in basinal waters tend to occur either as neutral hydrocarbon molecules (alkanes, aromatics, etc.) or as charged anionic species such as the carboxylic acids (acetic acid, oxalic acid, etc.). Most neutral hydrocarbons (with the exception of methane, CH_4) have only limited aqueous solubility and this decreases exponentially as the hydrocarbon molecule gets bigger and more complex. Organic acid anions, on the other hand, may have high solubilities in water and the dominant aqueous anion, acetate (CH_3COO^-), can exist at concentrations up to $2000\,mg\,l^{-1}$ (Hanor 1994). Organometallic complexes (i.e. those in which a metal is covalently bonded to the carbon of an adjacent hydrocarbon molecule) only rarely occur in nature, whereas metal–organic complexes,

Figure 3.15 Predicted solubilities of (a) Ag in vapor and (b) Au in vapor for a situation where hydrated AgCl and AuCl clusters, respectively, are in solution in a low-density gas-like phase (containing 0.1 vol.% HCl, density = 0.2 g cm^{-3} and fO$_2$ buffered by Ni/NiO). Source: After Migdisov and Williams-Jones (2013).

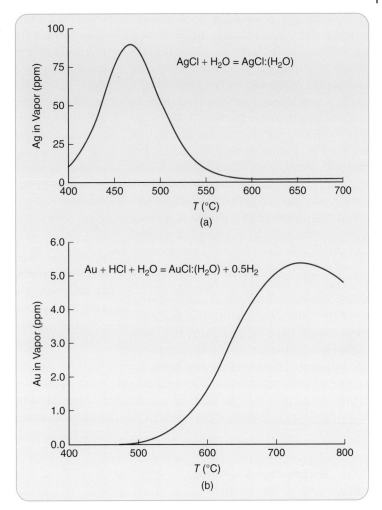

in which a metal cation is attached to the electron-donor atom (such as oxygen, sulfur, or nitrogen) of an organic ligand, are more common in basin-derived aqueous solutions, as well as in associated ore fluids (Giordano and Kharaka 1994; Giordano 1994).

In ore fluids metal–organic complexes will only exist if they are sufficiently stable to be able to compete with the normal inorganic complexes discussed previously. Modeling has shown that Pb and Zn can be transported as metal–organic complexes involving acetate or oxalate $(C_2O_4^{2-})$ ligands (e.g. $Zn(CH_3COO^-)_2$ and $Pb(C_2O_4^{2-})_2^{2-})$ in fluids that are relatively oxidized and have low total sulfide concentrations (Giordano 2002). In more saline fluids, such as those applicable to red-bed related stratiform base metal deposits, Pb and Zn would be transported predominantly as metal–chloride complexes, but Pb– and Zn–acetate complexes could still account for up to 5% of metal transport (Giordano and Kharaka 1994).

3.5 Precipitation Mechanisms for Metals in Solution

The previous section has shown that it is possible to stabilize many different metal–ligand complexes under wide ranging conditions in natural hydrothermal solutions, thereby creating the kind of solubility levels required to make effective ore-forming fluids. Once a metal is in solution, however, it then needs to be extracted from that fluid and concentrated in a portion of the Earth's crust that is sufficiently accessible and restricted in size to make an economically viable ore body. It is obvious that a wide range of precipitation mechanisms are likely to be effective since any mechanism that will destabilize a metal–ligand complex and, therefore, reduce the metal solubility, will cause it to be deposited in the host rock through which the hydrothermal solution is passing. The following sections examine the physical and chemical controls on metal precipitation, as well as adsorption and biologically mediated processes. The latter is important as it is now widely accepted that the geochemical pathways by which both metals and non-metals are distributed in the Earth's crust are affected to varying degrees by the action of microorganisms. The concept of biomineralization is becoming increasingly important to the understanding of ore-forming processes and this topic is likely to be the subject of much more research in the future.

3.5.1 Physico-Chemical Factors Affecting Metal Precipitation

The basic principles that control ore deposition have been summarized by Barnes (1979b) and Seward and Barnes (1997). At shallow crustal levels ore deposition will take place by open space filling, whereas deeper down, where porosity is restricted, replacement of existing minerals tends to occur. Decrease in temperature is the factor that, intuitively, is regarded as the most obvious way of promoting the precipitation of metals from hydrothermal fluids. At depth, however, temperature gradients across the structures within which fluids are moving tend to be minimal and in such cases metal precipitation will be neither efficient nor focused within a particular trap zone. In areas where temperature gradients are minimal, deposition of metals may be achieved more effectively by

changing the properties or composition of the hydrothermal fluid. If ore solution occurs by metal–chloride complexing then precipitation could occur by increasing the pH of the ore fluid (see Figure 3.13). An example is provided by the reaction of an acidic ore solution with a carbonate host rock and where precipitation is promoted by digestion and replacement of the host. Two coupled reactions illustrate the process in terms of calcite dissolution by an originally acidic, Zn^{2+}–chloride-bearing ore solution:

$$CaCO_{3(SOLID)} + 2H^+ \Leftrightarrow Ca^{2+} + H_2CO_{3(AQUEOUS)}$$
(3.2)

and subsequent sphalerite precipitation caused by increase of pH in the fluid as hydrogen ions are consumed:

$$ZnCl_n^{2-n} + H_2S_{(AQUEOUS)} \Leftrightarrow ZnS_{(SOLID)} + 2H^+ + nCl$$
(3.3)

It is apparent from reactions (3.2) and (3.3) that precipitation of metal–chloride complexes could also take place by increasing the H_2S concentration of the fluid (perhaps by fluid mixing or bacterial sulfate reduction) or by decreasing the Cl^- concentration (again perhaps by mixing of the ore fluid with dilute groundwater). As another option to calcite dissolution, an increase in fluid pH could also be brought about by boiling of the fluid, which results in acidic volatiles being partitioned preferentially into the vapor phase. There are, therefore, a number of ways of reducing the solubility of metal–chloride complexes and promoting metal precipitation.

The factors that promote precipitation of metal–sulfide complexes may be somewhat different to those for chloride complexes (Barnes 1979b). Although fluid mixing (or dilution) and temperature decreases will promote metal precipitation, it is oxidation of the ore fluid (see Figure 3.13) that is more likely to be effective in decreasing the solubilities of metal–sulfide complexes. Oxidation (or loss of electrons) causes a decrease in the pH and also the total sulfide concentration, thereby promoting metal precipitation, as shown in reaction (3.4):

$$Zn(HS)_3^- + 4O_{2(AQUEOUS)} \Leftrightarrow ZnS_{(SOLID)} + 3H^+ + 2SO_4^{2-}$$
$$(3.4)$$

Precipitation of ores from metal–sulfide complexes is controlled by oxidation, decreases in pH (or acidification), and decreases in sulfide concentration in the fluid, all tendencies that are opposite to those controlling metal–chloride dissociation. A more detailed discussion of the actual geological processes that affect ore deposition controls is presented below, together with different scenarios where they might apply. An understanding of the geological factors relating to metal deposition is obviously very important to the exploration geologist who needs to have a firm grasp of the concepts that control where and why hydrothermal ore bodies form.

3.5.1.1 Temperature

Since the solubilities of many metal–ligand complexes increase as a function of temperature (Figure 3.12) it is clear that cooling a fluid will generally have the effect of promoting metal deposition. Temperature decrease is particularly effective for destabilizing metal–chloride complexes because their solubilities are much more sensitive to temperature changes than are those of equivalent sulfide complexes. As mentioned previously, however, at deep crustal levels, where fluid and rock are in equilibrium and exist at the same temperature, thermal gradients are minimal and cooling is a slow and ineffectual process. Under these circumstances the concentration of metals and the formation of hydrothermal ores are generally not mediated by temperature decreases.

In the near surface environment, however, rapid decreases in the temperature of an ore fluid are much more likely to happen and in such environments cooling can play an important role in ore formation. The ocean floor undoubtedly represents the best example of where a dramatic reduction in the temperature of ore-forming fluids plays the dominant role in controlling metal deposition. VMS deposits typically form at sites where hot (up to 350 °C) brines with high metal contents are vented onto the ocean floor as black smokers (see Section 3.8.1 and Box 3.3). Precipitation of base metals (mainly Cu and Zn transported as chloride complexes; Scott 1997) and, in certain cases, precious metals (Au transported as a gold–bisulfide complex; Scott 1997) is virtually instantaneous as the ore fluids mix with an essentially infinite volume of cold (2–4 °C) sea water. In this environment it is the very rapid cooling of the fluid that causes a highly efficient precipitation of metal in the immediate vicinity of the exhalative vent. Black smokers are probably the most dramatic and efficient example of hydrothermal ore deposition anywhere on Earth.

The volcanic settings that give rise to the deposition of epithermal Au–Ag deposits (see Chapter 2, Section 2.11 and Box 2.4) are also characterized by steep geothermal gradients and high conductive and convective heat loss by fluids to the surrounding rocks. Cooling may, therefore, play a role in controlling metal solubilities although in such environments it is more likely to be the decrease in pressure and associated phase separation (i.e. boiling) that plays the dominant role in ore deposition.

3.5.1.2 Pressure

Pressure variations do not markedly affect the solubilities of metal–ligand complexes although it is clear that pressure increases will lead to a volume reduction which, in turn, promotes the dissociation of complexes to ionic species (Seward 1981). In effect a decrease in pressure tends to favor an increase in solubility and, therefore, works in the opposite sense to temperature. Pressure reduction is not a process that, on its own, is typically associated with ore formation. A major exception occurs, however, when pressure reduction is accompanied by an event such as boiling or effervescence. These specific processes are very important in ore deposition and are generally promoted by decreases in pressure. In this sense pressure only has an indirect influence on ore formation. Pressure reduction in fluids will accompany the rapid uprising of fluids in the Earth's crust and also the episodic seismic activity along fault zones, as discussed in Section 3.3.3.

3.5.1.3 Phase Separation (Boiling and Effervescence)

Although one might intuitively expect that boiling is promoted by an increase in temperature, there is no doubt that, in the upper levels of the Earth's crust, the transformation from liquid to vapor usually occurs because of a decrease in fluid pressure. At deeper levels in the crust a decrease in fluid pressure could also be accompanied by effervescence, or the transition from a single phase H_2O–CO_2 mixture to one where H_2O and CO_2 unmix (see Figure 3.5c). Both these processes are potentially important as mechanisms of precipitating metals from ore-forming solutions because they change the prevailing conditions under which metal–ligand complexes are stable.

In Section 3.3.3 the relationships between fluid flow and deformation were discussed and the fault-valve and suction-pump models were described with reference to the formation of mesothermal lode-gold and epithermal Au–Ag deposits respectively. The metamorphic fluids generally attributed to formation of mesothermal lode-gold deposits are characterized by single phase H_2O–CO_2 mixtures and it is possible that the episodic decreases in fluid pressure accompanying crack–seal processes in a fault or shear (Figure 3.9a and c) could promote precipitation of ore components. This is because there is a tendency for the H_2O–CO_2 solvus to move to higher temperatures with decreasing pressure (Brown 1998). A typical mesothermal lode-gold fluid comprising $H_2O + CO_2$ at 350 °C and with 2.6% NaCl, for example, will initially exist as a single homogeneous liquid just above the relevant solvus (shown as point A; Figure 3.5b). A pressure decrease will have the effect of moving the solvus plateau to higher temperatures, so that the fluid enters the immiscible, or two phase, field segregating CO_2 and H_2O. CO_2 effervescence will undoubtedly affect the prevailing fluid properties and likely promote precipitation of both gangue and ore components, as suggested in the fault-valve model. In addition, it is also suggested that fluxing of CO_2 through a magmatic-hydrothermal system in which metals are being transported in the vapor phase will strongly decrease the metal solubilities in the latter, providing an additional precipitation mechanism related to H_2O–CO_2 unmixing (Van Hinsberg et al. 2016).

Similarly, in the near surface environment where en echelon strike slip faulting results in a fault jog, the resulting rupture-related dilation will instantaneously reduce pressure (Figure 3.9c) to such an extent that a fluid could transform from a liquid to the vapor state. Boiling will, of course, only occur if the pressure reduction traverses the boiling point curve for the relevant hydrothermal fluid composition (refer to phase diagrams in Figure 2.2). The boiling away of an aqueous vapor from a hydrothermal fluid will result in the residual enrichment of solute, as well as partitioning of other volatile species into the gas phase, with probable increase of pH in the remaining fluid. Such processes will dramatically modify prevailing fluid properties and the pH increase will, as mentioned previously, be particularly effective in precipitating metal–chloride complexes from solution. When accompanied by temperature decreases, prolonged boiling will likely bring about solubility reduction of most dissolved species. As suggested by Sibson (1987) with respect to epithermal ore-forming environments, boiling is likely to be a very efficient mechanism for ore deposition at shallow crustal levels, resulting in high-grade mineral deposits.

3.5.1.4 Fluid Mixing/Dilution

The mixing of two fluids is regarded as another important mechanism for reducing solubility in ore-forming solutions and promoting metal precipitation. This is particularly the case when a relatively hot, metal-charged ore fluid mingles with a cooler, more dilute solution. Mixing of the two fluids will result in cooling of the hotter with modification of the prevailing ore fluid properties and destabilization of existing metal–ligand complexes. There are several examples that point to the importance of fluid mixing during ore deposition. In a classic stable isotope study, Sheppard et al. (1971) demonstrated that precipitation of chalcopyrite occurs at the interface between the potassic and phyllic alteration zones in porphyry copper deposits (see Box 2.1, Chapter 2), which also coincides with the zone where fluids of dominantly magmatic origin mingled with those derived from a meteoric source. A similar situation is known to exist with respect to precipitation of Au and Ag ores in certain epithermal deposits (Box 2.4, Chapter 2).

Hedenquist and Aoki (1991) showed that magmatic fluids venting directly from a volcano without any mixing or dilution below the surface are less conducive to the formation of economically viable deposits since ground preparation (or alteration) is minimized and the resultant ore accumulations at the surface have a low preservation potential. By contrast, systems characterized by mixing of magmatic waters with a meteoric water carapace 1–2 km below the surface will more likely be associated with the formation of economically viable mineralization. This is because mixing of a hot, saline, metal-charged magmatic fluid with a cooler, more dilute meteoric water promotes acid leaching of the host rocks, increases their permeability, and forces the fluids to condense and precipitate their dissolved metal solute. Hedenquist and Aoki (1991) suggested that interaction of magmatic fluids with a meteoric fluid blanket near the surface of active volcanic regions may be an important prerequisite to the formation of viable epithermal Au–Ag mineralization.

Box 3.1 Fluid Mixing and Metal Precipitation: The Olympic Dam Iron Oxide–Copper–Gold (IOCG) Deposit, South Australia

The Olympic Dam deposit, some 500 km north of Adelaide in South Australia, was discovered in 1975 beneath 350 m of sedimentary cover. It represents one of the most spectacular discoveries of modern times, containing well in excess of 2 billion tons of Cu, U, Ag, and Au ore (Solomon and Groves 1994). The ore body is hosted in a breccia complex within the apical portions of the 1588 Myr old (Creaser and Cooper 1993) Roxby Downs granite, the latter being a K-, U-, and Th-enriched A-type or anorogenic granite (Figure 3.1.1). Mineralization is generally attributed to mixing of two entirely different fluids, one of probable magmatic derivation, and the other an oxidized meteoric fluid (Reeve et al. 1990; Oreskes and Einaudi 1992; Haynes et al. 1995; Johnson and McCulloch 1995).

The breccias that host the ore comprise a variety of fragments, the majority of which are either granitic or hematitic, together with volcanic fragments, sedimentary rocks, and massive sulfide ore. The breccias are extremely iron rich and themselves comprise a huge iron resource made up of hematite and minor magnetite. Polymetallic mineralization, in the form of chalcocite, bornite, chalcopyrite, pitchblende, argentite, gold, and a variety of rare earth element (REE) minerals, is most closely linked to the hematitic breccia (Oreskes and Einaudi 1990). All breccias are cut by mafic and felsic dykes that are approximately the same age as

(Continued)

Box 3.1 (Continued)

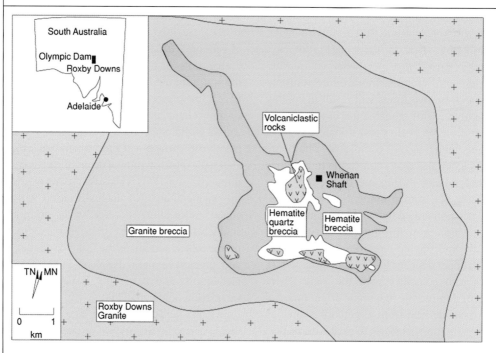

Figure 3.1.1 Simplified, subsurface geological map of the Olympic Dam ore deposit. Source: After Haynes et al. (1995).

the Roxby Downs granite, indicating that magmatism, brecciation, and mineralization were broadly coeval (Johnson and Cross 1991).

Ore genesis at Olympic Dam (Solomon and Groves 1994) was initiated by the high level intrusion of fertile A-type granite, with accompanying fluid exsolution and boiling, which gave rise to explosive volcanic activity and an early phase of brecciation. The fluid involved here is inferred to be of magmatic origin and moderately oxidizing. It was also clearly Fe-rich and precipitated an early generation of magnetite as well as REE-bearing minerals. Cooling of the magmatic system subsequently led to draw-down of groundwater (perhaps derived from a saline playa lake above the deposit; Haynes et al. 1995) that in turn led to phreatic explosive activity, further brecciation, and hematite precipitation. The spatial association between hematite breccia and polymetallic mineralization indicates that a low temperature, highly oxidizing, meteoric fluid was implicated in the dissolution and transport of substantial amounts of Cu, Fe, and U, possibly from the granite, or overlying volcanic rocks, or both. This fluid encountered and mixed with the less oxidized magmatic solutions (containing reduced sulfur species), destabilizing metal–ligand complexes and causing precipitation of metal sulfides together with iron oxides. Figure 3.1.2 illustrates a model showing the existence of two fundamentally different fluid types at different crustal levels, with ore precipitation occurring at the interface between the fluids. It is envisaged that fluid circulation and mixing continued episodically for a considerable period of time, perhaps in part stimulated by the high radioelement (K, U, Th) content of the granite and the resultant heat generated by radioactive decay. The longevity of the ore-forming process is consistent with the complexity and enormous size of the ore deposit.

Box 3.1 (Continued)

Figure 3.1.2 Geological model for the formation of the Olympic Dam deposit. The geological setting and the distribution of a lower, hotter, moderately oxidizing fluid of possible magmatic derivation, and an upper, saline, highly oxidizing meteoric fluid, is shown in (a). Drawdown of meteoric fluid together with uprising plumes of hotter water promote mixing of the fluids and metal precipitation, as shown in (b). Source: After Haynes et al. (1995).

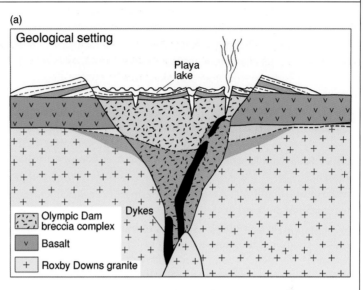

(a)

Geological setting

Playa lake

Dykes

Olympic Dam breccia complex

Basalt

Roxby Downs granite

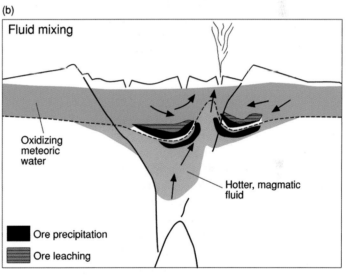

(b)

Fluid mixing

Oxidizing meteoric water

Hotter, magmatic fluid

Ore precipitation

Ore leaching

In another example demonstrating the importance of mixed fluids in ore formation, it has been suggested that sediment-hosted ore deposits associated with continental rift environments are the product of the mixing of meteoric and connate or basinal fluids. Using the active mineralizing systems of the Salton Sea geothermal system (see Section 3.7 below) as an analogue, McKibben et al. (1988) have suggested that the mixing of upwelling high-salinity brines of essentially basinal origin with descending meteoric waters of lower temperature and salinity may be a viable mechanism of promoting metal precipitation from the former. It has been suggested that this type of model might apply to the formation of several sediment-hosted ore deposits such as the stratiform Cu–Co ores of the central African

Copperbelt, and the sedimentary exhalative (SEDEX) Zn–Pb ores of Mount Isa and Broken Hill in eastern Australia (McKibben et al. 1988; McKibben and Hardie 1997).

Mixing of two fluids with distinct temperature, pH, and redox characteristics is a process that is likely to have widespread applicability to ore genesis in a variety of crustal settings. Another important example is the Olympic Dam deposit described in Box 3.1.

3.5.1.5 Fluid/Rock Reactions (pH and Eh Controls)

A widespread manifestation of hydrothermal mineralization processes is the development of alteration mineral assemblages in and around the fluid conduit. Alteration is caused by the reaction between fluids and their wall rocks and is a complex process that has been extensively studied, both in ore genesis research and for the understanding of metamorphic and metasomatic (mass transfer) processes in the crust. The interaction that occurs between a fluid and its wall rock promotes metal precipitation because it is yet another process that changes the ambient fluid properties, especially in terms of acidity (pH) and redox state. Reactions describing the factors controlling metal precipitation were presented in Section 3.5.1 above. Quantification of the effects of changes in pH and redox state on, for example, gold solubility is demonstrated in Figure 3.13, where calculated gold solubility contours for the dominant $Au(HS)_2^-$ complex, as well as for the less important $AuCl_2^-$ complex are shown. A reduced, neutral fluid like that at A, subject to an oxidation event akin to the one in reaction (3.4) above, would have its $Au(HS)_2^-$ solubility dramatically reduced, as shown by the oxidation trend in Figure 3.13. This would almost certainly result in significant precipitation of Au from the fluid. Similarly, a decrease in fluid pH from around 7 to 3 would also result in a reduction of gold solubility and deposition of the metal. It is pertinent to note that orogenic gold deposits are typically associated with relatively reduced, low salinity

H_2O–CO_2 fluids in which gold was transported as a $Au(HS)_2^-$ complex. The significance of ubiquitous CO_2 in this system has often been queried, although it has now been suggested (Phillips and Evans 2004) that its principal role was as a chemical buffer that maintained pH at neutral levels, thereby optimizing gold solubility (Figure 3.13).

As described earlier, the precipitation of ores from metal–chloride complexes may occur under different conditions altogether. In Figure 3.13 it is clear that the precipitation of Au from an oxidized, acidic fluid like that at B, in which the $AuCl_2^-$ complex is stable, will best be achieved by increasing pH (perhaps by consumption of H^+ during alteration) or by reduction (perhaps caused by sulfate reduction to sulfide).

The relationship between alteration and hydrothermal ore-forming processes is a very important one and for this reason the topic is covered in more detail in Section 3.6 below. In this section the link between fluid/rock reactions and the precipitation of metals from ore fluids is also discussed.

3.5.2 Adsorption

Although metal precipitation is generally considered to have been instigated by changes to the prevailing fluid properties and accompanying reduction in equilibrium solubility, it is evident that some ores form by adsorption of metal onto an existing mineral surface. Metal deposition by adsorption can occur from fluids whose concentrations are below their saturation levels and the process may, therefore, be important in certain ore-forming environments. Adsorption is defined as the adherence of an ion in solution to the surface of a solid (or mineral) with which it is in contact. The process occurs because a mineral surface may contain charge imbalances created by the fact that metal cations will not always be fully coordinated with anions such as O^{2-} or S^-. Sites of high charge density (either positive or negative) on a mineral surface represent the locations where adsorption of oppositely charged ions is likely to occur. Zones of high charge density are represented by features such as lattice defect sites, fracture planes, and trace element substitution sites. In general, a mineral surface in contact with an acidic solution (i.e. a high activity of H^+) will contain an abundance of positive charge and is likely to adsorb anionic complexes. Conversely, a mineral surface in contact with an alkaline solution (with a high activity of OH^-) will show a tendency for a surplus negative charge and will adsorb cations.

Experiments have shown that adsorption of different metals onto the surface of a range of silicate, oxide, and sulfide minerals can be an efficient way of forming an ore deposit. A variety of parameters affect adsorptivity and the most important of these are the surface area and surface properties of the adsorbent,

as well as the pH of the solution (Rose and Bianchi-Mosquera 1993). Temperature tends to have an inverse effect on adsorption and for this reason it is typically a low temperature phenomenon. Fine grained clay particles, as well as other materials with a high surface area to volume aspect, such as diatomaceous earth, are very efficient adsorbents and are widely used for filtration and removal of toxic matter from industrial effluent. With respect to ore deposition, however, it is the oxide and sulfide minerals that appear to play an important role in the adsorption of metals from hydrothermal fluids. Since most metals form cationic species in nature, it follows that adsorption will be promoted by an increase in solution pH. This trend is demonstrated in Figure 3.16a, where the experimentally derived adsorption efficiency of Cu onto goethite (FeO(OH)) is plotted for oxidizing

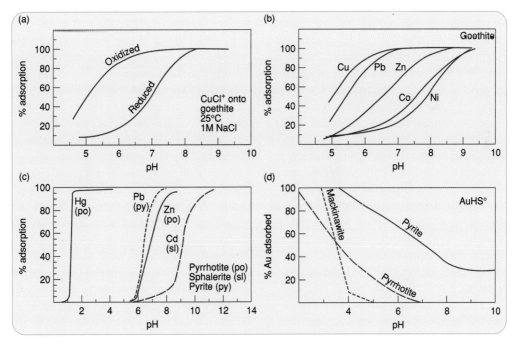

Figure 3.16 Plots of percentage adsorption versus pH for (a) copper as $CuCl^+$ onto goethite under oxidizing and reduced conditions and (b) a variety of transition metals onto goethite. Solutions in both cases are at 25 °C and 1 M NaCl, and data are from Rose and Bianchi-Mosquera (1993). (c) Plots of percentage adsorption versus pH for different metals with respect to various mineral sulfide substrates. Source: After Jean and Bancroft (1986). (d) Percentage adsorption for gold as AuHS onto different sulfide minerals as a function of pH. Solutions are at 25 °C and 0.1 M NaCl. Source: After Widler and Seward (2002).

and reducing conditions. Under oxidizing conditions, such as those applicable to the formation of red-bed hosted stratiform copper deposits (see Box 3.6), Cu will be adsorbed very efficiently from a CuCl$^+$-bearing solution at pH greater than 6 (Rose and Bianchi-Mosquera 1993). Under more reducing conditions Cu is only efficiently adsorbed under more alkaline conditions. Similar behavior to Cu is exhibited by other metals such as Pb, Zn, Co, and Ni (Figure 3.16b), although it is clear that the onset of efficient adsorption takes place at different pH values. Under the specific oxidizing conditions of the experiment, metals adsorb in the order Cu, Pb, Zn, Co, Ni with increasing pH. In a hydrothermal solution that is evolving along an acid neutralization trend caused, for example, by wall-rock alteration (see Section 3.6 below), sequential adsorption could result in metal zonation in an ore deposit.

Metal adsorption efficiency is clearly sensitive to pH, which is, in turn, mediated by the presence of surface OH$^-$-related charge distribution. Experimental data suggest the adsorption efficiency can increase sharply over a narrow range of pH increase. Adsorption efficiency is not, on the other hand, much affected by the type of mineral surface. In addition to goethite, which is a good adsorbent under oxidizing conditions (Figure 3.16a,b), sulfide minerals also attract metals to their surfaces in more reducing environments. Sulfide minerals are typically characterized by an intrinsic negative surface charge usually attributed to the presence of S^{2-} (Shuey 1975). Jean and Bancroft (1986) demonstrated that comprehensive adsorption of metals like Hg, Pb, Zn, and Cd can take place under alkaline conditions onto phases such as pyrite, pyrrhotite, and sphalerite (Figure 3.16c). Mercury is, in fact, readily adsorbed onto sulfide mineral surfaces over almost the entire pH range.

Interestingly, adsorption of gold (as AuHS) from solution by sulfide minerals exhibits a converse pattern of behavior to the base metals. Widler and Seward (2002) showed that the most efficient adsorption of Au onto phases such as pyrite and pyrrhotite occurs under acidic conditions, with a marked decrease in adsorptivity at neutral to alkaline pH ranges (Figure 3.16d). The reason for this trend is that the AuHS complex is itself only stable at low pH, whereas under neutral to alkaline conditions the negatively charged Au(HS)$_2^-$ complex is stable. Since a negatively charged complex will be repulsed by a negatively charged mineral surface, adsorption will decline as pH increases.

The actual process by which metal adsorption occurs is complex. Jean and Bancroft (1985) and Knipe et al. (1992) have suggested a two-stage process that is schematically illustrated in Figure 3.17. It is envisaged that adsorption is initiated by physical adsorption (or physisorption) where the metal–ligand complex is loosely held at the mineral surface by weak van der Waals forces. There is no charge transfer at this initial stage. The second stage of the process is chemical (chemisorption) and is promoted by reduction, or electron gain, and formation of covalent-like bonds between the metal ion and mineral substrate. Adsorbed metal atoms may diffuse across the mineral surface to form clusters at sites of high charge density (Figure 3.17). Further growth of the cluster could occur either by electrochemical precipitation or by conventional precipitation around the earlier formed metal nuclei.

Many sulfide minerals are good metal adsorbents and this fact clearly has an important bearing on the formation of different ore deposit types. Adsorption as a metal-depositing process in ore-forming environments may have been widely underestimated in terms of its importance. Adsorption of toxic metals onto a variety of different substrates is also likely to become increasingly important as an environmental tool in combating metal contamination and for this reason warrants more detailed study in the future.

As a final point, it should be noted that the charge imbalances on mineral surfaces give rise to the fact that many natural materials act as semiconductors. Sulfide mineral surfaces,

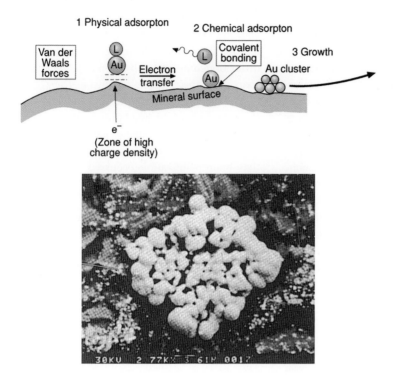

Figure 3.17 Diagram showing the steps involved in adsorbing metal ions onto a mineral surface. This case considers a gold–ligand (Au–L) complex that adheres initially by physical adsorption to a negatively charged sulfide mineral surface by weak Van der Waals forces and subsequently by chemical adsorption and covalent-like bonding to form a stronger attachment. Adsorbed gold ions can diffuse along the surface to form clusters such as that shown in the adjacent SEM image (inset). Further growth at the site of metal accumulation can result in the formation of discrete gold particles that adhere onto sulfide grains such as pyrite or arsenopyrite. Source: After Jean and Bancroft (1985), Knipe et al. (1992). Inset after Jean and Bancroft (1985).

for example, exist as both n-type (negative or electron saturated) and p-type (positive or electron depleted) semiconductors (Shuey 1975) and this influences metal precipitation onto these surfaces in much the same way as described above.

3.5.3 Biologically Mediated Processes of Metal Precipitation

It is a well-known fact that the evolution of all higher life forms since the end of the Precambrian is characterized by the ability of cells to interact chemically with inorganic metals such as silicon, calcium, and phosphorus, to form shells, teeth, and bones. The interaction of micro-organisms with a variety of other

metals is also important for the formation of authigenic mineral phases during diagenetic and hydrothermal processes. The concept of biomineralization is increasingly recognized as a significant process in ore formation, especially with respect to deposit types such as banded iron-formations, VMS-SEDEX ores, phosphorites, and supergene enrichments of Cu and Au. The following section briefly considers the basic principles of biomineralization and its significance to ore-forming processes.

The fundamental subdivision of life into prokaryotes and eukaryotes came about with an improvement in the understanding of cell microbiology in the mid-twentieth century (Patterson 1999). Prokaryotes are the more primitive microorganism and the ancestral life form on Earth, developing at some stage between 4650 Ma and about 3500 Ma when they first appear in the fossil record. They are single celled microorganisms that can be subdivided into two kingdoms, namely Bacteria and Archaea. Archaea can exist in extreme environments, such as volcanic hot-springs and ocean floor hydrothermal vents (thermophiles) or hypersaline evaporitic waters (halophiles), and utilize unusual metabolic pathways, involving the energy derived from hydrogen- and sulfur-based chemical reactions, to nourish themselves. Bacteria include a huge variety of microorganisms that occupy virtually every ecological niche on Earth. Bacteria perform many different functions: some are fermentative (*Escherichia coli*) whereas others either contain the enzymes for oxygen metabolism (i.e. they are aerobic) or lack such enzymes (anaerobic).

Eukaryotes evolved from prokaryotes and contain much more complex cellular structures, including a discrete nucleus, chromosomes, and characteristic cycles of cell division (mitosis and meiosis) and sexual reproduction. They probably appeared during Paleoproterozoic times although they could also have formed somewhat earlier (Patterson 1999). They are almost all aerobic and metabolize oxygen in discrete packages within the cell called mitochondria, which are semi-autonomous chemical entities containing DNA that is different from that of its own cell nucleus and more akin to prokaryotic DNA. This has led to the suggestion that more advanced cellular structures evolved by symbiotic inclusion of aerobic bacteria (i.e. mitochondria) into primitive eukaryotic entities. All higher life forms (the Animalia, Plantae, Fungi, and Protista kingdoms) consist essentially of eukaryotic cells.

3.5.3.1 Biomineralization

A capacity to chemically interact with metals is an inherent characteristic of most microorganisms. Mediation by micro-organisms in the formation of a variety of authigenic mineral phases is referred to as "biomineralization" and the process has important implications for ore formation. The major animal phyla incorporated biomineralization as a "lifestyle" soon after their formation during the Cambrian explosion. It is this process that allowed the more advanced life forms to develop an exoskeleton and teeth and, eventually, internal bones (the vertebrates). It is, however, apparent that biomineralization occurred long before the Cambrian Period and it has been suggested that the formation of magnetite in bacteria (the so-called magnetotactic bacteria) might represent an ancestral template for unraveling the genetics of this process in higher life forms (Kirschvink and Hagadorn 2000). It is the bacterial kingdom that is implicated in many of the processes linked to biologically mediated precipitation of metals during ore formation.

Biomineralization occurs in two ways (Lowenstam 1981; Konhauser 1998, 2003). The first is termed "biologically induced biomineralization" and refers to situations where minerals form as a consequence of the activity of microorganisms on their immediate environment. The second process is "biologically controlled biomineralization," in which the microorganism directly controls the uptake of inorganic material, either within the cell (where elements diffuse through the cell wall) or at the interface between the cell and its surrounds. These two processes are briefly described below with respect to iron, which is particularly relevant to many ore forming environments.

Biologically induced biomineralization is controlled by the same equilibrium principles that govern metal precipitation in inorganic chemical systems. The metabolism of microorganisms effects the properties (i.e. pH, Eh) of aqueous solutions in their immediate environment and can promote the precipitation of authigenic minerals as a function of their solubility products. More importantly, ionized cell surfaces can induce mineral nucleation in much the same way as discussed in the previous section on adsorption. A wide range

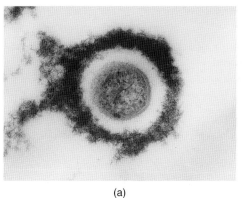

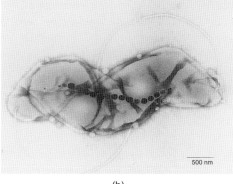

500 nm

(a)   (b)

Figure 3.18 Microphotographs of (a) bacterial cell with amorphous ferric hydroxide replacing the cell walls, and (b) magnetotactic spirillum bacteria showing a chain of magnetite crystals developed along the longitudinal axis of the microorganism. Source: Photographs courtesy of Kurt Konhauser and Dennis Bazylinski.

of different minerals can be biologically induced because the phases that form reflect the composition of the aqueous solution in which the microbe exists. Ferric hydroxide (or ferrihydrite $Fe(OH)_3$) is one of the most common biominerals, forming in a variety of different environments, and is often upgraded during diagenesis to more stable phases such as goethite and hematite. Precipitation of iron occurs in progressive stages, from Fe-staining of the cell walls (Figure 3.18a), to nucleation of small (<100 nm) ferric hydroxide grains and eventual replacement of bacterial colonies by the inorganic authigenic phase (Konhauser 2003). There are a number of bacteria that passively induce ferric iron precipitation onto their cell walls to form ochreous accumulations, a common one being *Leptothrix ochracea*, as well as the *Crenothrix, Clonothrix*, and *Metallogenium* genera. Marine hydrothermal vent sites are characterized by abundant iron-depositing microbial populations. It has also been suggested that Precambrian oceans were characterized by bacteria capable of binding both ferric iron and silica and that this biologically mediated process contributed to the formation of banded iron-formations (see Chapter 5). In the present-day surface environment, the bacterium *Thiobacillus ferrooxidans* is capable of oxidizing ferrous iron, creating

low pH solutions that have environmental implications during the formation of acid mine drainage from sulfide mineral-bearing ore deposits and mines. The fundamental problem associated with acid mine drainage is the production of sulfate and toxic labile metals in very acid solutions, as expressed in terms of reaction (3.5):

$$FeS_2 + 8H_2O \leftrightarrow Fe^{2+} + 2SO_4^{2-} + 16H^+$$

$$(3.5)$$

This problem can, however, also be remediated by bacterial activity, as discussed in the section on sulfate-reducing bacteria (SRB) below.

Biologically controlled biomineralization is characterized by the creation of an area internal to the cell into which ions specific to the requirements of the microorganism are diffused, thereby creating localized conditions where concentrations are increased until saturation is reached. Minerals are thus able to form even though the aqueous milieu surrounding the microbe would not be able to precipitate the same phase. This process is responsible for the formation, for example, of linear arrays of magnetite crystals within magnetotactic bacteria (Figure 3.18b), which enables them to navigate to preferred redox environments. The initial uptake of metal in this case is as ferric iron, with subsequent reduction to the ferrous state within the cell, but prior to transport into the magnetite bearing membrane. Fossil magnetotactic bacteria have been recognized from the 2000 Ma Gunflint Iron Formation, and this represents the oldest evidence for controlled biomineralization.

In terms of ore formation, the role that many bacteria play is to catalyze the oxidation of soluble metals to insoluble oxide phases. One particular group of anaerobic bacteria, the so-called "sulfate-reducing bacteria" (SRB), contribute to ore formation in a different way. SRB, such as *Desulphovibrio desulphuricans*, oxidize organic molecules by using sulfate as an electron acceptor, a process that generates dissolved sulfide (HS^- or H_2S). On diagenesis the sulfide reacts with other metals or minerals to form sulfide phases that are either directly implicated in mineralization, or form the precursors to later ore-forming processes. These low temperature processes can result in the formation of many different sulfide minerals, including pyrite (via mackinawite and greigite) as well as nickel, lead, and zinc sulfides (Konhauser 2003). In a modern setting SRB can be useful in the microbial treatment of pollution and acid mine drainage. With reference to reaction (3.5) above, SRB would catalyze the reduction of the sulfate to form sulfide, which in turn reacts with and stabilizes labile toxic metals. A bicarbonate ion (HCO^{3-}) is also commonly produced during these reactions, which has the effect of neutralizing the previously acid waters. This type of biological remediation of aqueous pollutants forms the basis for many operations that have to treat both acidic and metal toxic waters.

Biomineralization is a process that has considerable applicability to many other ore-forming environments in addition to those involving iron. Manganese is another redox-sensitive element that is biogenically precipitated in much the same way as iron. Manganous oxides such as birnessite and vernadite are the main phases formed in the

presence of bacteria. Bacterial concentrations of Mn and Fe on the ocean floor have been implicated in the formation of manganese nodules (see Chapter 5). The formation of phosphate-rich sediments has also been linked, at least in part, to biogenic mediation, with apatite overgrowths having been observed replacing the remains of cyanobacteria. Large-scale phosphate deposits (phosphorites; see Chapter 5) forming along the eastern seaboard of Australia have been attributed to the progressive assimilation of sea water phosphorus by benthic bacteria (Konhauser 2003). The amorphous silica that forms the characteristic siliceous sinters above epithermal gold deposits is also attributed to the presence of a diverse range of bacterial genera, the filaments of which act as nucleation sites for silicification. The siderophile metals (Au and Pt) are also amenable to biogenically mediated concentration, particularly with respect to the formation of nuggets in placer and lateritic environments (Watterson 1991; see Chapter 4). One of the more impressive demonstrations of the role of bacteria in the concentration of metals is derived from sulfur and lead isotopic determinations of sphalerite and galena ores from the world-class Navan deposit in Ireland. Although also characterized by replacement and epigenetic textural characteristics, the observation of pyritized worm tubes from a related deposit at Silvermines (Boyce et al. 1999) has shown that the Irish-type Zn–Pb–Ba ores are SEDEX types and related, at least in part, to syngenetic hydrothermal venting on the sea floor (see Section 3.8.2). Isotopic studies indicate that some 90% of the sulfides at Navan were derived from the bacterial reduction of Carboniferous sea water sulfate (Fallick et al. 2001). It is argued that without an ample supply of reduced sulfur the Navan deposit might not have developed to its current size, emphasizing the importance of biologically mediated ore-forming processes.

Bacterial biomineralization is an important process that has implications for the origin and evolution of life itself, and also for the formation of a wide variety of authigenic minerals in diagenetic and hydrothermal environments. In terms of ore-forming processes its role has probably been underemphasized and this is likely to be a fruitful area of research in the future.

3.6 Fluid–Rock Interaction – Introduction to Hydrothermal Alteration

The passage of hot fluids through the Earth's crust is invariably accompanied by hydrothermal alteration. Alteration is marked by the development of a new mineral assemblage that reflects the original rock composition, as well as the properties and amount of fluid that has traversed the system. Zones of alteration mark the pathways of hydrothermal fluids through the crust and may represent useful guides for the exploration of many ore deposit types. The nature of hydrothermal alteration also provides an indication of the fluid properties associated with ore formation and there is commonly a close relationship between the chemical reactions involved in alteration and those responsible for metal deposition. It is, therefore, appropriate that a summary of the characteristics of hydrothermal alteration accompanies any description of ore-forming processes. The subject is, however, a complex one and the following section is necessarily brief. More detailed accounts documenting the evolution of ideas regarding hydrothermal alteration are provided in the relevant chapters of the three editions of H.L. Barnes's *Geochemistry of Hydrothermal Ore Deposits* (namely Meyer and Hemley 1967; Rose and Burt 1979; Reed 1997).

Traditionally, alteration has been regarded as a process that, in its simplest form, involves both the hydrothermal fluid itself (essentially water with its dissociated components H^+ and OH^-) and the dissolved constituents of the aqueous solution. The low to moderate degrees of pervasive, isochemical alteration that characterize almost any rock evolving in the presence of a fluid can largely be explained by *hydrolysis* or H^+ *ion metasomatism* (see also Chapter 4, Section 4.2.2). This process can be depicted in terms of simple chemical equilibria, or reactions, although it must be remembered that these are likely to be simplifications of actual alteration processes. For example, one of the most common forms of alteration recognized in nature is the reaction of K-feldspar with water to form muscovite or sericite. This reaction can be expressed as follows:

$$\underset{\text{(k-fel)}}{{}^3\!/_2 KAlSi_3O_8} + H^+ \Leftrightarrow \underset{\text{(musc)}}{{}^1\!/_2 KAl_3Si_3O_{10}(OH)_2}$$
$$+ \underset{\text{(qtz)}}{3SiO_2} + K^+ \quad (3.6)$$

The hydrolysis of K-feldspar to muscovite in terms of this reaction requires nothing more than the presence of H^+ ions in an aqueous solution. The reaction is essentially isochemical in the sense that no new ingredients need to be added to the system. If the reaction proceeds to the right, H^+ ions are consumed and the fluid will become less acidic, continuing to do so until the K-feldspar is used up. The reaction will also produce quartz as part of the alteration assemblage, as well as K^+ ions dissolved in the aqueous solution.

If the products of reaction (3.6) are permitted to further react with H^+ ions in solution and the system undergoes an increase in its fluid/rock ratio, then muscovite will react to form kaolinite, as follows:

$$\underset{\text{(musc)}}{KAl_3Si_3O_{10}(OH)_2} + H^+ + {}^3\!/_2 H_2O \Leftrightarrow$$
$$\underset{\text{(kaol)}}{{}^3\!/_2 Al_2Si_2O_5(OH)_4} + K^+ \quad (3.7)$$

This reaction illustrates the way in which increasing fluid/rock ratios will change the alteration mineralogy by continued reaction with the rock after one set of mineral buffers has broken down (i.e. by consumption of K-feldspar in the case of reaction (3.6)). Alteration does not, however, necessarily result in the formation of only one mineral reactant, but generally forms an assemblage of phases. Plagioclase, for example, occurring in the same rock that originally contained the K-feldspar, might also react with H^+ ions in solution to form pyrophyllite, as follows:

$$2NaAlSi_3O_8 + 2H^+ \Leftrightarrow$$
$$\underset{(plag)}{} $$
$$\underset{(pyroph)}{Al_2Si_4O_{10}(OH)_2} + \underset{(qtz)}{2SiO_2} + Na^+ \quad (3.8)$$

Thus, reaction of a simple mineral assemblage with nothing but water could result in a multicomponent alteration assemblage, even if the system remains largely isochemical.

If consideration is now given to an open chemical system where the dissolved constituents of an aqueous solution are also involved in alteration, then the processes and their products will be still more varied. The reactions above ((3.6), (3.7), and (3.8)), describing hydrogen ion metasomatism, pertain to situations where reactants and products have similar bulk compositions. All three reactions also reflect acid–base exchange because an aqueous H^+ ion is consumed and replaced by cations from the reactants. The accumulation of aqueous base cations in the evolving fluid will have an effect on the nature of downstream alteration, as they may themselves react with wall-rock. This type of alteration process is known as *cation metasomatism*. A wide variety of alteration products can form by cation metasomatism, examples of which are:

$$\underset{(plag)}{2NaAlSi_3O_8} + 4(Mg, Fe)^{2+} + 2(Fe, Al)^{3+}$$
$$+ 10H_2O \Rightarrow \underset{(chl)}{(Mg, Fe)_4(Fe, Al)_2Si_2O_{10}(OH)_8}$$
$$+ \underset{(qtz)}{4SiO_2} + 2Na^+ + 12H^+ \quad (3.9)$$

and

$$\underset{(dol)}{3CaMg(CO_3)_2} + 4SiO_2 + 6H^+ + 4H_2O \Rightarrow$$
$$\underset{(talc)}{Mg_3Si_4O_{10}(OH)_2} + 6H_2CO_3 + 3Ca^{2+}$$
$$(3.10)$$

In reaction (3.9) the presence of Mg, Fe, and Al cations in a fluid which equilibrates with plagioclase will result in the chloritization of that mineral, with accompanying precipitation of quartz and significant H^+ production. This type of reaction will result in the fluid becoming more acidic during progressive alteration, which is contrary to the case for hydrolysis reactions. Reaction (3.10) demonstrates the process that takes place when a cation-bearing aqueous solution interacts with a carbonate mineral, such as dolomite, to form a silicate mineral, such as talc. This process, known as silication, is relevant to the formation of skarn ore deposits where fertile granites intrude carbonate country rocks (see Section 3.6.1 below).

More recent considerations of alteration processes have suggested that they should be regarded not only in terms of chemical equilibria, but also as dynamic systems that evolve with time. Given the time scale of most geological events, the attainment of equilibrium between rock and fluid is generally feasible, at least on a local scale. In dynamic systems, however, it is likely that kinetic considerations predominate over those of steady state equilibrium. The mineral assemblages that form as a result of hydrothermal alteration vary as a function of five factors (Reed 1997), namely:

- temperature;
- pressure;
- host rock composition;
- fluid composition; and
- the ratio of fluid to rock during the alteration process (termed the *fluid/rock ratio*).

In nature these five factors interact to create alteration assemblages that reflect the variations over time of all of them. The stabilities of mineral assemblages are dictated largely by temperature and pressure, whereas host rock and fluid compositions control the chemical make-up of a particular alteration assemblage. Fluid composition and temperature have a strong influence on the nature and extent to which the host rock is dissolved by the aqueous solution, as well as the nature and concentration of new constituents introduced into the system by the fluid. The fluid/rock ratio is considered to be the "master variable" (Reed 1997) in alteration processes because it dictates the extent to which reactions will move to completion.

Reed (1982, 1997) has shown that, because hydrothermal fluids migrate along pathways that represent evolving chemical gradients, alteration should be viewed in a dynamic sense as a function of changing fluid/rock ratio. This concept is schematically illustrated in Figure 3.19a, showing a fracture along which a hydrothermal fluid is flowing and adjacent to which alteration is taking place by diffusion-related reactions between fluid and country rock. At any instant in time, alteration pinches out downstream along the fracture because of the progressively diminishing ability of the fluid to react with the country rock as its enthalpy and its dissolved ingredients are effectively "used up" (e.g. an originally acidic fluid may become progressively neutralized as its H^+ ions are consumed by acid–base exchange; see below). The sequence of alteration minerals outwards or downwards of the fracture or vein reflects the sequence of reactions that relate to decreasing fluid/rock ratios (Figure 3.19a). The upstream portion of the vein has effectively "seen" more fluid (and is, therefore, characterized by a higher fluid/rock ratio) than the distal sections of the system, and this is reflected in a variable alteration assemblage.

The evolution of alteration assemblages in a dynamic hydrothermal system can be modeled by reacting a fluid of known initial composition, pH, and temperature with a rock, using constraints imposed by established thermodynamic, chemical, and equilibrium data (Reed 1997). The fluid composition and pH are then changed as the constituents of chemical reactions are consumed (i.e. as a mineral buffer system breaks down). Figure 3.19b shows how the alteration assemblage changes in a rock of andesitic composition subject to

reaction with a hydrothermal solution. The scenario is modeled in terms of a system with high initial fluid/rock ratios and an acidic (low pH) fluid, evolving to lower fluid/rock ratios and alkaline (high pH) conditions. The plot shows that an acidic fluid circulating with high fluid/rock ratios through the andesite will be characterized by an alteration assemblage comprising quartz + pyrophyllite + alunite, also referred to as an *advanced argillic* alteration assemblage (Figure 3.19b). As the alteration systematics evolve, the same rock, now infiltrated by a more alkaline fluid circulating at a lower fluid/rock ratio, would be altered, initially to a sericite + chlorite assemblage and then to a chlorite + epidote + muscovite + albite assemblage. The latter assemblage is termed *propylitic* alteration and also matches the typical greenschist facies metamorphic assemblage. The expected distribution of ore minerals in the evolving system is shown in Figure 3.19c.

Viewing alteration assemblages in terms of dynamic fluid/rock interaction offers many advantages to the understanding of alteration processes, as well as for explaining the relationship between fluid–rock interactions and ore deposition. The incorporation of time into the models of Figure 3.19 provides a good explanation for zonation and paragenetic sequence, two hallmark characteristics of hydrothermal ore deposits, discussed in more detail below. An evolving alteration system that moves from high to low fluid/rock ratios (i.e. from left to right in Figure 3.19) reflects a change from cation to hydrogen ion metasomatism, and from fluid-dominated to rock-dominated environments. The evolving system also dictates the compositional changes that occur in the fluid itself, such as the stepwise increase in fluid pH as H^+ ions are consumed by wall-rock reaction and stabilization of successive mineral assemblages (Figure 3.19d). The redox state of the fluid also changes as the alteration system evolves. Figure 3.19d shows that with decreasing fluid/rock ratio the concentration

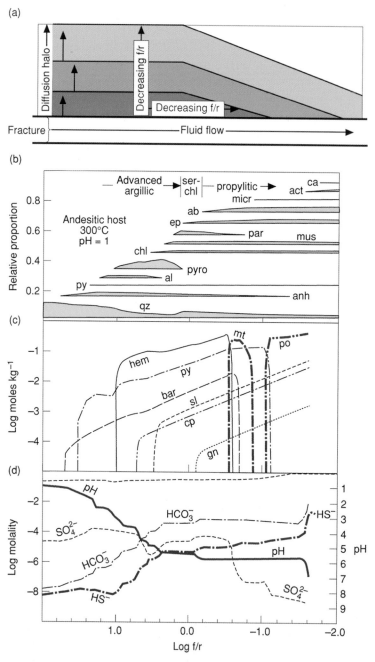

Figure 3.19 Models illustrating the dynamic nature of fluid/rock interaction and hydrothermal alteration as a function of changing fluid/rock ratio. (a) Conceptual model showing one side of a diffusion-related alteration halo around a fracture through which fluid is passing from left to right. As fluid reacts with its host rock it changes composition and, through time, alteration assemblages evolve both along and outward of the fracture system (i.e. with decreasing fluid/rock ratios). (b) The occurrence and relative proportions of the principal alteration minerals forming when a relatively oxidized aqueous solution (of magmatic derivation at 300 °C and with initial pH = 1) reacts with a rock of andesitic composition, as a function of changing fluid/rock ratio (ser, sericite; chl, chlorite; ca, calcite; act, actinolite; micr, microcline; ab, albite; ep, epidote; par, paragonite; mus, muscovite; pyro, pyrophyllite; al, alunite; py, pyrite; anh, anhydrite; qz, quartz). (c) The precipitation sequence of some of the major ore minerals as a function of changing fluid/rock ratio for the same setting as in (b) (po, pyrrhotite; mt, magnetite; hem, hematite; bar, barite; sl, sphalerite; cp, chalcopyrite; gn, galena). (d) The evolution of fluid pH, as well as concentrations of certain key ligands in the fluid, as a function of changing fluid/rock ratio (also for the same setting as b). Source: After Reed (1997).

of SO_4^{2-} in the fluid decreases as the HS^- content increases. This is because SO_4^{2-} is reduced by reaction with iron-rich wall-rocks, resulting in the precipitation of hematite in the alteration zone and fractionation of sulfide into the evolving aqueous solution (Reed 1997). Sulfate reduction is an important source of sulfide in many hydrothermal ore-forming systems and promotes base metal sulfide precipitation. It is the changes in variables such as pH and redox state of the fluid that are so important in promoting the precipitation of metals from ore fluids.

There are many examples of situations where reaction between a hydrothermal fluid and its wall rocks is the main cause of ore precipitation. This process can be recognized, for example, in samples that reveal textures illustrating the replacement of primary minerals in the host rock by ore minerals. Such a texture is distinct from situations where ore minerals are precipitated into open spaces such as fractures or vugs, where other mechanisms of ore deposition, such as fluid mixing or boiling, might have prevailed. Skarn deposits (see Section 2.10, Chapter 2) are a particularly good example of where alteration of carbonate rocks by acidic, metal-bearing solutions yields Ca–Mg silicate minerals accompanied by significant polymetallic mineralization. Carbonate rocks have the ability to neutralize acidic solutions and the pH increases that arise from fluid/rock reactions in this environment promote efficient precipitation of metal sulfides (Reed 1997). Likewise, greenstone belt hosted mesothermal lode-Au deposits are typically accompanied by extensive zones of quartz–albite–carbonate–muscovite–pyrite alteration in and around the shear zone or fault along which fluid flow was focused (see Section 3.9.1 below and Box 3.2). A common Au precipitation mechanism in these environments is the reaction of a $H_2O–CO_2$, sulfide-rich fluid with ferrous host rocks (commonly basalt or banded iron-formation) to form pyrite (Phillips 1986). This causes destabilization of the dominant $AuHS_2^-$ complex and precipitation of native Au together with the sulfide minerals.

The intimate association between alteration and hydrothermal ore-forming processes is well established. The recognition of alteration styles and mineral parageneses is now a standard tool in the array of techniques utilized by exploration geologists (Thompson and Thompson 1996) and has important bearing on the nature of the ore deposit itself.

3.6.1 Types of Alteration and Their Ore Associations

This section outlines some of the characteristics and ore associations of the main alteration styles associated with a variety of hydrothermal environments.

3.6.1.1 Potassic Alteration

Potassic (or K-silicate) alteration is characterized by the formation of new K-feldspar and/or biotite, usually together with minor sericite, chlorite, and quartz. Accessory amounts of magnetite/hematite and anhydrite may occur associated with the potassic alteration assemblage. It typically represents the highest temperature form of alteration (500–600 °C) associated with porphyry Cu-type deposits, forming in the core of the system and usually within the granite intrusion itself. A degree of K^+ (cation) metasomatism in addition to hydrolysis is considered to have taken place during the development of potassic alteration assemblages because studies of bulk compositions on a volume basis suggest that potassium has been added to the system. A variation of potassic alteration involving substantial addition of Na and Ca (called sodic and calcic alteration and characterized by abundant albite, epidote, and actinolite) is documented in some porphyry systems such as Yerington, Nevada, USA.

Box 3.2 Alteration and Metal Precipitation: Golden Mile, Kalgoorlie, Western Australia – An Archean Orogenic Gold Deposit

---

The "Golden Mile" is a zone of exceptional gold mineralization located immediately to the east of the twin cities of Kalgoorlie–Boulder in Western Australia. Gold was discovered here in 1893 and well over 1200 tons of gold has been won from the area since this time. Until 1975 mining was carried out in several underground operations, a situation which led to a steady decline in production due to increasing depths and decreasing grades. This situation was reversed in the late 1980s when ownership was amalgamated and plans initiated to develop a huge, low-grade open-pit operation known as the Fimiston or Super Pit. This is now a substantial gold mining operation in Australia, producing well over 800 000 oz of gold per year.

The Golden Mile is located in the Norseman–Wiluna greenstone belt of the Yilgarn Craton. Auriferous mineralization is structurally controlled and hosted mainly in steeply dipping shear zones or lodes that are centered on the Golden Mile Fault (Figure 3.2.1a). Innumerable subsidiary lodes are concentrated in and around this system and are hosted by a 2675 Myr old ferruginous tholeiitic intrusion (sill) called the Golden Mile dolerite. Although the initial high-grade, largely underground, workings were concentrated on the individual shear zones themselves, the system is accompanied by an extensive alteration halo of low grade mineralization which now forms the bulk of the ore in the Super Pit. It is the alteration of doleritic host rock by mineralizing, aqueo-carbonic hydrothermal solutions that is believed to be the main cause of the pervasive gold mineralization in this deposit (Phillips 1986) and this is discussed in more detail below.

A schematic indication of the regional geology of the Golden Mile, and the approximate position of the Super Pit, is shown in Figure 3.2.1a. The nature of the alteration halo around quartz–pyrite–gold mineralized shear zones (such as the East Lode at the Lake View mine, which has now been subsumed by the Super Pit) is shown in Figure 3.2.1b. The Golden Mile dolerite was regionally metamorphosed to a dominantly actinolitic assemblage prior to and during folding and ductile deformation of the Norseman–Wiluna greenstone belt. Metamorphism was superseded by introduction of a reduced, low-salinity aqueo-carbonic fluid that percolated along numerous shear zones, cutting preferentially through the relatively competent Golden Mile dolerite and progressively altering the host rock in and around them. Alteration is markedly zoned around individual shear zones and is characterized by a paragenetic sequence that commences with early (metamorphic) actinolite and then progresses successively to chlorite-, carbonate-(siderite), and eventually pyrite-dominated assemblages (Phillips 1986). Chlorite (with lesser carbonate, albite, and quartz) alteration forms a pervasive halo extending for several hundred meters around zones of shearing, and overprints the regional metamorphic assemblage. Carbonate assemblages reflect a more intense manifestation of alteration (i.e. higher fluid/rock ratio) and occur closer to individual shear zones (i.e. in halos extending up to 5 m away from the shear) or in more ferruginous parts of the differentiated dolerite sill. The shear zone itself is demarcated by a pyrite–quartz–muscovite–carbonate assemblage with which high-grade gold mineralization occurs. Phillips (1986) has argued that the alteration paragenesis is a product of a single metamorphic fluid that introduced CO_2, K, S, and Au into the system, and that alteration and mineralization were synchronous. Precipitation of gold is considered to have been facilitated by the interaction of a reduced auriferous fluid with Fe-rich host rocks, such as in the following reactions:

$$FeCO_3 + 2H_2S + 0.5O_2 \Rightarrow FeS_2 + CO_2 + 2H_2O$$

Box 3.2 (Continued)

(i.e. siderite replaced by pyrite) and

$$FeCO_3 + Au(HS)_2^- \Rightarrow FeS_2 + CO_2 + H_2O + Au$$

(gold bisulfide complex destabilized by pyrite formation)

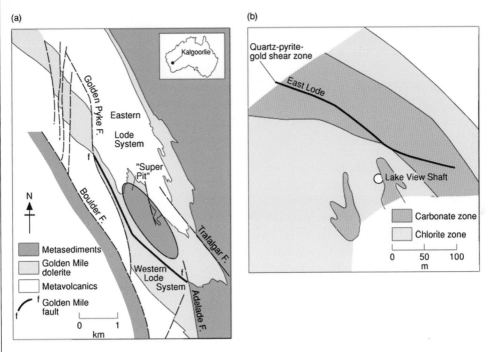

Figure 3.2.1 (a) General geological outline of the Golden Mile, east of Kalgoorlie, Western Australia, and the approximate position of the "Super Pit." (b) The nature of hydrothermal alteration in the Golden Mile dolerite. The East Lode was originally mined as a high grade underground deposit through the Lake View Shaft. The entire alteration zone now forms the bulk of the ore for the high tonnage–low grade operation in the Super Pit. Source: Both diagrams are after Phillips (1986).

This model emphasizes the role that host rock alteration plays on metal precipitation, as described in Section 3.5.1 of this chapter. Although the process is undoubtedly important in the formation of many hydrothermal ore deposits, it seldom takes place in isolation. At Golden Mile, for example, it is now recognized that the late-stage, high grade gold–tellurium–vanadium ores that characterize very rich vein systems, such as the Oroya shoot, were the product of the mixing of a (possibly magmatic) oxidized, SO_2-bearing fluid with a more reduced metamorphic fluid that was responsible for the pervasive alteration described above (Walshe et al. 2006). Fluid mixing processes also appear to be broadly coincidental with the transition from a ductile to a brittle deformational regime, a changeover that is commonly associated with significant mineralization in orogenic gold systems.

3.6.1.2 Phyllic (or Sericitic) Alteration

This alteration style is very common in a variety of hydrothermal ore deposits and typically forms over a wide temperature range by hydrolysis of feldspars to form sericite (fine-grained white mica), with minor associated quartz, chlorite, and pyrite (see reaction (3.6) and Figure 3.19). Phyllic alteration is associated with porphyry Cu deposits, but also with mesothermal precious metal ores and VMS deposits in felsic rocks.

3.6.1.3 Propylitic Alteration

Propylitic alteration is probably the most widespread form of alteration in that it is essentially indistinguishable from the assemblages which form during regional greenschist metamorphism. It comprises mainly chlorite and epidote, together with lesser quantities of clinozoisite, calcite, zoisite, and albite. It is a mild form of alteration representing low to intermediate temperatures (200–350 °C) and low fluid/rock ratios (see Figure 3.19). This style of alteration tends to be isochemical and forms in response to H^+ metasomatism. It characterizes the margins of porphyry Cu deposits as well as epithermal precious metal ores.

3.6.1.4 Argillic Alteration

This alteration style is commonly subdivided into intermediate and advanced categories depending on the intensity of host mineral breakdown. Intermediate argillic alteration affects mainly plagioclase feldspars and is characterized by the formation of clay minerals such as kaolinite and the smectite group (mainly montmorillonite). It typically forms below about 250 °C by H^+ metasomatism and occurs on the fringes of porphyry systems. Advanced argillic alteration represents an extreme form of base leaching where rocks have been stripped of alkali elements by very acidic fluids active in high fluid/rock ratio environments (see Figure 3.19). It is characterized by kaolinite, pyrophyllite, or dickite (depending on the temperature) and

alunite together with lesser quartz, topaz, and tourmaline. It is commonly associated with near surface, epithermal precious metal deposits where alteration is associated with boiling fluids and condensation of volatile-rich vapors to form extremely acidic solutions.

3.6.1.5 Silication

Silication is the conversion of a carbonate mineral or rock into a silicate mineral or rock (see reaction (3.10) above) and necessarily involves introduction of additional components into the system (cation metasomatism). It is the main process which accompanies the prograde stage in the formation of polymetallic skarn deposits, the latter occurring when a fertile, acidic, magmatic fluid (usually from a granite) infiltrates a carbonate host rock. Carbonate rocks are a particularly efficient host for metal deposition from hydrothermal solutions because of their ability to neutralize acidic fluids and their "reactivity," which enhances permeability and fluid flow. Skarn reactions and the formation of "calc–silicate" mineral assemblages are complex and develop over an extended temperature range. Fluid/rock ratios also develop to very high values in skarn systems and these factors result in the precipitation of a diverse, polymetallic suite of ores (see Section 2.10).

3.6.1.6 Silicification

Silicification should not be confused with silication and refers specifically to the formation of new quartz or amorphous silica minerals in a rock during alteration. Minor silicification develops in the alteration halos associated with many different ore deposit types and is usually a by-product of isochemical hydrolysis reactions where Si is locally derived (see reactions (3.6) and (3.8) above). The majority of fractures through which hydrothermal fluids have passed are at least partially filled with quartz to form veins. The Si in these settings is usually derived by leaching of the country rocks through which the fluids are circulating. Intense silicification, however, forms as a result of cation metasomatism, where Si^{4+} in solution is added to the system. This type of alteration is characteristic of the sinter zones in high level epithermal precious metal ore deposits.

3.6.1.7 Carbonatization

Carbonatization refers to the formation of carbonate minerals (calcite, dolomite, magnesite, siderite, etc.) during alteration of a rock and is promoted by fluids characterized by high partial pressures of carbon dioxide (P_{CO_2}) and neutral to alkaline pH. Archean greenstone belt related orogenic gold deposits represent an ore deposit type where carbonate alteration is virtually ubiquitous and is accompanied by an assemblage comprising quartz, muscovite, biotite, albite, and chlorite. It forms when reaction occurs between a low-salinity, CO_2-rich fluid and its host rock. The carbonate mineral that forms is a function of the composition of the host rock and could be dolomite in association with amphibolite, siderite in a banded iron-formation, or calcite in a granitic host.

3.6.1.8 Greisenization

The formation of a greisen is specific to the cupola zones of highly differentiated (S-type) granites that contain Sn and W mineralization, as well as significant concentrations of other incompatible elements such as F, Li, and B. Greisens represent an alteration assemblage comprising mainly quartz, muscovite, and topaz, with lesser tourmaline and fluorite, usually forming adjacent to quartz–cassiterite–wolframite veins.

3.6.1.9 Hematitization

Alteration that is associated with oxidizing fluids often results in the formation of minerals with a high Fe^{3+}/Fe^{2+} ratio and, in particular, hematite with associated K-feldspar, sericite, chlorite, and epidote. In the magmatic-hydrothermal environment, occurrences such as the Olympic Dam IOCG-type Cu–Au–Fe–U deposit in South Australia (see Box 3.1) are characterized by this style of alteration. The alteration may be related to redox processes where saline, oxidizing fluids come into contact with a more reduced host rock environment or mix with more reduced fluids.

3.7 Metal Zoning and Paragenetic Sequence

A characteristic feature of many hydrothermal ore deposits is the occurrence of a regular pattern of distribution, or *zoning*, of metals and minerals in space. Zoning can be observed at many different scales, ranging from regional patterns of metal distribution (at a scale of hundreds of kilometers), through a district scale, to individual ore-body related variations and even down to the level of a single vein or hand specimen (Smirnov 1977; Guilbert and Park 1986; Pirajno 1992; Misra 2000). A good example of regional metal zonation is the Andes metallogenic province, where the distribution of ore deposits is related to continental scale subduction along the western margin of South America. At a district scale, however, the distribution of metals is more tightly constrained and better defined, and commonly characterized by a consistent pattern that is reproduced in other ore-forming environments. District scale zoning is well exemplified by the Sn–W–Cu–Pb–Zn–Ag–Sb–U sequence of ores associated with the Cornubian batholith in southwest England (see Box 2.2, Chapter 2), a pattern that is reproduced in other granite-related ore-forming systems such as the very fertile Mole granite in New South Wales, Australia (Audétat et al. 2000). Zonation at this and smaller scales is clearly related to the systematic evolution of a hydrothermal fluid (caused by cooling and alteration) and the sequential precipitation of its contained metal budget by processes such as those discussed in Section 3.6 above. Although the processes controlling metal precipitation are complex and varied, the relatively consistent pattern of metal zonation in many different ore-forming environments means that the process should be amenable to rational and unified explanation.

Although there is still some ambiguity in its use, the term *paragenesis* is now widely applied to the association of minerals and metals that characterizes a particular ore type and, therefore, has a common origin (Guilbert and Park 1986; Kutina et al. 1967). Because both space and time need to be considered in the understanding of zonation, the phrase *paragenetic sequence* is now used to describe the distribution in time of a set of genetically related minerals or metals. The recognition and significance of parageneses and paragenetic sequences in hydrothermal ore deposits comes from early observations of ore-forming environments by some of the founders of modern economic geology, notably Waldemar Lindgren (1933). On the basis of these empirical observations Emmons (1936) devised the concept of a "reconstructed vein." This was an idealized composite vein that extended from deep in the crust up to the surface and contained a paragenetic sequence of ore components that is typical of the zoning patterns observed in many different hydrothermal ore deposit types. The "Emmons reconstructed vein," revised by Guilbert and Park (1986), is summarized in Table 3.3.

Observations suggest that metals such as Mo, W, and Sn tend to precipitate early (or at deep levels) from a high temperature hydrothermal solution. Such solutions are sometimes referred to as *hypothermal*. They are followed, in the ideal precipitation sequence, by Cu, and then Zn, Pb, Mn, and Ag, as the fluid infiltrates upwards in the crust and cools to form *mesothermal* solutions. The precious and volatile metals such as Au, Sb, and Hg are typically observed to represent the latest stages of the sequence, precipitating from still cooler *epithermal* solutions circulating near the surface. Numerous empirical studies of hydrothermal ore deposits leave little doubt that the Emmons sequence shown in Table 3.3 has general applicability. Many exceptions,

however, do exist and the scheme can be much improved and extended in the light of modern observations and theory.

The wide ranging applicability of the Emmons sequence suggests that the secular precipitation of metals might simply be related to the decline in solubility that generally accompanies temperature decrease. Sections 3.5 and 3.6 above, however, have shown that metal solubilities and precipitation are controlled by a variety of interdependent variables that are related to both the fluid itself and its surrounds. Barnes (1975) showed that if metals are considered to have been transported as *metal–sulfide* complexes then the relative stabilities of these complexes, when corrected for differing concentration of metals in solution, closely match the Emmons precipitation sequence. On this basis he identified the following paragenetic sequence:

$$Fe–Ni–Sn–Cu–Zn–Pb–Ag–Au–Sb–Hg$$

By contrast, if metal transport occurs by *metal–chloride* complexation then the above sequence no longer applies, and a subset of these metals is more likely to be precipitated according to the sequence:

$$Cu–Ag–Pb–Zn$$

This suggests that paragenetic sequences are likely to be controlled by more than just a single variable. Susak and Crerar (1981) demonstrated that paragenetic sequences need to be considered in terms of classes of metals that are grouped in terms of mineral stoichiometry and the valence state of the aqueous metal species. They went on to show that the precipitation sequence of metals from *within one particular class* will be controlled essentially by the Gibbs free energies of the associated minerals. The precipitation of metals from different classes would, however, be controlled by the environment within which the metal-bearing solution occurs, and this could be variable. Thus, a simple and reproduceable zonation pattern or paragenetic sequence is likely to

Table 3.3 The "Emmons Reconstructed Vein" concept showing empirical observations that relate to typical patterns of zonation and paragenetic sequences in many hydrothermal ore deposits.

Vein	Metal	Ore mineralogy (in bold), and related alteration assemblages
Shallow (epithermal)	Barren	Chalcedony, quartz, barite, fluorite, and carbonate minerals
	Mercury	**Cinnabar**: chalcedony, quartz, barite, fluorite, and carbonate minerals
	Antimony	**Stibnite**: quartz
	Gold–silver	**Gold, electrum, acanthite**: quartz, chalcedony, adularia, alunite, carbonate minerals, silicification, some potassic, phyllic, and propylitic alteration
	Barren	Quartz and carbonate minerals
	Silver–manganese	**Acanthite, rhodochrosite**: quartz and carbonate minerals, some phyllic, argillic, and propylitic alteration
Intermediate (mesothermal)	Lead	**Galena**: quartz with minor carbonate minerals
	Zinc	**Sphalerite**: quartz with occasional carbonate minerals, advanced argillic alteration
	Copper–arsenic–antimony	**Chalcopyrite, tennantite–tetrahedrite**: quartz, phyllic, propylitic, and argillic alteration
	Copper	**Chalcopyrite**: quartz, phyllic alteration
	Molybdenum–tungsten–tin	**Molybdenite, huebnerite, scheelite, cassiterite**: quartz, potassic alteration, greisens
Deep (hypothermal)	Barren	Potassic alteration, anhydrite, carbonate minerals

Source: After Guilbert and Park (1986).

apply only to hydrothermal ore-forming environments that consist of metals from the same zoning class. Any exceptions and reversals to the typical Emmons sequence are likely to be explained in terms of either disequilibrium, or precipitation of metals from different classes.

The predicted precipitation sequence for metals in specifically defined zonation classes is illustrated in Figure 3.20. The precipitation sequence for metals in solution from the same class can be read on this diagram in terms of decreasing Gibbs free energy values for a specific temperature. Thus, the precipitation sequence for metals in the MS (q = 2) zoning class is Mn–Zn–Cd–Pb–Hg for any temperature, but this sequence would not necessarily prevail under conditions where a different stoichiometry and valence state prevails. Whereas this scheme may be applicable to systems where metal solubilities control the paragenetic sequence, it should be emphasized that other factors might also control patterns of zonation. It was pointed out in Sections 2.6 and 3.4.2 that certain metals may fractionate into the vapor phase of an ore fluid undergoing phase separation. An example of this is documented in the extensively mineralized and zoned (Sn–W–Cu–Pb–Zn) Mole granite where more volatile species such as Cu, B, Li, As, and S preferentially partition into a vapor phase resulting in metal zonation patterns that cannot be interpreted simply by referring to relative metal solubilities (Audétat et al. 2000).

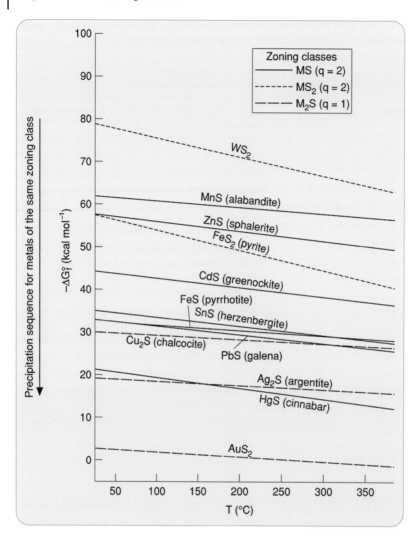

Figure 3.20 Plot of Gibbs free energy versus temperature for a variety of sulfide minerals categorized according to their stoichiometry (where M is metal and S is sulfide) and the valence state of the dissolved metal species (q). The precipitation sequence for metals in solution from the same zoning class is predicted in terms of decreasing values of free energy. Source: After Susak and Crerar (1981).

3.7.1 Replacement Processes

In the previous section the precipitation of minerals and metals was discussed in the context of aqueous solutions passing through open spaces in a rock, especially vugs and veins at shallow crustal levels. Different conditions apply to metals that are precipitated during the *replacement* of minerals deeper in the Earth's crust. Replacement occurs when the original minerals in a rock are dissolved and near simultaneous precipitation of secondary minerals occurs (Seward and Barnes 1997). In contrast to the more permeable near-surface environment where ore deposition takes place by open space filling, metal precipitation deeper in the crust occurs as fluids percolate along poorly interconnected microfractures and pore spaces (see Section 3.3.4). Thus, the process of replacement will only persist if porosity is maintained and if the accompanying fluid–rock reactions are characterized by a reduction in the molar volume of the mineral being replaced. An increase in molar volume would effectively block the through-flow of fluid as microfractures and pores are filled by secondary mineral precipitates whose volumes are greater than the material being replaced.

Comparison of molar volumes shows that calcite, for example, can be replaced by most calc–silicate and sulfide minerals, a feature that is commonly observed in skarn deposits. By contrast, oxide minerals such as magnetite and ilmenite are less likely to be completely sulfidized since the relevant reactions will involve an increase in molar volume. At best, only partial replacement of oxide by sulfide would occur since the volume increase, together with the accompanying decrease in microporosity that accompanies the reaction, would limit fluid flow. The sequence of precipitation of secondary or ore minerals during replacement processes can best be predicted in terms of decreasing molar volumes of the respective solids involved, as discussed by Seward and Barnes (1997). The mineral sequence shown below, with relevant molar volumes, is useful for the understanding of replacement processes and the prediction of replacement sequences. Replacement will only be effective when minerals of smaller volume replace those with larger volume. In the sequence presented below, minerals will only comprehensively replace those occurring to their left. In addition, mineral replacement will presumably occur more readily the greater the volume deficit between original and secondary mineral.

microcline ← scheelite ← anhydrite ← calcite
  [54]   [47]   [46]   [37]

← kaolinite ← galena ← molybdenite ← siderite
  [33]   [31]   [30]   [29]

← cinnabar ← muscovite ← arsenopyrite
  [28]   [28]   [26]

← bornite ← pyrite ← sphalerite ← chalcopyrite
  [25]   [24]   [24]   [22]

← pyrrhotite ← ilmenite ← hematite
  [18]   [16]   [15]

← magnetite ← chalcocite
  [15]   [14]

(This sequence is from Seward and Barnes (1997); [molar volumes] in $cm^3 \, mol^{-1}$.)

3.8 Modern Analogues of Ore-Forming Processes – The VMS–SEDEX Continuum

The term "volcanogenic massive sulfide," or VMS, refers to a large family of mainly Cu–Zn (occasionally with minor Pb and Au) deposit types that formed during episodes of major orogenesis throughout Earth history (see Chapter 6). They are also referred to as VHMS deposits, where the acronym stands for "volcanic hosted massive sulfide." VMS deposits occur in a variety of tectonic settings but are typically related to precipitation of metals from hydrothermal solutions circulating in volcanically active submarine environments. SEDEX deposits, by contrast, are dominated by a Zn–Pb (with lesser Cu, but commonly Ba and Ag) metal association and are also related to hydrothermal fluids venting onto the sea floor, but without an obvious or direct link to volcanism. Many of the large SEDEX deposits of the world are Proterozoic in age, although several examples, such as Red Dog (see Box 3.4) also formed in Phanerozoic times. Although there is generally no spatial or temporal link between SEDEX and VMS deposits, it is widely held that they represent a continuum and are conceptually linked by the fact that they formed by the same basic processes (Gilmour 1976; Plimer 1978; Guilbert and Park 1986; Kirkham and Roscoe 1993; Misra 2000). These processes are active and can be studied in modern day environments, as discussed below.

The notion of a continuum between VMS and SEDEX deposit types is contentious and readers should be wary of over-interpreting the genetic link between the two deposit types. The principal VMS and SEDEX deposits apparently formed at different times of Earth history and in different tectonic settings (Chapter 6). In addition, studies have shown that some of the stratiform base metal deposits, previously thought to be syngenetic and related to exhalative venting of hydrothermal solutions onto the sea floor, are more likely to be the product of replacement processes and, therefore, epigenetic in origin. A reinterpretation of this type has, for example, recently been made with respect to the Mount Isa deposits in Australia (Perkins 1997). With this cautionary proviso in mind, it is nevertheless convenient to discuss VMS and SEDEX deposits as a conceptual continuum, at least insofar as the principal ore-forming processes are concerned. Both deposit types can be discussed in terms of modern analogues, which is advantageous in that the ore-forming processes can be studied directly. VMS deposits are discussed in the light of the spectacular discoveries on the ocean floors of "black smokers," whereas SEDEX deposits are considered in terms of the rift-related hydrothermal activity in the Red Sea and also around the Salton Sea in California.

3.8.1 "Black Smokers" – A Modern Analogue for VMS Deposit Formation

Even though it had been known since the 1960s that warm brines vented onto the ocean floor, it was the discovery in 1977 of "black smoker" hydrothermal vents in the Galapagos Rift (Ballard 1977) and subsequently the report of massive Cu–Zn sulfide chimneys and mounds at 21°N on the East Pacific Rise (Francheteau et al. 1979), that provided fascinating insights into the nature of submarine hydrothermal activity and its potential impact on the understanding of ore deposition in such environments. Black smokers are described as hot (up to 400 °C), metal charged, reduced, and slightly acidic hydrothermal fluids that vent onto the sea floor, usually in zones of extension and active volcanism along mid-ocean ridges (Figure 3.21). The fluids originate essentially from cold (2 °C), alkaline, oxidizing, and metal-deficient sea water. They circulate through the basaltic

ocean crust and, in so doing, scavenge metals to form the hydrothermal fluids now observed at innumerable black smoker sites in the Pacific, Atlantic, and Indian Oceans as well as the Mediterranean Sea (Scott 1997). Black smoker fluids usually vent through tube-like structures, called chimneys (Figure 3.21), that are built out of a mixture of anhydrite, barite, and sulfide minerals such as pyrite, pyrrhotite, chalcopyrite, and sphalerite, as well as gangue opaline silica.

Metal-charged fluids venting on the ocean floor point to the hydrothermal exhalative processes by which most, if not all, VMS and SEDEX can be explained. Black smokers venting along mid-ocean ridges represent a direct analogy for one specific type of the VMS family of deposits, namely the ophiolite-hosted Cu–Zn deposits such as those of the Troodos Massif, Cyprus (Box 3.3). An ophiolite represents a section of a mid-ocean ridge or back-arc spreading center that is preserved by obduction onto continental material and is characterized by the preservation of a sheeted dyke complex, pillowed basalt, and pelagic sediment. Where venting has occurred, a lens of massive sulfide ore, comprising mainly pyrite, chalcopyrite, and sphalerite, is located at the ocean floor interface and this is underlain by a pipe-like zone of disseminated sulfides and intense chloritic alteration representing the conduit along which fluids passed on their way to the ocean floor (Figure 3.22a). The geometry of this ore deposit type is fairly typical of most other VMS deposits (Figure 3.22b) even though they may form in other tectonic settings such as island arcs, back-arc basins, and fore-arc troughs. Other well-known Cyprus-type VMS deposits include Lokken in Norway and Outokumpu in Finland. Well-known examples of VMS deposits formed in other tectonic settings include the greenstone belt hosted Kidd Creek and Mattabi deposits in the Archean Superior Province of Canada, Besshi and the numerous Kuroko-type deposits of the Japanese island arc, Rio Tinto and Neves Corvo in the Iberian

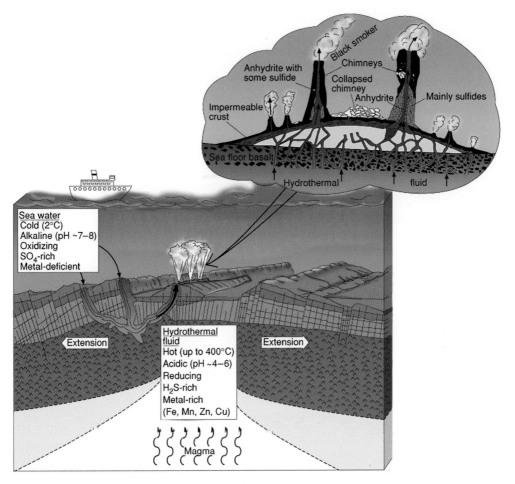

Figure 3.21 Conceptual diagram illustrating the fluid characteristics and circulation pattern in mid-ocean ridge environments that give rise to the formation of "black smokers" on the sea floor. Inset is a cross section of an exhalative vent site showing the construction of anhydrite–sulfide chimneys on top of a mound of massive sulfide mineralization. Source: After Lydon (1988), Scott (1997).

Pyrite Belt of Spain and Portugal, the Buchans and Bathurst deposits of Newfoundland, Jerome in Arizona, and the Mount Lyell and Hellyer deposits in Tasmania.

There is now general consensus that submarine venting represents a good analogue for the processes that applied during the formation of most of the actual VMS deposits preserved or fossilized in the Earth's crust. Excellent reviews of the characteristics of actual VMS deposits and the nature of their ore-forming processes are provided by Franklin et al. (1981, 2005), Hekinian and Fouquet (1985), Lydon

(1988), Large (1992), Scott (1997), Galley et al. (2007), and Hannington (2014). Studies of many actual deposits confirm that the fluids involved are derived dominantly from sea water, although fluid inclusion and stable isotope studies suggest that in certain cases a minor magmatic fluid component may have been incorporated into the circulating fluid system. The main source of metals is believed to have been the volcanic rocks through which the sea water was percolating, and evidence in support of this comes from mass balance considerations and the fact that metal assemblages

(a)

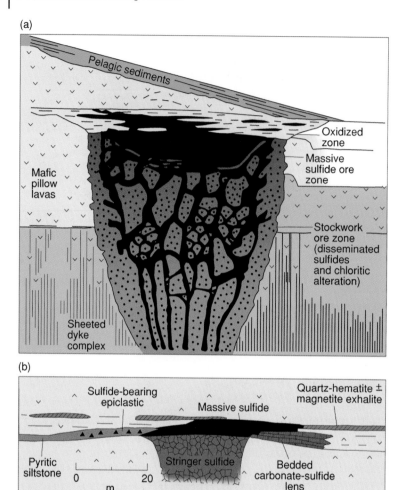

(b)

Figure 3.22 (a) Section through a typical ophiolite-hosted, Cyprus-type VMS deposit. Footwall rocks may consist of a sheeted dyke complex and the associated volcanic rocks are often pillowed with a tholeiitic composition. Source: After Hutchinson and Searle (1971). (b) Section showing the characteristics of VMS deposits other than the ophiolite-hosted type. Associated volcanic rocks may be intermediate or felsic in composition and a closer lateral link to chemical and epiclastic sediments is often apparent. The lens of massive sulfide ore formed on the ocean floor, and underlain by a stockwork zone of disseminated sulfides and intensely altered volcanic rock, is typical of VMS deposits in general. Source: After Large (1992).

in different VMS deposits are consistent with the expected metal contents and ratios in the associated primary igneous rock. For example, Cyprus-type VMS deposits, which reflect leaching of a dominantly mafic volcanic source rock, are typified by a Cu + Zn metal association (Franklin 1993) and this is explained by the fact that mafic volcanics are characterized by much higher Cu and Zn contents than their felsic equivalents. However, the reverse is true of Pb (see Table 1.2, Chapter 1) and this helps to explain why the Kuroko deposits of Japan, for example, described as a Zn–Pb–Cu variant of the VMS family (Pirajno 1992; Franklin 1993; Misra 2000) and associated with arc-related bimodal (i.e. felsic and mafic) volcanism, have a different metal association. By contrast, the source of sulfur is essentially from the sulfate component of the sea water itself, with reduction of sulfate to sulfide occurring during fluid–rock interaction prior to venting (Ohmoto et al. 1983). This is demonstrated in the compilation by Large (1992), which shows similar trends in sulfur isotope variations of VMS sulfide ores and sea water derived sulfate minerals over time (at least for Phanerozoic deposits).

Box 3.3 Exhalative Venting and "Black Smokers" on the Sea Floor: The VMS Deposits of the Troodos Ophiolite Complex, Cyprus

The 91 Myr old Troodos ophiolite complex in Cyprus represents an obducted and exposed fragment of oceanic crust that has been very well studied. It is also famous as a site of copper mining since the Bronze Age and contains numerous small to intermediate sized Cu–Zn VMS deposits hosted in pillowed basalts, mainly along the northern edge of the complex (Figure 3.3.1). The ophiolite is well exposed and comprises a cross section of the ocean crust, from lower mafic and ultramafic cumulates (gabbro and harzburgite that contain podiform chromitite deposits; see Chapter 1), overlain by a sheeted dyke complex and pillowed basalts (Figure 3.3.2). The largest deposits are Mavrovouni (15 million tons) and Skouriotissa (6 million tons), although several others have been mined in the recent past (Herzig and Hannington 1995).

Figure 3.3.1 Generalized geology of the Troodos ophiolite complex in Cyprus and the distribution of VMS style Cu–Zn mineralization.

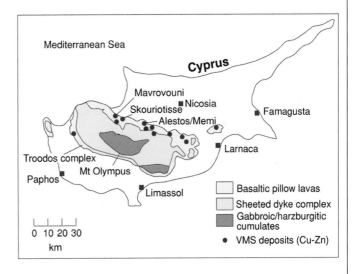

Mineralization in the Cyprus deposits comprises massive pyrite and chalcopyrite, with lesser sphalerite, hosted in pillow lavas and associated with zones of intense silicification and chlorite alteration (Figure 3.3.3; Constantinou and Govett 1973). The lavas are overlain by shales and cherts that are Fe- and Mn-rich (referred to as umber) and may also contain concentrations of Au. The hydrothermal fluid responsible for mineralization in the Troodos VMS ores was modified sea water that discharged through black smoker vents located along ridge axis parallel faults at temperatures of 300–350 °C (Spooner 1980). The characteristics of the Cyprus deposits have been compared directly with the TAG site, an active exhalative vent system on

(Continued)

Box 3.3 (Continued)

(a)   (b)

Figure 3.3.2 Photos of pillow lavas (a) and sheeted dykes (b) exposed in the Troodos ophiolite complex, Cyprus.

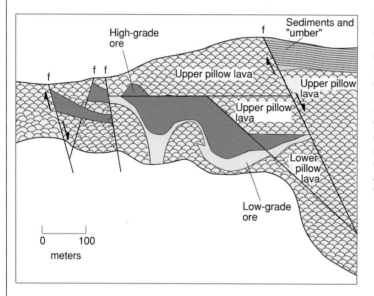

Figure 3.3.3 The nature and geometry of VMS mineralization in typical Cyprus deposits. Mineralization can be concentrated at the interface between pillowed basaltic lava flows, as shown, or it can occur between the upper pillow lavas and the umber sediments. Source: After Constantinou and Govett (1973).

the mid-Atlantic ridge (Eddy et al. 1998). It is also evident, however, that VMS mineralization can be associated with magmatic activity that is removed from that occurring along the ridge axis. The intrusion of discrete gabbroic plutons, and the formation of seamounts and sea floor calderas, for example, are features that promote fluid circulation and VMS styles of mineralization in areas that can be well away from the ridge axis. This observation has obvious exploration significance, and the Alestos and Memi deposits in Cyprus are two examples of classic VMS type mineralization associated with off-axis mineralization (Eddy et al. 1998).

VMS deposits are characterized by well developed metal zonation patterns defined by a typical sequence from Fe to Fe–Cu to Cu–Pb–Zn to Pb–Zn–Ba in an upward and lateral sense. As a first approximation it appears that this sequence reflects the variable solubilities of these metals and their sequential precipitation as the ore fluids cool. In detail, however, the development of this zonation is more complex and reflects the evolution of fluids and the growth mechanism of the massive sulfide mound with time, as described in Figure 3.23a (Large 1992). In this model it is envisaged that the temperature of the ore fluid increases with time as the deposit grows. Low temperatures would not be able to dissolve much in the way of base metals, although such fluids could transport sulfate complexes and precipitate anhydrite or barite on mixing with sea water (the so-called "white smokers" seen associated with low temperature vents). At temperatures approaching 250 °C however, the solubilities of Pb and Zn as chloride complexes would be high, reaching 100 ppm under the conditions applicable to Figure 3.23b. Copper would be poorly soluble, but some Au could be transported as a bisulfide complex. Discharge of such fluids would result in precipitation of barite/anhydrite together with sphalerite and galena, as well as minor gold (Stage 1, Figure 3.23a). As fluids evolve to higher temperatures (250–300 °C) they are now capable of containing significant Cu as a chloride complex and chalcopyrite will be precipitated in the footwall stockwork zone and at the base of the massive sulfide mound (Stage 2, Figure 3.23a). In the latter area it has been observed that chalcopyrite replaces sphalerite. Since the solvent capacity of a 250 °C fluid with respect to both Zn and Pb is high, it is feasible for the earlier formed sulfide minerals to be dissolved and reprecipitated further up in the mound, or distal to it. This zone-refining type of process accounts for the development of high grade massive sulfide cappings seen in certain VMS deposits. At still higher temperatures (300–350 °C) both Cu

and Au are in solution as chloride complexes, ultimately giving rise to further chalcopyrite precipitation, together with Au and pyrite. In this model the sequence Ba–Zn–Pb–Cu–Fe is the reverse of the paragenetic sequence predicted simply in terms of solubility variations with declining temperature. This scheme also predicts that VMS deposits could grow by input of metals from below and not necessarily by sulfide particles sedimenting out from the sea water, or aggregation of collapsed chimneys. Finally, it should be emphasized that biologically mediated precipitation mechanisms, promoted by the existence of sulfate reducing bacteria in exhalative vent environments, are likely to be important contributing factors during the formation of both VMS and SEDEX deposits. Large-scale precipitation of sulfides by bacterial reduction of sea water sulfate has, for example, been suggested for the Irish Zn–Pb–Ba deposits (see Section 3.5.3 above; Fallick et al. 2001).

Not all VMS deposits are characterized by a classic lens shaped sulfide mound and a well defined footwall alteration zone. Some have flat or tabular morphologies and appear to be distal from a vent, showing little or no evidence of footwall alteration/mineralization. Others exhibit indications of massive sulfide lenses stacked one on top of the other, or significant metal deposition below the sea water–rock interface (Large 1992). The morphology of atypical VMS deposits may largely be a function of the physical properties of the ore fluids themselves as they infiltrate up through the footwall volcanics and then vent onto the sea floor. Sato (1972) has suggested that ore fluids could be either more or less dense than sea water as a function of their temperature, salinity, and degree of mixing with cold sea water. Figure 3.24a (Type I) shows a situation where an ore solution is denser than sea water, giving rise to concentrated metal precipitation close to the site of venting. If the ore solution and sea water have similar densities metals will also be precipitated close to the site or in nearby topographic depressions (Type II,

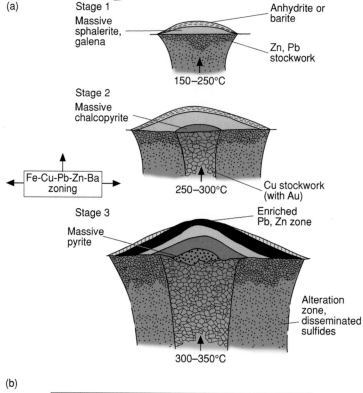

(a)

Stage 1
Massive sphalerite, galena

Anhydrite or barite

Zn, Pb stockwork

150–250°C

Stage 2
Massive chalcopyrite

Fe-Cu-Pb-Zn-Ba zoning

250–300°C

Cu stockwork (with Au)

Stage 3

Massive pyrite

Enriched Pb, Zn zone

Alteration zone, disseminated sulfides

300–350°C

Figure 3.23 (a) Secular model explaining the evolution of fluids, growth of massive sulfide mounds, zoning, and paragenetic sequence for VMS deposits. (b) Plot of Cu, Pb, Zn, and Au solubilities versus temperature, appropriate to fluid conditions during VMS ore deposition (pH = 4; 1 M NaCl; $aH_2S = 0.001$; $SO_4/H_2S = 0.01$). Source: Both diagrams after Large (1992).

(b)

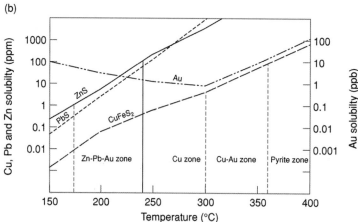

Cu, Pb and Zn solubility (ppm)

Au solubility (ppb)

ZnS

PbS

CuFeS$_2$

Au

Zn-Pb-Au zone | Cu zone | Cu-Au zone | Pyrite zone

Temperature (°C)

Figure 3.24a). In cases where the black smoker fluids were less dense than sea water (Type III, Figure 3.24a), buoyant metal-rich plumes would form dispersing metals into more distal marine sediments. Although this is not likely to be a particularly efficient process it remains a possible mechanism for explaining distal, low grade deposits associated with a higher proportion of ocean floor sediment.

Another factor that may be important in considerations of exactly where in the system metals accumulate relates to whether or not the hydrothermal solutions undergo boiling. The phase change from liquid to vapor is pressure-dependent and even though some black smoker fluids have been observed venting at temperatures as high as 400 °C, they appear on the sea floor as liquids because of the

(a)

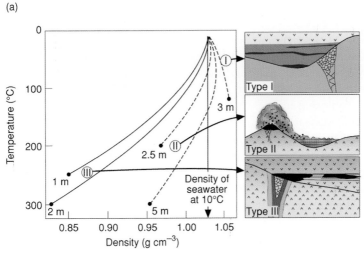

(b)

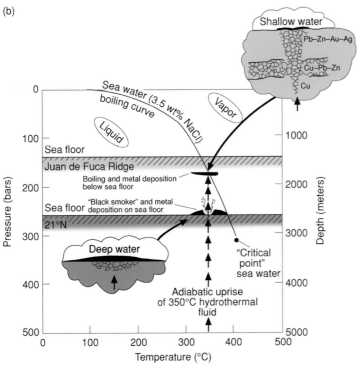

Figure 3.24 (a) Plot of temperature versus density for a variety of possible hydrothermal solutions (where 1 M, 3 M, etc. refer to the NaCl molality or the salinity of the solution) relative to sea water at 10 °C. Different solutions may be either less or more dense than sea water. At one extreme they would form buoyant plumes capable of distributing metals some distance away from the site of venting (Type III), whereas at the other they would precipitate metals proximally to the vent (Type I). Source: After Sato (1972). (b) Pressure (or depth) versus temperature plot showing the sea water (with 3.5 wt% NaCl) boiling curve relative to deep (21° N East Pacific Rise) and shallow (Juan de Fuca ridge) sea floor. Because of the strong pressure control on boiling, a 350 °C hydrothermal fluid rising adiabatically in the crust would intersect a deep ocean floor as a liquid and would vent as a black smoker precipitating metals into the sea. The same fluid venting into a shallower ocean would boil before intersecting the floor such that only a vapor phase would reach the surface. Metal precipitation in such a case will likely occur beneath the ocean floor. Source: After Delaney and Cosens (1982).

high confining pressures that exist deep in the oceans. This is demonstrated in Figure 3.24b, where the boiling point curve for sea water is plotted relative to deep (represented by 21°N East Pacific Rise) and shallow (represented by the Juan de Fuca ridge) ocean floor settings. Any 350 °C hydrothermal solution rising adiabatically through the crust would intersect the deep ocean floor as a liquid since it would not yet have intersected the boiling point curve. Metal precipitation on the sea floor from the black smoker would, in this case, occur by rapid quenching of the ore solution as it mixed with a virtually limitless volume of cold sea water. By contrast, the same fluid moving up toward a shallower sea floor is likely to intersect the boiling point curve before it vents to surface (Figure 3.24b). Boiling below the ocean floor is likely to promote brecciation of the footwall and local metal precipitation, possibly resulting in extensive stockwork mineralization. Classic VMS deposition is unlikely to occur and ore bodies may appear stacked one on top of the other. Although there are many variants in the family of VMS ores the processes applicable to the submarine exhalative model appear capable of explaining many of the features of this very important class of deposits.

3.8.2 The Salton Sea and Red Sea Geothermal Systems – Modern Analogues for SEDEX Mineralization Processes

SEDEX deposits contain more than half of the world's known resources of Pb and Zn and are generally represented by bigger and richer deposits than the VMS category, although there may be fewer of them. They are typically formed within intracratonic rift basins and are hosted by marine clastic or chemical sediments with little or no direct association with volcanic rocks. Most of the world's giant SEDEX deposits are Paleo- to Mesoproterozoic in age and well-known examples include HYC (McArthur River), Mount Isa and Broken

Hill in eastern Australia, Sullivan in British Columbia, Aggeneys and Gamsberg in South Africa, and Rajpura-Dariba in India. A significant group of deposits is Paleozoic in age, and these include Meggen and Rammelsberg in Germany, Navan in Ireland, Red Dog in Alaska, and Howard's Pass in Canada.

Modern analogues for the genesis of SEDEX deposits have been described from the Salton Sea and Gulf of California (Figure 3.25) and also the Red Sea. The study of these hydrothermal systems reveals that both syn-SEDEX and replacement processes apply and can, to a certain extent, be used to discount the controversial views that prevail regarding the origin of sediment-hosted stratiform deposits.

3.8.2.1 Salton Sea Geothermal System

The East Pacific Rise intersects the North American plate at the head of the Gulf of California (Figure 3.25a). At this point the prevailing stresses along this mid-ocean ridge are transferred onto the continent to form the essentially dextral San Andreas fault system. The fault cuts through a dry intermontane region called Imperial Valley, which is characterized by active, rift-related sedimentation. Early settlers discovered a huge salt pan in this valley, named the Salton Sink, from which salt was exploited for a few years in the late 1800s. In 1905, however, a canal, constructed to bring irrigation water from the Colorado River to early settlers in Imperial Valley, was breached by spring floods and the entire volume of this large river was channeled into the Salton Sink, at that time close to 100 m below sea level. It was two years before the breach was plugged and the Colorado River reverted to its natural course. By that time the salt pan had been transformed into the Salton Sea, a huge inland lake that covered an area of several hundred square kilometers. The Salton Sea soon became brackish as the lake waters dissolved the salt on its floor. Evaporation also increased the salinity of Salton Sea waters to the $Na–Cl–SO_4$ dominated composition that is observed today (McKibben and Hardie 1997).

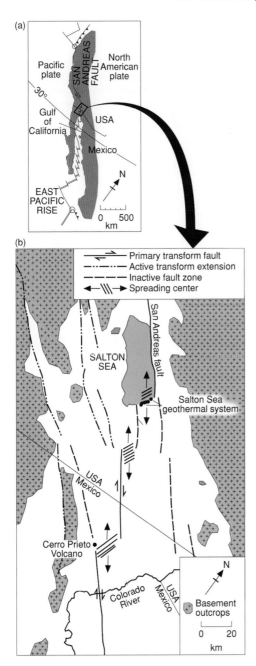

Figure 3.25 The Salton Sea Geothermal System (SSGS) and its location (inset) relative to the Gulf of California transform fault region and the dextral San Andreas fault system. Source: After McKibben and Hardie (1997).

Over the past century water from the Salton Sea has been circulating along fault-related pathways and through earlier formed lacustrine sediments to form an extensive hydrothermal system from which geothermal power is now generated. In addition, hot metalliferous brines, also derived directly from Salton Sea waters and heated by a combination of the high prevailing geothermal gradient and active volcanism in the region, vent from numerous sites, and it is these that are of particular interest to the study of ore-forming processes (White 1963; Skinner et al. 1969). The combination of well understood fluid compositional and circulation pathways, set in an active seismic and volcanic, intracontinental rift environment, has made the Salton Sea geothermal system a classic example of an ore-forming environment whose evolution can be observed, and from which many inferences about ancient sediment-hosted ore-forming processes can be made. In addition to its obvious applicability to the understanding of SEDEX ore-forming processes, the processes observed in this environment might also be relevant to the formation of rift-related, stratiform Cu deposits such as the Kupferschiefer ores of central Europe and the Central African Copperbelt (McKibben et al. 1988; McKibben and Hardie 1997).

Although the Salton Sea geothermal system does not exist in a submarine environment, the geometry of the system and the processes involved are considered to be useful for the understanding of sediment-hosted hydrothermal base metal deposits in which fluids circulated during and soon after deposition. A variety of conditions need to prevail in order to make SEDEX ore-forming processes viable. These include an abundant fluid source, a rich source of metals and complexing agents, active rifting, and deformation induced fluid flow. Detailed study of geothermal wells around the Salton Sea has shown that at depths of

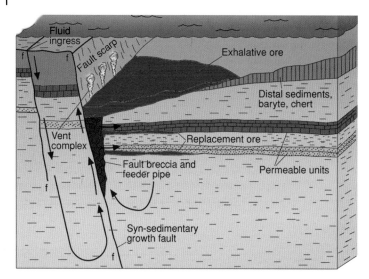

Figure 3.26 Diagram illustrating the setting for the formation of SEDEX-type Pb–Zn ores and a scenario which incorporates both exhalative and replacement concepts for the formation of these ores. Source: After compilations by Goodfellow et al. (1993), Misra (2000).

1000–3000 m a hot (350 °C), dense (up to 26 wt% dissolved solids), Na–Ca–K–Cl brine exists (McKibben et al. 1988). This fluid also contains significant concentrations of metals, in particular Fe, Mn, Pb, Zn, and Cu, derived from local lacustrine sediments. Precipitation of these metals has been induced by mixing of the hot metal charged brine with cooler, dilute surface waters about 1000 m below the surface, to form sediment-hosted base metal sulfide veins and associated chloritic alteration. This is clearly not an exhalative system and the ores that form will appear epigenetic in character. Ores that form in this environment could nevertheless have a stratiform geometry, with metals being deposited by replacement of pre-existing phases such as diagenetic cements or easily dissolvable detrital minerals. In the Salton Sea setting it is also conceivable, especially in the event of energetic fluid circulation, that the metal-charged brine could intersect the lake floor and vent ore-bearing solutions in much the same way that black smokers are exhaled along the mid-ocean ridges. It seems feasible, therefore, for SEDEX ore bodies to appear both syngenetic and exhalative in sediments that were deposited at the time of fluid

venting, but also distinctly epigenetic relative to the footwall sedimentary sequence (Goodfellow et al. 1993). A model illustrating the main features of SEDEX deposits and incorporating the concept that syngenetic exhalative and epigenetic replacement type ores could be coeval and form part of the same system is presented in Figure 3.26.

3.8.2.2 The Red Sea and the VMS–SEDEX Continuum

The extension of the East Pacific Rise (mid-ocean ridge) into continental North America to form the San Andreas fault system (Figure 3.25) provides support for the view that a conceptual continuum exists between VMS and SEDEX styles of mineralization. Present day ocean floor black smokers at 21 °N on the East Pacific Rise and the active, metal-charged geothermal systems of the Salton Sea perhaps represent extremes in the range of VMS and SEDEX ore-forming systems. The Guaymas Basin in the Gulf of California is also known to contain terrigenous sediment-hosted styles of VMS mineralization, also known as Besshi-type ores, from the type location in Japan. These intermediate styles of mineralization also provide support for the view that VMS and SEDEX styles of base/precious metal mineralization can be linked in terms of both tectonic setting and hydrothermal systems. Perhaps the best example of the conceptual link between VMS and SEDEX deposit types, however, is provided by the Red Sea rift, separating Africa and Arabia. The Red Sea is an ocean basin in the early stages of its development, containing active geothermal venting and deposition of base metal deposits in clayey muds accumulating in depressions on the sea floor (Bischoff 1969). Mineralization is particularly well developed at a site known as the Atlantis II Deep where muds are estimated to contain around 50 million tons of rich Zn–Cu–Pb–Ag–Au mineralization. Although the geological and tectonic setting of the Red Sea mineralization is analogous to a mid-ocean ridge VMS setting, the absence of black smokers and the presence of both clastic and carbonate sediment hosts to the mineralization are features perhaps more akin to SEDEX ores (Pirajno 1992).

Box 3.4 Sedimentary Exhalative (SEDEX) Processes: The Red Dog Zn–Pb–Ag Deposit, Alaska

The Red Dog mine in northwest Alaska is presently the world's largest producer of zinc concentrate, contributing about 6% of annual global supply. It is a very rich deposit and in 2002 had reserves of close to 100 million tons of ore grading 18% zinc, 5% lead, and 85 g ton^{-1} silver (www.teck.com). It was discovered in 1968 by an astute bush pilot who, from the air, noticed a color anomaly in the rocks caused by oxidation of the sulfide ores.

The Red Dog ores are hosted in Carboniferous black shales and limestone of the Kuna Formation in the De Long mountains of the Brooks Range, northern Alaska. The host succession has been deformed into a sequence of thrust allochthons during a Jurassic orogeny. Figure 3.4.1 shows an interpretive section through a Devonian–Carboniferous sequence of sediments in the Brooks Range prior to thrusting. The sequence of sediments, including limestones and turbiditic assemblages, is consistent with deposition in a restricted ocean basin, with no indication of volcanism in the immediate environs. Abrupt lateral facies changes in the succession point to the presence of syn-sedimentary growth faults that would have facilitated fluid circulation through the basin.

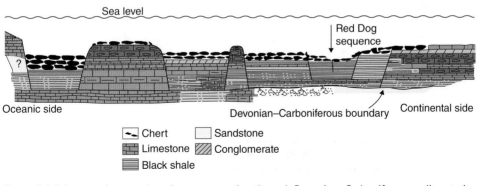

Figure 3.4.1 Interpretive, pre-thrusting, cross section through Devonian–Carboniferous sediments in the Brooks Range, illustrating the depositional setting of host rocks at the Red Dog mine. Syn-sedimentary growth faults and abrupt lateral facies variations are also a feature of the regional setting. Source: After Moore et al. (1986).

(Continued)

Box 3.4 (Continued)

Mineralization is stratabound and, despite the thrusting, is relatively flat-lying (Moore et al. 1986). It is marked by laterally extensive zones of silicification (chert and chalcedony) within which poorly bedded sulfide minerals (sphalerite, pyrite–marcasite, and galena) occur. Barite is closely associated with the ore and tends to cap the mineralized zones. An interesting feature of the ore zones is the occurrence of sinuous cylindrical structures interpreted to be worm tubes, similar to biotic structures observed in present day exhalative vent environments. This evidence supports the notion that silica and sulfide precipitation took place at the rock–water interface on the ocean floor and was, therefore, coeval with the deposition of the sediments. There is also, however, evidence for sulfide replacement of sediments at Red Dog, indicating that the dual processes of exhalation and replacement were taking place (see Section 3.8.2 of this chapter).

Figure 3.4.2 presents a simplified model of the ore-forming environment at Red Dog. Fluids circulating through Carboniferous organic-rich marine sediments scavenged metals from these host rocks and were vented onto the ocean floor at sights created by syn-sedimentary growth faults. Mineralization at Red Dog is massive, brecciated, and poorly bedded, suggesting that sulfides were precipitated fairly close to the vent sites (Moore et al. 1986). Variable sulfur isotope signatures in sulfide minerals suggest that precipitation was a product of mixing a buoyant hydrothermal plume with sea water that was becoming progressively more oxic over time. It is suggested that mineralization ceased when the sea water became sufficiently oxidizing to favor barite precipitation over sulfides (Moore et al. 1986).

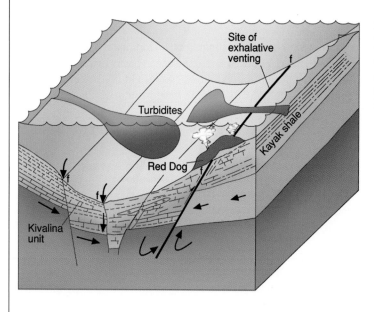

Figure 3.4.2 Simplified block diagram showing the environment of ore formation in the Carboniferous for the Red Dog deposit. Source: After Moore et al. (1986).

3.9 Mineral Deposits Associated with Aqueo-Carbonic Metamorphic Fluids

A significant proportion of the world's gold deposits are genetically linked to the formation of metamorphic fluids that are typically characterized by low salinity, near neutral pH, and mixed H_2O–CO_2 compositions. These gold deposits, of which there are many different types that formed over the entire span of Earth history, are associated with regionally metamorphosed terranes resulting from compression at convergent plate margins and have been termed *orogenic gold deposits* (Groves et al. 1998). They are also widely referred to, especially in the older literature, as "mesothermal" or "lode-gold" deposits, as they commonly precipitated at around 300 °C, forming quartz-vein or fracture dominated ore systems. Although orogenic gold deposits are commonly associated with

Archean granite–greenstone terranes (see Chapter 6), important examples of orogenic gold ores are also found hosted in a variety of Proterozoic and Phanerozoic settings. The dominant and characteristic genetic features that link all orogenic gold deposits are a synchroneity with major accretionary or collisional orogenic episodes and the production of metamorphic – and in some cases magmatic – fluids that precipitate metals at various crustal levels along deep-seated shear and fracture zones (Figure 3.27). It is widely recognized that these deposits were formed over a range of P–T conditions, are hosted in rocks subjected to a variety of metamorphic grades from granulite to greenschist facies, and subjected to structural regimes ranging from ductile through to brittle (Colvine 1989; Groves 1993).

In the sections that follow, a variety of orogenic gold deposits formed throughout geological time are described. In addition,

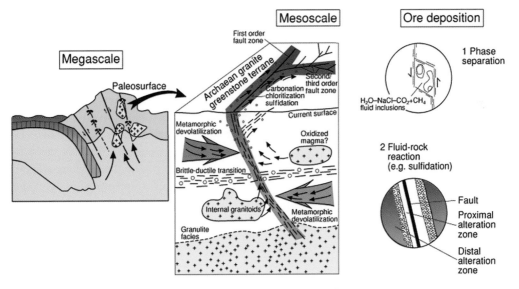

Figure 3.27 Schematic illustrations showing the principal features of Archean orogenic gold deposits – many of which also apply to Proterozoic and Phanerozoic examples. The main features at a megascale are that the deposits are associated with a convergent plate margin where metamorphism results in the production of fluids that are focused along major structural discontinuities. At a mesoscale, major shear zones within or along the margins of greenstone belts represent district-scale hosts to mineralization that are accompanied by broad zones of intense alteration. Individual deposits occur along second- and third-order structures. Ore deposition is a function of fluid rock reaction and/or H_2O–CO_2 phase separation. Source: After Hagemann and Cassidy (2000).

Carlin-type deposits, as well as the hydrothermal component of mineralization in quartz pebble conglomerate hosted gold ores such as the Witwatersrand and Tarkwaian deposits, are described, as they share several common features with orogenic gold deposits.

3.9.1 Orogenic Gold Deposits

There are many different types of orogenic gold deposits and their classification into a single category is bound to be controversial. From a descriptive and organizational point of view, however, the classification is appropriate and recognition of the different subtypes that undoubtedly do exist can perhaps best be made in terms of geological time and crustal evolution.

3.9.1.1 Archean

The Archean was a major gold metallogenic epoch and many important gold deposits occur in the granite–greenstone terranes of the world, examples of which include the Superior Province of Canada, the Yilgarn Craton of Western Australia, and the Zimbabwe Craton. A feature of Archean orogenic gold deposits is that they can be hosted in a variety of lithologies, including mafic volcanics, clastic metasediments, and banded iron-formations, as well as in associated felsic plutonic rocks. Despite the range of host rock lithologies, they are invariably found associated with zones of high strain that are manifest as brittle, brittle–ductile, or ductile deformation, depending on the crustal depth at which fluids were circulating. Reviews of the nature and characteristics of Archean deposits can be found in Witt (1991), McCuaig and Kerrich (1998), Mikucki (1998), and Hagemann and Cassidy (2000). Figure 3.27 illustrates the main features of Archean orogenic gold deposits. At a megascale the energy required to form crustal-scale zones of fluid flow, alteration, and mineralization, such as those preserved in the Abitibi (Superior) and Norseman–Wiluna (Yilgarn) greenstone belts, is ultimately derived from orogenic processes. Crustal thickening, deformation, metamorphism, and synorogenic magmatism all play important roles in the ultimate origin of fluids and their focused flow upwards into the crust. A major part of the fluids implicated in mineralization appears to have been derived from the dehydration that accompanies regional prograde metamorphism. These fluids are fairly reduced (fO_2 around the quartz–fayalite–magnetite or QFM buffer) and gold is preferentially transported as the $Au(HS)_2^-$ complex (Hagemann and Cassidy 2000; Ridley and Diamond 2000). Derivation of mineralizing fluids from oxidized granitic magma is also considered likely, although such fluids are probably restricted to environments where a spatial link between mineralization and granite intrusion is evident, or an association between Au and metals such as Mo, W, or Cu occurs (Burrows and Spooner 1987). Gold mineralization is broadly synchronous with the regional peak of metamorphism, although this peak may be in the lower crust rather than at the site of deposition in the upper crust. Fluids tend to be focused along major structural discontinuities that have strike lengths of tens to hundreds of kilometers and extend to considerable depths in the crust. Examples of well mineralized structural systems include the Larder Lake–Cadillac fault zone in the Abitibi belt and the Boulder–Lefroy fault zone in the Yilgarn craton. Individual deposits and mines are typically found along second- and third-order structures within these major discontinuities. Movement of fluid along these structures is explained by the fault-valve model, as discussed in Section 3.3.3 above. The sites of mineralization are marked, at a regional scale, by zones of intense carbonation and chloritization that can extend for tens of kilometers around the conduit (see Box 3.2). Alteration plays a major role in metal precipitation, and this is particularly the case for intense wall-rock alteration and sulfidization. H_2O–CO_2 phase separation is also considered to be an important process in gold deposition and may explain the siting of "bonanza" accumulations of gold in quartz veins. The latter mechanism may also explain the common association of gold mineralization with the brittle–ductile transition, as this zone coincides broadly with the relevant P–T conditions at which H_2O–CO_2 unmixing might occur (see Figure 3.5b). Mixing of metamorphic with meteoric fluids may explain gold precipitation at higher levels in the crust.

3.9.1.2 Proterozoic

The Proterozoic Eon is not as important as either its Archean or Phanerozoic counterparts in terms of gold mineralization (see Chapter 6). There are, nevertheless, several important examples of orogenic gold deposits from this period of time and these include Ashanti–Obuasi and several other deposits associated with the Birimian orogeny of West Africa, Telfer in Western Australia, Homestake in South Dakota, USA, Omai on the Guayanian craton of South America, and the Sabie–Pilgrim's Rest goldfield in South Africa. All these examples are associated with periods of major orogenesis and largely involve metamorphic fluids similar in character to those applicable to the formation of Archean orogenic gold deposits. Gold deposition is generally late orogenic and is hosted in major, high-angle thrust faults (Partington and Williams 2000). Granite intrusions, and possibly a magmatic fluid component, are associated with many of the Proterozoic deposits and, in cases such as Telfer (Rowins et al. 1997) and Sabie–Pilgrim's Rest (Boer et al. 1995), a polymetallic association of Cu, Co, and Bi, in addition to Au, is evident. In general, however, the characteristics of Proterozoic orogenic gold deposits are similar to those of the Archean.

3.9.1.3 Phanerozoic

Orogenic gold deposits of the Phanerozoic Eon were formed during two mineralization episodes, one in the Silurian–Devonian at around 450–350 Ma and the other in the Cretaceous–Paleogene at around 150–50 Ma

(Bierlein and Crowe 2000). These gold deposits are associated with convergent plate margins and hosted in compressional to transpressional shear zones that typically cut through thick marine shale sequences that have been metamorphosed to greenschist facies grades. They are often referred to in the older literature as "slate-belt hosted" gold deposits. There are many important examples of Phanerozoic orogenic gold deposits and these include, as Paleozoic examples, Bendigo, Stawell, and Ballarat in the Lachlan orogen of southeastern Australia, Haile in the Carolina slate belt, USA, Muruntau in Uzbekistan, the Urals of central Russia, and the Meguma terrane of Nova Scotia, Canada. Younger, Mesozoic examples include the Juneau gold belt of southern Alaska and other portions of the Cordilleran orogen of Canada and the western USA, including the Mother Lode of California (Bierlein and Crowe 2000).

3.9.2 Carlin-Type Gold Deposits

Although several of the features of Carlin-type ores mirror those of orogenic gold deposits, they are generally classified separately because of one notable difference. While linked to orogenic activity and hosted along major, deep-crustal structures, they are not related to compression but formed after the onset of *extensional* forces that followed earlier, subduction-related processes (Hofstra and Cline 2000). Carlin-type ores are located in Nevada, USA, and formed over a short period of time in the Paleogene Period, between 42 and 30 Ma. In this sense they are regarded as an essentially unique deposit type, although it is likely that similar ores are located elsewhere, such as in China. In Nevada they are hosted in Paleozoic carbonate rocks that have been subjected to severe crustal shortening and west to east directed thrusting (Figure 3.28a and Box 3.5). The carbonate sequences are structurally overlain by siliciclastic rocks that acted as aquifers along which metamorphically derived fluids flowed. During mid-Miocene

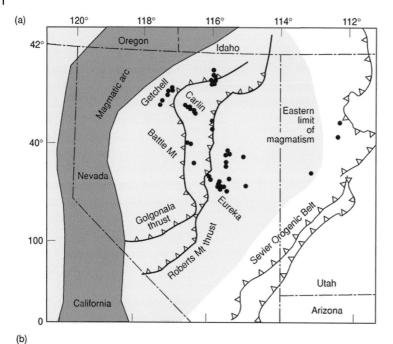

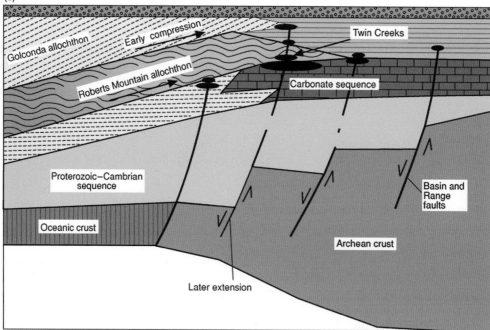

Figure 3.28 (a) Map showing the distribution of Carlin-type gold deposits in Nevada and their geological framework. (b) Cross section illustrating the fundamental controls behind the location of Carlin-type gold deposits. Source: After Hofstra and Cline (2000).

times host rocks were subjected to uplift and extension, which thinned the crust and gave rise to the Basin and Range faulting and associated fluid circulation along the newly formed dilatant structures.

Fluids associated with Carlin deposits are similar to the metamorphic fluids implicated in orogenic gold deposits, comprising low salinity, moderately acidic, reduced H_2O–CO_2 solutions in which gold is transported as either $Au(HS)_2^-$ or $Au(HS)$. Fluid temperatures are cooler than those associated with orogenic ores and fall in the range 150–250 °C, which has led to the suggestion that they may be meteoric rather than metamorphic in origin. Gold deposition occurs where normal faults intersect a less permeable cap rock, usually at a shale/limestone contact and in the crests of fault-propagated anticlines (Figure 3.28b). The precipitation mechanism is believed to be related to neutralization of the ore fluid during carbonate dissolution, although argillic alteration, silicification, and sulfidization also played a role in the ore-forming process. In Carlin-type deposits the gold occurs as micrometer-sized particles or in solid solution in arsenic rich pyrite, marcasite, or arsenopyrite.

Box 3.5 Circulation of Orogeny-Driven Aqueo-Carbonic Fluids: Twin Creeks – A Carlin-Type Gold Deposit, Nevada

Twin Creeks mine, formed in 1993 after the merger of the Chimney Creek and Rabbit Creek mines, is one of the largest Carlin-type deposits in Nevada, with reserves in 1999 of some 90 million tons of ore at an average gold grade of 2.5 g ton^{-1}. Most of the gold at the mine is extracted from the Mega Pit which produces some 280 000 tons of ore per day. Although gold at the Mega Pit is hosted essentially in limestones of the Ordovician Comus Formation, mineralization elsewhere on the mine (i.e. the Vista Pit) also occurs in overlying sequences that form a package of thrusted allochthons (see Figure 3.28a and b) emplaced in Carboniferous times. Gold mineralization occurs mainly in the arsenic-rich rims to pyrite grains formed during hydrothermal alteration (i.e. decalcification, dolomitization, silicification, and sulfidization) of the Comus Formation limestones (Thoreson et al. 2000).

In the Mega Pit the mineralized limestones are capped by the Roberts Mountain thrust which folded the underlying limestones and also possibly acted as a cap that limited further egress of hydrothermal fluids and concentrated mineralization within the Comus Formation. Figure 3.5.1 shows a cross section through the Mega Pit, illustrating the relationships between the early thrusting and folding of Comus limestones, and the later extensional faults along which mineralizing fluids are thought to have been focused. In this section line gold mineralization is concentrated largely within the core of a prominent eastward verging recumbent antiform.

Mineralization at Twin Creeks, as elsewhere in the Carlin district of Nevada, is considered to have been a result of fluid circulation between 42 and 30 Ma (Eocene–Oligocene) during the onset of extensional (Basin and Range) tectonism and mantle plume activity (Hofstra and Cline 2000). Gold deposition, therefore, took place long after deposition of the host rocks (in the Ordovician) and their subsequent deformation (thrusting and folding) in Carboniferous times. Although subduction-related magmatism took place in the region between 43 and 34 Ma, no intrusives are directly implicated in the formation of any Carlin-type deposit and most workers refer to a metamorphic fluid as the mineralizing agent. This fluid was a moderately acidic, reduced, H_2O–CO_2 solution at around 150–250 °C in which gold was transported as bisulfide complexes (Hofstra and Cline 2000). Gold precipitation is believed to have accompanied cooling and neutralization of the fluid as it interacted with reactive country rocks such as the Comus Formation limestones. Circulation of mineralizing fluids was focused essentially along

(Continued)

Box 3.5 (Continued)

---

major extensional faults, which is the main reason why major deposits are located along well defined structural trend lines. Individual deposits tend to occur beneath a less permeable, often siliciclastic, cap rock (such as the Roberts Mountain allochthon at Twin Creeks) causing the fluids ascending along the faults to flow laterally into more reactive sequences such as the Comus

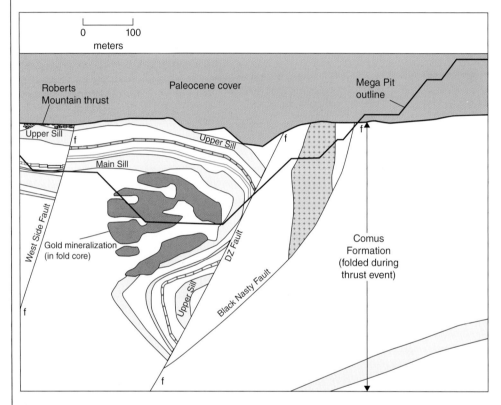

Figure 3.5.1 Cross section looking north of the Mega Pit at the Twin Creeks mine. Source: After Thoreson et al. (2000).

Formation. Carlin-type deposits are unique in the broader family of orogenic gold deposits and are perhaps intermediate in character between more typical compression-related orogenic gold ores and low-sulfidation epithermal gold deposits (see Chapter 2). There is, however, no reason why similar styles of epigenetic, sediment-hosted deposits, containing "invisible" or micrometer-sized gold in sulfides, should not be located in other parts of the world besides Nevada.

3.9.3 Quartz Pebble Conglomerate Hosted Gold Deposits

Quartz pebble conglomerates of Archean and Paleoproterozoic ages represent a very important source of gold, although production over the past century has been dominated by the Witwatersrand Basin in South Africa (Kirk et al. 2002). The seven major goldfields associated with the Witwatersrand sequences have produced in excess of 50 000 tons of gold metal – some 35% of all the gold produced in the history of mankind – since 1886 (Phillips and Law 2000). Other conglomerate-hosted gold deposits such as those of the Tarkwa (Ghana) and Jacobina (Brazil) basins are relatively minor by comparison. Irrespective of their production statistics, quartz pebble conglomerate hosted gold deposits are characterized by considerable controversy regarding the origin of their gold mineralization. Arguments revolve essentially around whether gold was introduced as detrital particles during sedimentation (a placer process – see Chapter 5; Kirk et al. 2001; Robb and Hayward 2014) or precipitated from hydrothermal solutions circulating through the sediments at some stage after sediment deposition. There is little doubt, however, that in all three cases (i.e. the Witwatersrand, Tarkwa, and Jacobina) the host sediments underwent burial and regional metamorphism and that the associated circulation of metamorphic fluids resulted in a clearly epigenetic component of mineralization. Fluid inclusion studies indicate that certain of these fluids have a mixed H_2O–CO_2 composition and have some similarities to those involved with orogenic gold deposits.

One view of the origin of gold mineralization in the Witwatersrand sediments is that it occurs along major unconformities, represented by conglomerate deposition, which represent zones of intense fluid flow and alteration (Figure 3.29a). Alteration is the product of fluid–rock reactions involving progressive neutralization of an acid fluid (see Section 3.5 above), with the most intense alteration being characterized by the assemblage pyrophyllite–chloritoid–muscovite–chlorite (Barnicoat et al. 1997; Philips and Law 2000). In the Witwatersrand Basin, however, gold mineralization is characterized by a long-lived paragenetic sequence (Robb et al. 1997; England et al. 2001; Frimmel and Minter 2002) and is likely to be related to more than just one-pass fluid–rock interaction. The existence of a prominent alteration halo around zones of pervasive fluid flow is nevertheless one of the reasons why the Witwatersrand

gold deposits are so large. At the scale of a single deposit similar effects can be seen, as exemplified by the João Belo reef in the Jacobina Basin of Brazil (Figure 3.29b). Here gold mineralization coincides with a section of the conglomerate horizon that is cut by a number of bedding-parallel shear zones along which fluids have flowed resulting in fuchsite (Cr–mica)–tourmaline–rutile–muscovite–pyrite alteration (Milési et al. 2002). Outside this alteration zone the conglomerate is reported to be barren.

The nature of the ore-forming fluid associated with the Witwatersrand, Jacobina, and Tarkwa auriferous conglomerates has been described as mixed H_2O–CO_2, fairly reduced, near-neutral, and with low salinity (Philips and Law 2000). Barnicoat et al. (1997) preferred a more oxidized, acidic fluid for the Witwatersrand Basin whereas Klemd et al. (1993) suggested that the ore fluids associated with the Tarkwaian conglomerates were dominantly high density carbonic with significant contained methane and nitrogen. Detailed fluid inclusion studies (Drennan et al. 1999; Frimmel et al. 1999) have indicated that in the Witwatersrand Basin different fluids are implicated in at least two distinct episodes of mineralization, and these include mixed H_2O–CO_2 compositions as well as aqueous, high-salinity fluids. The origin of these fluids is also contentious and options range from devolatilization of greenschist to amphibolite facies mafic rocks beneath the sedimentary basins, to derivation by prograde metamorphic reactions within the basin itself (Stevens et al. 1997 and Figure 3.4). The hydrothermal ore-forming processes of the Witwatersrand Basin in particular are long-lived and complex and this is the reason why there has been, and still is, so much debate on the genesis of these ores. It is, for example, now reasonably well established that the Witwatersrand basin was struck by a large meteorite at 2025 Ma and that the catastrophic exhumation of the gold-bearing sedimentary host rocks, as well as the likely circulation of related meteoric

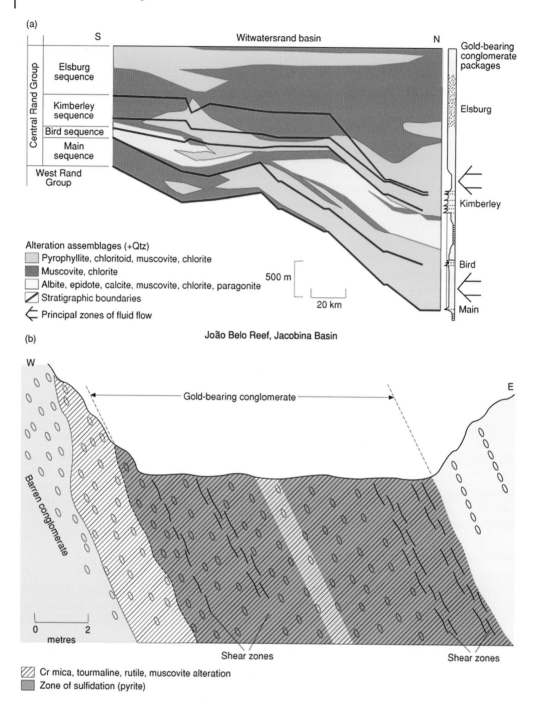

Figure 3.29 (a) Simplified north–south section showing the pattern of alteration at a regional scale associated with gold-bearing conglomerate packages in the Central Rand Group of the Witwatersrand Basin. Source: After Barnicoat et al. (1997). (b) Cross section through the João Belo reef in the Jacobina Basin showing the relationship between shear zone related fluid conduits, alteration, and gold mineralization at a deposit scale. Source: After Milési et al. (2002).

fluids, must also have played an important role in the ore formation process (Gibson and Reimold 1999).

3.10 Ore Deposits Associated with Basinal Fluids

Several sediment-hosted base metal deposits are genetically linked to the circulation of basinal or connate fluids during diagenesis. In these deposit types metal transport and deposition is generally restricted to the sedimentary sequence through which the basinal fluids circulate. Ores included in this category are the important stratiform sediment-hosted copper deposits (abbreviated to SSC, but also called red-bed copper deposits) and the family of Pb–Zn ores, usually associated with carbonate sediments (although some are sandstone-hosted), known as Mississippi Valley type (or MVT) deposits. Although the two deposit types are different in several respects, they both owe their origins to circulating basinal brines.

Sverjensky (1989) has suggested that the major differences between SSC and MVT deposits relate mainly to the type of sediment through which the fluids have traveled prior to ore deposition and also to the contrasting properties of Cu, Pb, and Zn in hydrothermal solutions. It is suggested that a basinal brine channeled into an *oxidized* (red-bed) aquifer would scavenge all three base metals from fertile detritus in the basin, but that the resulting ore fluid would be saturated only with respect to Cu and undersaturated relative to Pb and Zn. The tendency in such an environment would be for fluids to preferentially deposit copper sulfide minerals while the higher solubility metals remain in solution. Accordingly, SSC deposits tend to be characterized by a Cu > (Pb + Zn) metal association (Figure 3.30). By contrast, the same original fluid channeled through a relatively reduced carbonate aquifer would evolve to Zn and Pb saturated conditions in the presence of dissolved sulfur species and ore deposition would yield sphalerite > galena ≫ chalcopyrite, as is typical of most MVT deposits. In the situation where the original basinal brine passed through a quartz-rich sandstone aquifer, and the oxidation state of the fluid was buffered by hematitemagnetite, Pb would be particularly soluble and resulting ores would be galena-rich, consistent with the fact that sandstone-hosted MVT deposits tend to have Pb > Zn ≫ Cu metal ratios. It is interesting to

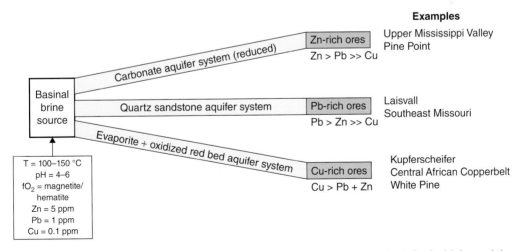

Figure 3.30 Diagrammatic representation showing the relationship between a single basinal brine and the conditions under which it might form SSC, as well as carbonate- and sandstone-hosted, MVT deposits. Source: After Sverjensky (1989), Metcalfe et al. (1994).

note that in the Central African Copperbelt, for example, Cu(–Co) mineralization is typically hosted in clastic sediments such as arkose and shale, whereas Zn–Pb mineralization is found associated with carbonate rocks.

Some of the processes relevant to the formation of sediment-hosted SSC and MVT deposits are discussed in more detail below.

3.10.1 Stratiform Sediment-Hosted Copper (SSC) Deposits

SSC deposits rank second only to porphyry copper deposits in terms of copper production and they represent the most important global source of cobalt, as well as containing resources of many other metals such as Pb, Zn, Ag, U, Au, PGE, and Re (Gustafson and Williams 1981). There are several deposits of this type around the world, but these are overshadowed by the huge resources of two regions, the Neoproterozoic Central African Copperbelt (Box 3.6) and the Permo-Triassic Kupferschiefer (or "copper shale") of central Europe. Other important deposits include White Pine in Michigan, Udokan and Dzhezhkazgan in Kazakhstan, Corocoro in Bolivia, and the Dongchuan district of China (Misra 2000). The formation of major SSC deposits seems to coincide with periods of supercontinent amalgamation such as Rodinia in the Neoproterozoic and Pangea in the Permo-Triassic (see Chapter 6).

SSC deposits are typically hosted within an intracontinental, rift-related sedimentary sequence. The early part of the sequence was either deposited originally as an oxidized (red-bed), eolian to evaporitic assemblage, or was rapidly oxidized during burial and diagenesis. This sequence is overlain by a shallow marine transgression that deposited a more reduced assemblage of shales, carbonates, and evaporites. Basin-derived fluid circulation was promoted by the high heat flow conditions accompanying rapid rifting and subsidence, and the relatively permeable environment created by porous clastic sediments deposited adjacent to active growth faults. The fluids circulating at this stage of basin evolution were saline, relatively oxidized, and pH neutral. Metals, in particular copper, are believed to have been leached from detrital minerals such as magnetite, biotite, hornblende, and pyroxene which themselves may have been derived from erosion of a fertile basement assemblage (a Paleoproterozoic magmatic arc in the case, for example, of the Central African Copperbelt; Rainaud et al. 1999; and Box 3.6). The aqueous transport of copper under these conditions was as a cuprous–chloride complex, probably $CuCl_3^{2-}$ (Rose 1976). Metal deposition occurred at a redox interface where the oxidized, metal-charged basinal fluids intersected overlying or laterally equivalent reduced sediments or fluids. The ore deposits are generally zoned, usually at a district scale, and typical zonation is characterized by the sequence barren/hematite – native copper – chalcocite – bornite – chalcopyrite – Pb/Zn/Co sulfides – pyrite.

SSC deposits can be subdivided into two subtypes (Kirkham 1989). The first, less important, subtype is hosted in continental red-beds that were probably originally oxidized, with ores precipitated around locally developed zones of reduction (for example, Dzhezhkazgan in Kazakhstan and Corocoro in Bolivia). The second subtype, represented by the much larger Kupferschiefer and Copperbelt deposits, is characterized by metal deposition in more reduced shallow marine sequences that were partially oxidized subsequent to deposition, and where oxidized (red-bed) sediments occur stratigraphically beneath the ore zone. In both cases, however, ore fluids are considered to have interacted with the dominantly clastic, oxidized sedimentary sequence and it is this feature that determines the nature and properties of the hydrothermal solutions. Such fluids are likely to have been characterized by low temperatures (<150 °C), a neutral range of pH (5–9) constrained by carbonate or silicate equilibria (K-feldspar/illite), oxidized conditions (hematite and other ferric-oxyhydroxides stable), and moderate to high salinities

Figure 3.31 Eh–pH plot for the system Cu–O–H–S–Cl at 25 °C (with $\Sigma S = 10^{-4}$ m and $Cl^- = 0.5$ m). Also shown are the stability fields for hematite, various sulfur species, and the cuprous–chloride complex for conditions compatible with the formation of SSC deposits. Source: After Rose (1976).

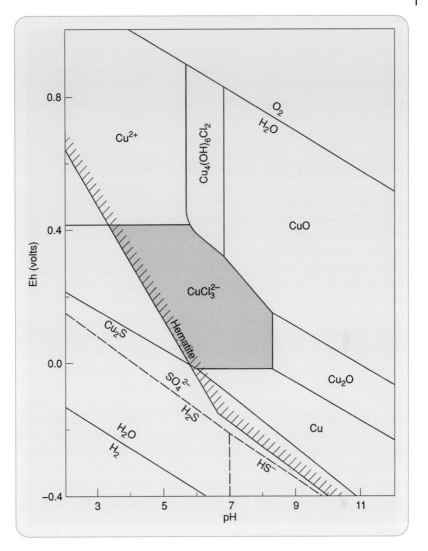

(typically up to 20 wt% NaCl equivalent), reflecting interaction of the fluids with evaporitic rocks. Rose (1976) showed that under these conditions the $CuCl_3^{2-}$ complex would have been particularly stable (Figure 3.30), although other aqueous species, such as $CuSO_4$ or $Cu(OH)_2^{2+}$ could also have existed. Solubility of copper as a chloride complex under such conditions has been estimated to be as high as 35–100 ppm (Haynes 1986a). In Figure 3.31 it is evident that the stability field of the $CuCl_3^{2-}$ complex largely coincides with that of hematite stability, and that the aqueous sulfur species would be SO_4^{2-} rather than H_2S or HS^-. Rose

(1976) points out that the conditions described above are particularly well suited to copper mobilization. By comparison, copper is relatively insoluble in normal meteoric waters and it is also unlikely to be easily leached from more reduced sediments where copper sulfides are stable. An extensive red-bed sedimentary environment where fertile detrital minerals readily break down to form a source of copper metal, and where stable $CuCl_3^{2-}$ complexes result in high copper solubilities, is, therefore, ideal for the formation of $Cu > (Pb + Zn)$ SSC deposits.

The formation of SSC deposits has, in the past, been a controversial topic and models supporting syngenetic as opposed to diagenetic origins have been much debated (Garlick 1982; Fleischer 1984; Jowett et al. 1987; Annels 1989; Sweeney et al. 1991). The syngenetic view, in which sulfides were believed to have precipitated directly out of anoxic sea water in processes akin perhaps to those giving rise to Mn deposition in the Black Sea at present times (see Chapter 5), have now largely been superseded by models that view ore genesis in terms of either diagenetic processes or epigenetic, hydrothermal fluid circulation (Sillitoe et al. 2017). A diagenetic origin is substantiated by many of the prominent features of SSC deposits, especially the broadly transgressive nature of ore zones relative to bedding, the clear relationship of metal deposition to a redox front, and replacement textures in the sulfide mineral paragenetic sequence. There are, nevertheless, several processes that can be called upon to explain ore deposition and metal zonation in SSC deposits, and in the Central African Copperbelt, for example, a hydrothermal remobilization of ores during the c. 500 Ma old Lufilian orogeny is clearly evident.

It is evident in the Kupferschiefer deposits, for example, that metal deposition is spatially associated with a reddish oxidized zone (termed the "Rote Fäule") that transgresses different lithologies and comprises hematite-replacing early diagenetic sulfides. The oxidized zone represents a chemical front in the footwall of the ore zone and against which a zoned sequence of base metal sulfides (chalcocite–bornite–chalcopyrite–galena–sphalerite–pyrite) is deposited. Much of the metal concentration is nevertheless concentrated within an organic-rich shale (i.e. the Kupferschiefer itself) since it is this unit that contained concentrations of early diagenetic framboidal iron sulfides (such as marcasite or pyrite) formed by bacterial reduction of sulfate to sulfide. Sawlowicz (1992) has documented evidence of copper sulfide replacement of framboids in the Kupferschiefer, and the process whereby chalcocite replaces biogenic sulfides has been applied more widely to

the formation of SSC deposits by Haynes (1986b). These features suggest low temperature processes and are consistent with metal deposition during the early stages of diagenesis of the sedimentary succession. A schematic representation of the zonal distribution of metals against the Rote Fäule in the Kupferschiefer ores is shown in Figure 3.32a, together with a model by Metcalfe et al. (1994) that demonstrates the effect of changing the redox state of the ore fluid on sequential metal precipitation (Figure 3.32b). In the latter model, metal precipitation is achieved by mixing the oxidized ore-bearing fluid with a reduced fluid, rather than reacting the fluid with a reduced sediment. The effect is nevertheless the same and metals are precipitated in the sequence Cu–Pb–Zn as a function of increased mixing of the two fluids. This is exactly the sequence observed in the Kupferschiefer ores and confirms that this type of zoning could be a result of sequential metal precipitation from mixed basinal fluids.

There are other ways of explaining the patterns of metal zoning and paragenetic sequence in SSC-type deposits. Alternative proposals prefer a late diagenetic origin for metal precipitation in SSC deposits where sulfate reduction to sulfide occurs at higher temperatures and is not bacterially induced. This could be achieved by fluid mixing (as suggested in Figure 3.32b) or simply by interaction between oxidized fluid and reduced country rock. Ripley et al. (1985) have proposed a scheme in which the pattern of zoning in SSC type ores is explained in terms of either fluid mixing or fluid-rock reaction. In a plot of Cu^+ versus Fe^{2+} activities (Figure 3.33) it is evident that a fluid with a high initial Cu content will become progressively more dilute as copper sulfides precipitate. This would produce a paragenetic sequence and zonation pattern that reflects a decreasing Cu content in the sulfide ores, such as chalcocite – bornite – chalcopyrite – pyrite.

Another feature of SSC deposits is that they contain appreciable amounts of other metals, such as Ag, Pb, and Zn (Kupferschiefer) and

(a)

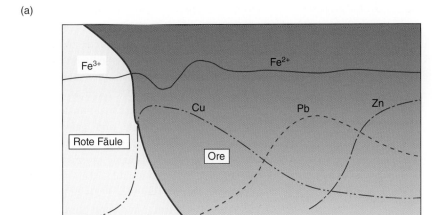

(b)

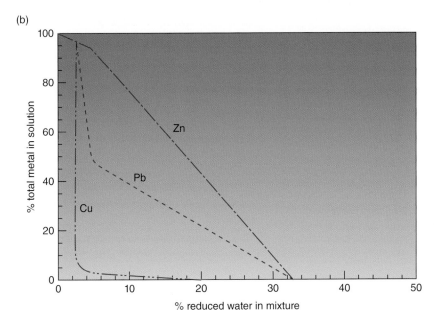

Figure 3.32 (a) Section showing metal zonation in typical Kupferschiefer ores in relation to the transgressive, oxidized "Rote Fäule" zone. Source: After Jowett et al. (1987). (b) Model showing the effects on metal precipitation of mixing a typical SSC-type ore fluid with a more reduced fluid. The sequence of deposition is the same as the zonal pattern observed in the Kupferschiefer. Source: After Metcalfe et al. (1994).

Co (Copperbelt), all of which form part of a broadly zonal arrangement of ores. Such patterns may be difficult to explain in terms of solubility contrasts when metal concentrations are well below their saturation levels, and in such cases a mechanism such as adsorption may be relevant (see Section 3.5 above). Figure 3.16b, for example, shows the variations in adsorption efficiency as a function of pH. It is feasible that variable adsorption efficiencies could control the uptake of different metals from red-beds into basinal fluids, as well as their subsequent deposition into nearby reducing sediments. This could explain regional zonation patterns in SSC ores (Rose and Bianchi-Mosquera 1993) although the details

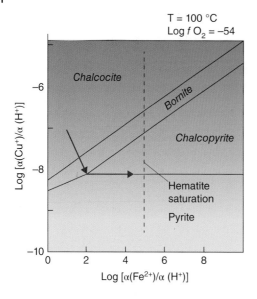

Figure 3.33 Plot of Cu$^+$/H$^+$ versus Fe$^+$/H$^+$ activity ratios to explain the zoning and paragenetic sequence that is often observed in SSC deposits when a Cu-bearing fluid reacts with a pyritic shale. In this situation the original copper content of the fluid is high and then decreases as a function of Cu-sulfide precipitation to form the paragenetic sequence chalcocite–bornite–chalcopyrite–pyrite. Source: After Ripley et al. (1985).

of how such a process relates more generally to an understanding of paragenetic sequences clearly requires more study.

3.10.2 Mississippi Valley Type (MVT) Pb–Zn Deposits

MVT deposits, like SEDEX and SSC ores, owe their origin to fluid circulation and metal transport/deposition within sedimentary basins.

Unlike the syngenetic to diagenetic time frame of SEDEX and SSC deposit formation, however, MVT ores are distinctly epigenetic and metals can be deposited long after sediment deposition. MVT deposits form from relatively low temperature fluids (<150 °C) and are broadly stratabound, mainly carbonate-hosted, and dominated by sphalerite and galena with associated fluorite and barite. Most MVT deposits contain significantly more sphalerite than galena, with the notable exception of the Viburnum Trend in southeast Missouri, where the reverse applies. Sandstone-hosted MVT deposits (such as Laisvall and some deposits in southeast Missouri) represent a conceptually intermediate category between SSC and carbonate-hosted MVT ores (Figure 3.30) and are characterized by Pb > Zn ≫ Cu. MVT deposits contain a significant proportion of the world's Zn and Pb reserves and there are many such deposits around the world in addition to the numerous mines in the "type area" of the southeastern USA. In the latter region the main metal-producing districts include the Viburnum Trend of southeast Missouri (which is mainly a Pb-producing region), the Tri-State (i.e. Oklahoma, Kansas, Missouri) region, the Illinois–Wisconsin region and the Mascot–Jefferson City area of east Tennessee. Other well-known MVT districts or deposits elsewhere in the world include the Pine Point and Polaris deposits of northern Canada, the Silesian district of Poland, Mechernich in Austria, the Lennard Shelf district, Sorby Hills and Coxco in Australia, and Pering in South Africa (Misra 2000).

Box 3.6 Circulation of Sediment-Hosted Basinal Fluids: 1 The Central African Copperbelt

The stratiform, sediment-hosted Cu–Co deposits of the Neoproterozoic Katangan sequence of Zambia and the Democratic Republic of the Congo (DRC) host the Central African Copperbelt, undoubtedly one of the great metallogenic provinces of the world (Fleischer et al. 1976). Several world class deposits (such as Nchanga, Konkola, and Tenke Fungurume), together with dozens of smaller mines, combine to make this region one of the foremost copper-producing areas of the world, as well as its most important Co supplier. Although figures for the entire

Box 3.6 (Continued)

Central African Copperbelt are unreliable, it has been estimated that the region still contains resources in excess of 150 million tons of Cu metal and 8 million tons of Co metal (Misra 2000).

The Katangan sediments were deposited in an intracratonic rift formed on a Paleoproterozoic basement comprising 2050–1870 Myr old granite–gneiss, and meta-volcanosedimentary sequences (Figure 3.6.1; Unrug 1988). The basement is thought to represent a deformed but fertile magmatic arc that contained the ultimate source of Cu (in porphyry type intrusions; Rainaud et al. 1999), and possibly Co too. Initiation of rifting and sediment deposition followed soon after intrusion of the Nchanga granite at 880 Ma. Initially sedimentation was dominated by siliciclastic deposition, but this was followed by playa lake and restricted marine incursions to form evaporites, shales, and carbonate rocks. A major glacial diamictite (the "grand

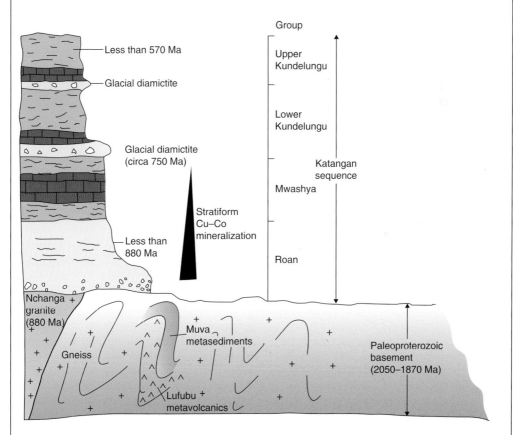

Figure 3.6.1 Simplified stratigraphic profile through the rocks of the Central African Copperbelt showing the relationship between the Paleoproterozoic granite–gneiss and volcano-sedimentary basement and the overlying Katangan sequence. Stratiform Cu–Co mineralization is largely hosted in a variety of sedimentary lithotypes in the lower part of the Roan Group, but does also occur higher up in the succession and up into the Mwashya Group. Source: The age constraints used in this compilation are after Rainaud et al. (2002a), Master et al. (2005), and Key et al. (2001).

(Continued)

Box 3.6 (Continued)

Figure 3.6.2 Folded metasediments hosting stratiform Cu–Co mineralization at the Chambishi Mine, Zambia. The bedding parallel ore is cut by an axial planar cleavage that formed during deformation. Source: Photograph courtesy of Lynnette Greyling.

conglomerat") occurs at the top of the Mwashya Group, the age of which (circa 750 Ma) makes it a likely correlative of the mid-Neoproterozoic "Snowball Earth" or Sturtian glaciation (Robb et al. 2002). The uppermost parts of the Katangan sequence (i.e. the Upper Kundelungu Group and its correlatives) now appear to be substantially younger (i.e. less than 565 Ma) than the underlying portions and are regarded as a separate basin that formed as a foreland to the Pan-African orogeny (Wendorff 2001).

Mineralization in the Central African Copperbelt is a much debated and contentious issue. Ideas on ore genesis in the 1960s and 1970s were dominated by syngenetic models, where metals were accumulated essentially at the same time as the sediments were being deposited (Fleischer et al. 1976). More recent work has shown that mineralization is more likely to be related to later fluids circulating through the sediments, either during diagenesis and com-paction of the sequence (Sweeney et al. 1991) or significantly later during deformation and metamorphism (Unrug 1988). Although much work is still required, it seems likely that metals were sequentially precipitated by processes related to reduction–oxidation reactions between sedimentary host rocks and a fertile basinal fluid. A highly oxidized, saline, and sulfate-rich fluid, derived by interaction with evaporites in the Roan Group, is regarded as the agent that scavenged Cu and Co from the sediments themselves. The Cu may have been introduced into the basin in Cu-rich magnetite or mafic mineral detritus eroded from the fertile Paleoprotero-zoic magmatic arc in the hinterland (S. Master, personal communication). Cobalt, on the other hand, may have been derived from penecontemporaneous sills that intrude the upper Roan Group (Annels 1984). Cu and Co would be highly soluble as chloride complexes in a basinal fluid of this type (see Section 3.10.1 of this chapter) and it is likely that very efficient leaching of these metals during diagenesis and oxidative alteration of the host rocks would have occurred. Precipitation of these metals would occur as the fluid reacted, either with more reduced strata (containing organic carbon or diagenetic framboidal pyrite), or with a second, more reduced,

---

Box 3.6 (Continued)

---

fluid. Metal zonation, which is a feature of mineralization throughout the Copperbelt, could, at least in part, be a product of the variable reduction potentials of metals such as Cu, Co, Pb, and Zn. An early or late diagenetic model for the origin and timing of stratiform mineralization in the Central African Copperbelt is consistent with the fact that it pre-dates the major period of compressive deformation that affected the entire region during the Lufilian orogeny (590–510 Ma; Rainaud et al. 2002b). Figure 3.6.2 shows the folded nature of the host rock strata at the Chambishi Mine in Zambia; the stratiform Cu–sulfide ore here is also folded and cut by an axial planar cleavage.

The majority of MVT deposits worldwide formed in the Phanerozoic Eon, and more specifically, in Devonian to Permian times; some deposits, such as Pine Point and the Silesian district, however, formed during the Cretaceous–Paleogene period (Leach et al. 2001). The former, more important, metallogenic epoch coincides with the assembly of Pangea when major compressional orogenies were active over many parts of the Earth (see Chapter 6). The later period is more specifically related to the Alpine and Laramide orogenies and, likewise, associated with compressional tectonic regimes. The timing of MVT deposits in relation to associated orogenic activity provides an important clue as to the causes of fluid flow and hydrothermal activity involved in their formation. It is generally accepted that topographically driven fluid pathways were critical to the development of large MVT ore districts, and that the carbonate host rocks maintained a hydrological connection to orogenic belts active during the period of ore deposition (Rickard et al. 1979; Oliver 1986; Duane and De Wit 1988). Other features believed to conceptually link most MVT deposits include a low-latitudinal setting, where high rainfall ensured an adequate fluid reservoir, and the presence, somewhere in the fluid flow system, of an evaporitic sabkha environment that contributed to the high fluid salinities necessary for enhanced metal solubility and transport (Leach et al. 2001).

Figure 3.34 illustrates the main characteristics of MVT deposits. Regional fluid flow is stimulated by a compressional orogeny that results in thrust faulting and uplift which, in turn, creates a topographic head and fluid flow down a hydrological gradient. Fluid flow in these tectonic settings occurs over distances of hundreds of kilometers and such fluids are implicated in the migration of hydrocarbons as well as metals (Rickard 1976; Oliver 1986). At the site of metal deposition MVT ores are sometimes bedded and focused along conformable dolostone/limestone interfaces, but are more commonly associated with discordant, dissolution-related zones of brecciation. The hydrothermal fluids that are linked to MVT deposit formation are typically low-temperature (100–150 °C), high-salinity (>15 wt% NaCl equivalent) brines, with appreciable SO_4^{2-}, CO_2, and CH_4, as well as associated organic compounds and oil-like droplets (Gize and Hoering 1980; Roedder 1984). These compositional characteristics are not unlike oil-field brines and reflect protracted fluid residence times in the sedimentary basin.

In many MVT deposits ores occur as cement between carbonate breccia fragments suggesting that brecciation occurred either before (perhaps as a karst related feature) or during metal deposition (Anderson 1975). Hydrothermal dissolution of carbonate requires acidic fluids and could occur by a reaction such as:

$$CaMg(CO_3)_2 + 4H^+ \Leftrightarrow$$
$$Ca^{2+} + Mg^{2+} + 2H_2O + 2CO_2 \quad (3.11)$$

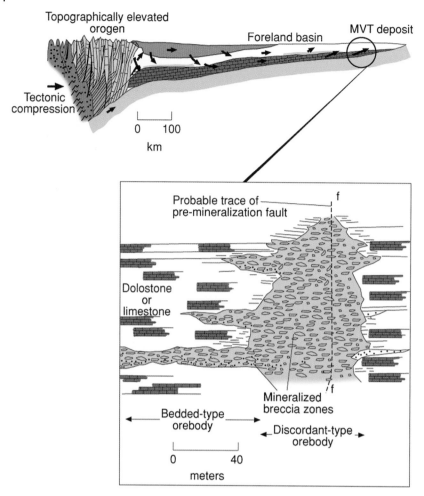

Figure 3.34 Diagram illustrating the concept of hydrological continuity between a compressional orogenic belt and a foreland sedimentary basin through which orogenically and topographically driven fluids flow, and within which MVT Zn–Pb deposits form. Inset shows characteristics of bedded and discordant mineralization at a deposit scale. Mineralization may be bedded and associated with dolostone/limestone sequences, or in discordant carbonate breccia zones. Source: After Garven et al. (1993).

The fluid itself might originally have been acidic, or hydrogen ions were produced by precipitation of metal sulfides according to the reaction:

$$H_2S + Zn^{2+} \Leftrightarrow ZnS + 2H^+ \qquad (3.12)$$

The production of additional Ca^{2+}, Mg^{2+}, and CO_2 in the fluid as a result of carbonate dissolution provides the ingredients for later precipitation of calcite and dolomite, a feature that is observed in some MVT deposits where secondary carbonate gangue minerals cement previously precipitated ore sulfides (Misra 2000).

The rather unusual nature of the ore fluid associated with MVT ore formation raises some interesting questions concerning metal transport and deposition (Sverjensky 1986). Zn and Pb are borderline Lewis acids (see Section 3.4.1 above) and can complex with both chloride and bisulfide ligands. At high temperatures and under acidic conditions the metal–chloride complex tends to be more stable, whereas at lower temperatures and neutral

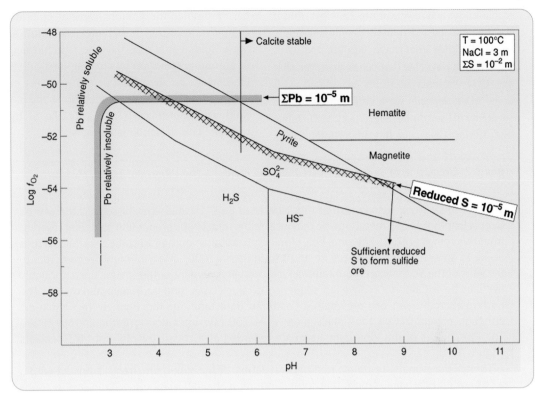

Figure 3.35 fO_2–pH plot showing key mineral and aqueous species stabilities under the conditions shown. Solubility contours of $\Sigma Pb = 10^{-5}$ m and reduced sulfur $= 10^{-5}$ m are also shown. Higher solubilities lie to the left of and above the Pb contour, and below the sulfur contour. Source: After Anderson (1975).

to alkaline pH the metal–bisulfide complexes dominate. There seems little doubt, given the prevailing conditions, that metal–chloride complexes predominated during MVT ore formation (Sverjensky 1986). Anderson (1975) pointed out that viable Zn and Pb solubilities are, however, only achieved when fluids are either relatively oxidized (i.e. the sulfur in the fluid is transported as SO_4^{2-} and not as H_2S or HS^-), or at low pH. Figure 3.35 shows that if a fluid was sufficiently oxidized for SO_4^{2-} to have been stable, then high Pb solubilities could have occurred over a range of pH. That the sites of MVT ore deposition are characterized by sulfide minerals however, implies that metal precipitation must have been accompanied by the reduction of sulfate to sulfide. In some MVT deposits, such as the Pine Point district, such a process is feasible

and ore deposition has been attributed to the mixing of a sulfur-deficient ore fluid with another carrying significant reduced sulfur (i.e. H_2S or HS^-) at the site of ore deposition (Beales and Jackson 1966). Alternatively, the ore fluid might originally have contained significant SO_4^{2-}, with reduction and metal precipitation occurring, again either by mixing with a reduced fluid, or simply by reaction of the fluid with organic matter. In many MVT deposits however, textures and geochemistry are incompatible with a fluid mixing model and ores appear to have precipitated over a protracted period of time. If fluid mixing and/or sulfate reduction did not occur then the ore fluid itself would have needed sufficient reduced sulfur to form sulfide minerals (i.e. in excess of 10^{-5} m; see Figure 3.35). In a reduced fluid metal–chloride complexes will only be stable under acidic conditions (low pH) and it has been suggested that this could not apply to MVT fluids that have equilibrated with carbonate host rocks. It appears, however, that the CO_2 contents of fluids associated with some MVT deposits were sufficiently high to form low pH solutions (Sverjensky 1981; Jones and Kesler 1992). Such acidic fluids would have promoted the stability of metal–chloride complexes resulting in high metal solubilities. Such fluids could also have been implicated in ground preparation by creating dissolution breccias at the site of ore deposition. This in turn resulted in metal precipitation as the ore fluid was neutralized (Eq. (3.11) above) by reaction with the carbonate host rock.

Box 3.7 Circulation of Sediment-Hosted Basinal Fluids: 2 The Viburnum Trend, Missouri

The Viburnum Trend, also known as the "New Lead Belt," in southeast Missouri is the world's largest lead-producing metallogenic province. In the mid-1980s production plus total reserves from this remarkable, 65 km long, belt of mineralization amounted to some 540 million tons of ore at an average grade of 6% Pb and 1% Zn (Misra 2000). The Viburnum Trend occurs to the west of the St. Francois Mountains, a Precambrian basement outlier that is unconformably overlain by a Cambro–Ordovician sequence of siliciclastic and carbonate sediments. The host rocks to Pb–Zn ore are mainly Cambrian dolostones of the Bonneterre Formation, although mineralization took place more than 200 Myr later during Carboniferous times. The Pb–Zn ores of the Viburnum Trend are regarded as a classic example of MVT deposits (Figure 3.7.1).

The galena–sphalerite dominated mineralization of the Viburnum Trend is hosted within a sequence of carbonate rocks that formed during a shallow marine incursion over a continental shelf (Larsen 1977). A stromatolitic reef developed around the northern and western margins of a positive feature (the St. Francois Mountains) on this shelf, resulting in the formation of biostomal dolomites that are underlain by the porous Lamotte sandstone and overlain by the relatively impervious Davis shale (Figure 3.7.2). Accurate age dating of the actual mineralization in the Viburnum Trend, and indeed for much of the related Pb–Zn MVT styles of mineralization throughout the southeastern USA, indicates that it formed during a relatively brief interval of time in the late Carboniferous (i.e. between about 330 and 300 Ma; Leach et al. 2001). This interval coincides with a period of Earth history characterized by major compressional orogeny, during which time the amalgamation of the supercontinent Pangea was taking place (see Chapter 6). In the southeastern USA this event is recorded by the Ouachita collisional orogen and it is suggested that it was this event that stimulated fluid flow throughout the region. Orogeny- or topography-driven fluid flow (see Sections 3.3.1 and 3.10.2 of this chapter) focused hydrothermal solutions into connected paleo-aquifers such as the Lamotte sandstone, with subsequent metal precipitation occurring in response to features such as fluid interaction and dissolution of carbonate rocks. An interesting feature of many of the Viburnum Trend deposits is the close association between basement topography, overlying sedimentary strata pinch-outs, and mineralization. Figure 3.7.2 shows a west–east section through the Fletcher Mine where mineralization is constrained to a zone in the Bonneterre dolostones that is directly above a basement topographic high. Stratal pinch-outs against this buttress suggest that fluids, originally flowing through underlying successions such as the Lamotte sandstone (where some mineralization does occur) or the lower stromatolitic reefs, might have been forced upwards into the clastic carbonates, and that this was an important factor promoting metal precipitation.

Box 3.7 (Continued)

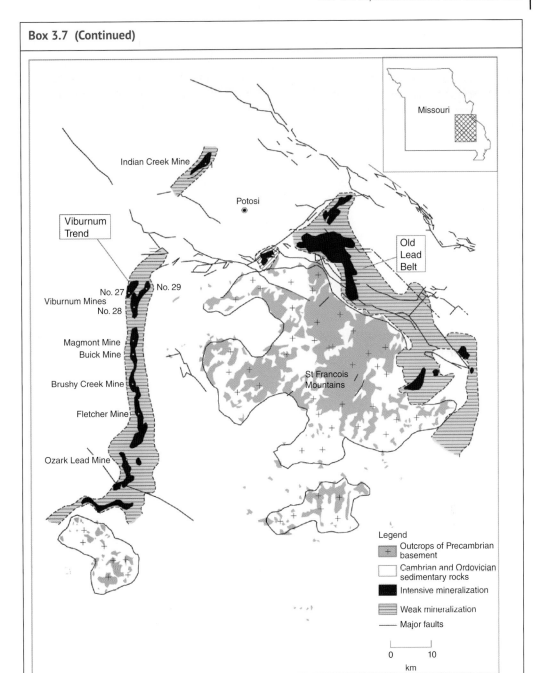

Figure 3.7.1 Geological setting of the Viburnum Trend Pb–Zn mineralization and the major mines in this belt. Source: After Kisvarsanyi (1977).

(Continued)

Box 3.7 (Continued)

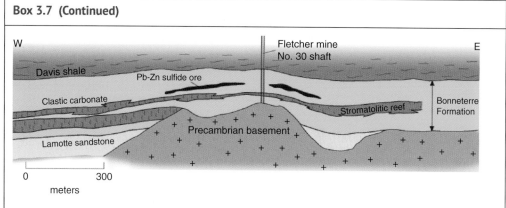

Figure 3.7.2 West–east section through the Fletcher Mine, Viburnum Trend, showing the close association between mineralization, basement topographic highs, and stratal pinch-outs. Source: After Paarlberg and Evans (1977).

3.11 Ore Deposits Associated with Near Surface Meteoric Fluids (Groundwater)

Few hydrothermal ore deposits are linked to the circulation of ambient meteoric fluids, or groundwaters, in the near surface environment since metals are generally not transported in these types of fluids. A notable exception is the family of sediment-hosted uranium deposit types associated with low temperature meteoric fluid flow (Nash et al. 1981). One category of these deposit types is the calcrete-hosted, uranium–vanadium ores that form in arid regions such as parts of Western Australia and Namibia. These deposits are related to high rates of evaporation in surficial environments and, consequently, are discussed in Chapter 4. The other category, the sandstone-hosted uranium ores, is related to groundwater flow and redox precipitation mechanisms, and is discussed below.

3.11.1 A Brief Note on the Aqueous Transport and Deposition of Uranium

In nature uranium occurs in two valence states, as the uranous (U^{4+}) ion and the uranyl (U^{6+}) ion. In the magmatic environment uranium occurs essentially as U^{4+} and in this form is a highly incompatible trace element that occurs in only a few accessory minerals (zircon, monazite, apatite, sphene, etc.) and is concentrated into residual melts. The uranous ion is, however, readily oxidized to U^{6+} in meteoric waters. In most natural aqueous solutions uranous complexes are insoluble, but U^{6+} forms stable complexes with fluoride (under acidic conditions at pH < 4), phosphate (under near neutral conditions at 5 < pH < 7.5), and carbonate (under alkaline conditions at pH > 8) ligands. The oxidized form of uranium is, therefore, easily transported over a wide range of pH conditions, whereas the reduced form of the metal is far less soluble. Figure 3.36, an Eh–pH diagram for the system U–O_2–CO_2–H_2O at 25 °C, shows that for most meteoric waters in the near neutral pH range, the dominant aqueous species are likely to be U^{6+}–oxide or –carbonate complexes. Uranium can be precipitated by lowering Eh of the fluid to form a uranous oxide, in the form of either uraninite or pitchblende.

Consequently, many low temperature uranium deposit types are a product of oxidation–reduction processes where the metal is transported as U^{6+} and precipitated by reduction. In the presence of high

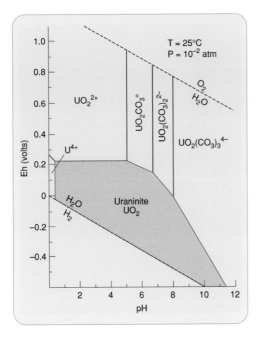

Figure 3.36 Eh–pH diagram showing relevant aqueous uranium species for the conditions specified. For most meteoric waters in the near neutral pH range, the dominant aqueous species are likely to be U^{6+}–oxide or –carbonate complexes. These will be precipitated from solution by reduction (i.e. a decrease of Eh) to form U^{4+}–oxide, or uraninite. Source: After Langmuir (1978).

concentrations of other ligands, uranium can be transported as a variety of different complexes, such as UO_2Cl^+, which is stable at higher temperatures ($>100\,°C$) and low pH (Kojima et al. 1994). In addition to redox controls, uranium precipitation may also be promoted by fluid mixing, changes in fluid pH, and adsorption. A more complete overview of uranium ore-forming processes is presented in Nash et al. (1981).

3.11.2 Sandstone-Hosted Uranium Deposits

Some of the world's best-known sandstone-hosted uranium deposits occur in the USA. These deposits occur mainly in three regions, the Colorado Plateau region (around the mutual corners of Utah, Colorado, New Mexico, and Arizona states), south Texas, and the Wyoming–South Dakota region. Smaller, generally non-viable, deposits are known from elsewhere, such as the Permian Beaufort Group sandstones of the Karoo sequence in South Africa and the Lodève deposits in France. The deposits are generally associated with Paleozoic to Mesozoic fluvio-lacustrine sandstones and arkoses and are stratabound, although different geometrical variants exist. Most of the deposits in the Colorado Plateau region are tabular and associated with either organic material or vanadium (Northrop and Goldhaber 1990). Deposits in Wyoming and south Texas, by contrast, are sinuous in plan and have a crescent shape in cross section. Although still stratabound these ores are discordant to bedding and are typically formed at the interface between oxidized and reduced portions of the same sandstone aquifer (Reynolds and Goldhaber 1983; Goldhaber et al. 1983). It is the latter type that is commonly referred to as a roll-front uranium deposit. The genesis of the two subtypes is quite different and they are discussed separately below.

3.11.2.1 Colorado Plateau (Tabular) Uranium–Vanadium Type

Important features of tabular sandstone-hosted deposits are that they can occur stacked one on top of the other, and that the ore zones are bounded above and below by sandstones containing a dolomitic cement. The generally accepted model for the formation of these deposit types is based on work done in the Henry Basin of Utah, where stacked U–V ore bodies are considered typical of the Colorado Plateau region as a whole (Northrop and Goldhaber 1990). The ore bodies of this area occur predominantly within the Jurassic, sandstone-dominated Salt Wash Member of the Morrison Formation, although on a regional scale they are slightly discordant with respect to bedding (Figure 3.37a). The ores are believed to have formed at the interface between two discrete, low temperature

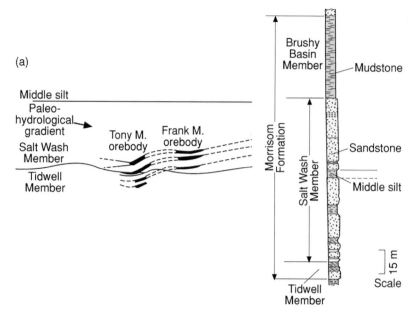

Figure 3.37 (a) The orebody geometry and nature of host rocks for the tabular uranium–vanadium deposits of the Henry Basin area of Utah, considered typical of the Colorado Plateau region as a whole. (b) Secular genetic model for the tabular, stacked orebody geometries of Colorado Plateau type deposits, involving an upward-migrating mixing zone between a lower stagnant brine and an overlying, ore-bearing, aquifer-focused meteoric fluid. Source: After Northrop and Goldhaber (1990).

meteoric fluids. One of these fluids, a relatively stagnant but saline basinal brine containing Na^+, Mg^{2+}, and Ca^{2+} cations and Cl^- and SO_4^{2-} anions, underlies a low salinity, meteoric fluid that flows readily along aquifer horizons bringing with it highly soluble metal species such as $UO_2(CO_3)_2^{2-}$ and VO^+. Mixing of these two fluids results in precipitation of metals in subhorizontal tabular zones reflecting the fluid interface (Figure 3.37b). The main uranium ore mineral in these deposits is a uranous silicate, coffinite $(U(SiO_4)_{1-x}(OH)_{4x})$, which is associated either with syn-sedimentary organic (plant) debris, or together with a vanadium-rich chlorite in the intergranular sandstone pore spaces.

Fluid mixing and metal precipitation occurred within a few hundred meters of the surface such that fluid temperatures were unlikely to have exceeded 30–40 °C. The stagnant basinal brine is considered to have been locally derived, but interacted with evaporitic sequences, which explains its high Mg/Ca

ratios. The overlying meteoric fluid was essentially groundwater that scavenged U^{6+} and V^{4+}, as well as other oxide-soluble metals such as Cu, Co, As, Se, and Mo, from overlying, uranium-enriched tuffaceous rocks, or from breakdown of detrital Fe–Ti oxide minerals in the aquifer sediments through which it flowed (Reynolds and Goldhaber 1978). Mixing of the two fluids and associated diagenetic reactions ultimately gave rise to ore formation, the details of which are presented in Northrop and Goldhaber (1990). One of the products of fluid mixing in this environment was a dolomite cement that precipitated from pH-related (Ca, Mg)CO_3 oversaturation within the basinal brine. The pH changes in the mixing zone appear to have been a product of the formation of vanadium-rich clays (smectite–chlorite) during compaction and sediment diagenesis. Vanadium is incorporated into the dioctahedral interlayer hydroxide sites (as $V(OH)_3$) of the clay minerals, which reduces the [OH^-] activity of the fluid and decreases its pH.

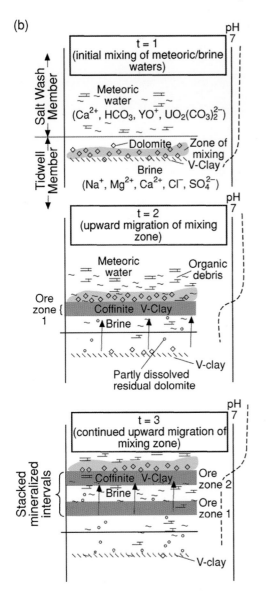

Figure 3.37 (*Continued*)

solubility which promotes the adsorption of UO_2^+ molecules onto the surface of quartz grains. Once the fluid is reduced (probably by bacterially induced SO_4 reduction and production of H_2S) to stabilize U^{4+} complexes, bonding between the uranous ion and silica leads to the formation of coffinite (Northrop and Goldhaber 1990). It should be noted that organic matter played an important role during the ore-forming process, as is evident from the very close association between coffinite and plant debris. In addition to the fact that organic matter is itself a reductant, it supplies a source of nutrients for sulfate reducing bacteria that in turn supply biogenic H_2S for reduction of U^{6+} and V^{4+} to their insoluble lower valency forms. In the Grants mineral belt of the Colorado Plateau region uranium mineralization is associated with organic matter in the form of humate that was introduced epigenetically into the sediments by organic acid rich fluids.

The model illustrated in Figure 3.37b is particularly applicable to the tabular, stacked ores of the Colorado Plateau region in which there is a uranium–vanadium association. It suggests that the stacked nature of the tabular ore bodies could be the product of a two-fluid interface that episodically migrates upwards with time. The dominant control on ore formation is pH, although adsorption, bacterial mediation, and redox reactions are also important processes. Evidence for the migration of the low pH zone is provided by the fact that the footwall zones to each ore horizon retain evidence of ore-related dolomite crystals partly dissolved by the acidic fluid as it moved upwards through the sequence.

The same chemical controls are also implicated in the formation of coffinite ore, although the actual processes are more complex and involved two stages. The uranyl–dicarbonate complex most likely to have been dissolved in this meteoric fluid is stable over only a limited pH range (around 7–8 in the conditions applicable to Figure 3.36). A decrease in fluid pH caused by fluid mixing and V-clay formation causes a lowering of $UO_2(CO_3)_2^{2-}$

3.11.2.2 Roll-Front Type

Although somewhat similar to tabular deposits, roll-front ores form as a consequence of different chemical processes. In its simplest form the ore genesis model for roll-front deposits that best accommodates their geometry, is one where an oxidized, meteoric fluid transporting soluble uranyl–carbonate complexes flows along a sandstone aquifer and

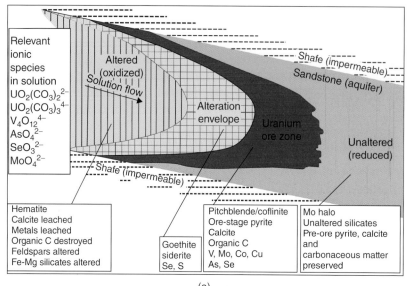

(a)

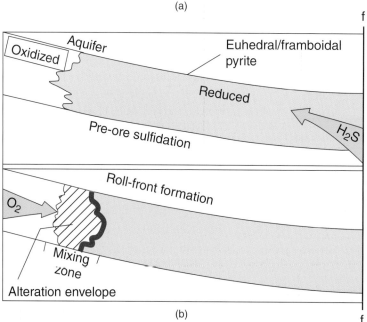

(b)

Figure 3.38 (a) Cross section through a typical roll-front type sandstone-hosted uranium deposit, such as those in Wyoming, South Dakota, and south Texas. Uranium mineralization occurs at the interface (or redox front) between the altered/oxidized and unaltered/reduced portions of the same aquifer. Source: After De Voto (1978), Misra (2000). (b) A two-fluid mixing model for roll-front deposits. Initially a reduced fluid migrates up a syn-depositional fault (f) and sulfidizes the sandstone (top box). This is followed by down-dip migration of an oxygenated, uranium-bearing meteoric fluid and the precipitation of uranium ore at the redox front formed in the mixing zone (bottom box). Source: Model is after Goldhaber et al. (1978).

precipitates uranium ore at a redox front (Figure 3.38a). Up-dip of the roll-front, the sandstones are altered (detrital silicate minerals are altered to clays, Fe–Ti oxide phases are leached) and oxidized (secondary hematite has formed, organic carbon is biodegraded). The redox front itself is characterized by an inner alteration envelope (marked by goethite, siderite, pyrite, or marcasite) and an outer ore zone (comprising pitchblende and/or coffinite with pyrite and some organic carbon). Down-dip of the redox front, the sandstone is relatively unaltered and contains a more reduced assemblage of pre-ore pyrite, calcite, and organic matter, with detrital mineral phases relatively intact (Figure 3.38a). In this model the redox front can be regarded as the position in space and time where the meteoric fluid has lost its capacity to oxidize the sandstone through which it is percolating. Eh of the fluid changes abruptly at the redox front and soluble uranyl–carbonate complexes are destabilized with precipitation of uranous oxide or silicate minerals. Other soluble metal–oxide complexes also present in the fluid (such as V, As, Se, Mo, Cu, Co) are likewise reduced and precipitate as various minerals, either before or after uranium depending on their relative solubilities as a function of Eh. The roll-front does not remain static and it will migrate down the paleoslope as the meteoric fluid is recharged and the process of progressive, down-dip oxidation of the sandstone evolves with time. The down-dip migration of the redox front can be equated to the process of zone-refining and the ore body is "frozen in" only once the paleohydrological regime changes and fluid flow ceases.

A modification of this scheme has been proposed by Goldhaber et al. (1978, 1983), who noted that some roll-front deposits are spatially associated with syn-depositional faults, and that the uranium ores are superseded by paragenetically later sulfide minerals. A fluid mixing model was proposed to accommodate these features and it is envisaged that a reduced fluid migrated up the fault and sulfidized the sandstone aquifer for several hundred meters on either side (Figure 3.38b). This stage of ground preparation created the reduced environment responsible for locating the ore-related redox front. Evidence for this additional stage in the mineralization process is provided by euhedral pyrite (in deposits that do not contain organic matter) or framboidal pyrite (in those deposits that do). This stage was followed by, or was largely coincidental with, down-dip migration of an oxidized, uranium-bearing meteoric fluid and the subsequent precipitation of uranium ore at a redox front represented by the zone of fluid mixing (Figure 3.38b). In certain cases, such as the deposits of south Texas, continued flow up the fault of sulfide bearing solutions from deeper within the basin resulted in post-mineralization sulfidization and the growth of late-stage pyrite and marcasite that overprinted even the altered up-dip sections of the sandstone aquifer.

3.12 Summary

There are many different types of fluids circulating through the Earth's crust and these have given rise to a wide variety of hydrothermal ore deposit types that exist in virtually every tectonic setting and have formed over most of Earth history. Juvenile fluids exsolve from magmas, in particular those with felsic compositions, and give rise to granite related ore deposits that include porphyry Cu–Mo, skarn and greisen related Sn–W, intrusion linked Fe oxide–Cu–Au, and high sulfidation Au–Ag ores (discussed in Chapter 2). Metamorphic fluids, derived from volatiles liberated during prograde mineral reactions, are typically aqueo-carbonic in composition and are associated worldwide with orogenic gold deposits that are particularly well developed in Archean and Phanerozoic rocks. Basinal fluids formed during diagenesis interact with either reduced (forming Pb–Zn dominant ores) or oxidized (forming Cu dominant ores)

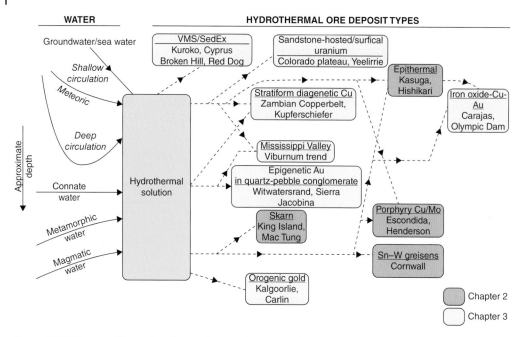

Figure 3.39 Diagram illustrating the relationship between different fluid types and various hydrothermal ore deposit types. Source: The diagram is relevant to both Chapters 2 and 3, and is modified after Skinner (1997).

sedimentary environments. These fluids are implicated in the formation of MVT (Pb–Zn) and stratiform sediment-hosted copper ores. Circulation of near surface meteoric waters can dissolve labile constituents such as the uranyl ion, giving rise to a variety of different sediment-hosted uranium deposits. Finally, sea water circulating through fractured oceanic crust is vented onto the sea floor as black smokers, providing a modern analogue for the formation of Cu–Zn-dominated VMS deposits. Similar exhalative processes also occur in different tectonic settings and with different metal assemblages, giving rise to sediment-hosted, Zn–Pb-dominated SEDEX type deposits.

The various hydrothermal deposits described in Chapters 2 and 3, and their relationships to the different fluid types discussed in these sections, are summarized in Figure 3.39.

Formation of hydrothermal ore deposits is linked not only to the generation of significant volumes of fluid in the Earth's crust, but also its ability to circulate through rock and be focused into structural conduits (shear zones, faults, breccias, etc.) created during deformation. The ability of hydrothermal fluids to dissolve metals provides the means whereby ore-forming constituents are concentrated into this medium. Temperature and composition of hydrothermal fluids (in particular the presence and abundance of dissolved ligands able to complex with a variety of metals), together with pH and fO_2, control the metal-carrying capability of any given fluid. Precipitation of metals is governed by a reduction in solubility which can be caused by either compositional (fluid-rock interaction, fluid mixing), or physical (P and T) changes in the fluid itself. Economically viable hydrothermal ore deposits occur when a large volume of fluid with a high metal-carrying capacity is focused into a location that is accessible, and where efficient precipitation mechanisms can be sustained for a substantial period of time.

Further Reading

For those readers wishing to delve further into hydrothermal ore-forming processes, the following is a selection of references to books and journal special issues.

Barnes, H.L. (ed.) (1967). *Geochemistry of Hydrothermal Ore Deposits*, 1e. Holt, Rinehart and Winston Inc.

Barnes, H.L. (ed.) (1979). *Geochemistry of Hydrothermal Ore Deposits*, 2e. Wiley.

Barnes, H.L. (ed.) (1997). *Geochemistry of Hydrothermal Ore Deposits*, 3e. Wiley.

Barrie, C.T. and Hannington, M.D. (eds.) (1999). *Volcanic-Associated Massive Sulfide Deposits: Processes and Examples in Modern and Ancient Settings*, *Reviews in Economic Geology*, vol. 8. El Paso, TX: Society of Economic Geologists, 408 pp.

Gill, R. (1996). *Chemical Fundamentals of Geology*. London: Chapman & Hall, 290 pp.

Goodfellow, W.D. (ed.) (2007). *Mineral Deposits of Canada*, Special Publication 5. Geological Association of Canada, 1061 pp.

Guilbert, J.M. and Park, C.F. (1986). *The Geology of Ore Deposits*. London: W.H. Freeman and Co, 985 pp.

Hagemann, S.P. and Brown, P.E. (eds.) (2000). *Gold in 2000*, *Reviews in Economic Geology*, vol. 13. El Paso, TX: Society of Economic Geologists, 559 pp.

Hedenquist, J.W., Thompson, J.F.H., Goldfarb, R.J., and Richards, J.P. (eds.) (2005). *One Hundredth Anniversary Volume*. Society of Economic Geologists, 1136 pp.

Misra, K.C. (2000). *Understanding Mineral Deposits*. Dordrecht: Kluwer Academic Publishers, 845 pp.

Pirajno, F. (1992). *Hydrothermal Mineral Deposits*. New York: Springer-Verlag, 709 pp.

Richards, J.P. and Larson, P.B. (eds.) (1998). *Techniques in Hydrothermal Ore Deposits Geology*, *Reviews in Economic Geology*, vol. 10. El Paso, TX: Society of Economic Geologists, 256 pp.

Ridley, J. (2013). *Ore Deposit Geology*. Cambridge University Press, 409 pp.

Wolf, K.H. (ed.) (1976). *Handbook of Strata-Bound and Stratiform Ore Deposits*, vol. 1–14. New York: Elsevier.

Part III

Sedimentary/Surficial Processes

4

Surficial and Supergene Ore-Forming Processes

TOPICS

Principles of chemical weathering
    dissolution and hydration
    hydrolysis and acid hydrolysis
    oxidation–reduction
    cation exchange
Formation of lateritic soil/regolith profiles
Bauxite ore formation
Nickel, gold, and PGE in laterites
Clay deposits
    HREE and ion adsorption clays
Calcretes and surficial uranium deposits
Supergene enrichment of copper and other metals

CASE STUDIES

Box 4.1 Lateritization: Bauxite – The Los Pijiguaos Deposit, Venezuela
Box 4.2 Supergene Processes: Supergene and "Exotic" Mineralization in Porphyry Copper Deposits

---

4.1 Introduction

Once metals have been concentrated in the Earth's crust and then exposed at its surface, they are commonly subjected to further concentration by chemical weathering. The relationship between weathering and ore formation is often a key ingredient that leads to the creation of a viable deposit and many ores would not be mineable was it not for the fact that grade enhancement commonly occurs in the surficial environment. There are several deposit types where the final enrichment stage is related to surficial weathering processes. Some of these deposit types are economically very important and contain ores, such as bauxite, that do not occur in any other form. Ore-forming processes in the surficial environment are intimately associated with pedogenesis, or soil formation, which is a multifaceted process reflecting local climate and the chemical interactions between rock and atmosphere. Soil itself is an extremely valuable resource since the world's food production is largely dependent on its existence and preservation. This chapter examines the concentration of metals in soil, as well as in the associated regolith (i.e. all the unconsolidated material that rests on top of solid, unaltered rock). Many different metals are enriched in the surficial environment, the most important of which, from a metallogenic viewpoint, include Al, Ni, Mn, Fe, Cu, Au, Pt, REE (Rare Earth Elements), and U. Emphasis is placed here on laterites, with their associated enrichments of Al, Ni, Au, and platinum group elements (PGEs). In addition, clay deposits, uranium-enriched calcretes, and the supergene enrichment of Cu in porphyry-type deposits, are also discussed. Surficial enrichment of metals takes place in many other settings too, and is a key process, for example, in the formation of viable iron ores associated with banded iron-formations. The latter are described in more detail in Chapter 5 (Box 5.2).

4.2 Principles of Chemical Weathering

From a metallogenic point of view chemical weathering can be subdivided into three processes:

- Dissolution of rock material and the transport/removal of soluble ions and molecules by aqueous solutions. Some of the principles applicable to this process at higher temperatures were discussed in Chapters 2 and 3.
- Production of new minerals, in particular clays, oxides and hydroxides, and carbonates. Again, this topic was discussed briefly in the section on hydrothermal alteration in Chapter 3.
- Accumulation of unaltered residual material such as silica, alumina, and low solubility metals such as gold.

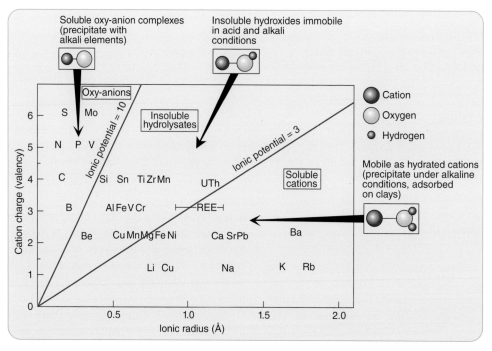

Figure 4.1 Simplified scheme on the basis of ionic potential (ionic charge/ionic radius) showing the relative mobility of selected ions in aqueous solutions in the surficial environment. Source: Modified after Leeder (1999).

The main chemical processes that contribute to weathering include dissolution, oxidation, hydration, hydrolysis, and acid hydrolysis (Leeder 1999). Once weathering has commenced and fine clay particles have been produced, cation exchange further promotes the breakdown of minerals in the weathering zone. Each of these processes is relevant to ore formation in the surficial environment and is discussed briefly below.

4.2.1 Dissolution and Hydration

Certain natural materials such as halite (NaCl) and other evaporitic minerals, as well as the carbonate minerals (calcite, siderite, dolomite, etc.), tend to dissolve relatively easily and completely in normal to acidic groundwaters.

This type of dissolution contrasts with the breakdown of most rock-forming silicate minerals which dissolve less easily and do so incongruently (i.e. only certain components of the mineral go into solution). The relative solubilities of different elements in surface waters depends on a variety of factors, but can be qualitatively predicted (Figure 4.1) in terms of their ionic potential (or the ratio of ionic charge to ionic radius). Cations with low ionic potentials (<3) are easily hydrated and are mobile under a range of conditions, although they will precipitate under alkaline conditions and are readily adsorbed by clay particles. Similarly, anions with high ionic potentials (>10) form soluble complexes and dissolve easily, but will precipitate together with alkali elements. Ions with intermediate

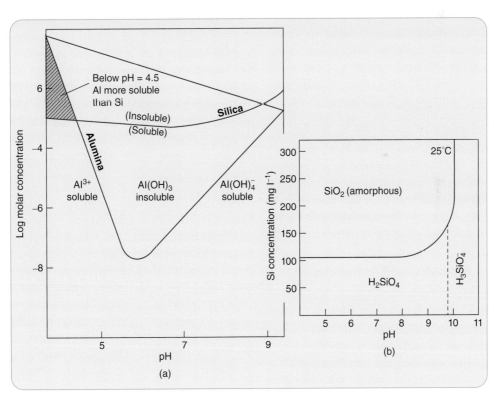

Figure 4.2 (a) The solubility of Si and Al as a function of pH. Source: After Raiswell et al. (1980). (b) The solubility of amorphous silica at 25 °C. Source: After Bland and Rolls (1998).

values (ionic potentials between 3 and 10) tend to be relatively insoluble and precipitate readily as hydroxides.

Over the neutral pH range (5–9) at which most groundwaters exist, silicon is more soluble than aluminum (Figure 4.2) and consequently chemical weathering will tend to leach Si, leaving behind a residual concentration of immobile Al and ferric oxides/hydroxides. This is typical of soil formation processes in tropical, high rainfall areas and yields lateritic soil profiles, which can also contain concentrations of bauxite (aluminum ore) and Ni. Lateritic soils will not, however, form under acidic conditions (pH <5) because Al is more soluble than Si (Figure 4.2) and the resultant soils (podzols) are silica-enriched and typically depleted in Al and Fe.

Another process that contributes to the dissolution of minerals in the weathering zone is hydration. In aqueous solutions water molecules cluster around ionic species as a result of their charge polarity and this contributes to the efficacy of water as a solvent of ionic compounds. Hydration of minerals can also occur directly (Bland and Rolls 1998), good examples of which include the formation of gypsum ($CaSO_4 \cdot 2H_2O$) from anhydrite ($CaSO_4$) and the incorporation of water into the structure of clays such as montmorillonite. Mineral hydration results in expansion of the lattice structure and assists in the physical and chemical breakdown of the material.

4.2.2 Hydrolysis and Acid Hydrolysis

Hydrolysis is defined as a chemical reaction in which one or both of the O—H bonds in the water molecule is broken (Gill 1996). Such reactions are important in weathering. One example occurs during the breakdown of aluminosilicate minerals such as feldspar, and also the liberation of silicon as silicic acid into solution, as shown by the reaction:

$$Si^{4+} + 4H_2O \Leftrightarrow H_4SiO_4 + 4H^+ \quad (4.1)$$

Although quartz itself is generally insoluble, it is feasible to mobilize silica over a wide range of pH (see Figure 4.2b) by hydrolysis reactions that result in the formation of relatively soluble silicic acid (H_4SiO_4). Another example relates to the hydrolysate elements such as Fe and Al (Figure 4.1) that are relatively soluble in acidic solutions, but will precipitate as a result of hydrolysis. The hydrolysis of aluminum, yielding an aluminum hydroxide precipitate, is illustrated by reaction (4.2):

$$Al^{3+} + 3H_2O \Leftrightarrow Al(OH)_3 + 3H^+ \quad (4.2)$$

It is this type of process that results, for example, in the concentration of aluminum (as gibbsite $Al(OH)_3$) and ferric iron (as goethite $FeO(OH)$) in lateritic soils.

Acid hydrolysis refers to the processes whereby silicate minerals break down in the weathering zone. In this process activated mineral surfaces (i.e. those with a net charge defect) react with acids (H^+ ions) in solution and in so doing displace metal cations from the crystal lattice (see Section 4.4 below), which then either go into solution or precipitate. The process is most active at surfaces exposed by fractures, cleavages, and lattice defect sites.

4.2.3 Oxidation

Oxidation (and reduction) refers essentially to chemical processes that involve the transfer of electrons. In the surficial environment, oxygen, present in either water or the air, is a common oxidizing agent. The element most commonly oxidized in the surficial environment is probably iron, which is converted from the ferrous (Fe^{2+}) to the ferric (Fe^{3+}) valence state by oxidation (loss of electrons). An example of the role of oxidation in chemical weathering is provided by the relative instability of biotite compared to muscovite. Biotite has the formula $K^+[(Mg^{2+}, Fe^{2+})_3(Si_3Al)O_{10}(OH)_2]^-$ and is much more easily weathered than muscovite $K^+[(Al_2)(Si_3Al)O_{10}(OH)_2]^-$ because of the ease with which the ferrous iron can be oxidized to ferric iron. Weathering and the resulting oxidation of Fe in biotite leads to a charge imbalance that destabilizes the mineral, a process that is less likely to happen in muscovite since it contains no iron in its lattice. The presence of iron in minerals such as olivine and the orthopyroxenes is one of the main reasons why they are relatively unstable in the weathering zone.

4.2.4 Cation Exchange

Clay particles are often colloidal in nature (i.e. they have diameters <2 μm) and are characterized by a net negative surface charge brought about, for example, by the replacement of Si^{4+} by Al^{3+} in the clay lattice. The negative charge is neutralized by adsorption of cations onto the surface of the colloids (see Figure 3.16, Chapter 3 and Section 4.4.2 below). The adsorbed cations may be exchanged for others when water passes through weathered material containing clay colloids and this has an effect on mineral stabilities as well as the nature of leaching and precipitation in regolith profiles. A simplified scheme (after Bland and Rolls 1998) illustrating the effect of cation exchange at the surface of a colloidal particle is shown in reaction (4.3) below. In this example cation exchange promotes the solution of Ca in groundwaters:

$$\boxed{COLLOID} - Ca^{2+} + 2H^+ (aq) \rightarrow$$
$$\boxed{COLLOID} - H^+ + Ca^{2+}(aq) \quad (4.3)$$
$$\underset{H^+}{\overset{|}{}}$$

4.3 Lateritic Deposits

4.3.1 Laterite Formation

Laterite is the product of intense weathering in humid, warm, intertropical regions of the world, and is typically rich in kaolinitic clay as well as Fe- and Al-oxides and oxyhydroxides.

Laterites are generally well layered, due to alternating downward percolation of rainwater and upward movement of moisture in the regolith during seasonal dry spells – they are also often capped by some form of duricrust. Laterites are economically important as they represent the principal environment within which aluminum ores (bauxite) occur (Retallack 2010). They can also contain significant concentrations of other metals such as Ni, Mn, and Au, as well as Cu and PGE. Laterites form on stable continental land masses, over long periods of time. In parts of Africa, South America, and Australia, laterization has been ongoing for over 100 million years (Butt et al. 2000), and, consequently, laterites are characterized by thick regolith profiles (up to 150 m) in which intense leaching of many elements has occurred such that the soils no longer reflect the rock compositions from which they were derived (Tardy and Roquin 1992). As with most soil/regolith profiles the main processes involved in lateritization can be subdivided into those of *eluviation*, where clays and solutes are removed from a particular horizon, and *illuviation*, where material is accumulated, usually at a lower level. A generalized lateritic regolith profile is shown in Figure 4.3, together with an indication of the pattern of horizontally orientated leaching and retention of different elements in the regolith zones (Butt et al. 2000).

The base of a lateritic regolith profile is characterized by the saprolith zone (saprock), which is highly weathered rock where the primary texture and fabric is still preserved. Given that the fluids involved in this type of weathering are typically oxidizing and slightly acidic, the lowermost saprock zone (Figure 4.3) is characterized by the destabilization of sulfides and carbonates and the associated leaching of most chalcophile metals and alkaline/alkaline earth elements. The lower saprolite zone is also characterized by the destruction of feldspars and ferromagnesian minerals, with Si and

Al retained in clay minerals (kaolinite and halloysite). Fe oxides and oxyhydroxides also form in this zone with partial retention of some transition metals in phases such as hematite and goethite. The mid- to upper-saprolite zone sees alteration of all but the most resistant of minerals as well as destruction of earlier formed secondary minerals such as chlorite and smectite. Only minerals such as muscovite and talc tend to survive intact through this zone. The upper part of the regolith profile, the pedolith zone, is characterized by complete destruction of rock fabric and leaching of all but the most stable elements. This zone is dominated compositionally by Si, Al, and ferric Fe occurring mainly as kaolinite, quartz, and hematite/goethite. A ferruginous residual zone may be well developed over mafic/ultramafic bedrock, whereas kaolinite is more abundant over felsic lithologies. The zone is also characterized by accretionary and dissolution textures such as pisoliths and nodules, or kaolinite replacement by gibbsite or amorphous silica. Some metals (Au, Cr, V, Sc, Ga) tend to be associated, by adsorption, with Fe oxides/oxyhydroxides in the ferruginous residuum, whereas other elements are retained in the upper zones simply because their mineral hosts are particularly stable (i.e. Zr, Hf in zircon, Cr in chromite, Ti in rutile, etc.).

Laterites can be subdivided into ferruginous (ferricretes) and aluminous (alcrete or bauxite) varieties. Ferricretes are characterized by a ferruginous residuum in the upper zone, where Fe_2O_3 contents can be as high as 20%. They tend to form under specific climatic conditions where rainfall is less than about 1700 mm per year but average temperatures are high (Tardy 1992). Under higher rainfall conditions ferricrete dissolution tends to occur and aluminum hydroxide (gibbsite $Al(OH)_3$) accumulates.

4.3.2 Bauxite Ore Formation

Bauxitic ore, in the form of the minerals gibbsite/boehmite and diaspore, is the principal source of aluminum metal, demand for which

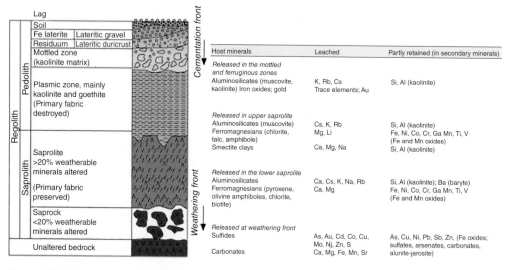

(a) Generalized lateritic regolith profile   (b) Element mobility in typical lateritic profile

Host minerals	Leached	Partly retained (in secondary minerals)
Released in the mottled and ferruginous zones		
Aluminosilicates (muscovite, kaolinite) Iron oxides; gold	K, Rb, Cs Trace elements; Au	Si, Al (kaolinite)
Released in upper saprolite		
Aluminosilicates (muscovite)	Cs, K, Rb	Si, Al (kaolinite)
Ferromagnesians (chlorite, talc, amphibole)	Mg, Li	Fe, Ni, Co, Cr, Ga Mn, Ti, V (Fe and Mn oxides)
Smectite clays	Ca, Mg, Na	Si, Al (kaolinite)
Released in the lower saprolite		
Aluminosilicates	Ca, Cs, K, Na, Rb	Si, Al (kaolinite); Ba (baryte)
Ferromagnesians (pyroxene, olivine amphiboles, chlorite, biotite)	Ca, Mg	Fe, Ni, Co, Cr, Ga Mn, Ti, V (Fe and Mn oxides)
Released at weathering front		
Sulfides	As, Au, Cd, Co, Cu, Mo, Nj, Zn, S	As, Cu, Ni, Pb, Sb, Zn, (Fe oxides; sulfates, arsenates, carbonates,
Carbonates	Ca, Mg, Fe, Mn, Sr	alunite-jarosite)

Figure 4.3 (a) A generalized lateritic regolith profile showing the different horizons and the terminology used in their description. (b) Generalized pattern of element mobility in lateritic regoliths. Source: After Butt et al. (2000).

Figure 4.4 Eh–pH diagram showing conditions relevant to the formation of laterites and bauxite ore. The solubility contours are in mol l^{-1} and assume equilibrium of the solution with gibbsite and hematite/goethite. Source: After Norton (1973).

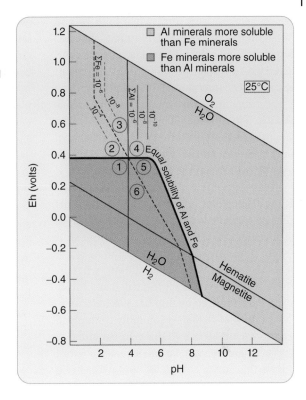

increased dramatically in the latter half of the twentieth century. The accumulation of an alumina-rich residuum, as opposed to one enriched in iron, in the upper zone of a lateritic profile is a function of higher rainfall, but also lower average temperatures (around 22 °C rather than 28 °C for ferricretes) and higher humidity (Tardy 1992). Actual alumina enrichment in the upper parts of laterite profiles is due, at least in part, to relatively high Si mobility compared to Al, and probably reflects near neutral pH conditions (between 4.5 and 9; see Figure 4.2a). This results in incongruent dissolution of minerals such as felspar and kaolinite, where Si is leached in preference to Al, yielding a gibbsite-like residue. This process (after Bland and Rolls 1998) is described simplistically as:

$$\text{feldspar} - (\text{loss of Si}) \rightarrow \text{kaolinite}$$
$$- (\text{loss of Si}) \rightarrow \text{gibbsite} (\text{Al(OH)}_3) \quad (4.4)$$

Seasonal climatic variations are also considered important to the formation of bauxitic ores as the alternation of wet and dry spells promotes fluctuation of groundwater levels and, hence, dissolution and mass transfer. Variations in bauxitic profiles, as well as transformation from hydrated gibbsite to its relatively dehydrated equivalent, boehmite, or to diaspore (AlO(OH)), result from such fluctuations. The mineralogical profiles in zones of bauxite mineralization may be quite variable. In humid, equatorial, laterite zones, hydrated minerals such as gibbsite and goethite predominate, whereas in seasonally contrasted climates the ores are relatively dehydrated and boehmite–hematite assemblages form (Tardy 1992).

The redistribution of iron, and the segregation of Al and Fe, is a necessary process in bauxite formation because ferruginous minerals tend to contaminate the ore. High quality bauxitic ores form when both Fe and Si are removed, but not alumina, whereas ferricretes and conventional laterites are characterized by different combinations of element leaching. The interplay of Eh and pH is critical to the formation of high quality bauxitic ores as discussed in some detail by Norton (1973). The essential features of Norton's model are shown on an Eh–pH plot in Figure 4.4. Si is leachable by hydrolysis (see Eq. (4.1) above) and is increasingly soluble at higher pH, whereas Al is most soluble at either very low or high pH (Figure 4.2a). Fe is most mobile as ferrous iron at low Eh and pH. In lateritic environments where concentrations of both Fe and Al occur, it is the rather special conditions whereby these two metals are segregated that provide the means for high quality bauxite formation. The plotting of Fe and Al solubility contours in Figure 4.4 allows one to draw an isosolubility curve where Al and Fe are equally soluble. Above this curve, at high Eh and pH, Al-rich minerals such as gibbsite will be more soluble (and Al preferentially leached) than below this curve where Fe minerals such as hematite or goethite will be more soluble (and Fe preferentially leached). Using these constraints, several situations can be described that are relevant to laterite formation. The area of Eh–pH space encompassing fields 1 and 2 on Figure 4.4 will be characterized by leaching of both Fe and Al and will not result in laterite formation, but more likely Si-enriched podzol soils. Fields 3 and 4 are characterized by restricted solubility of Fe and Al and laterites and bauxites will only form if the bedrock composition itself has high iron or aluminum contents. In field 3 bauxites are unlikely to form because of preferential leaching of Al from the soil, although over a protracted time period and

given the right climatic conditions, high Fe laterites will form under these conditions. Groundwater solutions forming in field 5 will contain more Fe than Al and laterites will form in this environment. The optimum conditions for bauxite formation are provided in field 6 where groundwater solutions will preferentially remove Fe. In this field Al hydrolysates are stable, especially at pH between 5 and 7, and gibbsite will accumulate. This model shows that both the initial and subsequent Eh and pH of the groundwater solutions involved will impose important controls on the relative mobilities of Al and Fe and, consequently, on the dissolution and precipitation of relevant minerals during laterite and bauxite formation. It should be noted that disequilibria and the role of organic acids in element mobility might complicate some of the generalizations made above and could account for the variability that is sometimes observed in actual weathering profiles.

The huge bauxite deposits of Jamaica represent a good example of the controls involved in the formation of aluminum ores. Thick deposits of bentonitic volcanic ash were laid down on a limestone bedrock and the former are considered to be the ultimate source of residual alumina in the bauxite deposits. Desilication of volcanic glass and other silicate minerals by rapid and efficient drainage through the underlying karst limestone is considered to have been responsible for the gibbsite-dominated ores (Comer 1974). In Venezuela, by contrast, lateritic bauxites are derived by deep weathering of a granitic bedrock (Meyer et al. 2002) during prolonged uplift and erosion of the Guyana Shield starting about 35 million years ago (see Box 4.1).

Box 4.1 Lateritization: Bauxite – The Los Pijiguaos Deposit, Venezuela

Approximately 90% of the world's bauxite, the principal source of alumina, is obtained from laterites formed by deep weathering of continental planation surfaces exposed to hot humid climates. Global production of bauxite has risen significantly in recent years and is presently at around 125 million tons per annum (Meyer et al. 2002). One of the largest bauxite deposits in the world is at Los Pijiguaos in Venezuela, where annual production is about 5 million tons from a total resource that could be as much as 6 billion tons.

The Los Pijiguaos bauxite deposit formed from a Mesoproterozoic, metaluminous Rapakivi-type granite (the Parguaza granite) that was deeply eroded and weathered during Cretaceous–Paleogene times to form the Nuria planation surface (Figure 4.1.1). This profound weathering event gave rise to the development of thick lateritic soil cover in which economic concentrations of bauxite (around 50 wt% Al_2O_3) form a mineable zone that averages 8 m thick. The complete laterite profile comprises an upper concretionary zone (bauxite), underlain by a mottled or saprolitic layer and then the granitic bedrock. The bauxite layer itself can be subdivided into four zones comprising various concretionary and pisolitic textures (Figure 4.1.2). The deposit as a whole is subdivided into blocks that are defined by fluvial channels that cut through the deposit subsequent to its formation (Figure 4.1.1).

The bauxite at Los Pijiguaos is made up principally of gibbsite with lesser quartz and rare kaolinite. The original granite textures in the bauxite layer have been almost entirely destroyed, with both feldspar and quartz being largely replaced by gibbsite (Figure 4.1.2 inset). An increasing degree of lateritization is marked by systematic compositional changes in the rock. This is demonstrated in the Al_2O_3–SiO_2–Fe_2O_3 ternary plot of Figure 4.1.3 where the process of lateritization is characterized by an increase in the alumina content.

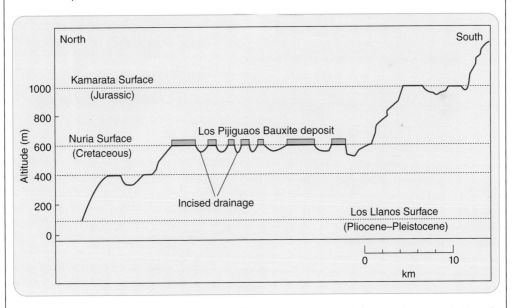

Figure 4.1.1 Schematic profile through the Los Pijiguaos bauxite deposit showing the concentration of laterite on a dissected late Cretaceous planation surface. Source: After Meyer et al. (2002).

(Continued)

Box 4.1 (Continued)

Figure 4.1.2 Photograph of typical bauxitic ore at Los Pijiguaos (white rod is 1 m); the upper portion is unconsolidated and comprises gibbsite and loose pisoliths whereas the lower portions are cemented and concretionary. Inset – photomicrograph showing the replacement of quartz by gibbsite in bauxite from Los Pijiguaos. Source: Both photographs courtesy of Michael Meyer.

Box 4.1 (Continued)

Figure 4.1.3 Ternary Al$_2$O$_3$–SiO$_2$–Fe$_2$O$_3$ plot showing the residual increase in alumina content with progressive lateritization in the Los Pijiguaos bauxite ores. PG is the composition of the unaltered granite. Source: After Meyer et al. (2002).

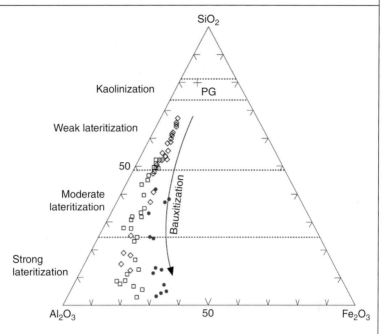

Quartz and gibbsite are inversely correlated in the bauxite ore and iron contents remain relatively constant, with an increase in Fe$_2$O$_3$ in the most highly lateritized samples. The bauxite at Los Pijiguaos is a product of desilication, hydration and residual Al$_2$O$_3$ concentration of the Parguaza granite, and involves a mass loss of some 60% (Meyer et al. 2002). These processes produce a high grade and high purity gibbsitic ore from which alumina can be extracted by low temperature digestion, making Los Pijiguaos one of the most commercially profitable bauxite deposits in the world.

4.3.3 Nickel Laterites

Laterites that are developed on serpentinized ultramafic rocks originally containing abundant olivine and orthopyroxene and, hence, elevated Ni contents (i.e. 0.1–0.3%; Figure 4.5 and Chapter 2), commonly develop concentrations of Ni silicate or oxide that are up to 10 times the original concentration. Given the relative ease of mining this type of ore compared to hard-rock magmatic sulfide ores, as well as improvements in extractive metallurgical processes, Ni laterites are widely exploited and account for up to 60% of global nickel supply (Butt and Cluzel 2013). The principles by which these ores form are relatively

straightforward and basically involve eluviation of Ni from the uppermost lateritic residuum and concentration into underlying saprolitic illuvium as Ni is adsorbed onto goethite, or as newly formed nickeliferous talc, serpentine, chlorite or clay (smectite) minerals (Figure 4.5).

The primary olivine and orthopyroxene minerals that are the source of the Ni form the main ingredients of an ultramafic rock that may originally have been part of an obducted ophiolite complex (such as the New Caledonia deposits in the south Pacific or the deposits of Moa Bay in Cuba; Aiglsperger 2015) or a layered mafic intrusion (such as at Niquelandia in Brazil). These minerals are serpentinized, in

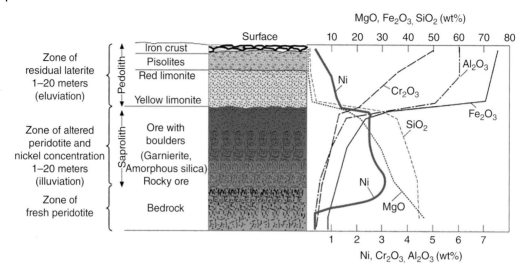

Figure 4.5 Descriptive profile and Ni distribution in a lateritic regolith typical of the New Caledonian deposits. The chemical profile clearly distinguishes the ferruginous/aluminous residual zone where Si, Mg, and Ni are leached, from the saprolith where illuviation has resulted in concentration of Ni. Source: After Troly et al. (1979), Guilbert and Park (1986).

most cases prior to lateritization, by interaction with sea water or during low-grade metamorphism and alteration. The alteration of olivine is by hydration to amorphous silica, serpentine, talc, and goethite, with the serpentine able to accommodate as much or more of the Ni as the original olivine. Most nickel laterites form on bedrock that is already largely altered to serpentinite (Golightly 1981). In lateritic regoliths the groundwaters become progressively less acidic with depth (pH increases downwards from around 5 to 8.5; Golightly 1981) and bicarbonate is the main anion in solution. Lateritization proceeds under suitable tropical or sub-tropical climatic conditions to form clays and Fe- and Al-oxide/oxyhydroxide minerals (see Section 4.3.1 above) as illustrated in reaction (4.5) below (after Golightly 1981) where olivine breaks down to smectite clays (saponite–nontronite) and goethite:

$$4\,(Fe_2, Mg_2)SiO_4 + 8H^+ + 4O_2 \Leftrightarrow$$
$$\underset{\text{olivine}}{}$$
$$(Fe_2, Mg_3)Si_4O_{10}(OH)_2$$
$$\underset{\text{smectite}}{}$$
$$+ 6FeO(OH) + 5Mg^{2+} \qquad (4.5)$$
$$\underset{\text{goethite}}{}$$

Once smectite, and also serpentine and talc, are present in the regolith, further concentration of Ni takes place largely by cation exchange, mainly with Mg^{2+}. This results in the formation of a variety of unusual Ni enriched clay and phyllosilicate minerals, such as kerolite (Ni–talc), nepouite (Ni–serpentine), and pimelite (Ni–smectite). An example of such an ion exchange reaction, where nepouite forms from serpentine, is shown in Eq. (4.6) below (after Golightly 1981):

$$Mg_3\,Si_2O_5(OH)_4 + 3Ni^{2+}(aq) \Leftrightarrow$$
$$\underset{\text{serpentine}}{}$$
$$Ni_3\,Si_2O_5(OH)_4 + 3Mg^{2+}(aq) \qquad (4.6)$$
$$\underset{\text{nepouite}}{}$$

Classification of Ni laterite ores suggests that they can be subdivided mineralogically into three types, namely, oxide ore, hydrous Mg–silicate ore and clay silicate ore (Butt and Cluzel 2013). Oxide ore forms at a relatively early stage of lateritization where Ni is residually concentrated by adsorption onto goethite, possibly at neutral to slightly alkaline pH as discussed previously in Chapter 3 and shown in Figure 3.16b. Oxide ores may also contain

Mn-oxide minerals that are enriched in Co as well as Ni. A more advanced degree of lateritization results in the formation of new hydrous silicate minerals that are typically highly nickeliferous, such as Ni lizardite or nepouite (a serpentine), nimite (a chlorite) and kerolite (a talc). A third type of ore, typically found in areas of low relief (Butt and Cluzel 2013), forms when Ni smectites (nontronite and saponite) are preferentially developed over serpentinized peridotite. The formation of Ni laterite deposits as a function of an evolving weathering and erosional regime is illustrated in Figure 4.6 (after Butt and Cluzel 2013). Humid, warm, sub-tropical climates, where weathering rates are greater than erosional rates, create an environment whereby Ni released during the breakdown of olivine or serpentinized olivine is retained in the lateritic profile by adsorption onto goethite or smectite, while other elements such as Mg and Si are leached (Figure 4.6a). Ni enrichment is by residual concentration rather than active neo-precipitation and Ni grades in these oxide and clay hosted ores are typically low. In accretionary environments where uplift and fluvial drainage is more active, adsorbed Ni on early formed goethite or smectites may be leached and re-concentrated lower in the lateritic profile to form new Ni-rich serpentine, talc and chlorite phases (Figure 4.6b,c). This multi-stage evolution commonly results in higher grade deposits that have formed in uplifted and eroded plateaus and then subsequently preserved, on aridification, beneath a duricrust.

The collective term applied to Ni–phyllosilicate ore minerals in lateritic environments is "garnierite," after Jules Garnier who discovered the enormous Ni laterite ores of New Caledonia in the nineteenth century. Enormous resources of Ni ore are located in ultramafic rooted lateritic regoliths in many parts of the world. The thickest laterite and the richest concentrations of garnierite tend to occur where the bedrock is characterized by closely spaced joints or fractures, as this is where maximum groundwater circulation and fluid–rock interaction takes place. The best Ni laterite concentrations also appear to display a topographic control and tend to occur either beneath a hill or on the edge of a plateau or terrace (Golightly 1981).

4.3.4 Gold in Laterites

It is well known that gold occurs in the upper soil dominated portions of laterite weathering zones in many parts of the world, including Brazil, West Africa, and Western Australia (Wilson 1984; Nahon et al. 1992). This gold takes many forms, ranging from large, rounded nugget-like particles and gold dendrites in cracks and joints, to small crystals in pore spaces. In the Yilgarn Craton of Western Australia, for example, the recovery of gold by metal detector from lateritic profiles is a popular activity and some spectacular nuggets, occasionally over 1 kg in mass, have been found. Whereas the primary source of gold in these environments is high in silver (strictly a Au–Ag alloy with typically 5–10% Ag as well as minor Hg and other impurities), the gold that is concentrated in lateritic profiles has been chemically purified. This suggests that differential mobilization and decoupling of Au and Ag takes place as meteoric fluids percolate through the weathering zone (Figure 4.7).

It is only under certain conditions in the near surface environment that Au and Ag can be substantially mobilized. Actual measurements of near surface groundwaters in lateritic environments indicate that aqueous solutions are relatively acidic (pH <5), oxidizing, and in Western Australia have low to moderate salinities because of a marine influence on rainfall (Mann 1984). With increasing depth, groundwater pH tends to increase and the fluids become more reducing as the ferric/ferrous ratio decreases (Figure 4.8). The process of lateritization, described in Section 4.3.1 above, is controlled essentially by hydrolysis as groundwaters react with the regolith and iron is oxidized. Two reactions reflect the nature of the process and indicate why

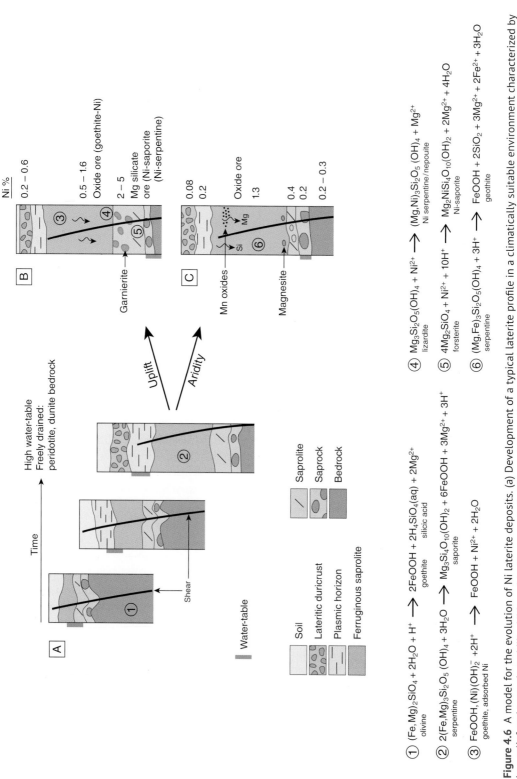

① $(Fe,Mg)_2SiO_4 + 2H_2O + H^+ \rightarrow 2FeOOH + 2H_4SiO_4(aq) + 2Mg^{2+}$
   olivine   goethite   silicic acid

② $2(Fe,Mg)_3Si_2O_5(OH)_4 + 3H_2O \rightarrow Mg_3Si_4O_{10}(OH)_2 + 6FeOOH + 3Mg^{2+} + 3H^+$
   serpentine   saporite

③ $FeOOH, (Ni)(OH)_2^- + 2H^+ \rightarrow FeOOH + Ni^{2+} + 2H_2O$
   goethite, adsorbed Ni

④ $Mg_3Si_2O_5(OH)_4 + Ni^{2+} \rightarrow (Mg,Ni)_3Si_2O_5(OH)_4 + Mg^{2+}$
   lizardite   Ni serpentine/nepouite

⑤ $4Mg_2SiO_4 + Ni^{2+} + 10H^+ \rightarrow Mg_2NiSi_4O_{10}(OH)_2 + 2Mg^{2+} + 4H_2O$
   forsterite   Ni-saporite

⑥ $(Mg,Fe)_3Si_2O_5(OH)_4 + 3H^+ \rightarrow FeOOH + 2SiO_2 + 3Mg^{2+} + 2Fe^{2+} + 3H_2O$
   serpentine   goethite

Figure 4.6 A model for the evolution of Ni laterite deposits. (a) Development of a typical laterite profile in a climatically suitable environment characterized by low relief and tectonic stability. Ni is residually concentrated in goethite and smectitic clays according to reactions 1 and 2. (b) Development of hydrous Mg silicate type ores following tectonic uplift and downward leaching of lateritic Ni and the precipitation of newly formed "garnierite" minerals lower in the profile. (c) Onset of more arid climatic conditions with the resulting precipitation of Si (as opalescent quartz) and MnO minerals (within which Co and Ni concentrations may occur) Source: After Butt and Cluzel (2013).

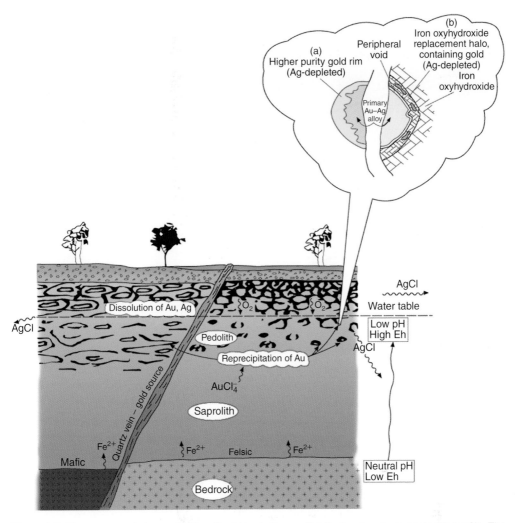

Figure 4.7 The nature and characteristics of gold and silver redistribution in a typical lateritic profile. The inset diagram shows two mechanisms whereby a primary Au–Ag alloy particle could be broken down during lateritization and silver preferentially removed. Source: Modified after Mann (1984), Nahon et al. (1992).

groundwaters become more acidic upwards in lateritic profiles. Reaction (4.7) describes the initial breakdown, by hydrolysis, of sulfide phases such as pyrite in the lower sections of the weathering zone close to bedrock, and the production of hydrogen ions:

$$2FeS_2 + 2H_2O + 7O_2 \Leftrightarrow$$
$$2Fe^{2+} + 4SO_4^{2-} + 4H^+ \quad (4.7)$$

Reaction (4.8) illustrates the oxidation of ferrous iron in the vicinity of the water table, to form goethite in the ferruginous pedolith,

with further production of hydrogen ions (after Mann 1984):

$$2Fe^{2+} + 3H_2O + O_2 \Leftrightarrow 2\underset{goethite}{FeOOH} + 4H^+$$

$$(4.8)$$

These reactions illustrate some of the processes active during lateritization and explain why groundwaters are likely to be acidic in such environments. Laterites formed over felsic rocks are prone to be more acidic than those over a mafic substrate, as the latter will

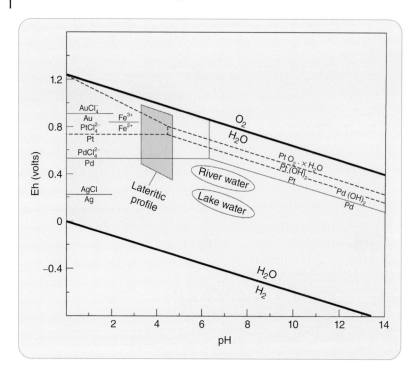

Figure 4.8 Eh–pH diagram showing the characteristics of a lateritic profile in relation to lacustrine and riverine waters. Also shown are the redox potentials and stabilities of Au, Ag, Pt, and Pd as chloride complexes, illustrating the range of Eh that controls solubility of the various metals. Source: Information compiled from data in Fuchs and Rose (1974), Mann (1984), Bowles (1986).

have pH buffered by bicarbonate ions formed during alteration (Mann 1984).

Experimental work (Cloke and Kelly 1964; Mann 1984) indicates that at low pH and high Eh, and in the presence of Cl^- ions, gold in the surficial environment can go into solution as an $AuCl_4^-$ complex (Figure 4.8). Substantial Au dissolution only occurs, however, at relatively high Eh when Fe^{2+} is completely oxidized in the presence of free oxygen. By contrast, silver will go into solution more readily, and under more reducing conditions, as several complexes, including AgCl (the most stable) and $AgCl_2^-$ or $AgCl_3^{2-}$ (Figure 4.8). Accordingly, particles of Au–Ag alloy in the surficial environment will be more likely to be leached of their Ag over a broader range of conditions.

A model for gold concentration in laterites is shown in Figure 4.7 and is discussed below in terms of gold– and silver–chloride speciation, as well as the likely Eh–pH conditions of a typical lateritic profile, shown in Figure 4.8. Using these constraints it is apparent that primary Au–Ag alloy particles in the near surface regolith above the water table will

be dissolved in any acidic, oxygenated, and moderately saline groundwaters prevailing in this environment. Mann (1984) has shown that the breakdown of the particle and dissolution of Au and Ag actually takes place by replacement of the alloy by an iron oxyhydroxide phase (such as goethite). The goethite itself, however, can be seen to contain tiny particles of high purity gold (Figure 4.7 inset b) suggesting that the latter metal is rapidly reprecipitated as goethite forms, whereas silver remains in solution and is transported away. This is explained by the fact that high gold solubilities are only maintained at high Eh and that the metal will precipitate with only a slight decrease in Eh under mildly reducing conditions, as described by the following reaction:

$$AuCl_4^- + 3Fe^{2+} + 6H_2O \Leftrightarrow Au + 3FeOOH + 4Cl^- + 9H^+ \quad (4.9)$$

A decrease in $AuCl_4^-$ solubility and precipitation of gold is, therefore, promoted by interaction of the fluid with a more reducing regolith in which Fe^{2+} and Mn^{2+} contents are higher with increasing depth in the profile and below the water table. Redox reactions, therefore, control the precipitation of gold in this environment and explain its association with minerals such as goethite and manganite (MnO(OH)). It is clear in Figure 4.7 that under the same conditions at which gold precipitates, silver as AgCl remains in solution and will be transported away from the site of gold deposition.

Primary Au–Ag alloy particles that are not closely associated with an iron or manganese oxyhydroxide tend to show a preferential depletion in silver around their edges and the formation of higher purity gold halos (Figure 4.7 inset a). Such features are a product either of silver diffusion from the outer margins of the particle, or of complete dissolution of the entire alloy and in situ reprecipitation of the gold. It should also be noted that the efficacy of chloride ion ligands in promoting gold and silver dissolution is particularly appropriate to laterites that presently occur in arid climates, such as Western Australia. In high-rainfall, tropical regions chloride ion activity is reduced by dilution and other compounds, such as organic humic and fulvic acids, may play a more important role in gold dissolution (Nahon et al. 1992).

Finally, it should be noted that microorganisms may play an important role in the concentration of gold, as well as other metals such as Cu, Fe and U, in the weathering environment (Sillitoe et al. 1996; Southam and Saunders 2005; Brugger et al. 2013; Zammit et al. 2015). Secondary gold, in the form of the nuggets found either in laterites or in other environments where placer gold deposits occur, are sometimes characterized by spherical nano-morphologies consistent with the size and shape of bacteria (Watterson 1991). Southam and Beveridge (1994) have shown experimentally that *Bacillus subtilis*, a common bacterium in soils, is capable of accumulating gold by diffusion across the cell wall and precipitation within the cytoplasm. It is suggested that gold was stabilized internally as an organo-gold complex and then precipitated in colloidal form once a critical concentration

had been reached. Subsequent diagenesis of sediment containing such gold-enriched microorganisms would result in recrystallization and coalescence of gold to form nugget like shapes.

4.3.5 A Note on Platinum Group Element (PGE) Enrichment in Laterites

In addition to gold in laterites, it is now widely appreciated that PGEs can also be concentrated in weathering profiles. Laterites in Sierra Leone, for example, are known to contain PGE deposits (Bowles 1986) and significant concentrations of these precious metals also occur in non-lateritic soils such as over the Stillwater Complex in Montana and in Quebec (Fuchs and Rose 1974; Wood and Vlassopoulos 1990). Evidence for the mobilization of PGE in the weathering zone is provided by the existence of well formed, crystalline Pt–Fe or Os–Ir–Ru alloys in pedolith, and in some cases PGE-sulfide minerals that are compositionally different, and often larger, than those in the source rock (Bowles 1986). In the case of laterite-hosted PGE concentrations, it is considered likely that the controls on Pt and Pd solubility were similar to those affecting Au and Ag and that these precious metals go into solution as chloride complexes in acidic, oxidized environments (Bowles 1986). Figure 4.8 shows that $PdCl_4^{2-}$ is stable over a broader range of Eh and pH than $PtCl_4^{2-}$, a feature that is consistent with observations that palladium is more readily dissolved and concentrated in groundwaters than is platinum (Wood and Vlassopoulos 1990). Like gold and silver, platinum and palladium tend to become decoupled in the weathering environment, with Pd going into solution and Pt concentrating in soils and sediment. Similarly, redox reactions control the precipitation of PGE in this environment, with Pt, like Au, precipitating before Pd as Eh is reduced. In non-lateritic environments, where groundwaters are near neutral to alkaline and more reducing, PGE will not be transported as chloride complexes but more likely as neutral hydroxide complexes ($PdOH_2$ and $PtOH_2$). Decoupling of Pt and Pd still occurs in non-lateritic weathering profiles, although it is suggested that whereas Pd is dispersed in solution, Pt, together with Au, is less easily mobilized, and concentrated perhaps as colloidal particles (Wood and Vlassopoulos 1990). Finally, it should be noted that there is evidence pointing to an effective interaction between PGE and organic acids, in particular humic acid, in weathering zones (Wood 1990; Bowles et al. 1994). Pt and Pd are reasonably soluble in natural solutions containing humic acid, although the exact nature of the complex involved is not known. Humate is also thought to promote PGE transport as colloidal suspensions, whereas solid organic debris may also be responsible for precipitation of precious metals.

In general, and in spite of their similar chemical properties, Au and the PGE differ in terms of their mobility in the surficial environment, a feature that has been attributed in part to microbiological processes (Brugger et al. 2013). However, a spectacular example of where Au and PGE have both been concentrated into a deeply weathered lateritic profile is the Serra Pelada deposit in the Carajás mineral province of Brazil. This deposit was discovered and exploited exclusively by artisinal miners who took advantage of the rich and easily extractable laterite-hosted ore. Average grades are reported to have been around 15 g ton^{-1} Au, 4 g ton^{-1} Pd, and 2 g ton^{-1} Pt, with the primary ore body thought to resemble other Fe oxide–Cu–Au deposits (such as the Olympic Dam deposit; see Box 3.1, Chapter 3) found in the Carajás province (Grainger et al. 2002). Weathering has resulted in a lateritic profile dominated by kaolinite, manganese oxides, and Fe oxides/oxyhydroxides. Amorphous carbon is present as the weathered residue of an originally carbonaceous siltstone. Most of the ore constituents have been extensively remobilized during lateritization, with base metals occurring in Mn oxides and Au–Pt–Pd occurring together with amorphous carbon

and Fe–Mn oxides/oxyhydroxides (Grainger et al. 2002). Some of the ore also occurs as nuggets of Au and PGE alloy, some of which are reported to weigh in excess of 6 kg.

4.4 Clay Deposits

Clay minerals are volumetrically the most abundant product of weathering and either occur in situ or have been transported to a site of deposition. They are also economically important and used for a variety of industrial applications including paper, ceramics, filtration, and lubricants. Clay formed during weathering reflects both the nature of the source material and the weathering conditions. Nevertheless, similar clay mineral assemblages are ultimately derived from both felsic and mafic rocks (Figure 4.9). Temperature and rainfall in particular, as well as local Eh–pH conditions, determine the nature and rates of clay-forming processes. Some of the more important clay minerals considered here include kaolinite, illite, and the smectite group (including montmorillonite). Kaolinite is formed under humid conditions by acid hydrolysis of feldspar-bearing rocks (e.g. in laterites), whereas illite forms under more alkaline conditions by weathering of feldspars and micas. The smectite clays commonly weather from intermediate to basic rocks under alkaline conditions and are highly expandable, containing intracrystalline layers of water and exchangeable cations. It should be noted that clay minerals are not only the result of weathering processes but can also form as the products of low temperature hydrothermal alteration (see Chapter 3).

Clay formation is initiated during chemical weathering by acid hydrolysis (see Section 4.2.2 above and Figure 4.10). A typical reaction that describes the process is shown below:

$$(Ca, Na)Al_2Si_2O_8 + 2H^+ + H_2O \Leftrightarrow$$
$$\underset{\text{plagioclase}}{}$$

$$\underset{\text{kaolinite}}{Al_2Si_2O_5(OH)_4} + Na^+ + Ca^{2+} \quad (4.10)$$

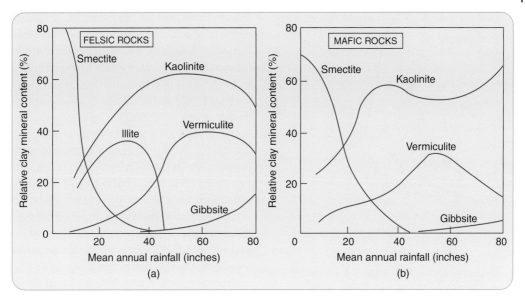

Figure 4.9 The effect of rainfall on clay mineral formation in (a) felsic and (b) mafic rocks. Source: After Barshad (1966).

In the above reaction plagioclase feldspar is broken down during weathering to form kaolinite, with the release into solution of Na and Ca ions. The actual mechanisms of weathering are undoubtedly more complex and may involve several incremental stages during the formation of clays. A schematic diagram illustrating the various stages involved during the weathering and breakdown of K-feldspar by acid hydrolysis is shown in Figure 4.10. Alteration by acid hydrolysis is most relevant at acidic pH and the process is kinetically less effective under more alkaline conditions, where other processes such as Si-hydroxylation may be more efficient.

The mobility of different elements in the surficial environment varies considerably (Figure 4.1) and this has led to the recognition of a hierarchy of mobility, as follows:

$$Ca > Na > Mg > Si > K > Al \approx Fe \quad (4.11)$$

The alkaline and alkali earth elements are typically the most soluble in groundwaters, with Al and ferric Fe being relatively immobile. The combination of kinetic and thermodynamic constraints on clay mineral formation during weathering results in

a well defined sequence of reaction pathways that can be related, for example, to climatic conditions. The effects of rainfall on the formation of clay mineral assemblages is shown in Figure 4.9, for both felsic and mafic parental rock compositions. The general pattern is that smectite clays tend to form under relatively arid to semi-arid conditions, whereas kaolinite tends to dominate in wetter climates. This pattern may also reflect a paragenetic sequence of clay formation, with smectites forming relatively early on and being superseded, or replaced, by kaolinite or vermiculite.

4.4.1 The Kaolinite (China Clay) Deposits of Cornwall

The Cornubian batholith of southwest England, well known for its polymetallic magmatic-hydrothermal mineralization (Box 2.2 in Chapter 2), also hosts some of the largest and best quality kaolinite (or china clay) deposits in the world. Of considerable demand in the paper and ceramic industries, the kaolinite resource of Cornwall surpasses the base metal resource in value, and considerable reserves of high quality china clay still exist.

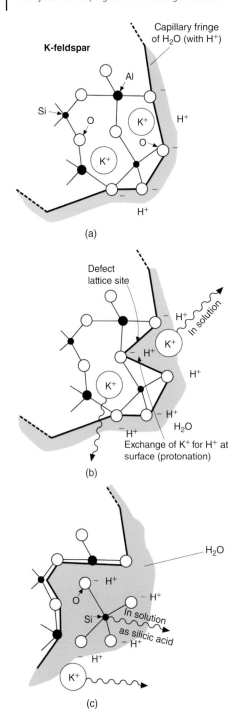

(a)

(b)

(c)

Figure 4.10 Representation of the stages during the progressive weathering and breakdown of K-feldspar by acid hydrolysis, leading to the formation of clay minerals. Source: Modified after Leeder (1999).

The formation of the Cornish kaolinite deposits is controversial. Some workers have argued in favor of a weathering-related origin (Sheppard 1977), but this has been countered by proponents of a hydrothermal process who maintain that clay formation represents the end stage of a long-lived paragenetic sequence initiated soon after granite emplacement and terminated with circulation of low temperature meteoric fluids during clay formation (Alderton 1993). The actual processes of kaolinite formation are essentially the same irrespective of the actual fluid origin, and on balance it seems likely that both a late hydrothermal process and a tertiary weathering event contributed to clay formation.

Zones of intense kaolinization are best developed in areas where greisen altered Sn–W ore veins and major NW–SE trending faulting have occurred. This suggests a genetic link to previous, higher temperature, hydrothermal processes, but could also point to groundwater circulation focused into sites previously accessed by high temperature fluids. Kaolinite itself is preferentially developed from the plagioclase feldspar component of the granite, while K-feldspar remains relatively stable except in zones of extreme alteration and weathering. This is consistent with the mobility hierarchy described in Eq. (4.11) above, where Ca and Na are more easily leached than K during acid hydrolysis. The process is also demonstrated in reaction (4.10) although, as mentioned previously, the actual mechanism is likely to be more complex. Many workers have described an intermediate stage of either muscovite or smectite formation prior to the ultimate formation of kaolinite (Exley 1976; Durrance and Bristow 1986). This, too, is consistent with the pattern of progressive clay mineral formation illustrated as a function of rainfall in Figure 4.9. It seems likely, therefore, that acid hydrolysis type reactions gave rise to early alteration of plagioclase feldspars to form sericite (fine grained muscovite), as well as incipient argillic alteration. Exhumation of the batholith during Mesozoic

Figure 4.11 Strongly kaolinitized granite at Treviscoe in Cornwall, being hydraulically mined for "china clay." Source: Photo is courtesy of Robin Shail.

times, together with further meteoric fluid circulation and deep weathering, continued the alteration process to form zones of intense kaolinization, particularly in areas where previous alteration had taken place. It appears likely that kaolinization was also accompanied by leaching of Fe from the altered granite, a process that contributed to the very high purity and "whiteness" of Cornish china clay (Figure 4.11).

4.4.2 "Ion-Adsorption" Rare Earth Element (REE) Deposits in Clays

The majority of the world's REE production comes from China where these critical metals are extracted, either from the huge Bayan Obo carbonatite deposit, or from numerous surficial deposits in the southeast of China where the REE are hosted in clay minerals (Wall 2014). Bayan Obo and other carbonatite related deposits are dominated by the light rare earth elements (LREE – comprising the series La to Sm) whereas the clay hosted deposits of SE China produce a larger proportion of heavy rare earth elements (HREE – comprising the series Eu to Lu). In fact the majority of the world's HREE production is derived from the surficial, clay hosted deposits of SE China – these are generally referred to as "*ion adsorption clay deposits*" because the HREE are preferentially adsorbed onto the surfaces of certain clay minerals, in particular kaolinite, halloysite and smectites (Kynicky et al. 2012; Bao and Zhao 2008).

The HREE bearing ion adsorption clay deposits of SE China have formed above intensely weathered granites that now have thick lateritic soil profiles developed over them (Bao and Zhao 2008; Kynicky et al. 2012). The key requirements needed to produce viable ion adsorption clay deposits include:

1) A parental granite that contains a variety of REE-bearing accessory mineral phases that can be dissolved wholly or partially during

lateritization. Highly resistant and stable HREE bearing minerals such as zircon tend to be preserved even during intense weathering and do not contribute their REE content to the clay deposits. The clay hosted deposits of SE China have formed over granites that contain REE bearing minerals such as allanite, monazite and bastnaesite that will break down under extreme weathering.

2) The host laterite must comprise clay minerals suited to the adsorption of REE, and under conditions where ambient pH controls the preferential concentration of the HREE by ion exchange processes.

3) The deposits are very low grade (typically 0.05–0.2% REE oxide contents) and

the REE must, therefore, be easily and cost-effectively extractable by leaching.

The physical and chemical processes by which the HREE are preferentially adsorbed relative to LREE in clay deposits are complex. The adsorption of REE, as well as other metallic cations, is related to the creation of negatively charged sites on the surface of clay minerals caused by cation exchange. Permanent charge defects are created when Si^{4+} and Al^{3+} are substituted by cations of lower charge; variable charge defects are created when the amphoteric OH^- radical combines with either Si or Al on crystal margins to form either a positive charge defect at low pH, or a negative charge defect at higher pH. Thus, REE adsorption on clays such as

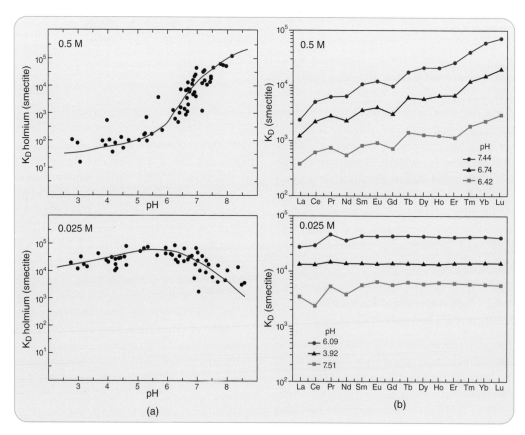

Figure 4.12 Experimental data for smectite clay that illustrates (a) the relationship between the sorption coefficient for holmium (K_D holmium) and pH at two different solution ionic strengths (0.5 M and 0.025 M) and (b) K_D's for all the REE (plotted with increasing atomic number) in smectite as a function of pH, again at two different solution ionic strengths. Source: After Coppin et al. (2002).

Figure 4.13
Experimental data for Eu adsorption onto kaolinite showing that the increase in the pH-dependent sorption coefficient (K_D) migrates toward lower pH with increasing temperature (from 25 to 150 °C) Source: After Tertre et al. (2006).

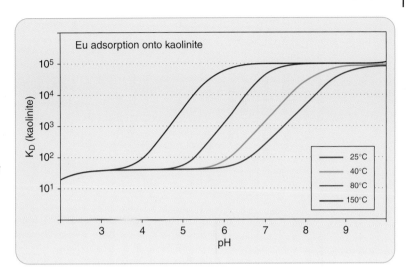

kaolinite and the smectites is controlled by the pH of fluids implicated in the weathering process, which is in turn also a function of fluid temperature (Bradbury and Baeyens 2002; Coppin et al. 2002; Moldoveanu and Papangelakis 2012; Tertre et al. 2006).

As a guide to the processes that give rise to ion adsorption clay deposits, the experimental work of Coppin et al. 2002 (Figure 4.12a) illustrates the variations in sorption coefficient (i.e. K_D – the solid/solution concentration ratio) on smectite as a function of pH for one of the HREE, holmium (Ho). For fluids with low ionic strength, sorption is independent of pH whereas at higher ionic strengths the HREE are much more strongly adsorbed than the LREE under more alkaline (higher pH) conditions. In Figure 4.12b experimental data for all the REE (i.e. lanthanum to lutetium) shows again that HREE are preferentially adsorbed over LREE when fluids have high ionic potential, and that this is not the case at lower ionic potentials. The experiments of Tertre et al. (2006) likewise demonstrate that the abrupt increase in sorption coefficient (K_D) with increasing fluid pH is also affected by temperature, such that the sorption cliff occurs at increasingly lower pH (i.e. more acidic conditions) with increasing temperature (Figure 4.13).

The experimental data discussed in this section indicate that adsorption of REE onto clay minerals is a process that is controlled by pH at high fluid ionic strengths, but is essentially pH independent at low ionic strengths. The relative adsorption of HREE over LREE is a function of both fluid ionic strength and pH. It is suggested (Coppin et al. 2002) that, in environments of low fluid ionic strength, adsorption of all REE occurs at the sites of permanent negative charge defect by *physical adsorption* (i.e. physisorption – where attraction is caused by the Van Der Waals force and the electronic structure of the crystal is relatively undisturbed; see Figure 3.17). By contrast, at higher ionic strengths, preferential HREE adsorption occurs at the sites of variable negative charge which are themselves created under high pH conditions. This attraction is by *chemical adsorption* (or chemisorption) where distinct covalent bonds form between REE cations and the clay mineral surface. Extracting the REE from clays is carried out by a leaching process during which a lixiviant, such as ammonium sulfate (($NH_4)_2SO_4$), readily strips away the physisorbed metals (Moldoveanu and Papangelakis 2012), but is less effective against chemisorbed elements that are more strongly bonded to the clay particles.

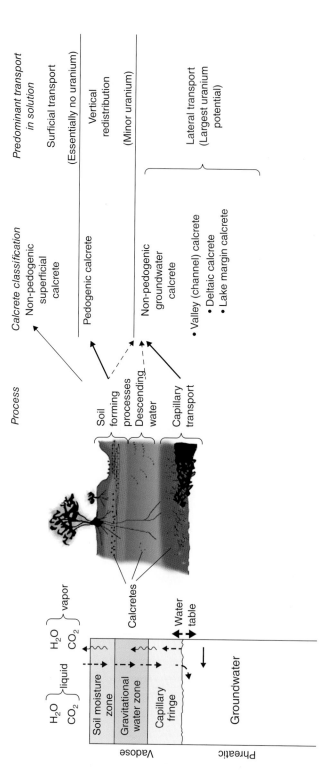

Figure 4.14 Simplified classification scheme for calcretes and a summary of the processes by which they form. An indication of the suitability of calcretes for hosting surficial uranium deposits is also provided. Source: After Carlisle (1983).

4.5 Calcrete-Hosted Deposits

Most of the laterally extensive surface or near-surface calcrete (also referred to as caliche or duricrust) layers that typify arid environments around the world are classified as "pedogenic calcrete" because they represent calcified soils (Klappa 1983). Calcrete is defined as an accumulation of fine grained calcite ($CaCO_3$) in the vadose zone (i.e. above the water table) that formed during a combination of pedogenic (soil forming) and diagenetic (lithification) processes. The solution and precipitation of calcite in the surficial environment can be represented by the following general equation (after Klappa 1983):

$$Ca^{2+} + 2HCO_3^- \Leftrightarrow CaCO_3 + CO_2 + H_2O$$

$$(4.12)$$

The solubility of $CaCO_3$ increases with decreasing temperature, with lower pH, and with increasing partial pressure of CO_2. Thus, extraction of CO_2 from an aqueous solution will precipitate $CaCO_3$. Precipitation of calcite may also result when water is removed from a soil profile during evaporation and evapotranspiration, resulting in an increase in the concentration of metal ions in the solution. A simplified classification of calcrete types is shown in Figure 4.14 where it is evident that, in addition to extensively developed pedogenic calcretes, a more locally distributed non-pedogenic variant, termed valley or channel calcrete, also occurs. It is the latter variety that is particularly important as the host rock for surficial uranium deposits.

4.5.1 Calcrete-Hosted or Surficial Uranium Deposits

Important uranium resources have been discovered in channelized calcretes from arid regions in Australia and Namibia. The Yeelirrie (Western Australia) and Langer Heinrich (Namibia) deposits represent well-known examples of surficial uranium ores formed by accumulations of a bright yellow potassium–uranium vanadate mineral, carnotite ($K_2(UO_2)_2(V_2O_8)\cdot3H_2O$), within calcretized fluvial drainage channels (Carlisle 1983). Since carnotite is a uranyl (U^{6+})–vanadate, reduction of hexavalent uranium is *not* the main process involved in the precipitation and concentration of uranium ores (see Chapter 3). Rather, the formation of this type of uranium deposit is related to high rates of groundwater evaporation and the resultant decrease of aqueous carbonate, vanadium, and uranium solubilities within a few meters of the surface. Calcretized fluvial channels represent the remnants of rivers from a previous higher rainfall interval. When such channels drained a uranium-fertile source region, they formed zones within which uranium was transported by focused groundwater flow, with subsequent mineral precipitation and concentration of uranium ore. In this process, uranium and other components are required to remain in solution until they reach a zone where carnotite can be precipitated by evaporation of the groundwater.

At Yeelirrie in Western Australia, fluvial channels started to incise the bedrock during Paleogene times, although the period of aridity commenced only in the late Pliocene. Active calcrete deposition and carnotite ore formation are thought to have occurred over the past 500 000 years (Carlisle 1983). Calcrete crops out along the axis of a paleochannel that can be traced for over 100 km and within which a carnotite bearing orebody, some 6 km long, 0.5 km wide, and up to 8 m thick, and comprising a resource of about 46 000 tons of U_3O_8, is defined (Figure 4.15a). The Langer Heinrich deposit in Namibia is remarkably similar in many respects to Yeelirrie (Hartleb 1988).

The main ingredients of carnotite, U, K, and V, are derived locally from weathered granites (for the uranium and potassium) and possibly more mafic rocks for the vanadium. In relatively carbonated groundwaters under

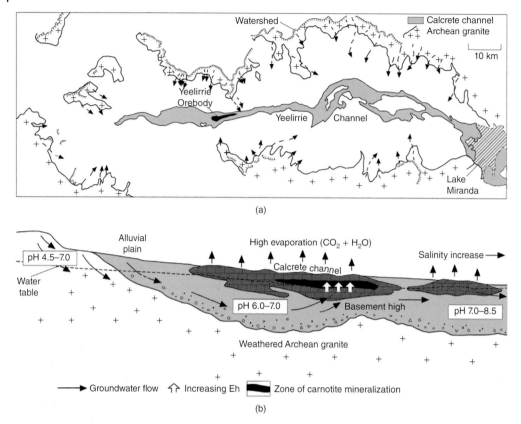

(a)

(b)

Figure 4.15 (a) Geological setting of the Yeelirrie carnotite deposit hosted in channelized calcrete, Western Australia. Source: After Mann and Deutscher (1978), Carlisle (1983). (b) Model depicting the setting and processes involved in the formation of carnotite deposits in calcretized channels. Source: After Carlisle (1983).

near neutral conditions, U^{6+} is transported as a carbonate complex (see Figure 3.36), whereas vanadium was probably in solution as V^{4+}, although the exact species and complexing agent is not known (Mann and Deutscher 1978). Focused groundwater flow introduced the ore-forming ingredients into the channel, with a combination of high evaporation rates and calcite precipitation ensuring that the solution evolved toward higher salinities and pH along the flow path (Figure 4.15b). In detail, however, the actual precipitation of carnotite is a complex process and appears to be related either to evaporation and decomplexation of uranyl–carbonate complexes, or to oxidation of V^{4+} to V^{5+}, or both (Mann and Deutscher 1978). A general equation for the

precipitation of carnotite can be written as follows (after Carlisle 1983):

$$2UO_2^{2+} + 2H_2VO_4^- + 2K^+ + 3H_2O \Leftrightarrow$$
$$\underset{\text{carnotite}}{K_2(UO_2)_2(V_2O_8)\cdot 3H_2O} + 4H^+ \quad (4.13)$$

Evaporation removes CO_2 from solution and drives equilibrium equations such as Eq. (4.12) to the right, promoting precipitation of calcite and probably also carnotite. Precipitation of carnotite, however, produces hydrogen ions (see Eq. (4.13)), which lowers pH and increases the solubility of calcite. Dissolution of calcite provides additional CO_3^{2-} to the solution, which in turn increases uranium solubility and favors dissolution of carnotite. The processes described in terms of Eqs. (4.12) and (4.13) appear, therefore, to be counter-productive – textual evidence for repeated dissolution and reprecipitation of ore and gangue minerals is indeed evident in these deposits. In reality, however, the very high rates of evaporation that occur in arid regions such as Australia and Namibia increase the Ca^{2+} and Mg^{2+} concentrations of the groundwater to such an extent that calcite (as well as dolomite) is readily precipitated in these environments. Formation of carbonate minerals, under conditions in which they would not otherwise have precipitated, also promotes the destabilization of uranyl–carbonate complexes and eventually results in carnotite formation coeval with calcretization. Mann and Deutscher (1978) have suggested that oxidation of vanadium may also play a role in carnotite formation since V^{4+} is soluble in mildly reducing waters at near neutral pH, but is precipitated as V^{5+} with increasing Eh. They envisage that vanadium might have been transported separately in deeper groundwaters flowing below the calcretized channel and that this fluid mixed with the surficial, oxidized, U- and K-bearing fluid to form carnotite in a mixing zone. The migration of the lower V-bearing fluid to the near surface environment is caused by positive relief in the basement topography (Figure 4.15b) which force groundwaters upwards into the evaporative zone to mix with oxidized fluids, or be directly oxidized by interaction with the atmosphere. This notion is consistent with the common observation that carnotite mineralization is spatially related to pinch-outs and constrictions in the fluid flow path.

4.6 Supergene Enrichment of Cu and Other Metals in the Near Surface Environment

The processes of weathering can be responsible for the in situ enrichment of Cu, as well as other metals such as Zn, Ag, and Au. In fact, many mineral deposits have their overall grades enhanced because of the localized enrichment of metals that occurs at or near the surface as a result of weathering (Reich and Vasconcelos 2015). The process is generally referred to as *supergene enrichment* and is an important criterion particularly in the assessment of low-grade porphyry copper deposits where the presence of an enriched, easily extractable supergene blanket of secondary copper ore minerals above the primary or *hypogene* ore may be the factor that makes them economically viable. The processes involved in the formation of supergene mineralization are similar to those discussed in Section 4.3.4 and effectively involve leaching of metals from primary ores by oxidation or hydrolysis, the transport of the metals to a site usually close by, and the precipitation of secondary minerals by a variety of means that could include reduction, bacterial mediation, adsorption, and cation-exchange (see Section 3.5). This section focuses specifically on copper and the formation of supergene enrichments in porphyry-type deposits, although the principles also apply to other deposit types.

4.6.1 Supergene Oxidation of Copper Deposits

Unlike the processes applicable to gold and nickel, copper enrichments are not specific to lateritic environments and supergene copper deposits can occur in any surficial environment where oxidized, acidic groundwaters are able to destabilize sulfide minerals and leach copper. The principles involved in this process are illustrated in Figure 4.16 and also summarized in terms of the reaction below:

$$4CuFeS_2 + 17O_2 + 10H_2O \Leftrightarrow$$
$$\text{chalcopyrite}$$
$$4Fe(OH)_3 + 4Cu^{2+} + 8SO_4^{2-} + 8H^+ \quad (4.14)$$
$$\text{goethite}$$

In most porphyry copper environments pyrite is the dominant sulfide mineral and its hydrolysis and oxidation dictates the production of hydrogen ions (i.e. the decrease in pH) in the weathering zone (see Eqs. (4.7) and (4.8) above). Sulfide mineral breakdown may also

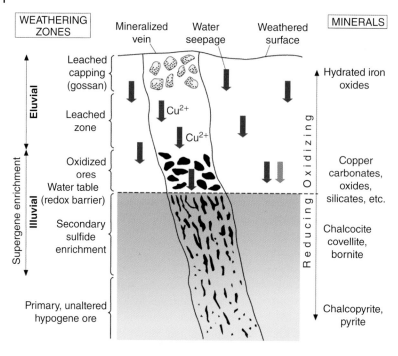

Figure 4.16 Schematic section through a copper deposit showing the typical pattern of an upper, oxidized horizon (the leached or eluvial zone) overlying a more reduced zone of metal accumulation (the supergene blanket or illuvial zone). The uppermost zone of ferruginous material, often containing the skeletal outlines of original sulfide minerals, is known as gossan. The redox barrier may be the water table or simply a rock buffer. Source: After Webb and Rowston (1995).

be accompanied by the formation of goethite in the regolith and the liberation of SO_4^{2-}. Chalcopyrite is the major copper–iron sulfide mineral and its breakdown, described in reaction (4.14), produces soluble cuprous ions that are dissolved in groundwater solutions. In the regolith profile, therefore, oxidizing, acidic groundwaters leach the primary, hypogene orebody of its metals, leaving behind an eluviated zone that may, if the intensity of leaching is not too severe, be residually enriched in iron, as hematite or goethite/limonite. The upper clay- and Fe–oxyhydroxide-rich capping, which also contains the skeletal outlines of the original sulfide minerals, is referred to as a *gossan*. Gossans are very useful indicators of the previous existence of sulfidic ore, not only in porphyry systems, but in many other ore-forming environments too. The soluble copper ions percolate downwards in the regolith profile and encounter progressively more reducing conditions, either as a function of the neutralization of acid solutions by the host rock, or at the water table. Copper is then precipitated as various secondary minerals, the compositions of which reflect the groundwater composition as

well as local pH and Eh in the supergene zone. The so-called "copper-oxide" minerals that precipitate here are actually compositionally and mineralogically variable and can include copper –carbonate, –silicate, –phosphate, –sulfate, –arsenate, as well as –oxyhydroxide phases (Chávez 2000; Dill 2015). In addition, Cu also replaces pre-existing sulfide minerals (i.e. pyrite and chalcopyrite) in more reduced zones where the latter minerals are still stable. Cu typically replaces the Fe in hypogene minerals such as pyrite and chalcopyrite, to form a suite of Cu-enriched sulfide phases including chalcocite (Cu_2S), covellite (CuS), and bornite (Cu_5FeS_4).

A more detailed illustration of the distribution of stable mineral assemblages in a typical supergene weathering profile over a porphyry copper deposit is shown in Figure 4.17a. This simulation integrates the effects of time (or more accurately the evolution of the chemical system in terms of fluid/rock ratio) in much the same way that alteration processes were described in dynamic terms as a function of evolving fluid/rock ratios in Section 3.6 of Chapter 3. The primary, hypogene ore is considered to be hosted in a granite and comprises pyrite + chalcopyrite that is weathered by a flux of slightly acidic rainfall. Within about 6000 years (period 1 in Figure 4.17a) the primary sulfides are totally dissolved from the leached zone and copper is reprecipitated as chalcocite and covellite by replacement of primary sulfides in the supergene blanket, as shown in reaction (4.15). Progressive replacement of hypogene pyrite by secondary copper sulfides is also illustrated in Figure 4.17b:

$$CuFeS_2 + 3Cu^{2+} \Leftrightarrow 2Cu_2S + Fe^{2+} \quad (4.15)$$
chalcopyrite · · · · · · · · · chalcocite

The stable assemblage in the leached (gossanous) zone during stage 1, and also into stage 2, is hematite/goethite, alunite, and quartz. The dissolution and/or replacement of primary sulfides is differential and chalcopyrite tends to disappear from both leached and supergene zones before pyrite, which is only removed completely from the upper zone by stage 2. With time, and as the fluid/rock ratio increases, the leached zone develops gibbsite, muscovite/sericite, and minor clay minerals (stage 3), eventually evolving into a unit dominated by sericite and kaolinite (stage 4). Alunite disappears and SO_4^{2-} is removed from the weathering zone. The pH of groundwaters in the leached zone starts off low (acidic) during the sulfide dissolution stage and then evolves to slightly more alkaline conditions with time. The supergene blanket initially develops an assemblage comprising covellite and chalcocite, as the primary sulfides are replaced, together with quartz and alunite. Once chalcopyrite is consumed from the leached zone, the downward migrating flux of Cu ions in solution is reduced and bornite develops later in the paragenetic sequence (i.e. from stage 2 onwards). Again, in this zone the clay minerals and muscovite/sericite are stabilized by progressive hydrolysis of primary feldspars and micas and, ultimately, the stable mineral assemblage is not unlike that of the overlying leached zone except that copper sulfide (bornite) is still stable. In the primary host rock weathering has a limited effect,

although over time relatively unstable minerals such as plagioclase, biotite, and magnetite will tend to disappear. It is interesting to note that sulfate (SO_4^{2-}), produced in abundance during the oxidation of primary sulfides (see reaction (4.14) above and Box 4.2), is generally transported away from the in situ weathering zone, and very little sulfate reduction (either organic or inorganic) tends to occur in the supergene blanket, or below the water table. This is another reason why low-S copper sulfides, such as covellite and chalcocite, form in this environment.

In addition to the formation of high metal/sulfur copper sulfide minerals at a redox barrier in the supergene blanket (Figure 4.16), it is apparent that a diverse suite of secondary copper minerals can form at the base of the oxidized zone (Figure 4.18). This suite includes oxides such as cuprite and tenorite (the former usually associated with native Cu), carbonates such as malachite and azurite, sulfates such as brochantite, antlerite, and chalcanthite, chlorides such as atacamite, silicates such as chrysocolla, and phosphates such as libethinite. The precipitation of these oxidized secondary minerals is generally due to direct precipitation from groundwaters that are saturated with respect to one or more of the various components that make up this suite of minerals. Figure 4.18 shows an Eh–pH diagram that identifies the stability fields of several of the copper minerals encountered in supergene enrichment zones above porphyry copper deposits. Under relatively low Eh conditions, below the water table, the high metal/sulfur copper sulfides are stable, whereas closer to the water table, cuprite and native Cu form. Above the water table the stability fields of a range of secondary copper minerals (note only a few are shown) occur and these are controlled essentially by pH. This is because it is essentially the pH that controls the solubilities of complexing ligands (such as CO_3^{2-}, OH^-, SO_4^{2-}, PO_4^{2-}, etc.) and determines which of the relevant copper complexes are likely to be saturated in any given environment.

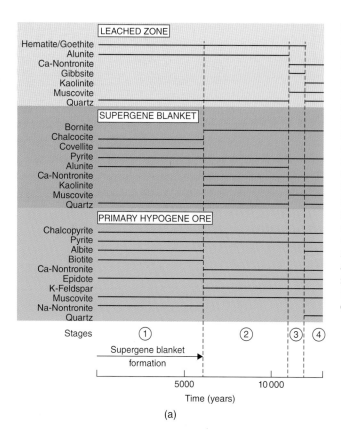

(a)

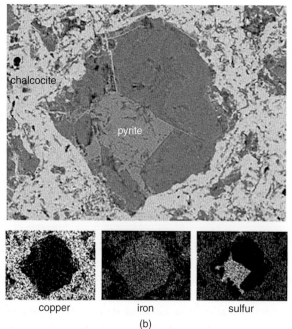

(b)

Figure 4.17 (a) Illustration of the stable mineral assemblages modeled as a function of time in the weathering zones above a porphyry copper deposit. The primary, hypogene ore (least altered) is compared to the overlying supergene blanket and the oxidized, leached zone. Note that with time (from stages 1 to 4) the mineral assemblages evolve as certain phases are consumed and others precipitated. Source: After Ague and Brimhall (1989). (b) Copper enrichment in a supergene profile – photomicrograph shows the breakdown of pyrite and its replacement by secondary copper sulfide minerals (chalcocite), with associated SEM back-scatter images showing the distribution of copper, iron, and sulfur. Source: Photos courtesy of Ryan Mathur.

Figure 4.18 Eh–pH diagram showing the stability fields of selected copper minerals at 25 °C and 1 atm. Source: After Guilbert and Park (1986).

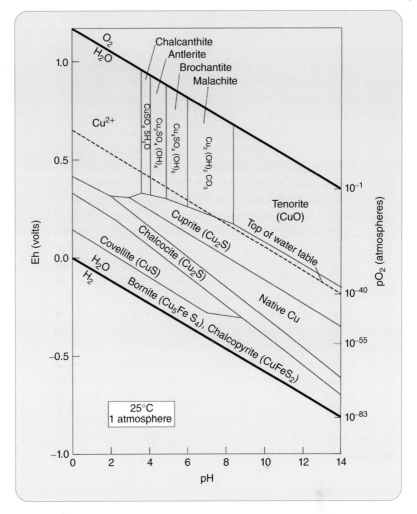

The extent to which leached and supergene zones develop, and the nature of the secondary minerals that form in them, depends to a certain extent on the quantity of primary, hypogene sulfides in the host rock and the acidity generated in the groundwaters by their oxidation/weathering. Weathering of a host rock with a low sulfide content will typically result in minimal acidity, limited mobility of Fe and other metals, and formation of secondary copper minerals that are stable in the near neutral to slightly acidic pH range (Chávez 2000). Conversely, weathering of protores with a high sulfide mineral content will result in more extreme acidity and the formation of secondary copper minerals stable only at lower pH. Local conditions also play an important role, however, and supergene enrichment in the proximity of a limestone, for example, will result in local groundwaters with a high CO_3^{2-} content, resulting in the stabilization of minerals such as malachite or azurite under neutral to alkaline conditions (Figure 4.15 and Eq. (4.16) below).

$$2Cu^{2+} + CO_3^{2-} + 2OH^- \Leftrightarrow \underset{\text{malachite}}{Cu_2(OH)_2CO_3}$$

(4.16)

Probably the most impressive examples of supergene enrichment are associated with the giant porphyry copper deposits of northern Chile, such as Chuquicamata, El Salvador, and El Abra. In addition to world-class hypogene orebodies at depth, these deposits have been subjected to climatic and geomorphic conditions during Paleogene – Neogene times that favored significant supergene enrichment (Sillitoe and McKee 1996). The supergene and "exotic" ores at Chuquicamata, Chino (Santa Rita) and El Salvador are described in more detail in Box 4.2.

4.6.1.1 A Note on Supergene Enrichment of Other Metals

Zones of supergene enrichment may be found in the surficial environment above any exposed metal orebody. In porphyry systems, molybdenum is normally removed from regolith profiles, although it can accumulate under very oxidizing conditions as ferrimolybdite, and under alkaline conditions as powellite (Ca–molybdate). Relative to Cu, the other base metals do not form secondary mineral enrichments as commonly. Zn is often dispersed in groundwaters, whereas Pb tends to be relatively immobile. These two metals do, however, form a range of sulfate and carbonate supergene minerals, such as anglesite ($PbSO_4$), cerrusite ($PbCO_3$), and smithsonite ($ZnCO_3$), under certain conditions. With new technologies available in extractive metallurgy, non-sulfide zinc ores, especially those comprising willemite (Zn_2SiO_4), now represent an important category of mineralization associated with oxidation of stratiform, sediment-hosted base metal deposits. Examples include the Skorpion Zn mine in Namibia and Vazante in Brazil. Silver tends to behave in much the same way as copper, and acanthite (Ag_2S) is readily oxidized in acidic groundwaters, with Ag being reprecipitated as the native metal, or as Ag–halides (AgCl, AgBr, AgI) in regions characterized by arid climates. In the Atacama desert of Chile there have been spectacular concentrations of supergene silver ores discovered, including a 20 ton aggregate of embolite (Ag[Cl,Br]) and native silver (Guilbert and Park 1986). The concentration of gold, with specific reference to lateritic regolith profiles, is discussed in Section 4.3.4 above.

Box 4.2 Supergene Processes: Supergene and "Exotic" Mineralization in Porphyry Copper Deposits

Porphyry coppers are high tonnage–low grade deposits. The average grade of primary sulfide, or hypogene, mineralization is typically so low that they are only marginally economic. However, many porphyry copper deposits are capped by a blanket of enriched ore formed by supergene processes. The supergene blanket, usually accessible during the early stages of mining, contributes significantly to the overall viability of the mining operation (Reich and Vasconcelos 2015). At Chuquicamata, Chile, for example (Figure 4.2.1a), one of the largest copper deposits in the world with a total reserve of some 11.4 billion tons of ore at 0.76 wt% copper, the supergene blanket comprised a major proportion of the ore body that has now largely been mined out. The supergene blanket is made up of a barren leached zone, an upper copper "oxide" zone of antlerite, brochantite, atacamite, and chrysocolla, and an underlying copper sulfide zone made up mainly of chalcocite (Ossandón et al. 2001; Figure 4.2.2). In addition to the in situ supergene ores, some of the giant porphyry copper deposits of northern Chile, such as Chuquicamata, El Salvador, and El Abra, also contain "exotic" copper oxide mineralization. This is secondary ore that has been transported laterally from the leached portion of the supergene blanket and precipitated in the drainage network surrounding the deposits. Again, at Chuquicamata exotic copper oxide mineralization is located in a gravel filled paleochannel, extending from the main pit to the South or Mina Sur deposit, a distance of about 7 km. Some 300 million tons of exotic ore was deposited in these gravels as chrysocolla and copper wad, the latter being a Cu-rich, K-bearing Mn oxyhydrate (Mote et al. 2001). Similar exotic copper

Box 4.2 (Continued)

Figure 4.2.1 (a) Cross section through the Chuquicamata mine showing the distribution of the supergene blanket (represented by the leached cap, the copper "oxide" zone, and supergene chalcocite ore) in relation to the primary hypogene ore. Source: After Ossandón et al. (2001). (b) Simplified map of the El Salvador district showing the distribution of gravel-filled paleochannels and exotic deposits extending radially away from the main pit. Source: After Mote et al. (2001).

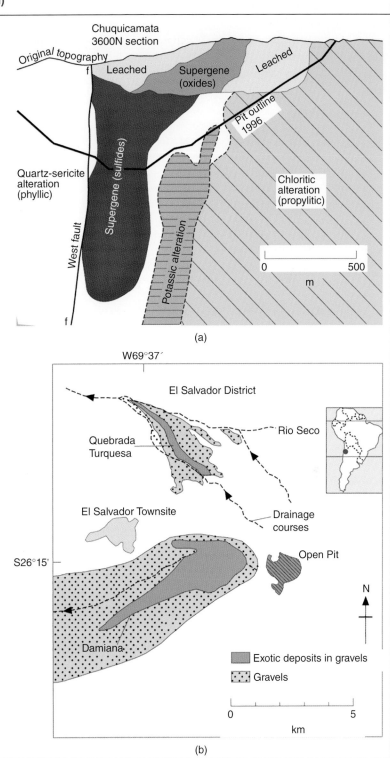

Box 4.2 (Continued)

Figure 4.2.2 Supergene profile exposed in the south wall of the Chino (Santa Rita) porphyry copper mine, New Mexico, USA. Source: Photograph courtesy of Ryan Mathur. The various zones are shown relative to the water table and are described below in terms of simple chemical reactions. Source: Reich and Vasconcelos (2015).

LEACHED ZONE – sulfide minerals broken down, Cu leached and goethite precipitated

$$4CuFeS_2 + 17O_2 + 10H_2O \rightarrow 4Fe(OH)_3 + 4Cu^{2+} + 8SO_4^{2-} + 8H^+$$

GREEN "OXIDIZED" ZONE – secondary copper minerals (i.e. copper carbonates, oxides, phosphates, silicates etc.) precipitated above the water table
See Eq. (4.16)
REDUCED SULFIDE ZONE – secondary copper sulfides (chalcocite) precipitated below the water table

$$2Cu^{2+} + HS^+ \rightarrow Cu_2S + H^+$$

mineralization occurs at El Salvador (the Damiana and Quebrada Turquesa deposits) and also at El Abra.

Supergene and exotic mineralization is related to the formation of acidic groundwaters that take up Cu into solution and reprecipitate it elsewhere. Secondary copper is either reprecipitated in the chalcocite rich supergene blanket beneath the leached zone (Figure 4.2.1a), or transported laterally by groundwaters moving through paleo-drainage channels to form the distal exotic ore bodies of chrysocolla and copper wad (Figure 4.2.1b). The high degree of

---

Box 4.2 (Continued)

preservation of secondary copper ores in northern Chile is related to episodic tectonic uplift in combination with the pattern of global climatic rainfall and watertable fluctuations (Mote et al. 2001). Accurate dating of copper wad and alteration assemblages in supergene mineralized zones of northern Chile indicates that surficial processes were long-lived and episodic. Some supergene processes commenced just a few million years after the hypogene ores had formed at the Eocene–Oligocene boundary (around 35 Ma). The main pulses of supergene ore formation, however, occurred toward the end of the Oligocene (20–25 Ma) and in the mid-Miocene (12–15 Ma). These pulses coincided broadly with periods of relatively high rainfall which promoted weathering, leaching of copper and groundwater flow, as well as the deposition of gravels in proximal drainage channels. The subsequent preservation of both supergene and exotic styles of mineralization is due to a decrease in erosion rate (i.e. reduction in uplift) and, more importantly, the onset of hyperaridity and reduced groundwater flow, as the Atacama desert formed from mid-Miocene times (Alpers and Brimhall 1989).

4.7 Summary

The chemical processes that contribute to weathering include hydration and dissolution, hydrolysis and acid hydrolysis, oxidation and cation exchange. Surficial and supergene ore-forming processes are related essentially to pedogenesis, which can be simplified into an upper zone of eluviation (leaching of labile constituents and residual concentration of immobile elements) and an underlying zone of illuviation (precipitation of labile constituents from above). Laterites are the product of intense weathering in humid, warm intertropical regions and are important hosts to bauxite ores, as well as concentrations of metals such as Ni, Au, and the PGE. Residual concentrations of alumina, and the formation of bauxitic ores, occur in high rainfall areas where Eh and pH are such that both Si and Fe in laterites are more soluble than Al. Ni enrichments occur above ultramafic intrusions in the illuviated laterite zone where the metal is concentrated in phyllosilicate minerals by cation exchange. Au and Pt enrichments also occur in laterites above previously mineralized lithologies, forming in the presence of highly oxidized, acidic, and saline groundwaters. Fixation of the precious metals occurs by reduction or adsorption, in the presence of carbonaceous matter or Fe and Mn oxyhydroxides. Oxidized, acidic groundwaters are capable of leaching metals, not only during laterite formation, but in any environment where such fluids are present. Supergene enrichment of Cu can be very important in the surficial environment above hypogene porphyry styles of mineralization. Enrichment of copper occurs in the illuviated zone, although in this case both Cu–sulfide minerals (in the relatively reduced zone beneath the water table) and Cu–oxide minerals (above the water table) form.

The formation of clay deposits, such as kaolinite (china clay), is a product of progressive acid hydrolysis, essentially of plagioclase feldspar. Clay itself is a valuable commodity, especially in the ceramic and paper industries, and certain clays that formed from highly weathered granites (such as in SE China) have preferentially adsorbed and concentrated the HREE and represent the principal global source of these critical metals. Calcrete is a pedogenic product of high evaporation environments but can also form in paleo-drainage channels in arid climatic regions. In uranium-fertile drainage systems, calcrete channels represent zones where concentrations of secondary uranium minerals precipitate together with calcite by evaporation-induced processes.

Further Reading

Bland, W. and Rolls, D. (1998). *Weathering: An Introduction to the Scientific Principles*, 271. London: Arnold.

Butt, C.R.M. and Cluzel, D. (2013). Nickel laterite ore deposits: weathered serpentinites. *Elements* 9: 123–128.

Guilbert, J.M. and Park, C.F. (1986). *The Geology of Ore Deposits*, 985. New York: W.H. Freeman and Co. (Chapter 17).

Leeder, M. (1999). *Sedimentology and Sedimentary Basins: From Turbulence to Tectonics*, 592. Oxford: Blackwell Science (Chapter 2).

Martini, I.P. and Chesworth, W. (1992). *Weathering, Soils and Paleosols. Developments in Earth Surface Processes 2*, 618. New York: Elsevier.

Reich, M. and Vasconcelos, P.M. (2015). Supergene metal deposits. *Elements* 11 (5): 305–342.

Williams, P.A. (1990). *Oxide Zone Geochemistry*, 286. New York: Ellis Horwood.

Wilson, R.C.L. (1983). *Residual Deposits: Surface Related Weathering Processes and Materials*, 258. London: The Geological Society of London and Blackwell Scientific Publications.

5

Sedimentary Ore-Forming Processes

TOPICS

Sedimentary basins and the ores within them
Clastic sedimentation and heavy mineral concentration – placer deposits
   sorting mechanisms relevant to placer formation
   application of sorting mechanisms to placer deposits
   sediment sorting in beach and eolian environments
Chemical sedimentation – iron-formations, phosphorites, and evaporites
   iron-formations and ironstones
   bedded manganese deposits
   ocean floor manganese nodules
   phosphorites
   black shales
   evaporites
Fossil fuels – oil/gas formation and coalification
   oil and gas formation (conventional)
   coalification
   unconventional hydrocarbons – shale gas, oil shales, and tar sands
   gas hydrates

CASE STUDIES

Box 5.1 Placer Processes: The Alluvial Diamond Deposits of the Orange River, Southern Africa
Box 5.2 Chemical Sedimentation – Banded Iron-Formations: The Mount Whaleback Iron Ore Deposit, Hamersley Province, Western Australia
Box 5.3 Evaporite Deposits: The Boulby Mine (Na and K), Cleveland, UK and Salar de Atacama (Li and K), Chile
Box 5.4 Fossil Fuels – Oil and Gas: The Arabian (Persian) Gulf, Middle East

---

5.1 Introduction

Sedimentary rocks host a significant proportion of the global inventory of mineral deposits and also contain the world's fossil fuel resources. Previous chapters have considered a variety of ore-forming processes that have resulted in the epigenetic concentration of metals and minerals in sedimentary rocks. This chapter will concentrate on processes that are syngenetic with respect to the host sediments and where the ores are themselves sediments or part of the sedimentary sequence.

Introduction to Ore-Forming Processes, Second Edition. Laurence Robb.
© 2021 John Wiley & Sons Ltd. Published 2021 by John Wiley & Sons Ltd.

Processes to be discussed include the accumulation of heavy, detrital minerals and the formation of placer deposits, the deposition of organic-rich black shales, and the precipitation mechanisms that give rise to Fe, Mn, and P concentrations in chemical sediments. In addition, there is an overview of the origins of oil and gas deposits, as well as of coalification processes, because the organic source material for these deposits reflects the local depositional environment and ore formation was broadly coeval with sediment accumulation. Although it might be argued that the migration of oil through sediments involves later fluid flow through the depository, such processes are generally early diagenetic in character and therefore warrant inclusion in the present chapter. By contrast, syngenetic, clastic sediment-hosted SEDEX Pb–Zn–Ag, sandstone-hosted U ores, and late diagenetic red-bed Cu–(Ag–Co) type deposits are discussed in Chapter 3, since they involve processes that are more hydrothermal in character, or have their metals originating from outside the sedimentary host rocks. These subdivisions are somewhat arbitrary and reflect more the organizational structure of this book than a rigorous genetic classification.

Modern trends in the application of basin analysis techniques to exploration have undoubtedly been set in the oil industry, where, in recent decades, major advances have been made in understanding the relationships between sediment deposition, tectonics, and the maturation of organic material. The evolution of oil and gas within this cycle can be effectively assessed in terms of organic chemistry and fluid flow patterns in the rock (Eidel 1991). This level of understanding requires a high degree of integration of numerous earth science disciplines (including paleontology, stratigraphy, sedimentology, structural geology, plate tectonics, geohydrology, organic geochemistry, and others), an approach that is increasingly being applied to all types of exploration in sedimentary basins (Force et al. 1991).

As with the fossil fuels, ore-forming processes that give rise to placer deposits, as well as metal enrichments associated with chemical sediments and diagenetic fluid flow, are intimately related to the origin, tectonic setting, and evolution of the sedimentary host rocks. Major syn-sedimentary ore deposits tend to occur in a limited range of basin types (Eidel 1991). Passive continental margin chemical sedimentary ores (iron-formations and ironstones, bedded Mn deposits, and phosphate ores), as well as shoreline and fluvial placer deposits (gold, cassiterite, diamonds, and zircon–ilmenite–rutile black sands), reflect a cratonic metallogenic setting where features such as protracted stability and a variety of sedimentological processes apply. Sediment-hosted ores in which late diagenetic or epigenetic processes are responsible for metal accumulation (such as red-bed Cu–(Ag–Co) deposits and some SEDEX Pb–Zn–Ag ores; see Chapter 3) tend to be associated with rift basins where rates of deposition are high and active faulting promotes the circulation of connate and meteoric fluids. It is also noteworthy that many oil and gas fields are also associated with rift-related cratonic basins.

The relationships between sedimentation and ore formation are multifaceted and only selected topics are discussed in this chapter. Additional information is available from the works listed at the end of the chapter.

5.2 Clastic Sedimentation and Heavy Mineral Concentration – Placer Deposits

A placer deposit is one in which dense (or "heavy") detrital minerals are concentrated during sediment deposition. They are an important class of deposit type and can contain a variety of minerals and metals, including gold, uraninite, diamond, cassiterite, ilmenite, rutile, and zircon. Well-known examples of placer deposits include the late Archean

Witwatersrand and Huronian basins (Au and U) of South Africa and Canada respectively. Geologically more recent occurrences include the diamond placers of the Orange River system (see Box 5.1) and the western coastline of southern Africa, the cassiterite (Sn) placers of the west coast of peninsula Malaysia, and the beach-related "black sand" placers (Ti, Zr, Th) of Western Australia and New South Wales, South Africa, Florida, and India.

The hydrodynamic processes involved in placer formation are invariably very complex. It is difficult to predict, for either exploration or mining purposes, where heavy mineral concentrations occur. This section will only touch briefly on the principles of hydrodynamics that are applicable to geomorphology, civil engineering, and flood mitigation, as well as to sedimentology and the formation of placer deposits. More detailed reviews of the principles involved in clastic sedimentation and placer formation can be found in Miall (1978), Slingerland (1984), Slingerland and Smith (1986), Force (1991), Pye (1994), and Carling and Breakspear (2006).

5.2.1 Basic Principles

The formation of placer deposits is essentially a process of sorting light from heavy minerals during sedimentation. In nature heavy mineral concentration occurs at a variety of scales, ranging from regional systems (alluvial fans, beaches, etc.), through intermediate features (the inner bank of a river bend or a point bar), to small-scale features (bedding laminae or cross-bed foresets). An experimental simulation of placer processes is illustrated in Figure 5.1, where a heavy mineral is fed via a "tributary" into the "trunk river." The confluence of two orthogonal flow streams results in the creation of a vortex or eddy downstream of the junction, and the accompanying erosion of the bedload with removal of small, light particles and retention of larger and heavier minerals.

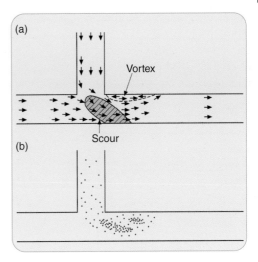

Figure 5.1 Results of a flume experiment simulating (a) fluid flow at the confluence of a tributary with its trunk river and (b) the resulting distribution of heavy minerals. Source: After Best and Brayshaw (1985).

The transportation and deposition of sediment in fluvial and related systems is a multifaceted process (Selley 1988; Friedman et al. 1992; Pye 1994; Allen 1997; Leeder 1999), the basic principles of which are essential to the understanding of placer formation. To complicate matters, sediments can also be sorted by the action of wind, and this process has applicability to the formation of placers in, for example, beach environments (Kocurek 1996).

One of the parameters used to quantify the conditions of fluid motion is the Reynolds number, which is a dimensionless ratio identifying fluid flow as either laminar and stable, or turbulent and unstable. The Reynolds equation is expressed as:

$$Re = UL\delta_f/\eta \qquad (5.1)$$

where Re is the dimensionless Reynolds number; U is the fluid velocity; L is the length over which the fluid is flowing; δ_f is the fluid density; and η is the fluid (molecular) viscosity.

For low Reynolds numbers, flow is laminar and vice versa for turbulent flow (Figure 5.2a). The behavior of a particle in a river channel will depend to a large extent on the energy of

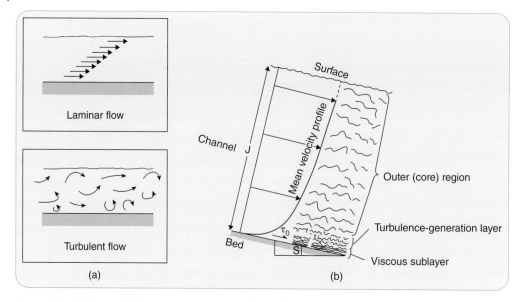

Figure 5.2 (a) Schematic illustration, using streamlines, of the nature of laminar and turbulent fluid flow. (b) Internal structure of turbulent fluid flow in a natural channel. τ_0 is the boundary shear stress imposed by the fluid on its bed and is a function of J (the flow depth) and S (the bed slope), as well as fluid density. Source: After Slingerland and Smith (1986).

the transporting fluid. Fluid flow in natural stream channels is predominantly turbulent, the detailed anatomy of which is shown schematically in Figure 5.2b. Three layers of flow can be identified. The bottom zone is the non-turbulent *viscous sublayer*, which is narrow and may break down altogether in cases where the channel floor is rough and turbulence is generated by the upward protrusion of clasts from the bed load. Above it is the *turbulence generation sublayer*, where shear stresses are high and eddies are generated. The remainder of the stream profile is the *outer or core layer*, which has the highest flow velocities. A shear stress (τ_0 in Figure 5.2b) is imposed on the bed load by the moving fluid and is a function of fluid density, the slope of the stream bed, and flow depth. The curved velocity profile of the channel section is the result of frictional drag of the fluid against the bed.

The types of fluid flow in water (or air) define the character and efficiency of mass (sediment) transport. A particle or grain will move through a fluid as a function of its size, shape, and density, as well as the velocity and viscosity of the fluid itself. In water, a particle

at any instant will move in one of three ways: the heaviest particles (boulders, gravel) roll or slide along the channel floor to form the bedload (or traction carpet); intermediate sized particles (sand) effectively bounce along with the current (a process known as saltation); while the finest or lightest material (silt and clay) will be carried in suspension by the current (Figure 5.3a). In air the types of movement are similar, but the lower density and viscosity of air relative to water dictate that moving particles are smaller, although their motion may be more vigorous (Figure 5.3b).

The type of particles moving in a fluvial channel by saltation and suspension is partly a function of the nature of the fluid flow (as defined by the Reynolds number), whereas bed load movement is determined by shear stress at the boundary layer (Figure 5.2b) and the characteristics of the particles themselves. Combining fluid flow (hydrodynamic) and physical mass transport parameters provides a useful semi-quantitative indication of the processes of sedimentation, a technique first presented diagrammatically by Hjulström (Figure 5.4; after Sundborg 1956). In this

Figure 5.3 Illustration of the different mechanisms of sediment transport in water (a) and air (b). Source: After Allen (1997).

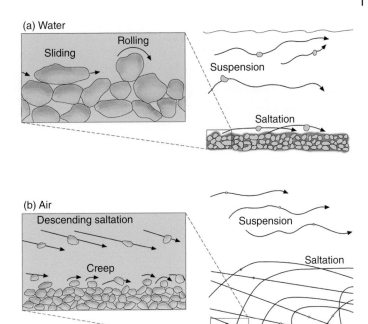

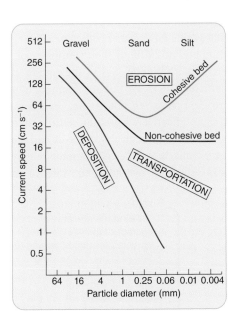

Figure 5.4 A Hjulström diagram showing how sedimentary processes can be assessed in terms of hydrodynamic (flow velocity) and physical (grain size) parameters. Critical conditions for deposition are shown, as well as those for erosion and transportation for two situations in which cohesive and non-cohesive channel beds apply. Source: After Sundborg (1956), Friedman et al. (1992).

diagram the conditions under which either erosion, transportation, or deposition will take place are shown as a function of flow velocity and grain size. Deposition occurs either as flow velocity decreases or grain size increases (or both) – parameters that are very relevant to the formation of placer deposits.

5.2.2 Hydraulic Sorting Mechanisms Relevant to Placer Formation

Slingerland and Smith (1986) divided the mechanisms of sorting into four types, namely:

- free or hindered settling of grains;
- entrainment of grains from a granular bed load by flowing water;
- shearing of grains in a moving fluidized bed;
- differential transport of grains by flowing water.

Each of these is discussed in more detail below with respect to their roles in the formation of placer deposits.

5.2.2.1 Settling

The settling velocity of a perfectly spherical particle in a low Reynolds number (non-turbulent) fluid can be determined using Stokes's Law, which is expressed as follows:

$$V = gd^2(\delta_p - \delta_f)/18\eta \qquad (5.2)$$

where V is the particle settling velocity; g is the acceleration due to gravity; d is the particle diameter; δ_p and δ_f are the densities of particle and fluid respectively; and η is the fluid (molecular) viscosity.

The relationship indicates that particle settling velocities in the same fluid medium are proportional to particle diameters (squared) and densities. Figure 5.5 shows that, in terms of Stokes's Law, different sized grains of quartz, pyrite, and gold may settle at the same velocity, a condition that is referred to as *hydraulic (or settling) equivalence*. This suggests that particles of differing size and density could conceivably reside together in the same sedimentary layer, to form a rock such as a conglomerate (Coetzee 1965). Settling equivalence is sometimes used to explain the concentration of small, heavy detrital minerals together with larger clasts in a coarse sedimentary rock but on its own is, in fact, an oversimplification of the dynamic sorting mechanisms taking place during placer formation. Hydraulic equivalence is, nevertheless, a fundamental concept that quantifies the relationships between particle size and density in a fluid medium.

There are several reasons why simple Stokesian settling is inadequate as an explanation for placer forming process. These include the fact that stream flow is generally turbulent (it has a high Reynolds number), particles are not spherical, and particle sizes may be either too big or too small for the relationship in Eq. (5.2) to be upheld. In addition, if the concentrations of grains in the fluid is high (>5%) then settling is no longer unhindered and settling velocities are retarded by grain–grain collisions and current counterflow.

The random instability of turbulent flow makes it virtually impossible to predict particle settling velocities and there is no completely satisfactory model for simulating this condition (Slingerland and Smith 1986). Likewise, particle shape has an important effect on settling velocity. A tabular biotite grain, for example, will settle between 4 and 12 times slower than a quartz grain of equivalent diameter. Big grains have large coefficients of drag in a fluid and accordingly their settling velocities vary as a function, not of the square of the diameter (Eq. (5.2)), but of the square root (Figure 5.5). Smaller grains are, therefore, much more effectively sorted by settling than are big grains. Although the ratios of particle diameters (quartz : pyrite : gold approximately 32 : 2 : 1) reflecting hydraulic equivalence in

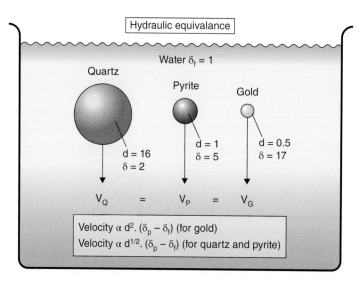

Figure 5.5 Illustration showing the principle of hydraulic equivalence for particles settling according to Stokes's Law. The settling velocities of quartz, pyrite, and gold, with radii and densities as shown, are the same, indicating that they would settle out of a non-turbulent column of water into the same sedimentary layer.

Figure 5.5 appear to be reasonably consistent with what one might expect in a gold-bearing conglomerate such as in the Witwatersrand Basin, the diameter of gold itself in these deposits is typically much too small to be explainable by Stokesian-type settling. For that component of Witwatersrand gold that is detrital, it is, therefore, likely that some other concentration mechanisms, such as entrainment (see below) might have applied (Frimmel and Minter 2002).

Particle settling in nature is a difficult parameter to quantify because of the wide range of variables likely to affect it. A more realistic expression of particle settling velocity, which takes into account the frictional drag that arises from different shapes, is provided by the following expression (after Slingerland and Smith 1986):

$$V = [4(\delta_p - \delta_f)gd/3\delta_f C_d]^{1/2} \qquad (5.3)$$

where V is the particle settling velocity; g is the acceleration due to gravity; d is the particle diameter; δ_p and δ_f are the densities of particle and fluid respectively; and C_d is the coefficient of drag and is defined as 24/Reynolds number.

Settling does play a role in the formation of placer deposits, but on its own is of little use in understanding the processes by which they form. The existence of a state of hydraulic equivalence will not explain how heavy detrital minerals are sorted or concentrated in dynamic river or beach systems. Hydraulic equivalence describes a condition of equal settling velocities and accounts for unsorted accumulation of heavy minerals in coarser grained sediment (conglomerate). By contrast, if settling velocities are not equal and settling occurs in a system that is flowing (i.e. a river), then heavy minerals will be segregated laterally downstream as a function of size and density (see section on transport sorting below). In any event it is the movement or flow of the fluid medium that has the dominant role to play in placer formation.

5.2.2.2 Entrainment

Entrainment sorting refers to the ability of a fluid in contact with bed load particles to dislocate certain grains from that bed and move them further downstream. As with settling, a considerable amount of effort has been spent by fluid dynamicists to quantify the criteria for entrainment in terms of variables such as the hydraulic/flow conditions, particle size, shape, and density. In the channel cross section shown in Figure 5.2b, the concept of a fluid force (referred to as τ_0, the boundary shear stress) acting on the bed load is demonstrated. It is clear that τ_0 must exceed the forces that keep any given particle in place (i.e. size, mass, shape, friction) before the particle can start to move. The critical shear stress required to initiate movement of any given particle is known as the Shields entrainment function, or simply the Shields parameter (θ), a detailed description and derivation of which is provided in Slingerland and Smith (1986). The Shields parameter is expressed as:

$$\theta = \tau_c/[(\delta_p - \delta_f)gd] \qquad (5.4)$$

where θ is the dimensionless Shields parameter; g is the acceleration due to gravity; d is the particle diameter; δ_p and δ_f are the densities of particle and fluid respectively; and τ_c is the critical boundary shear stress.

In relatively simple situations the Shields criterion can be shown to simulate the movement of bed load particles in a channel reasonably well, and Figure 5.6a illustrates the critical conditions that differentiate between entrainment of a particle into the channel and non-movement of the grain. For a uniform bed load, particle size is the dominant control and sand, for example, will have a lower entrainment threshold than gravel. Again, however, a uniform bed load seldom applies and there are many factors which complicate entrainment sorting. The bed load roughness, or, put another way, the extent to which a particle sits proud of the bed, will clearly have an effect on the entrainment threshold. In Figure 5.6a this variable is illustrated and

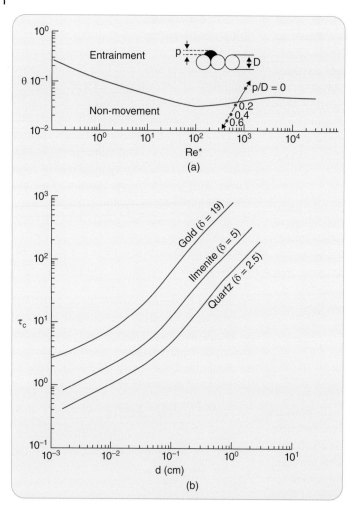

Figure 5.6 (a) Shields diagram showing the threshold conditions between entrainment and non-movement of a grain for a bed load of uniform size in terms of the Shields parameter (defined in Eq. (5.6)) and the grain Reynolds number (Re* or the Reynolds number as it pertains to a single particle). The effects of increasing the protrusion of a particle above the bed floor (i.e. increasing the ratio p/D) will have the effect of shifting the entrainment threshold: the higher a particle protrudes above the bed, the easier it will be to entrain. (b) Diagram showing the effects of grain density (δ) on the critical boundary shear stress (τ_c). Greater shear stresses are required to entrain denser particles. Source: Details after Slingerland and Smith (1986) and Reid and Frostick (1994).

quantified by the ratio p/D. An increase in this value equates with grains which jut out and this can be seen to have the logical effect of decreasing the entrainment threshold. Other factors which complicate entrainment sorting include the clast shape (sphericity tends to make entrainment easier), bed consolidation, and the range of grain sizes in any given bed load. The latter point is particularly important in the situation where sediment is made up of a bimodal assemblage of large and small grains. In this case small grains, even if they are relatively light, will not be entrained because they rapidly become entrapped within the much larger particles and are no longer available for entrainment.

The effects of entrainment on sorting and placer accumulation are shown in Figure 5.6b. In this diagram (similar to the Shields diagram but where θ is replaced by τ_c and Re* is replaced by grain diameter) the critical boundary shear stress for entrainment increases as a function of grain density. This indicates that, for a uniformly sized bed and with all other factors being equal, lighter particles will be effectively entrained at lower shear stresses and, therefore, winnowed away, leaving a residual concentration of heavier particles. It is, therefore, quite feasible to entrain quartz from the bed load of a stream and leave behind a residual accumulation of heavy minerals that could be preserved as a placer deposit. Thin laminae of heavy minerals in a fluvial setting, or along cross-bed foresets, could be the result of this type of sorting process. Entrainment processes also apply to wind-blown sedimentation and sorting, but the magnitude of the critical thresholds and the size and density of entrained particles will be quite different in the eolian environment.

5.2.2.3   Shear Sorting

Shear sorting of grains is a process that only applies to the concentrated flow of suspended particles in a fluidized bed. During the movement of suspended particles in a dense granular medium, grain collisions create a net force that is perpendicular to the plane of shearing and disperses the granules toward the free surface (i.e. upwards). Counter-intuitively, the dispersive pressure is greater on large grains within the same horizon of flow so that these grains migrate upwards relative to smaller and lighter particles. The same effect can be rationalized in terms of a concept known as kinetic sieving, where smaller grains simply fall between larger ones (Slingerland 1984).

The effects of shear sorting have been quantified for sediment populations of mixed sizes and densities by Sallenger (1979), who showed that two grains of different densities (δ) coming

to rest in the same horizon would have relative sizes (d) given by:

$$d_h = d_l(\delta_l/\delta_h)^{1/2} \qquad (5.5)$$

where d_h and d_l are the diameters of heavy and light fractions and δ_h and δ_l are the densities of heavy and light particles respectively.

An illustration of shear sorting in a concentrated granular mixture comprising quartz and magnetite, in the proportion 90 : 10, is shown in Figure 5.7. The horizon at a relative height of between 0.75 and 0.5 units from the surface is seen to contain magnetite concentrations up to double the initial concentration. The mechanism is, therefore, one of the few processes that will explain heavy mineral concentrations in an elevated horizon of the sediment strata and could also explain inversely graded particle concentrations. Shear sorting applies to concentrated suspended particle loads such as beach swash zones and wind-related dune formation. The process could be applicable to the concentration of Ti–Zr–Th black sand placers in these environments.

5.2.2.4   Transport Sorting

Transport sorting is the most important sorting process and the one most applicable to the broadest range of environments in which

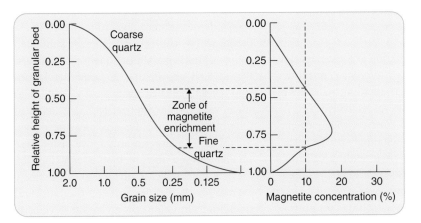

Figure 5.7 Quartz grain size and magnetite concentrations in a shear sorted granular dispersion comprising quartz and magnetite in the proportion 90 : 10. The relative depth of the granular bed is on the ordinate; original magnetite concentration (10%) is shown as the vertical dashed line. Source: After Slingerland (1984).

placer deposits form. The quantification of transport sorting is difficult, but conceptually it refers simply to the differential transport rates that exist during movement of particles in a flowing fluid medium. It is complex because it incorporates two distinct components, namely the varying rates of movement of grains both in the bed load (determined by entrainment), and in suspension (determined by settling and shear sorting).

The previous discussion of settling considered the concept only in terms of unhindered, non-turbulent flow and it was emphasized that it had limited use in understanding the dynamics of placer-forming processes. The concept of suspension sorting, however, is an extension to the rather simplistic considerations of the previous section and refers to the fractionation of grains with different settling velocities into different levels above the bed in a turbulent channel flow system. Subsequent to its deposition downstream, sediment sorted in this fashion can result in substantial heavy mineral enrichments. It is discussed here, rather than earlier, because it is one of the two components that contribute to the determination of transport sorting.

The concentrations of heavy (h) and light (l) particles that coexist at any point in the channel flow system can be quantified in terms of an equation derived in Slingerland (1984):

$$(C/C_a)_h = (C/C_a)_l^{V_h/V_l} \qquad (5.6)$$

where C is the concentration at a given level in the channel flow; C_a is a reference concentration; and V_h and V_l are the settling velocities of heavy and light particles respectively.

The above equation predicts that particles in a turbulent flowing channel will be sorted vertically according to their settling velocities which, according to Stokes's Law, are determined by their relative sizes and densities. The concept is pertinent to transport sorting because the effects of suspension sorting are only relevant downstream once deposition of a particular sediment horizon, with its possible

enrichment of a hydraulically equivalent mineral, has taken place. Note that the results of suspension sorting are contrary to those obtained by shear sorting in a concentrated fluidized bed.

The other component of transport sorting applies to the movement of bed load and is discussed previously in the section on entrainment. The entrainment threshold of a particle sitting on the bed is determined by the Shields parameter (Eq. (5.4)) but is also strongly influenced by bed roughness. Consequently, sediment transport rates would be expected to decrease as bed roughness increases. Slingerland (1984) has modeled transport sorting processes using an initial sediment comprising quartz and magnetite, in the proportion 90 : 10, and mean grain diameters of about 0.4 and 0.2 mm respectively. The results confirm that for a given shear velocity (another measure of the boundary shear stress, as shown in Figure 5.2b) all sizes of quartz and magnetite exhibit a decrease in transport rate as bed roughness increases. In addition, and as shown in Figure 5.8a, the proportion of magnetite in the moving sediment load increases as a direct function of shear velocity for a given bed roughness (trend a–a′) but decreases as a function of roughness for a given shear velocity (trend b–b′).

The effects of transport sorting have also been demonstrated experimentally in a flume, where sediment transport is simulated using simplified, scaled down parameters. Figure 5.8b shows that a positive relationship exists between transport velocity and shear velocity for any particle size and bed roughness. However, for any measure of bed roughness (K) larger grains move faster than smaller ones, a feature that is explained by the fact that smaller grains progress less easily over a rough bed because of trapping and shielding. Thus, for any given shear velocity and bed roughness, the largest grains have the fastest transport rates (trend c–c′–c″ in Figure 5.8b). Conversely, the slowest transport

Figure 5.8 (a) Computer simulation showing the effects of transport sorting in terms of bed roughness (measured by the height of clasts protruding from the bed in millimeters) and shear velocity U*. The plot shows the approximate limit for transport of magnetite and the magnetite concentrations (%) in the sediment load. Maximum enrichment of magnetite on the bed (i.e. the formation of a lag deposit) would occur above this limit under conditions of low shear velocity and high bed roughness. Source: After Slingerland (1984). (b) Results of a flume experiment showing the effects on grain transport velocities (V_g) of shear velocity (U*), bed roughness, and grain diameter. Source: After Meland and Norman (1966).

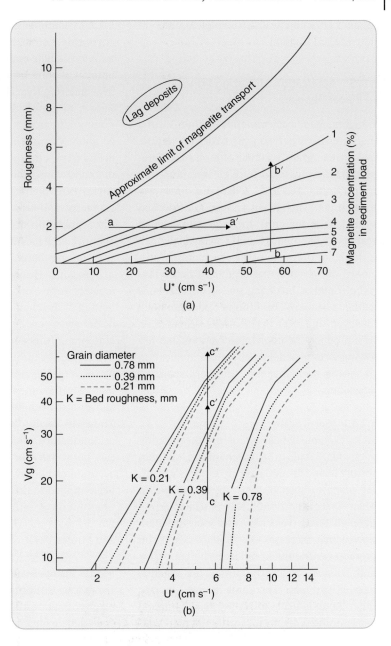

rates for a given shear velocity are a feature of the smallest particles (i.e. 0.21 mm in this particular experiment) moving over the roughest bed (K = 0.78 in Figure 5.8b). Clearly, transport sorting is a complex process which can have quite different outcomes from one part of a channel flow system to another. A bed of sediment that is being deposited at any given moment may undergo enrichment in

heavy minerals if the combination of shear velocity and the sizes of heavy and light particles relative to bed roughness mitigate against entrainment, thereby resulting in the formation of a lag deposit. If, on the other hand, the same particles are subsequently entrained into the sediment load under a different flow regime, they might then be subjected to suspension sorting which would result in heavy mineral enrichment at a completely different downstream location.

5.2.3 Application of Sorting Principles to Placer Deposits

Enrichment of heavy minerals by grain-sorting mechanisms occurs on all scales in nature, from single grain laminae on cross-bed foresets, to large regional scale concentrations that have accumulated in a particular sedimentary environment such as an alluvial fan or beach system. It is tempting to suggest that small scale systems might be more easily explained in terms of only one of the mechanisms discussed above, and that regional systems are more complex, involving several mechanisms acting in concert. This rationale does not hold, however, and even small scale systems can be a product of complex interactions; the following examples demonstrate the application of sorting mechanisms to different scales of deposition.

5.2.3.1 Small Scale

As an example of heavy mineral concentration at a small scale, the mechanisms of grain sorting in and around a dune or ripple migrating along the bed of a stream are shown in Figure 5.9a. The dune crest is characterized by high shear velocities and non-turbulent flow which increases the Shields factor and promotes entrainment of larger, lighter grains and residual concentration of the smaller, heavier particles. The dune foreset, on the other hand, is likely to receive heavy mineral concentrations by shear sorting of high concentration grain avalanches down the slope of the advancing dune. The trough or scour forming ahead of the dune is likely to receive heavy grain concentrations by settling sorting in a locally turbulent micro-environment. It is, therefore, evident that several sorting mechanisms apply even to the small scale heavy mineral concentrations that one often sees in and around dune and ripple features in river- or beach-related sediments.

5.2.3.2 Intermediate Scale

An example of heavy mineral concentration at an intermediate scale is provided by the development of point bars along the convex bank of a meandering river channel. Such sites are well known to miners dredging river channels for accumulations of minerals such as cassiterite or gold. Meandering river channels are also prime targets during the exploitation of placer diamonds along the Orange River in southern Africa (see Box 5.1). Figure 5.9b shows the geometry and flow patterns associated with the aggradation of a point bar in a meander channel and its migration toward the opposite concave bank, which is being subjected to erosion. Heavy mineral concentrations actually form in degraded scours along the bottom of the channel itself. Both components of transport sorting seem to contribute to the enrichment process, namely settling of grains in the highly turbulent environment formed by convergence of disparate flow orientations, and entrainment, because shear stresses are above the thresholds for most of the bed load. Transport sorting, therefore, results in placer accumulations which are then preserved as the point bar sediment migrates over the accumulated heavy minerals. This process has been applied to the concentration of gold and uraninite in point bars formed on the fluvial fan deltas of the Witwatersrand basin (Smith and Minter 1980).

5.2.3.3 Large Scale

Large scale sorting of heavy minerals over areas of tens of square kilometers has been described from deltaic fan conglomerates of the Welkom goldfield in the Witwatersrand basin, South Africa (Minter 1978). Figure 5.9c shows the relative distributions of gold and uraninite (represented by the Au/U ratio) in the composite Basal–Steyn placer with respect to the sediment entry point and major fluvial braid channels on the fan delta complex. Close to the entry point channel fill and longitudinal bars comprise coarse pebble conglomerate, with maximum clast sizes in the range 20–40 mm

Figure 5.9 Heavy mineral sorting processes at various scales and in different sedimentary environments: (a) at a small scale along laminae associated with dunes and ripples forming along a stream bed. Source: After Slingerland (1984). (b) At an intermediate scale in an aggrading point bar forming along the convex bank of a meander channel. Source: After Smith and Beukes (1983). (c) At a large scale on a large braided, alluvial fan complex such as the Welkom goldfield, Witwatersrand Basin. Source: After Minter (1978).

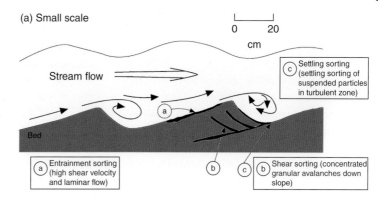

(a) Small scale

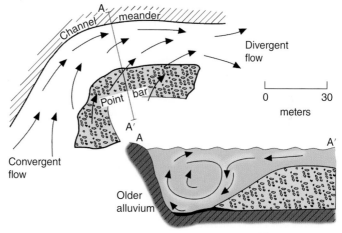

(b) Intermediate scale

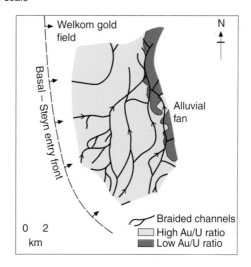

(c) Large scale

in diameter. By contrast, the contained gold particles are much finer and range between about 0.5 and 0.005 mm in size. Several kilometers downstream the fan delta is built of less deeply channeled quartz–arenite. The Au/U ratio decreases downstream and this is a function of a decrease in gold and an increase in uraninite contents down the paleoslope (Minter 1978). This has been interpreted to reflect transport sorting which results in net deposition of gold in more proximal locations of the fan, with the less dense uraninite being more effectively concentrated in distal parts. Transport sorting ensures that clast sizes (and also, therefore, bed roughness), as well as shear velocity, decrease markedly down slope. Uraninite grains are significantly larger, and also lighter, than gold and will exhibit higher grain transport velocities, and, therefore, be carried a greater distance down slope. In addition, gold will be transported less easily over the proximal, high bed roughness portions of the fan delta complex and will, therefore, be trapped more effectively in these areas compared to uraninite. Many of the laterally extensive Witwatersrand placers represent regional, low dip angle, unconformity surfaces where the fan delta complexes are being slowly degraded and the fluid regime is consistent over large areas. This type of setting represents an excellent site for the formation of very large placer deposits because significant volumes of sediment can be subjected to a consistent set of transport sorting mechanisms (Slingerland and Smith 1986; Carling and Breakspear 2006).

5.2.4 A Note Concerning Sediment Sorting in Beach and Eolian Environments

Much of the previous discussion applies to processes pertaining to unidirectional water flow and is, therefore, mainly relevant to fluvial environments. As mentioned previously, however, many important placer deposits are associated with sediments deposited in shoreline environments (Force 1991b) where sediment sorting is largely controlled by the dynamics of waves and by tidal fluctuations. It is also possible in such environments that sorting processes could be influenced by wind action and that the latter might interact with water-borne sediment dispersal patterns. In fact, heavy mineral concentrations in some of the beach-related "black sand" (Ti–Zr) placer deposits of the world, such as Richard's Bay in South Africa, are a likely product of both beach- and wind-related sorting processes. This section briefly considers the physical processes involved in such environments and emphasizes some of the differences that exist compared to the fluvial system.

Box 5.1 Placer Processes: The Alluvial Diamond Deposits of the Orange River, Southern Africa

Diamond is the hardest known natural substance. High quality diamonds (i.e. based on clarity, color and a lack of flaws) are used for jewelry, whereas flawed and poorly crystalline stones are used as abrasives in a wide range of cutting tools. The specific gravity (3.5) and hardness of diamond ensure that it is concentrated together with other heavy minerals in both fluvial and marine placer deposits. Over much of recorded history India was the only country to produce diamonds from alluvial sources, but in the 1860s diamonds were discovered in South Africa, specifically in gravels of the Orange River and its tributaries. This led to the discovery of the primary kimberlitic source of diamonds around the town of Kimberley, and then in 1908, of the huge beach placers along the west coast of southern Africa (Figure 5.1.1).

The alluvial and beach diamond placers of South Africa and Namibia are the product of deep erosion of kimberlites on the Kalahari Craton in Cenozoic times. The erosion of diamondiferous kimberlite liberated the diamonds onto the land surface for subsequent

Box 5.1 (Continued)

redistribution into the catchment of the Orange River drainage system and its precursors. This drainage flows westwards and exits into the south Atlantic at Alexander Bay–Oranjemund at the South Africa–Namibia border. Diamonds are trapped in gravel terraces that represent preserved sections of river sediment, abandoned by the present river as it migrates laterally or incises downwards (Lynn et al. 1998). Diamonds are also washed out into the ocean and redistributed by long-shore currents to be concentrated in gravels either beneath the wave base, or in remnant beaches reflecting previous fluctuations of sea level.

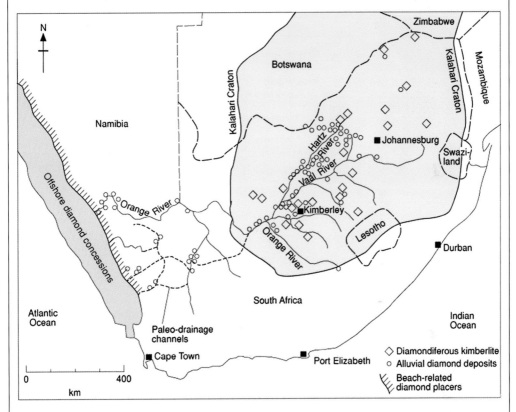

Figure 5.1.1 Map showing the Orange River drainage system in relation to the distribution of deeply eroded kimberlites on the Kalahari Craton. The distribution of alluvial diamond deposits along major river channels and paleo-channels, as well as the beach placers along the west coast of southern Africa, are also shown. Source: After Lynn et al. (1998).

   In the upper reaches of the Orange River system (which includes the Vaal River) diamonds are preferentially concentrated in areas where the rivers flow over resistant bedrock, such as the Ventersdorp lavas, where good gully and pothole trap sites can form. Diamonds are concentrated as residual lag deposits in such traps. Concentration is enhanced by the extreme hardness of diamond as other less resistant minerals undergo attrition and are more easily washed out of irregularities in the bedrock (Lynn et al. 1998). In the lower, more mature reaches of the Orange River, a number of gravel terraces occur (Figure 5.1.2). These contain large, low

(Continued)

Box 5.1 (Continued)

grade accumulations of diamonds and were deposited in lower Miocene to upper Pleistocene times, occurring at different elevations relative to the present day channel (Van Wyk and Pienaar 1986). As many as five terraces can be identified, each characterized by diagnostic fossil faunal and floral assemblages.

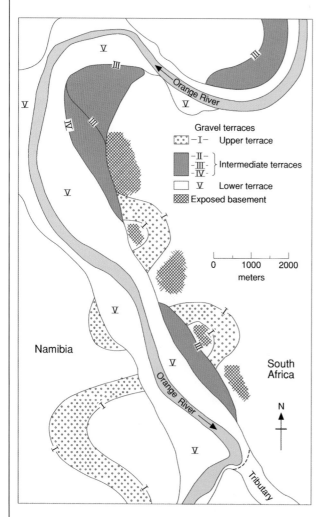

Figure 5.1.2 Map showing the distribution of gravel terraces in the vicinity of the Bloeddrift diamond mine along the lower reaches of the Orange River. Source: After Van Wyk and Pienaar (1986)

Diamonds derived from gravels of the lower reaches of the Orange River are 97% gem quality since it is the unflawed diamonds that typically resist the rigors of mechanical transport and abrasion. Most of the diamonds are between 0.85 and 1.30 carats in size, but occasionally very large stones (60–100 carats) are also recovered.

The single most important feature of the concentration of diamonds in terrace gravels is bedrock irregularity (Figure 5.1.3). The development of large potholes, plunge pools, and bedrock ribbing are all features that impact on grade distribution in the gravels. The lowermost sections of the gravel profile are typically the highest grade. Bedrock depressions can be

Box 5.1 (Continued)

extremely productive, with grades of over 40 carats per 100 tons of ore being reported (Van Wyk and Pienaar 1986). Large terraces are generally higher grade and contain bigger stones in their upstream portions, attesting to efficient trapping and sorting mechanisms prior to preservation.

Figure 5.1.3 Diamondiferous gravels related to the paleo-Orange River drainage system, deposited unconformably on irregular, potholed bedrock.

5.2.4.1 Beaches

The sorting of sediment in beach environments occurs at various scales, ranging from longshore and orthogonal processes (related to mass transport parallel and at right angles to the shoreline) to swash zone-related features attributed to the "to and fro" motion of waves breaking on the shore. A detailed overview of these various processes is presented in Hardisty (1994) and Allen (1997). Beach-related placer deposits appear to be mainly related to processes occurring in the swash zone and it is on this environment that the present discussion will focus.

The origin and nature of ocean waves is a quantifiable topic of study and the effects of wave action on sediment bed forms are well understood. Beaches represent the interplay between sediment supply, wave energy, and shoreline gradient. Sediment transfer on beaches is influenced largely by tides and currents, whereas sediment sorting is dominated by waves and the swash processes. Figure 5.10a illustrates the spectrum of wave-dominated shoreline types, emphasizing the differences between a low energy, shallow gradient beach comprising mud and silt, and a steep, highly energetic environment in which gravel beaches are deposited. An intermediate situation reflects the setting in which sand-dominated beaches are likely to form.

Although the type of beach that forms is important to whether a placer deposit is likely to develop, environmental considerations alone are unlikely to explain the dominant processes in beach placer formation. It is apparent from the research that has been carried out in this field that factors such as source fertility and swash processes are a more important consideration (Komar and Wang 1984; Komar 1989; Hughes et al. 2000). Source fertility is self-evident and dictates whether a placer deposit is likely to contain ilmenite–zircon or diamond concentrations, or

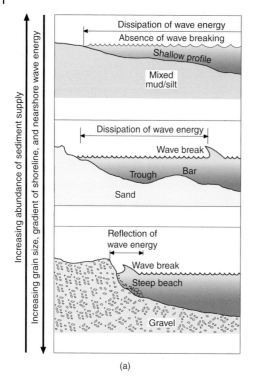

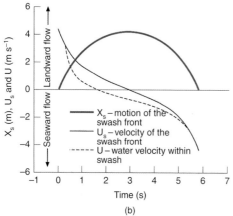

Figure 5.10 (a) Factors controlling the nature of wave dominated beaches and shorelines. Source: After Reading and Collinson (1996). (b) Mathematical model illustrating the dynamics of the swash zone created for a 0.5 m high breaker at the shoreline. Source: After Hughes et al. (2000).

none at all. By contrast, the flow dynamics of a swash zone are more difficult to evaluate. The application of these processes to the concentration of heavy minerals on a beach has been studied by Hughes et al. (2000). An illustration of modeled swash dynamics is presented in Figure 5.10b and shows that motion of the swash front is symmetric about the surge and backwash, but the water velocity is asymmetric because local flow reversal occurs before the front has reached its final landward advance. The dynamics of the swash zone have been used to evaluate mineral-sorting mechanisms in the beach placers of southeast Australia and show that neither settling nor entrainment is likely to be important in the concentration of heavy minerals. Settling is discounted because swash zones are typically not deep enough to allow effective settling of heavy minerals to occur. Entrainment, likewise, is ineffective as a sorting mechanism because the large bed shear stresses prevailing imply that most, if not all, the minerals would be in motion with little or no prospect of selective entrainment (Hughes et al. 2000). The dominant process appears rather to have been shear sorting, where the dispersive (upward) pressures acting on large, dense grains moving in a concentrated sheet flow were larger than those applicable to smaller, less dense grains. In a study of the heavy mineral placers of the Oregon, USA, coastline, Komar and Wang (1984) were able to discount grain settling as a sorting mechanism because the settling velocities of light and heavy mineral fractions were found to be similar. In contrast to the Australian situation, this study found that selective grain entrainment was likely to have been the most efficient concentration process and the one that best accounted for lateral variations observed in the distribution of heavy minerals. Figure 5.11a shows the variable distribution of heavy minerals as a function of distance offshore, a local characteristic that demands a sorting process capable of laterally sorting the mineral components. Figure 5.11b illustrates the calculated threshold stresses for the various minerals examined in the Oregon beach placers as a function of the measured concentration factors and shows that the densest and smallest grains were the most effectively concentrated. This may be due largely to the fact that these

Figure 5.11 (a) Plot showing the changes in heavy mineral contents along a profile of beach sediment in Oregon as a function of distance offshore. (b) Plot of concentration factors of the principal minerals in an Oregon beach placer deposit as a function of calculated threshold stress, illustrating the likelihood of selective entrainment processes during the formation of the deposit. Source: After Komar and Wang (1984).

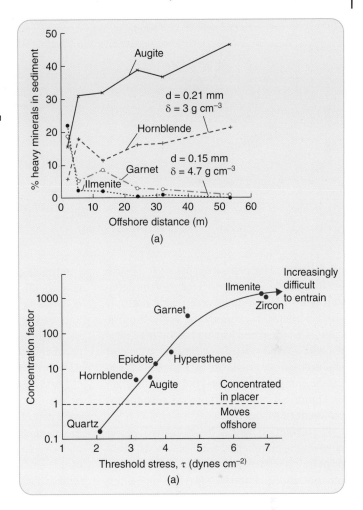

grains were the most difficult to entrain and, consequently, were residually concentrated in the near shore environment. The fact that entrainment sorting seems to apply to the Oregon placers, but not those formed along the Australian coastline, is possibly a function of the higher wave energy in the latter area and the development of conditions favoring shear sorting over entrainment (Hughes et al. 2000). Although the mechanics of the sorting processes are similar, the detailed patterns of heavy mineral distribution on beaches may be quite different to those in fluvial systems where transport sorting is the dominant process.

5.2.4.2 Wind-Borne Sediment Transport

As illustrated in Figure 5.3, the transport of sediment by wind, although similar in principle to that by water, differs in detail because of the much lower viscosity and density of air and the generally higher kinetic energies of eolian transport. Grains transported by wind are subjected to different entrainment criteria than those moved subaqueously and are also subjected to more dramatic ballistic effects due to their energetic motion. In Eq. (5.4), the critical shear stress required to initiate entrainment of a given particle is seen to vary linearly as a function of its density and size. Bagnold (1941) showed that for eolian sediment transport the critical shear stress required to initiate entrainment varied as a function of the square root of density and size. This implies that it is easier to move small particles by wind than it is to move them by water, which explains the abundance of widespread,

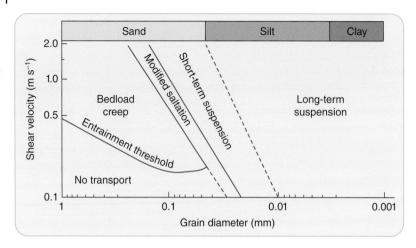

Figure 5.12 Wind blown (eolian) sediment transport characteristics as a function of shear velocity and grain size. Source: After Tsoar and Pye (1987).

coarse grained lag deposits in desert and beach environments. Such deposits form by the deflation of a surface initially comprising multiple sized particles, by winnowing away the fine grained material and leaving behind the gravel residue. Wind-related sediment transport characteristics are shown in Figure 5.12 and can be compared with the Hjulström diagram of Figure 5.4 which pertains to subaqueous sediment transport.

The extent to which beach-related placer deposits are modified by wind-blown sediment dispersal is a question that has not received much attention, despite the fact that such interactions are almost certain to have taken place (Chan and Kocurek 1988). At the Richard's Bay deposits, for example, a substantial proportion of the Ti–Zr heavy minerals are extracted from eolian dunes even though much of the original concentration took place in the swash zone of the adjacent beaches.

5.2.5 Numerical Simulation of Placer Processes

Computer simulations of natural placer processes have now been developed with a view to improving the prediction of heavy mineral distribution patterns in placer deposits. An example is the MIDAS (Model Investigating Density And Size sorting) code that predicts the transport and sorting of heterogeneous size–density graded sediment under natural

conditions (Van Niekerk et al. 1992; Vogel et al. 1992). It has been applied to the problem of grade control in the Witwatersrand basin and can predict gold distribution patterns in the host conglomerates with remarkable accuracy (Nami and James 1987; Nami and Ashworth 1993).

The conglomerates that host concentrations of detrital gold in the Witwatersrand basin were deposited in braided alluvial fan systems, the main elements of which comprise channel and overbank flow deposits. The geometry of this environment is shown in Figure 5.13a, which illustrates that fluid flow can be resolved into two components, a higher velocity flow contained within the channel, and a lower velocity flow in the overbank plain that is transverse to that in the channel. Hydrodynamic considerations suggest that fine grained detrital gold particles in the Witwatersrand deposits were transported in suspension as the prevailing hydraulic conditions exceeded those required to entrain and suspend gold in the channel. The model considers the distribution of gold, therefore, in terms of an interaction between flow in the channel where gold is being entrained, and flow on the overbank plain where suspended particles will tend to settle because of reduced transport capacity. The MIDAS code integrates the effects of flow velocity, sediment concentration, and the various parameters of transport sorting for both the longitudinal flow in the channel and the

Figure 5.13 (a) Geometry and flow characteristics of a braided fan delta complex used to simulate gold distribution patterns in Witwatersrand placer deposits. (b) Comparisons of computer simulations and actual gold distribution patterns in two different profiles across the Carbon Leader Reef. Channel edge gold concentrations are reasonably well modeled in terms of transport sorting. Source: After Nami and James (1987).

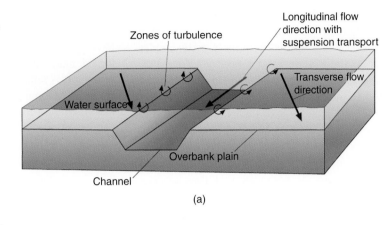

(a)

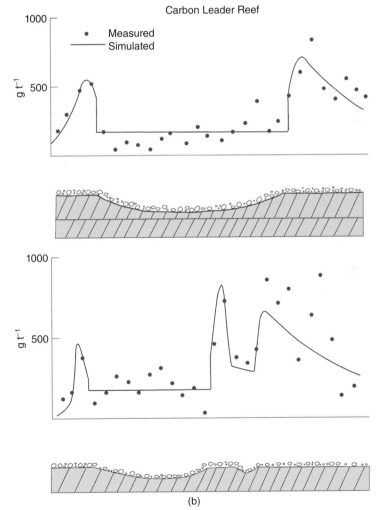

(b)

transverse flow in the overbank plain. The results of model simulations are compared to actual gold grade distributions in two examples from the Carbon Leader Reef (Figure 5.13b) and exhibit a high degree of correlation, even when the bed morphology is variable. The fact that the highest gold grades are often found along the edges of channels has long been known but has lacked a feasible explanation. The simulation suggests that the highest gold grades coincide with zones of flow interaction and that gold distribution patterns in the Witwatersrand deposits are a function, at least in part, of transport sorting. It should be emphasized, however, that this relationship exists despite the fact that it is well known that most of the presently observed gold in the basin has been remobilized and is secondary in nature. This conundrum may be explained if secondary gold has not been remobilized very far from the sites where it was originally deposited as placer grains (Frimmel 1997; Frimmel and Minter 2002).

5.3 Chemical Sedimentation – Iron-Formations, Phosphorites, and Evaporites

In contrast to the mechanical processes of sedimentation discussed above, where clastic sediment is sorted and deposited by water and wind, chemical sedimentation refers to the precipitation of dissolved components from solution, essentially out of sea water or brine. A wide variety of rocks are formed by the compaction and lithification of chemical precipitates and these include carbonate sediments (limestone and dolomite), siliceous sediments (chert) and iron-rich sediments (ironstones and iron-formations), as well as less voluminous accumulations of manganese oxides, phosphates, and barite. The majority of the world's Fe, Mn, and phosphate resources, all extremely important and strategic commodities, are the products of chemical sedimentation and are hosted in

chemical sediments. In addition, rocks such as limestone, comprising essentially $CaCO_3$, have great value as the primary raw material for the manufacture of cement. Furthermore, sediments known as evaporites, in which chemical precipitation is promoted by evaporation, contain the main economically viable concentrations of elements such as K, Na, Ca, Mg, Li, I, Br, and Cl, as well as compounds such as borates, nitrates, and sulfates, all of which are widely used in the chemical and agricultural industries.

Most chemical sediments form in marine or marginal marine environments. The continental shelves, together with intratidal and lagoonal settings, represent the geological settings where chemical sediments and associated deposits are generally located. The chemical processes by which ore concentrations form are complex and controlled by parameters such as solubility, oxidation–reduction and pH, as well as climate, tectonics, paleolatitude, and biological–atmospheric evolution. This section will first consider the processes associated with the formation of ironstones, iron-formations and bedded manganese oxide ores. This is followed by a discussion of phosphorites and also the formation of evaporites, carbonaceous "black shales," and oceanic manganese nodules.

5.3.1 Iron-Formations and Ironstones

The description and nomenclature of the iron-rich sediments from which most of the world's iron ore is extracted have been the subjects of much debate over the past several decades. It is now generally accepted that iron-formations are sedimentary rocks that contain >15% Fe in which an iron mineral is interlayered with chert or a carbonate mineral, and which formed episodically, but throughout, Earth history. The definition of iron-formations now also includes rocks termed ironstones, the latter referring specifically to iron-formations of Phanerozoic age, which have lower contents of chert and are often phosphorus rich (Bekker et al. 2010). Iron-formations may be bedded, laminated or granular and the iron minerals (usually hematite, magnetite or goethite) may be interbedded with chert or a carbonate mineral such as siderite or ankerite. It is the Precambrian iron-formations of the world that host a majority of currently viable iron ore deposits, the latter forming after iron-formations have been subjected to a process of alteration and up-grading (see below).

Iron-formations can be divided into several sub-types, summarized as follows:

1) In Precambrian successions[1] iron-rich sediments can be subdivided into banded iron-formation (BIF) and granular iron-formation (GIF). Although there is considerable overlap, BIFs are typically >2.3 Ga old, forming prior to the Great Oxidation Event (GOE), whereas GIFs were formed largely in the interval 2.3 to 1.8 Ga (Bekker et al. 2014; Ramanaidou and Wells 2014). BIFs and GIFs were also formed in Neoproterozoic sequences, largely in response to the surficial redox conditions prevailing during the "Snowball Earth" periods, and are known as Rapitan type iron-formations.

2) In the Phanerozoic Eon, iron-rich sediments formed in marginal marine settings are granular and referred to as ooidal ironstones, abbreviated herein as POI (Phanerozoic Ooidal Ironstones). Another important category of ooidal ironstones formed in continental, fluvial settings and are referred to as channel iron deposits (CID).

---

1 Precambrian iron-formations were previously classified on the basis of depositional setting (Gross 1980) as *Algoma-types* (typically >2.5 Ga – hosted within meta-volcanic settings and linked to exhalative processes on the sea-floor) and *Superior-types* (typically 2.5–1.8 Ga and formed in shallow, continental margin marine settings). In addition, *Rapitan-type* iron-formations are defined as Neoproterozoic in age and associated with redox fluctuations linked to global Snowball Earth events. However, with improved dating, and the recognition of their overlapping age characteristics, the distinction between Algoma and Superior types has blurred and the two terms are now best used as an indication of environment of deposition and proximity to sea-floor volcanic-exhalative activity (Bekker et al. 2014).

3) Recent (Holocene) accumulations of iron, typically forming in swampy, unconsolidated surficial settings, are referred to as bog iron ores.

Although each of these iron-formation sub-types has its particular features and modes of origin, collectively they are all associated with reduction–oxidation processes – their formation is also distinctly episodic and related to specific events and stages in Earth history. The three sub-types are described individually below, in order of increasing importance.

It is worth noting that mining of iron ore prior to the 1900s was dominated by exploitation of bog iron and POI deposits. Increased demand through the twentieth century, coupled with the discovery of the huge iron ore resources of the USA, Brazil, South Africa, and Western Australia, resulted in the fact that modern extraction is now essentially from iron ores hosted in Precambrian iron-formations. Currently, more than 2.5 billion tons of iron ore are mined globally each year (Simonson 2011), the major proportion of which is hosted in iron-formations.

5.3.1.1 Bog Iron Ores

Bog iron ores form principally in lakes and swamps of the glaciated tundra regions of the northern hemisphere, such as northern Canada and Scandinavia. The deposits are typically small and thin, and comprise concentrations of ferrihydrite and other limonitic minerals (Fe oxyhydroxides) associated with organic-rich shale. They formed in the recent geological past (Holocene), and in some places are still doing so at present (Stanton 1972). The principal concentration mechanism for iron in bog ores, as well as other iron ore deposits, is related to the fact that the metal occurs in two valence states, namely Fe^{2+} (the ferrous ion) which is generally soluble in surface waters, and Fe^{3+} (the ferric ion) which is less soluble and precipitates out of surface solutions. Concentration of iron occurs when aqueous solutions containing labile Fe^{2+}, in a relatively reduced environment, are oxidized, with the

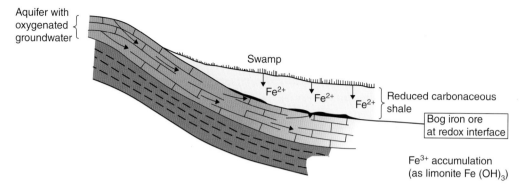

Figure 5.14 The development of limonitic concentrations in the formation of a bog iron ore where a reduced solution transporting ferrous iron interacts with oxidized groundwater flowing along an aquifer. Source: After Stanton (1972).

subsequent precipitation of Fe^{3+}. Figure 5.14 illustrates the principle with respect to bog iron ores accumulating at the interface between oxygenated surface waters flowing along an aquifer and the reduced iron-rich solutions percolating downwards through a swamp. At this interface ferrous iron in solution is oxidized to ferric iron which precipitates, usually as goethite. Bog iron ores precipitate in several different environments including lakes (such as the central African great lakes and in the Saint Petersburg area of Russia), glaciation related swamps and peat bogs (Europe and Scandinavia), and streams (Ramanaidou and Wells 2014).

Bog ores were exploited in ancient times and may be very high grade, although they are no longer considered a viable resource because they are small and sporadically distributed. In more recent times exploitation of bog iron ores supported the development of industrial activity, such as in the Pines River region of New Jersey, USA, where iron-based industrial activity spanned the eighteenth and nineteenth centuries (Ramanaidou and Wells 2014).

5.3.1.2 Phanerozoic Ooidal Ironstone (POI) Deposits

Although they have been documented in rocks as old as the Paleoproterozoic, ironstones are predominantly Phanerozoic in age and are widespread in occurrence, representing an important source of iron in the eastern USA and western Europe, particularly in the nineteenth and early part of the twentieth centuries (Young 1993). POI deposits are also referred to as Minette or Lorraine-type iron ores, well-known occurrences of which are found in the Jurassic sediments of England and the Alsace-Lorraine region of France and Germany. In North America the Silurian Clinton type iron ores of Kentucky and Alabama are analogues of the younger European deposits. Ironstones form typically in shallow marine and deltaic environments but are also known to have been deposited in continental settings – they consist of goethite and hematite that have been rolled into oolites or pellets, suggesting the action of mechanical accumulation and abrasion. The deposits are siliciclastic and contain little or no chert, but are commonly associated with Fe-rich silicate minerals such as glauconite and chamosite. The environment of POI deposition in shallow marine settings suggests that the iron was introduced from a continental source via a fluvial system in which the metal was either in solution as Fe^{2+} or transported as a colloid (see Chapter 3). Ironstones are often linked stratigraphically to organic-rich black shales and, in certain cases, develop in strata representing periods of major marine transgression and continental margin flooding.

They do not occur randomly in time and show distinct peaks in the Ordovician–Silurian and again in the Jurassic periods. Their formation appears, therefore, to be related to a pattern of global tectonic cyclicity and specifically to times of continental dispersal and sea-level highstand (see Chapter 6), as well as periods of warmer climate and increased rates of chemical sedimentation (Van Houten and Authur 1989).

The origin of ironstones is controversial and needs to account for both the Fe concentration processes and the formation of the ubiquitous oolites that typify these ores (Young and Taylor 1989). A model for the formation of Minette-type ironstones, after Siehl and Thein (1989), is presented in Figure 5.15. Iron is thought to have been concentrated initially on continental land masses that were subjected to deep weathering and erosion in a warm, humid climatic regime. Highly oxidized lateritic soils forming in such an environment would have been the sites of iron enrichment as insoluble Fe^{3+} remained in situ while other components of the regolith were leached away (see Chapter 4). In addition, laterites were also the sites where iron ooids formed in response to low temperature chemical and biologically mediated processes. The

lateritic soils and ooids were then transferred into a shallow marine environment, either by flooding during transgression or by erosion during regression, to be reworked and concentrated in fluvio-deltaic or littoral settings. Note that pedogenic ooids are different to those that arise from the mechanical abrasion of particles in the swash zones of shallow marine environments. Subsequent diagenesis of the ironstone accumulations resulted in further post-depositional modifications to the texture and mineralogy of the ores. Unlike the relatively simple processes invoked to explain the formation of bog ores, POI appear to have formed only in very specific environments and in response to a special set of processes (including oxidation–reduction, diagenesis, mechanical sedimentation, and microbial activity).

An interesting feature of several iron-producing districts, particularly in Kazakhstan and Western Australia, is the development of ooidal CIDs (Ramanaidou and Wells 2014). CIDs form in a continental setting and are geologically recent, having been deposited in response to Cenozoic weathering, erosion, and fluviatile drainage. The ooids are deposited in meandering fluvial channels that have incised into an older sequence comprising,

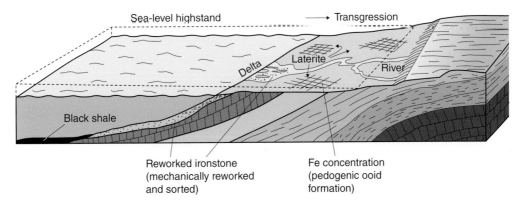

Figure 5.15 Simplified environmental model for the formation of oolitic ironstone ores. The model invokes initial Fe enrichment and pedogenic ooid formation in lateritic soils on the continental edge and transfer of this "protore" to a marginal marine setting with subsequent mechanical abrasion and reworking/concentration of ooids. Source: After Siehl and Thein (1989).

either significant iron-formations and iron ore deposits (such as the Hamersley Province), or a heavily lateritized ferruginous regolith. It has been suggested that iron-rich ooids formed in the pedogenic environment and were subsequently concentrated into fluvial channels during erosive cycles. The ooids were then cemented within the channels by goethitic material to form moderate to high grade iron ores of substantial size. CID mines in the Robe River area of the Hamersley Province, for example, comprise in excess of 30 Mt of iron ore at a grade of 57% Fe (Ramanaidou and Wells 2014). Although large in size, marine POI deposits are typically lower grade and regarded as uneconomic under current conditions – the Jurassic ironstones of the Cleveland district of north Yorkshire, UK, comprised in excess of 370 Mt of ore, but at an average grade of only 28% Fe.

5.3.1.3 Banded and Granular Iron-Formation – An Enigmatic Rock Type

The term BIF is somewhat controversial because of the connotation it has with respect to stratigraphic terminology. For this reason the term "iron-formation" is hyphenated and applies strictly to a bedded chemical sediment comprising alternating layers rich in iron and chert, and containing 15% or more iron (Klein and Beukes 1993; Bekker et al. 2010). BIFs and GIFs host the most important global source of iron ore and far outweigh the ironstone and bog iron ores in terms of reserves and total production. Whereas the latter occur predominantly in Phanerozoic rock sequences, BIFs are older and formed in essentially three periods of Archean and Proterozoic Earth history, namely 3500–3000 Ma, 2500–1800 Ma, and 1000–500 Ma (Figure 5.16). These three periods equate broadly with different geologic

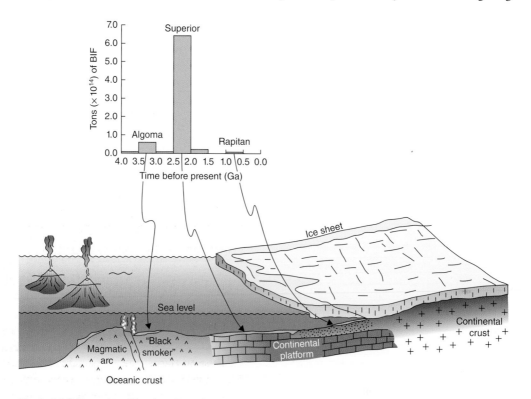

Figure 5.16 Tectonic and environmental model showing the depositional settings for Algoma, Superior, and Rapitan type BIFs. Source: After Clemmey (1985), Maynard (1991). The inset chart illustrates the approximate tonnages of BIF resource for each of the three major types as a function of time. Source: After Holland (1984).

and tectonic settings and are referred to as, respectively, Algoma, Lake Superior, and Rapitan types (see footnote 1 in Section 5.3.1). Algoma type BIFs are associated with volcanic arcs and are typically found in Archean greenstone belts. These deposits tend to be fairly small, but they are mined in places such as the Abitibi greenstone belt of Ontario, Canada. The majority of Lake Superior, or simply Superior, type BIFs are located on stable continental platforms and were mainly deposited in Paleoproterozoic times. They host by far the most important category of iron ore deposits (Figure 5.16) and most of the major currently producing iron ore districts of the world fall into this category. Examples include the Hamersley Basin of Western Australia (see Box 5.2), the Transvaal Basin of South Africa, the "Quadrilatero Ferrifero" of Brazil, the Labrador trough of Canada, the Krivoy Rog–Kursk deposits of the Ukraine, the Singhbhum region of India, and the type area in the Lake Superior region of the USA. Finally, the Rapitan type iron ores represent a rather unusual occurrence of iron ores associated with glaciogenic sediments formed during the major Neoproterozoic ice ages. The type occurrence is the Rapitan Group in the McKenzie Mountains of northwest Canada.

In addition to the environmental classification, BIFs have also been categorized in terms of the mineralogy of the associated iron phases. Although in most BIFs the iron mineral is an oxide phase (hematite or magnetite), carbonate (siderite), silicate (greenalite and minnesotaite), and sulfide (pyrite) iron minerals also occur, together with chert or carbonaceous shale. This observation led James (1954) to suggest a facies concept for BIF in terms of which the progression from oxide through carbonate to sulfide phases was considered to reflect precipitation of the relevant iron minerals in successively more reducing environments (iron silicate phases are stable over a wide range of Eh and do not, therefore, conform to this simple progression). A consideration of the stability of the main iron phases

as a function of Eh and pH confirms that the redox state of the depositional environment plays an important role in determining the mineralogy of BIFs, although the situation is undoubtedly more variable than the facies concept would suggest. Figure 5.17a shows that ferrous iron is stable in solutions that are acidic and reducing, but that for a given pH, oxidation (or an increase in the Eh) will stabilize hematite (Fe_2O_3 or ferric oxide), which is the principal iron phase over a wide range of geologically pertinent conditions. The phase diagram indicates that magnetite is only stable under reducing and alkaline conditions, but this field expands well into the range of neutral pH if the activities of carbonate and sulfur are lower than the prevailing conditions for this diagram. Siderite and pyrite are only stable under reducing conditions although these fields would expand if the activities of total carbonate or sulfur in solution were increased. It should be noted that most sulfide facies "iron-formations" are no longer considered true BIFs because they are typically hosted by carbonaceous shale (Bekker et al. 2010).

BIFs are enigmatic rocks whose origin is important for reasons other than that they contain significant resources of iron ore (Konhauser et al. 2011; Taylor and Konhauser 2011). They formed episodically in Earth history and identify specific stages in the tectonic, atmospheric, and oceanic evolution of the Earth system (Beukes and Gutzmer 2008; Bekker et al. 2010, 2014). They are chemical sediments in which the major components, Fe and Si, appear to have been derived from the ocean itself, rather than from a continental source, as is the case for ironstones. This is evident from the lack of aluminous and silicate mineral particulate matter in BIFs. Models for the formation of these rocks are controversial and hampered by a lack of modern analogues. Features which need to be explained include the origin of the Fe and Si, their transport and precipitation mechanisms, the cause of the delicate silica- and iron-rich banding at various scales, and their episodic formation

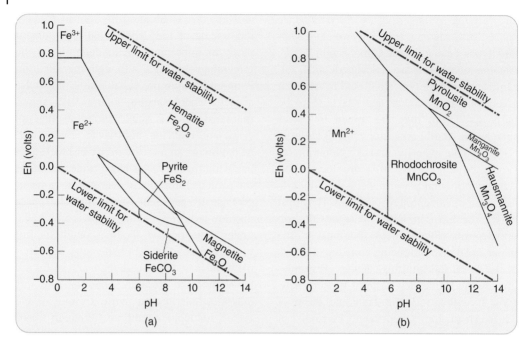

Figure 5.17 (a) Eh–pH diagram showing the stabilities of common iron minerals. The conditions that apply to this particular phase diagram are: T = 25 °C, P_{total} = 1 bar, molarities of Fe, S, and CO_3 are, respectively, 10^{-6}, 10^{-6}, and 1. (b) Eh–pH diagram showing the stabilities of common manganese minerals. Identical conditions apply, but with the molarity of Mn = 10^{-6}. Note that the manganese oxides (MnO_2 and Mn_2O_3) are stable at higher Eh than the equivalent ferric oxide (hematite), and would only form, therefore, under more oxidizing conditions. Source: Diagrams modified after Garrels and Christ (1965), Krauskopf and Bird (1995).

during specific time periods in Earth history, in particular at around 2500 and 1900 Ma.

The secular periodicity of iron-formations and the correlation between their deposition and, for example, the injection of mantle plumes and the formation of Cu–Zn rich volcanogenic massive sulfide (VMS) deposits, is illustrated in Figure 5.18. Regarding the source of iron, some, but not all, episodes of prolific BIF deposition coincide with periods of pronounced mantle plume activity, a relationship that has led to the suggestion that plume related magmatism resulted in an increase in the flux of Fe into the oceans, thereby promoting BIF deposition (Isley and Abbott 1999). Mantle plumes that intersect continental crust cause rifting and extrusion of continental flood basalt, whereas those intruding ocean basins form oceanic plateaus and seamount chains. Erosion of continental flood basalt increases

the flux of Fe into the oceans via weathering and riverine transport, while more voluminous oceanic volcanism is associated with enhanced hydrothermal and exhalative activity, which also augments the Fe flux into ocean waters. Thus, episodes of global plume activity are suggested to have provided ocean waters with an enhanced iron content that can be linked to the episodicity of BIF deposition through Precambrian times. The relationships between mantle plume activity, global tectonics, and formation of other deposit types such as VMS Cu–Zn ores, are discussed in more detail in Chapter 6.

5.3.1.4 Mechanisms by Which BIFs Are Deposited

It is generally agreed that the conditions that promote BIF deposition include (i) a reducing atmosphere and oceanic waters with a low oxidizing potential so that Fe^{2+} is readily

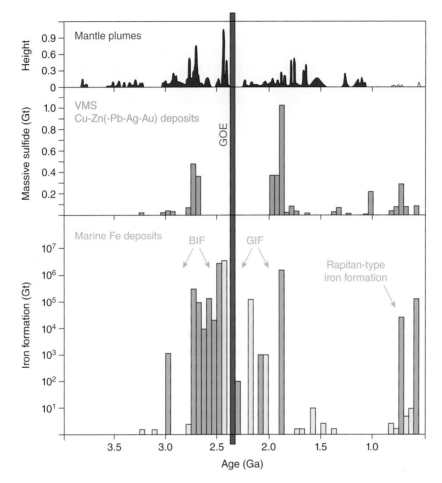

Figure 5.18 Episodic occurrences of VMS and iron deposits (BIF and GIF in darker and lighter yellow colors respectively) as a function of time, mantle plumes, and the Great Oxidation Event (GOE). Source: Modified after Bekker et al. (2010, 2014).

soluble; (ii) ocean waters with transient high Fe^{2+} contents; and (iii) an oxidative mechanism able to convert ferrous to ferric ions and cause the precipitation of iron oxyhydroxide particles on the ocean floor, as shown in Eq. (5.7) (after Bekker et al. 2014).

$$2Fe^{2+} + 0.5O_2 + 5H_2O \leftrightarrow 2Fe(OH)_3 + 4H^+$$
$$(5.7)$$

Pervasively reducing conditions would have prevailed on the Earth's surface prior to the Great Oxidation Event (GOE), thought to have occurred at around 2.32 Ga (Bekker et al. 2004; Figure 6.8). However, significant volumes of BIF were deposited for up to 400 Ma

after the GOE suggesting that reduced conditions persisted to maintain Fe^{2+} in solution in localized oceanic environments. If mantle plumes were the source of Fe enrichment in the oceans, whether through direct introduction of Fe^{2+} from hydrothermal exhalations (black smokers; see Chapter 3) on the sea floor or riverine input from erosion of continental flood basalts, the period between 2.7 and 1.9 Ga is characterized by the coincidence of three major plume events and the great majority of BIF and GIF deposition (Figure 5.18). Rare earth element patterns of BIFs suggest that hydrothermal venting and volcanic activity are likely to have contributed significantly to the source of Fe^{2+} in Algoma type BIFs, but that this influence diminished with time such that Rapitan ores reflect a normal ocean water character (Misra 2000). It is notable that Rapitan style BIF deposition in the Neoproterozoic is characterized by an absence of plume activity (Figure 5.18). Finally, the oceans themselves are also likely to be the source of silica in BIFs because the solubility of amorphous silica (as $Si(OH)_4$) is relatively high (about $120\,mg\,l^{-1}$) in ocean waters. It seems likely that the Archean–Paleoproterozoic oceans were episodically saturated with respect to silica, ensuring periodic precipitation and accumulation of siliceous matter on the ocean floor during this early stage of Earth history. This pattern is likely to have changed in later geological times when the organisms that use silica to build an exoskeleton evolved (i.e. siliceous protozoans such as radiolaria) and this element was extracted from the water column. The modern oceans, for example, typically contain $<10\,mg\,l^{-1}$ dissolved silica and are markedly undersaturated (MacKenzie 1975). The decrease of available Si in the oceans as a function of biological evolution contributes to an explanation for the changing character of iron ores, from BIFs to ironstones, with time.

The transport and precipitation mechanisms involved in BIF genesis are best explained in terms of the Fe^{2+} upwelling model (Klein and Beukes 1993; Bekker et al. 2010, 2014), analogous to the observations of phosphorus upwelling onto the continental shelves (see Section 5.3.3; Kazakov 1937, 1939). Figure 5.19 illustrates the main elements of the model. Ferrous iron from deep, reduced ocean levels is introduced to shallower shelf environments by upwelling currents similar to those seen along the western coastlines of continents such as Africa and the Americas today. These currents interact with the shallower waters and Fe^{2+} is oxidized, with subsequent hydrolysis and precipitation of ferric hydroxide ($Fe(OH)_3$). Three scenarios are envisaged being relevant to the formation of Precambrian BIFs and GIFs – (i) Oxidation of Fe^{2+} at a redox interface where

oxygen is biologically produced (Figure 5.19a). It is envisaged that cyanobacteria were the likely organisms producing photosynthetic oxygen in shallow-water environments and this mechanism would not have occurred prior to the evolution of oxygen photosynthesizing organisms; (ii) Direct biologically induced precipitation of Fe^{2+} by microbial organisms (Figure 5.19b). Metabolic microbial iron oxidation, for example, is a mechanism whereby certain organisms, such as the *Proteobacteria*, use Fe^{2+} as an electron donor, generating energy via the metabolic process of chemoautotrophy. There are several pathways whereby direct microbial oxidation of Fe^{2+} may take place and this mechanism is suggested to have been the most important for Precambrian BIF deposition (Bekker et al. 2014); and (iii) Oxidation of Fe^{2+} by ultraviolet radiation from the sun (Figure 5.19c). This mechanism might have prevailed in Archean times and prior to the evolution of oxygen producing organisms. It is possible that the flux of ultraviolet radiation at the Earth's surface prior to GOE might have been higher because of the lack of an ozone layer, supporting the view that photo-oxidation of Fe^{2+} was more prevalent in Archean times.

There are clearly several mechanisms that are likely to have contributed to BIF deposition and each of these may have prevailed at different times and in different environments. Other processes relevant to BIF deposition pertain to the formation of iron carbonate minerals such as siderite. In marine environments where CO_3 activities are sufficiently high, $FeCO_3$ might precipitate in response, again, either to production of oxygen by photosynthesizing organisms living in the photic zone, or to photo-oxidation of ferrous to ferric ions (Cairns-Smith 1978). Precipitation of silica is likewise problematical, but it has been suggested that evaporation from the ocean surface could promote local silica oversaturation and thereby promote precipitation as a gel which, on compaction, is transformed to chert. The explanation of banding in BIFs remains a problem and suggestions

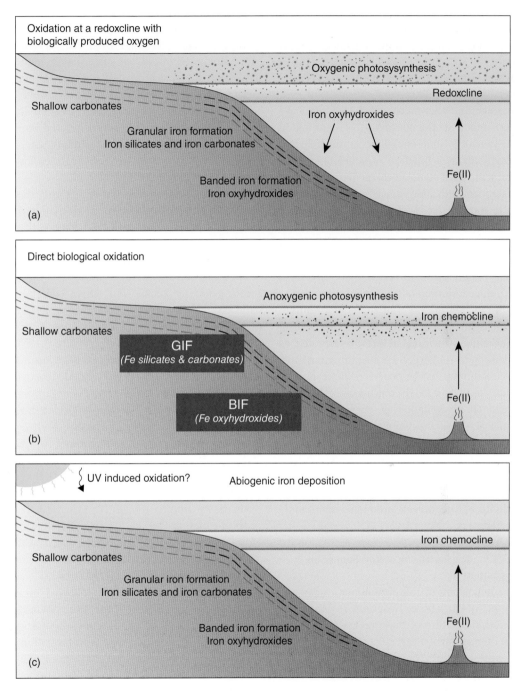

Figure 5.19 Models illustrating different scenarios in which the upwelling and oxidation of ferrous iron from an oceanic source can be used to explain the formation of shallow marine BIFs in Archean and Paleoproterozoic times: (a) oxidation of ferrous iron at shallow levels by biologically produced oxygen involving photosynthesizers such as cyanobacteria – not a likely scenario in Archean times; (b) oxidation of ferrous iron by direct biological precipitation involving metabolic pathways; and (c) photo-oxidation of ferrous iron by abiological processes such as ultraviolet radiation. Source: After Bekker et al. (2010, 2014).

for the formation of alternating Fe and Si rich bands range from diurnal cycles (where photosynthesis and photo-oxidation shut down at night) for very thin lamellae, to seasonal cycles (where summer seasons promote biological activity and enhance iron precipitation) for mesobands. Other mechanisms that may have contributed to the formation of banding in BIFs include differential flocculation of Fe and Si colloids as a function of water chemistry, and some form of biological mediation. Studies of biomineralization processes (see Chapter 3) in modern day environments have shown that the filamentous bacterium *Chloroflexus* is capable of binding either silica or iron, and occasionally both silica and iron, to the cell walls (Konhauser and Ferris 1996). This form of induced biomineralization is considered to apply to the formation of some Precambrian BIFs, a notion supported by the presence of spheroidal structures in BIFs interpreted as microfossils (LaBerge 1973). At the same time, oversaturation of the oceans with respect to silica could have led to continuous biologically mediated precipitation of chert, where cell walls acted as the templates for nucleation of silica (Konhauser and Ferris 1996). In this model, the banding in iron-formations is attributed either to episodic depletion of ferrous iron or to periodic (seasonal) diminished activity of the oxygen producing planktonic bacteria.

5.3.1.5 The Periodicity of Iron-Formation Deposition

The deposition of iron-formations in discrete periods of Earth history (Figure 5.18) is another intriguing feature of these rocks that reflects the interaction of a variety of processes related to atmospheric and biological evolution, as well as to the pattern of global tectonic cycles (Bekker et al. 2010; see Chapter 6 and Section 6.4).

In Archean times (>2.5 Ga) BIF deposition was relatively restricted in extent and frequency, but nevertheless occurred over more than 1000 million years of Earth history, from

as early as c. 3.8 Ga when they were deposited, for example, in the Isua greenstone belt of SW Greenland. Archean BIFs are typically of Algoma type and associated with mafic volcanics in greenstone belt settings. Several examples of Superior type BIFs are, however, also known to have formed as early as c. 3.0 Ga, such as those in the Witwatersrand and Pongola sequences of the Kaapvaal Craton, where deposition is associated with clastic to shallow marine platformal settings without any association to volcanism (Beukes and Cairncross 1991). Although the Archean crust may have been composed of smaller land masses that lacked the major continental free-board of later epochs, deposition of Archaean BIFs took place repeatedly in oceanic basins that were generally anoxic and where oxidation and precipitation of iron took place, either by UV-induced photo-oxidation (Figure 5.19c) or direct biological oxidation (Figure 5.19b). Typically, Archean BIFs may also have formed in smaller, more restricted oceanic basins, accounting for the smaller sizes of individual deposits compared to the huge, laterally extensive, Superior types. The peak of Archean BIF deposition took place at around 2.7 Ga, which coincides with the emplacement of a superplume (Figure 5.18; Isley 1995; Bekker et al. 2010).

In the Paleoproterozoic Era (2.5 Ga to 1.6 Ga) two distinct pulses of Superior type iron-formations appear to have taken place. The first, dominated by BIFs, occurred *prior* to the GOE between c. 2.6 Ga and 2.4 Ga – a quieter period then occurred followed by a second pulse of largely GIF deposition subsequent to the GOE at around 2.2 Ga to 1.8 Ga (Figure 5.18). Superior type iron-formations were deposited on shallow continental platforms during periods of sea-level highstand related to plume activity and enhanced input of Fe^{2+} into the oceans. The initial pulse was characterized by the development of major iron ore deposits on a global scale and coincides with a mantle superplume event that saw the emplacement of mafic magmas in both continental and marine environments – as much as 60% of all Precambrian BIF might have been deposited at this time (Isley and Abbott 1999). The later pulse, post-GOE, resulted in substantial iron-formation deposition around the margins of the Superior Craton and coincides with the emplacement of a plume-related large igneous province (LIP) in this region and at this time. Oxidation of Fe^{2+} during this pulse could increasingly have been the result of accumulation of free oxygen in the upper portions of the ocean water column, caused by the evolution of cyanobacteria and resultant oxygenic photosynthesis (Figure 5.19a).

Subsequent to 1.85 Ga, for a period of about 1100 million years, a significant decline in the deposition of iron-formation is evident (Figure 5.18). Major BIF deposition – the Rapitan type iron-formations – only recommenced at around 750 Ma with the advent of the Sturtian glacial period, which was the earliest of the Neoproterozoic "Snowball Earth" events (Hoffman et al. 1998). Periods of near-global glaciation in the late Neoproterozoic, and the associated anoxia of ocean waters caused by the isolation of seawater from the atmosphere by an intervening global ice cap, may have increased the concentration of Fe^{2+} in the oceans at these times (Canfield et al. 2008). The ensuing "hot-house" caused by the build-up of greenhouse gases in the atmosphere caused the relatively rapid melting of the global ice cap and the reinstatement of oxic conditions in the shallower parts of the ocean water column. Ferrous iron would have been oxidized in these regions to form Rapitan iron-formations, often in association with glacial diamictite deposition. Rapitan type BIFs and GIFs are typically less chert rich than Superior types and more phosphate enriched – this may reflect a generally higher phosphorus content in late Neoproterozoic ocean waters (see Section 5.3.3) as well as the start of silica extraction from seawater by marine microorganisms at around this time (Porter et al. 2003). Mantle plumes would appear not to have played a

major role in iron-formation deposition in the late Neoproterozoic (Figure 5.18). The Braemar BIFs and diamictic ironstones associated with Sturtian aged glaciogenic sediments of the Adelaide geosyncline of South Australia represent an example of rocks formed by these processes (Lottermoser and Ashley 2000).

Ooidal ironstones formed in marine settings were also deposited episodically with peaks of formation in two periods, initially during the mid-Paleozoic (Ordovician to Devonian) and again in the mid-to-late Mesozoic (Jurassic–Cretaceous). The ironstones of the Clinton Group in the central and southern Appalachians of the USA are Silurian in age, whereas the Minette ironstones of the Lorraine region of NE France were deposited in the early Jurassic (Ramanaidou and Wells 2014). Ooidal ironstone formation can be linked to periods of global ocean anoxia, sea-level highstand and preferential development of Cu–Zn VMS deposits, the latter reflecting enhanced development of ocean crust and sea floor exhalative activity. These relationships are discussed in more detail in Chapter 6 (Section 6.5.3).

5.3.1.6 Transformation of BIFs into Viable Iron Ore Deposits

BIFs typically comprise approximately equal proportions of iron minerals (such as hematite and/or magnetite) and chert. Iron contents in BIFs are typically in the region of 30%, a concentration that is not usually adequate to extract profitably. Accordingly, most ore-grade BIF has been naturally upgraded, usually by removal of silica, to yield an ore that is residually enriched in iron (typically >55%) relative to the parental material (Taylor et al. 2001). The mechanisms by which BIF is upgraded are complex and often long-lived, and several models have been proposed for this process (Tsikos et al. 2003; Ramanaidou and Wells 2014; Hagemann et al. 2016). Upgrading of BIF to iron ore was initially thought to have been dominantly a supergene process taking place during a weathering cycle, and somewhat akin to lateritization. It was subsequently realized that iron ores were often structurally controlled and that hypogene processes were dominant – the description of the Mount Whaleback iron ore deposit in the Hamersley Province of Western Australia in Box 5.2 discusses these processes in more detail.

Examination of BIFs in several major iron producing regions of the world indicates that the transformation to ore-grade material is a process facilitated by the passage of hypogene fluids along major structural discontinuities, usually during an orogenic event subsequent to BIF deposition. The fluids that interact with BIFs are variable in character (magmatic, basinal, or meteoric, but typically mesothermal, with moderate to low salinities and occasionally bicarbonate enriched; Taylor et al. 2001) and initially leach silica from the rock with an accompanying mass/volume reduction and porosity increase. Magnetite and carbonate minerals form at this early stage. Further alteration progressively oxidizes magnetite to hematite (to form martite) with further dissolution of silica and also the earlier formed carbonate minerals (Hagemann et al. 2016).

Supergene enrichment of BIFs may also, however, occur during exhumation and weathering of the sequence. If present, late supergene alteration replaces previous iron oxide minerals with iron oxyhydroxides (goethite) and is accompanied by formation of fibrous quartz and clay minerals. The transport of labile iron in an otherwise oxidized, surficial environment has been explained in terms of an electrochemical model (Morris et al. 1980; Morris and Kneeshaw 2011). Iron oxide minerals are good conductors of electrons and accordingly large electrochemical cells are set up within the BIF that transport Fe^{2+} in meteoric waters to BIFs in the anode area where it is oxidized and precipitated as Fe^{3+} – bearing goethite. Goethite replaces chert in these areas and magnetite is progressively oxidized to martite, forming the high-grade goethite-martite ores that are much sought after as a direct-shipping product (Ramanaidou and Wells 2014). The application of the electrochemical model to the Mount Whaleback deposit in Western Australia is discussed in more detail in Box 5.2.

Box 5.2 Chemical Sedimentation – Banded Iron-Formations: The Mount Whaleback Iron Ore Deposit, Hamersley Province, Western Australia

---

The Archean to Paleoproterozoic Hamersley Province of Western Australia contains huge volumes of Superior-type BIF that host some of the world's largest iron ore deposits. Western Australia is the world's largest exporter of iron ore, the bulk of which comes from enriched BIF in the Hamersley Province. Several large mines, including Mesa J, Tom Price, Paraburdoo, and Mount Whaleback, together with many smaller operations, combine to produce over 170 million tons of export quality iron ore annually, much of which goes to Japan, South Korea, and China (Kneeshaw 2002).

The Mount Whaleback deposit is the largest single iron ore deposit in Australia, with an original ore resource of some 1.8 billion tons. The deposit is located in the southeastern portion of the Hamersley Province in an area that was episodically deformed, with the most intense folding attributed to the Ophthalmian orogeny between 2400 and 2200 Ma (Figure 5.2.1). The ore is hosted in BIF and ferruginous shales of the Dales Gorge Member, which is part of the Brockman Iron Formation. The latter unit is the middle of three major iron-formation sequences in the Hamersley Province, forming between 2540 and 2450 Ma (Barley et al. 1997). This period of Paleoproterozoic Earth history was a time when similar chemical sediments were being deposited in continental margin or platformal settings elsewhere in the world, such as South Africa and Brazil (see Section 5.3.1 of this chapter). The BIF host rock in all these

Box 5.2 (Continued)

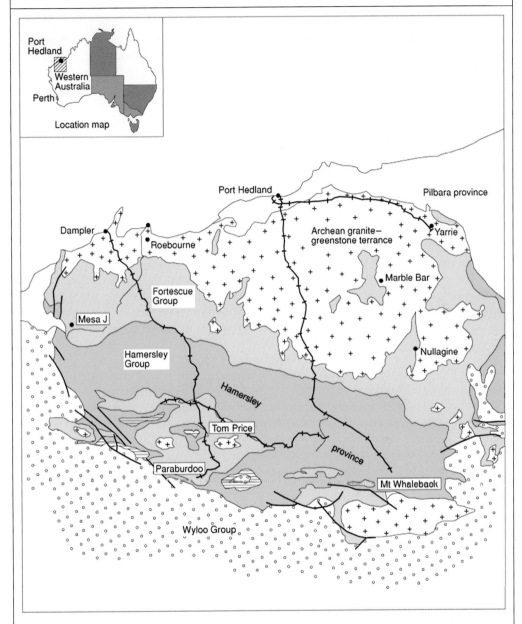

Figure 5.2.1 Simplified geology and location of some of the major iron ore deposits in the Hamersley Province, Western Australia. Source: After Kneeshaw (2002).

deposits was originally a bedded chert–magnetite/hematite chemical sediment that typically contained around 30% Fe. In order to make an economically viable ore deposit the BIF needed upgrading so that the Fe content is at least double this value, a process that is achieved by either selectively leaching or replacing the silica in the rock. The processes whereby BIF is enriched to form iron ore are multifaceted and both supergene and hypogene processes have been advocated.

(Continued)

Box 5.2 (Continued)

The conventional model of ore formation at Mount Whaleback, and indeed most other deposits of the Hamersley Province, is that BIF was exposed and subjected to supergene alteration after the main period of deformation of the host rock. This process involved in situ replacement of chert by goethite and oxidation of magnetite to hematite (a process called martitization). The actual processes involved are somewhat akin to those taking place during the formation of laterites (see Section 4.3 in Chapter 4), although rather than leaching of silica and residual enrichment of iron, the chert is largely replaced by Fe oxyhydroxides. After supergene alteration the resulting goethite–martite rock was then buried and metamorphosed to form microcrystalline hematite–martite ore. This model, envisaging a combination of supergene and metamorphic processes, was initially attributed to Morris (1985) and is believed to be consistent with the characteristics of many of the enriched BIF deposits of the region. These characteristics include the high Fe grades (65% Fe) of the enriched BIF, the lumpy character of the ore, and low phosphorous contents (around 0.05%), all features that are highly desirable in the marketing of the ore (Kneeshaw 2002).

Exposure of the BIF and supergene alteration apparently took place prior to 2200 Ma (Martin et al. 1998b), which is the maximum age of deposition of conglomerates adjacent to the Hamersley Province (i.e. the Wyloo Group; Figure 5.2.1) within which detrital fragments of hematite ore are found. Fluid inclusion and stable isotope data have also suggested that the fluids involved with the early phases of hematite–martite ore formation were hot (up to 400 °C), saline hydrothermal solutions that are inconsistent with a supergene or near-surface setting. These and other observations, such as the mineralogy of the ore assemblage at the Tom Price mine, have resulted in a number of workers suggesting a hypogene origin for the enriched BIF-hosted hematite–martite ores (Barley et al. 1999; Powell et al. 1999). One suggestion is that hot, oxidizing basinal fluids were focused into low angle thrust faults during the main stage of deformation and folding (i.e. "orogenic fluids"; see Chapter 3) and that it was these fluids that were initially responsible for silica dissolution and hematite recrystallization in the BIF.

One feature that appears to favor a hypogene origin for the Hamersley iron ores is that they can extend to considerable depths (over 500 m) below the surface. Although this feature on its own does not militate against supergene processes, work by Morris et al. (1980) at Mount Whaleback envisaged supergene alteration by a novel electrochemical process that might explain it (Figure 5.2.2). The high electrical conductivity of magnetite-rich layers in BIF enables oxidation potentials to extend to significant depths, with an electrochemical cell being completed by ionic conduction through groundwaters circulating along suitably positioned fault systems. An effective cathode is set up at surface where electrons are consumed in oxygenated groundwaters. The accompanying anode exists at depth where ferrous ions are oxidized to ferric ions and then precipitated as an iron oxyhydroxide such as goethite. This electrochemical process explains alteration processes at significant depths, but does not preclude the existence of normal supergene processes, or lateritization, in the near surface environment (Morris et al. 1980; Morris and Kneeshaw 2011).

Finally, it should be noted that the Hamersley Province is characterized by a significant component (about 35%) of much younger mineralization that is related to supergene processes active during a cycle of Paleogene erosion and weathering. These iron deposits are referred to as channel iron deposits (CID – see Section 5.3.1) and are located in mature river courses

Box 5.2 (Continued)

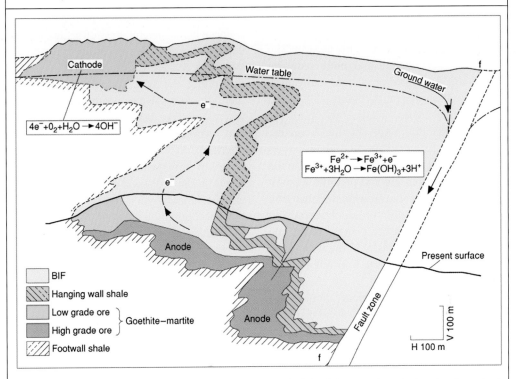

Figure 5.2.2 Simplified cross section through the Mount Whaleback iron ore deposit showing the extent of enriched BIF-hosted ore and the postulated nature of iron enrichment by an electrolytically mediated process. Source: After Morris et al. (1980).

and colluvial fans, both of which are spatially linked to bedded ores (Kneeshaw 2002). Very significant iron ore resources have now been documented in CIDs and it is likely that they will play an increasingly important role in future iron ore production from the region (Ramanaidou and Wells 2014). Although the origin of the enriched BIF ores remains controversial, the region has been subjected to a long and involved geological history and the ore-forming processes are expected to be complex and episodic.

5.3.2 Bedded Manganese Deposits

Most of the world's sedimentary manganese deposits formed in environments similar to those in which BIF and ironstone ores also formed. Superior type BIFs are sometimes closely associated with manganese ores, the prime example being the world's largest exploited Mn deposits in the Kalahari manganese field of South Africa (Miyano and Beukes 1987; Frakes and Bolton 1992; Cornell and Schütte 1995). In this case, enormous reserves of bedded manganese oxide ores were formed within the Transvaal Basin at stratigraphic levels above those where the equally substantial BIF-hosted iron ores occur. This is not unexpected given the remarkably similar chemical properties of Fe and Mn (Maynard 2014). The geochemical behavior of manganese, like iron, is controlled by oxidation potential and it exists as Mn^{2+}, which is soluble under reducing and acidic conditions, as well as Mn^{3+} and Mn^{4+}, which are less soluble and stabilize as manganese oxides under relatively oxidizing and alkaline conditions (Figure 5.17b). Pyrolusite, or MnO_2, is the dominant oxide phase at high Eh and over a range of pH, whereas rhodochrosite ($MnCO_3$) is stable under alkaline and more reducing conditions. Comparison of the manganese and iron phase diagrams (Figures 5.17a and b) shows that higher oxidation potentials are required to stabilize pyrolusite than hematite. The fact that Fe^{2+} oxidizes more readily than Mn^{2+} means that iron can precipitate while Mn remains in solution and provides a possible explanation for the observation that iron ores may be spatially (or stratigraphically) separated from manganese ores, as in the case of the Transvaal Basin. In the latter case, early precipitation of ferric iron out of solution to form BIFs (within which the Sishen and Thabazimbi deposits occur) would have depleted the water column of iron so that later precipitation, at a higher stratigraphic level and under more oxidizing conditions, resulted in the formation of Fe-depleted manganese oxide ores – evidence for this process is provided by an iron isotope study of the Mn ores which contain depleted $\delta^{56}Fe$ values reflecting the previous deposition of BIFs and extraction of Fe from the system (Tsikos et al. 2010).

Manganese deposits are located in sediments of variable age, from the Paleoproterozoic to recent. Some of the biggest deposits are Phanerozoic in age, such as the Cretaceous Groote Eylandt manganese oxide ores in northern Australia and the Molango district of Mexico, where the ores are made up dominantly of rhodochrosite ($MnCO_3$). An interesting modern day analogue for the formation of sedimentary manganese ores is provided by the Black Sea, where active sedimentation results in ongoing accumulation of MnO_2. The Black Sea is markedly stratified in terms of water density and composition and the waters below about 200 m depth are euxinic (i.e. highly reduced, with H_2S stable). Pyritic muds accumulate on the sea floor and this effectively depletes the entire water

column of iron. Mn by contrast is concentrated in the deeper waters, since it exists in solution as Mn^{2+}, but is depleted in shallower, oxidized waters where it precipitates and settles out. The depth–concentration profiles for Mn and Fe in the Black Sea are illustrated in Figure 5.20a, illustrating the zone of manganese enrichment that forms just below the redox interface where mixing between high-Mn euxinic and oxidized surface waters occurs. At the interface, Mn^{2+} is oxidized to Mn^{3+} and particulate Mn forms which settles at deeper levels where it can redissolve. However, where the enriched mixing zone intersects shallow sea floor, accumulation of either MnO_2 or $MnCO_3$ (depending on the local oxidation potential; see Figure 5.17b) takes place on the sediment substrate. Mineralization is further enhanced if the sites of accumulation are protected from dilution by clastic sediment input, as shown for the north-central portion of the Black Sea (Figure 5.20b).

5.3.3 A Note on Ocean Floor Manganese Nodules

Since their discovery during the 1872–1876 oceanographic expedition by *HMS Challenger*, it has become apparent that a vast resource of Mn, Fe, Cu, and Ni (as well as lesser amounts of Co, Zn, and other metals) is contained within the (ferro)-manganese nodules that occur in the deep, pelagic portions of most major oceans (Hein and Koschinsky 2014). Manganese nodules typically occur in parts of the ocean basins where sediment accumulation rates are very low (less than 7 m per million years; Heath 1981) and are absent from areas of rapid sedimentation, as well as from equatorial and high-latitude regions of high biological productivity and chemical sedimentation. They appear to be best developed in the Pacific Ocean, where exploration has identified large areas of prolific nodule formation (up to 100 nodules per square meter). These areas also contain nodules that have higher than normal metal contents, with up

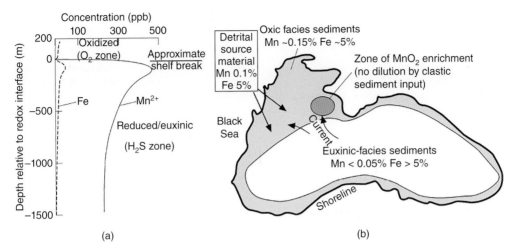

Figure 5.20 (a) Depth-concentration profiles for Fe and Mn in the Black Sea. Fe is depleted in the water column by pyrite accumulation on the sea floor, whereas Mn occurs to moderately high levels in solution in the dominantly reduced water column. Maximum enrichment of Mn occurs just below the redox interface where Mn^{2+} is oxidized to Mn^{3+} and precipitates. (b) The distribution of Mn in sea floor sediments of the Black Sea. Accumulation of Mn occurs at the intersection of the zone of Mn precipitation with the sea floor. Source: Diagrams modified after Force and Maynard (1991).

to 2 wt% combined Cu + Ni and substantial Co and Zn (Figure 5.21). One such area is the Clarion–Clipperton region, to the west of Mexico.

Manganese nodules are typically ovoid, a few centimeters in diameter, and have the appearance on the sea floor of a burnt potato. In cross section they are layered, often partly concentric, and represent concretions that have apparently nucleated around a clastic or biogenic fragment. They often exhibit internal radiating shrinkage cracks reflecting a change in mineralogy from hydrated Fe and Mn oxy-hydroxides to a variety of more compact, but chemically variable, Fe and Mn oxides such as vernadite, birnessite, todorokite, goethite, and hematite (Heath 1981). The formation of manganese nodules is still not fully understood. Metals are undoubtedly derived from the sea water itself, but ultimately come from various sources including submarine exhalative vents, clastic and volcanic input, and diagenesis of pelagic sediments. The formation of the nodules and the concentration of metals within them have been attributed to one or more of the following mechanisms:

- Settling of clay and biogenic (fecal) debris that contained either absorbed or ingested metals obtained from the sea water, with subsequent release of metals during dissolution and decomposition.
- Direct precipitation of metals from sea water onto a suitable substrate which forms the nucleus around which concretionary growth takes place.
- Upward diffusion of metals in ocean bottom sediment pore waters.
- Authigenic reactions in ocean bottom sediments during alteration and compaction.
- Bacterial activity and the oxidation of transition metals. In this regard, certain types of bacteria (such as *Metallogenium*) are known to be particularly prevalent in the oxidation and subsequent precipitation of labile Mn^{2+} and have been detected in manganese nodules from the Atlantic Ocean (Trudinger 1976).
- Preferential uptake of dissolved transition metals into specific Fe and Mn oxyhydroxide minerals, such as the affinity of Cu and Ni for todorokite (a multi-element Mn oxyhydroxide).

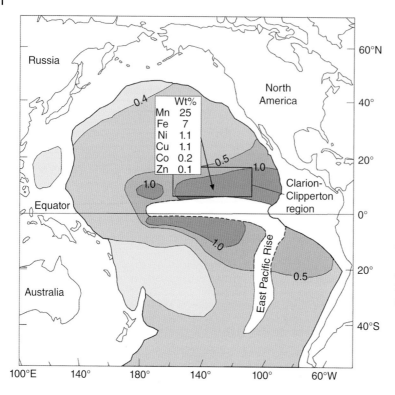

Figure 5.21 Map showing sub-equatorial zones of base metal enriched manganese nodule concentration in the Pacific Ocean, contoured with respect to the copper content (in wt%) of nodules. Nodule-free areas occur close to the continents, where sedimentation rates are high, as well as along the equator and East Pacific Rise, where biological productivity is high. The Clarion–Clipperton region, a particularly enriched zone, is boxed. Source: After Heath (1981).

The current level of uncertainty about the processes involved in the formation of deep ocean manganese nodules is reflected in the variety of suggested mechanisms listed above. Many of these suggestions are not mutually exclusive and it is likely that nodule formation is a long-lived process. The biggest problem with the exploitation of manganese nodules at present is not so much technical as it is political, environmental and economic. The technologies required for exploiting manganese nodules at depth in several kilometers of ocean water could be overcome, but the formulation of an internationally acceptable Law of the Sea is a much bigger and more contentious issue. Manganese nodules, nevertheless, represent an important metal resource for the future.

5.3.4 Phosphorites

Phosphorus is an important element that is essential for the growth and development of most living organisms. It is a basic ingredient of some proteins, including DNA and RNA, and is a building block for the bones and teeth of vertebrates, as well as the shells and chitinous exoskeletons of many invertebrates. Plants also require it for growth, and it is for this reason that phosphate is such an important ingredient of fertilizer. In order to sustain global food supply the manufacture of artificial fertilizers has become a major industry that requires approximately 250 million tons of phosphate rock each year (Heckenmüller et al. 2014). Phosphate ore, which is used to manufacture phosphoric acid and subsequently a variety of different fertilizer types, is obtained predominantly from phosphorus-rich sediments or phosphorites, the latter defined as a sediment containing more than 15–20% P_2O_5. The natural phosphorus cycle varies in response to patterns of global tectonic evolution and climate change, and more recently due to anthropogenic factors such as farming and fertilization. The input of phosphorus into ocean sinks, for example, is shown to have increased during periods of orogenesis (such as during the uplift of the Himalayan and Andean mountain chains) when enhanced physical and chemical weathering resulted in an increased flux of dissolved phosphorus and particulate phosphate minerals into the oceans (Filippelli 2008). Phosphorus in the oceans is then fixed, either biologically or during sedimentation and diagenesis on continental margins. The most common phosphate mineral is fluorapatite ($Ca_5(PO_4)_3F$) which is usually simply referred to as apatite and occurs as an accessory phase in many igneous rock types. In sedimentary environments a carbonate anion partially replaces the phosphate radical to form carbonate-fluorapatite (CFA) which is a microcrystalline or amorphous mineral that was previously referred to as either collophane or francolite.

Phosphorites generally form in much the same environmental niches as do BIFs, ironstones, and bedded manganese ores, namely along continental shelves and in shallow marginal marine settings such as lagoons and deltas. The processes of formation are, therefore, very similar to those for Fe and Mn deposits and essentially involve the upwelling of cold ocean water containing above-normal concentrations of metals, and their subsequent precipitation onto the continental shelves. In fact, one of the earliest published accounts of the upwelling hypothesis, attributed to the Russian scientist A.V. Kazakov in the 1930s (Kazakov 1937, 1939), applied to the formation of phosphate-rich sediments in a continental shelf setting, and the concept was only later applied to iron and manganese deposits. Deep sea drilling and ocean floor exploration have shown that phosphate concentrations are presently forming offshore along the coastlines of Namibia and Chile–Peru, and are related to upwelling of deep, cold currents carrying small amounts of the phosphorus oxyanion in solution. The same environments are also well known as prolific fishing grounds, a feature which substantiates the relationship between the supply of phosphatic nutrients and biological productivity.

When the idea of oceanic upwelling was first introduced, it was suggested that precipitation of phosphorus onto the sea floor in areas of upwelling was inorganic and related to decreasing solubility in the near shore environment. This notion is over-simplified, and it is now known that the phosphorus cycle in the oceans is largely controlled by organic processes. A role for microorganisms in the deposition of phosphate is also considered likely (Trudinger 1976). Phosphorus dissolves in sea water as either PO_4^{3-} (stable under very alkaline conditions), HPO_4^{2-} (the dominant anionic complex), or $H_2PO_4^-$ (stable under acidic conditions) complexes. These anionic complexes are absorbed by the organisms living in shallow marine environments to form their shells, bones, and teeth. The subsequent accumulation of phosphorus on the sea floor, therefore, is not related to a chemical redox reaction, as in the cases of Fe and Mn, but occurs after the host organism dies and settles to the ocean floor. Phosphorus is released from the decaying organism to form a calcium phosphate compound, which then converts to CFA. The concentration of calcium phosphate on the sea floor is described in terms of the reaction:

$$3Ca^{2+} + H_2PO_4 \Leftrightarrow Ca_3(PO_4)_2 + 2H^+$$

$$(5.8)$$

which suggests that its precipitation is dependent, among other things, on pH. Figure 5.22 shows apatite solubility in sea water as a function of pH and oxalate content of the solution (oxalic acid is $H_2C_2O_4$). It is apparent that apatite solubility is significantly higher under more acidic conditions and, on this basis, apatite formation is more likely to occur under the relatively alkaline, warm-water conditions that apply to continental shelves. Deeper, colder, and more acidic waters will promote phosphate solubility.

The formation of CFA in phosphatic sediments is, however, not a straightforward process (Piper and Perkins 2014) and there has been considerable debate as to whether

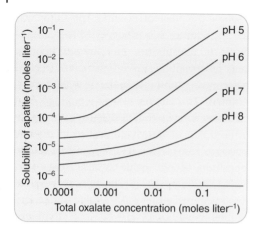

Figure 5.22 Estimates of apatite solubility in sea water (expressed in terms of phosphate concentration) as a function of solution pH and oxalate concentration. Source: After Schwartz (1971).

it develops as a primary precipitate at the sea water–sediment interface (i.e. where the organic material is decaying) or as a product of diagenesis during later compaction and dewatering of the sediment. This debate has bearing on the origin and formation of phosphorite ores and evidence seems to exist for both processes. Typical concentrations of dissolved phosphate in the deep portions of the present day oceans are low (about 50–100 ppb; Bentor 1980) and the shallow marine environment is even more strongly depleted because of biological uptake. Given the low present day concentrations of phosphate in solution, even in the deep ocean, it seems very unlikely that sea water is saturated with respect to apatite. This, in turn, suggests that the primary precipitation of apatite at the sea water–sediment interface is also unlikely. Although this situation applies generally to the present day oceans, it may not, however, have prevailed in previous geological eras when factors such as ocean chemistry and biological evolutionary development were different. It may also not apply to restricted, but biologically productive, environments (such as a lagoon) where high rates of organic decay can produce localized saturation of phosphate

in sea water, as in the case of the southern African example described below.

Sheldon (1980) has suggested that phosphate concentrations in the Precambrian oceans might have been higher than at present because the lower-O_2, higher-CO_2 atmospheric conditions prevailing at that time would have meant a more acidic sea-water composition and higher phosphate solubility. It is also apparent that oceans that existed before the development of phosphate-dependent organisms did not witness biological phosphorus depletion. Likewise, oceans at this time could not produce the environments in which biologically mediated deposition and concentration of CFA might occur. Substantial Precambrian phosphorite deposits, therefore, do not exist, and the few that do occur are likely to have formed from the direct precipitation of phosphate from sea water. As the atmosphere evolved, however, any event that coincided with a substantial proliferation of life, such as the Cambrian explosion, could have had the effect of removing CO_2 from the oceans and atmosphere, increasing pH in the shallow water environment and further promoting the direct precipitation of apatite from a saturated sea water column. A more efficient form of concentration, however, would have been through the decay of phosphate-dependent life forms and the formation of phosphorites in association with organic-rich sediments. This is probably the reason why most large phosphorite deposits are Phanerozoic in age. It is pertinent to note that the Precambrian–Cambrian boundary, for example, coincides with a period in Earth history when substantial deposition of phosphoritic sediments took place (Cook and Shergold 1984), good examples of which include the deposits of Mongolia, southeast Asia and Australia. By contrast, formation of Mesozoic and Cenozoic phosphorite deposits must have taken place essentially during the diagenetic stage of sediment evolution, since the phosphate concentration levels in the water column were too low to allow for direct CFA precipitation. In this scenario, the concentration of phosphate in organic-rich host sediments builds up progressively with time such that interstitial pore-fluids would eventually become highly enriched in phosphate (measurements of up to several thousand ppb HPO_4^{2-} in solution have been recorded; Bentor 1980). This represents an ideal environment for the formation of diagenetic apatite as the sediments are compacted and lithified. Such processes may pertain to deposits such as the Cretaceous–Paleogene deposits of Morocco and possibly also the oolitic phosphorites of Florida. In the latter case, however, subsequent mechanical abrasion and sedimentary transport of diagenetic aggregations resulted in deposits that are reworked and allochthonous (Riggs 1979).

Other factors, such as global climatic patterns and eustatic sea-level changes, are also known to have played a role in the formation of phosphorites (Cook and McElhinny 1979; Sheldon 1980). Figure 5.23a shows that there is a broad correlation between the number of phosphate deposits that formed and periods of sea-level highstand. The explanation offered is that elevated sea levels and flooding of the continental shelves enhanced circulation and upwelling. If enhanced upwelling occurred in an equatorially aligned sea way (Figure 5.23b) then phosphorite formation would have been promoted by the biogenic concentration and/or solubility decrease that accompanied the existence of such biologically productive environments. The late Mesozoic to Cenozoic phosphorite deposits of North Africa, deposited in the Tethyan sea way, might represent examples of this control mechanism. Figure 5.23a shows that there is also a correlation between glacial events and phosphogenesis. In this case the lack of vertical mixing in the oceans during stable periods results in a phosphate buildup in the deep water sink. The development of strong trade wind systems between glacial and non-glacial events in mature, longitudinally orientated oceans would promote the establishment of major oceanic gyrals which would,

in turn, result in strong upwelling along the western edges of continental land masses (Figure 5.23c). This mechanism could explain the development of the Permian Phosphoria Formation, probably the world's largest accumulation of phosphate rock, along the western margin of continental USA in Pangean times. This scenario also applies to the present day Atlantic Ocean and may account for the pattern of upwelling and phosphate deposition along the western margin of Africa (see Section 5.3.4.1 below).

5.3.4.1 A Model for Phosphogenesis Based on Present Day Deposition

Birch (1980) studied the nature and origin of phosphorites forming off the western and southern continental margins of South Africa and Namibia. Exploration of the continental shelf in this region reveals two types of phosphorite (Figure 5.24a). One variety, developed mainly along the Namibian coastline, comprises oolitic CFA ores derived by accretionary growth arising from the direct precipitation of phosphate from sea water. The other type, formed mainly along the South African coastline, consists of phosphatic replacement of fossiliferous limestone and is diagenetic in origin. Both types of phosphorite are geologically recent (Eocene to Miocene) in age and are believed to have formed at the same time. A model for this type of penecontemporaneous ore formation is shown in Figure 5.24b. Diffuse upwelling along the outer reaches of the shelf results in reduced nutrient supply and biological productivity, and consequently the rate of phosphate release during decay of organisms on the sea floor is low. Apatite is undersaturated and cannot precipitate directly from the sea water, but subsequent diagenesis can lead to replacement of limestone by calcium phosphate. By contrast, in the near shore environment very high biological productivity in restricted shallow embayments and lagoons is maintained by wind-induced nutrient upwelling. Mass mortality of life forms in these restricted

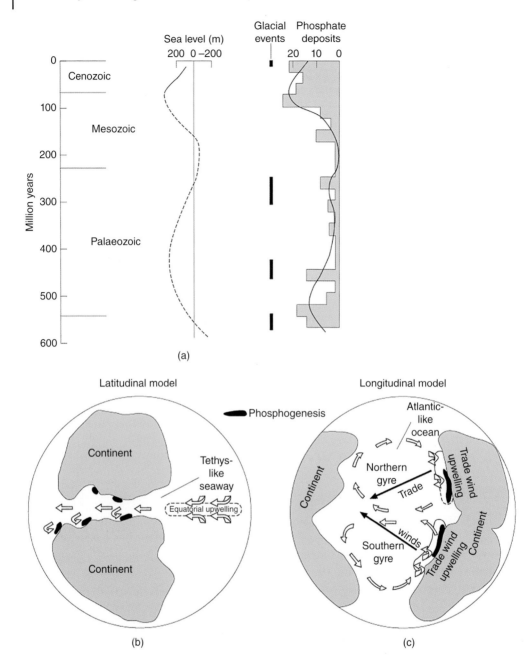

Figure 5.23 (a) Generalized correlations between the development of phosphate deposits, glacial events, and eustatic sea level changes in the Phanerozoic Eon. (b) Model for phosphate deposition related to equatorial upwelling in a latitudinally orientated seaway such as Tethys in the Mesozoic. (c) Model for phosphate deposition related to trade wind induced circulation patterns in a mature longitudinally orientated ocean. Source: After Sheldon (1980).

Figure 5.24 (a) Map showing the distribution of phosphate-rich sediments along the continental margins of South Africa and Namibia. (b) Model showing contrasting modes of phosphate deposition in the outer shelf (where diagenetic accumulation of apatite is taking place) and in restricted near-shore environments characterized by high biological productivity and seasonal mass mortality (where precipitation of apatite directly from sea water is occurring). Source: Diagrams after Birch (1980).

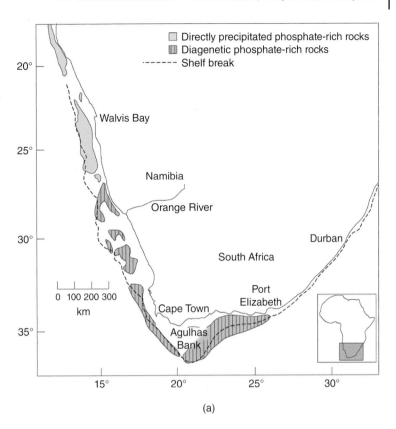

(a)

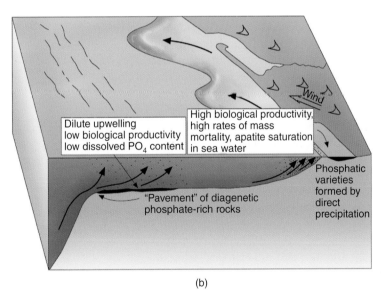

(b)

environments (especially during summer low tides) leads to the accumulation of abundant organic matter on the substrate with local increases in dissolved phosphate content to saturation levels. Varve-like layers of pure CFA, as well as accretionary oolitic growth of the phosphate mineral, provide the textural evidence for an origin by direct precipitation of phosphate at the sea water–sediment interface.

5.3.5 Black Shales

Shales that are rich in organic matter are economically important because they are often enriched in a large variety of metals, including V, Cr, Co, Ni, Ti, Cu, Pb, Zn, Mo, U, Ag, Sb, Tl, Se, and Cd. Although few metalliferous black shales are presently mined (with the exception of China where several small deposits are exploited) they represent a significant metal resource and probably also a source of ore components for younger hydrothermal deposits. As mentioned previously, many black shales are spatially associated with ironstones (Figure 5.15) and form in the same environments. At a global scale, they formed penecontemporaneously with ironstones, and their major periods of development can be linked to the first order eustatic sea-level highstands in the Ordovician–Devonian and Jurassic–Paleogene intervals (see Chapter 6). In this respect they are also linked to the Ocean Anoxic Events (OAEs) that are now well documented in the Phanerozoic Eon. Sea-level highstand resulted in continental margin flooding and transgression of muddy sediment across the shelf. The deeper water conditions that applied at these times resulted in poorer circulation, a reduction in oxygen levels, and widespread oceanic stagnation (Van Houten and Authur 1989), providing a depositional environment favorable for metalliferous black shale formation.

Well known examples of metalliferous black shales (Meyers et al. 1992) include the Cambro–Ordovician Alum shale of Scandinavia, which is particularly enriched in

uranium, and the Devonian New Albany shale of Indiana, USA, which has significant concentrations of Pb as well as V, Cu, Mo, and Ni. The euxinic shales forming on the floor of the Black Sea represent an example of where manganese, as well as Co, Cu, Ni, and V enrichment, is taking place in a present day sedimentary basin (Section 5.3.2; Holland 1984).

Black shales are characterized by significant quantities of organic carbon which is preserved from degradation/oxidation by the almost total lack of free oxygen in the immediate environment of deposition. This environment is not only anoxic but also euxinic (where reduced sulfur species are stable) and it is this combination, together with a lack of clastic dilution, that provides the optimal conditions for development of metalliferous black shales (Leventhal 1993). A sediment organic content of even 1% will bacterially and chemically deplete the oxygen content of the immediate environment to virtually zero. This microenvironment promotes the growth of sulfate-reducing bacteria which in turn produce HS^- or H_2S (depending on the local pH) and the development of euxinic conditions. The biologically mediated development of euxinic waters can be described in terms of the following reaction:

$$R(CH_2O)_2 + SO_4^{2-} \rightarrow R + 2HCO_3^- + H_2S \tag{5.9}$$

where $R(CH_2O)_2$ is an abbreviated representation of a complex (organic) carbohydrate molecule. The subsequent concentration of metals in this environment is brought about by the affinity of positively charged metals in solution in the sea water for a thio complex (such as HS^-), as demonstrated in the following simplified reaction:

$$HS^- + Me^{2+} \rightarrow MeS + H^+ \tag{5.10}$$

where Me is a metal. The above reactions imply a correlation between organic carbon, sulfur, and metal contents in black shales since the higher the content of sulfate-reducing bacteria, the more sulfide is produced and the

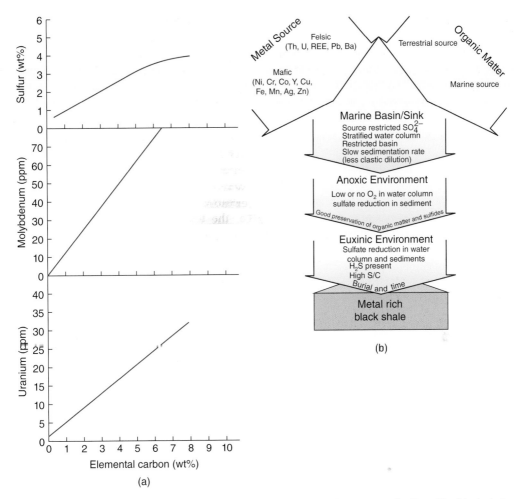

Figure 5.25 (a) Plots of sulfur, molybdenum, and uranium versus carbon content for Devonian black shales from the eastern USA. Source: After compilations by Holland (1984). (b) Schematic outline of the geological and environmental conditions required for the optimal formation of metalliferous black shales. Source: After Leventhal (1993).

greater the degree of extraction of metal from the sea water column. Such a correlation is demonstrated in Figure 5.25a, for Devonian black shales in the eastern USA, and lends support to the processes outlined above.

The source and metal content of black shales will obviously vary from place to place and this accounts for the observation that different horizons have characteristic metal signatures. In oceanic settings submarine exhalative vents, or black smokers, represent likely metal sources, whereas in continental margin settings (such as the Black Sea) metals are derived from terrigenous clastic input. Figure 5.25b summarizes the geological and environmental conditions necessary for the optimal development of a metalliferous black shale. This type of sedimentary rock represents a potentially important source of exploitable metals in the future.

5.3.6 Evaporites

The bulk of the world's production of rock salt (halite), as well as of borates, sodium carbonate, sodium sulfate, and potash for agricultural fertilizers, comes from rocks known as evaporites. These are chemical precipitates that form at or near the surface as a result of the solar evaporation of a brine, derived either from sea water or lacustrine waters rich in dissolved salts. There are two main environments in which evaporites form. The major one, in which the so-called "saline giants" of the world formed, is marginal marine in setting and represented by large lagoons or restricted embayments into which periodic or continual sea water recharge occurs. This type of deposit may be both laterally extensive and thick, but is characterized by a relatively limited range of mineral precipitants derived from sea water, which itself has remained fairly constant in terms of bulk composition over much of geological time. The second setting is intracontinental and lacustrine and results in much smaller, thinner deposits which are characterized by a more diverse range of mineral precipitates since continental fluvial input introduces a broader range of dissolved ingredients to the lake waters (Eugster 1980). Examples of deposits representing the former include the Permian Zechstein Formation, which formed in a shallow sea covering large areas of the United Kingdom, The Netherlands, Germany, and Poland, and also the Silurian Salina Formation of New York and Michigan states in the USA. Typical products mined from these deposits include halite, potash, and sulfates. Probably the largest known evaporite sequence occurs in late Miocene strata beneath the floor of the Mediterranean Sea, where deposits extend laterally for over 2000 km and may be up to 2 km thick. Much of this resource is obviously unexploitable, but is so large that its formation caused a significant reduction in oceanic salinity at the time (Kendall and Harwood 1996). Examples of intracontinental lacustrine deposits include the dry lakes, or salars, of Chile, Death Valley in California, and the Great Salt Lake of Utah. In addition to the products mentioned above, these deposits are also important producers of

borates (from California) and alkaline metals such as Mg, Br, and Li (from Chile and Utah). Two examples of evaporite deposits, Boulby in the UK and Salar de Atacama in Chile, representing marine and intracontinental lacustrine settings respectively, are described in Box 5.3.

The principles behind the formation of evaporite deposits are relatively simple, although actual deposits can exhibit a variety of different mineral salts and paragenetic sequences. As sea water or brine evaporates and water vapor is removed into the atmosphere, the salinity (i.e. the total content of dissolved salts in solution) of the residual solution increases and individual salts precipitate as their respective solubility limits are reached. The order or sequence of precipitation reflects the scale of increasing solubility at a given temperature, such that the salts with the lower solubilities precipitate first, followed consecutively by salts with progressively higher solubilities. The relative quantities of precipitated products in an evaporite deposit are also constrained by the solubility limits of the various salts dissolved in the brine solution. These considerations seem to suggest a relatively straightforward precipitation process and uniform paragenetic sequence. In fact, the chemistry of brine solutions is complex and factors such as convection dynamics, post-precipitation diagenesis, and equilibrium, as opposed to fractional crystallization, combine to ensure considerable diversity in the nature of these deposits.

Marine evaporites that formed from sea water that was well mixed and had a constant composition over much of geological time tend to be dominated by the same assemblage of major mineral precipitates, namely halite (NaCl), gypsum/anhydrite ($CaSO_4$/ $CaSO_4 \cdot 2H_2O$), and sylvite (KCl). Marine evaporite deposits can, however, also contain minor accumulations of other salts, especially those representing the late stage precipitation of high solubility compounds. The latter typically comprise hydrated, multi-element, Mg, Br, Sr, K chlorides and borates, and are called bitterns. The preservation of bitterns

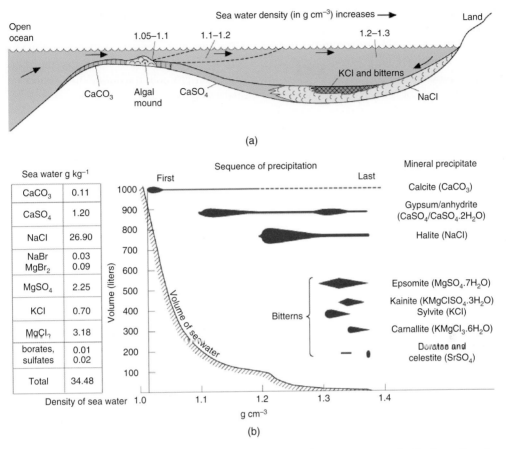

Figure 5.26 (a) Schematic cross section showing the important features necessary for the formation of large marine evaporite sequences. (b) Paragenetic sequence for an evaporite assemblage from typical sea water containing the ingredients shown in the left-hand column. The amount of sea water (per 1000 litre volume) that has to evaporate in order to consecutively precipitate the observed sequence of mineral salts is shown by the curve adjacent to the paragenetic sequence. Source: Diagrams modified after Guilbert and Park (1986).

in evaporite deposits appears to be controlled largely by geological factors. Figure 5.26a shows a schematic cross section through a shallow, marginal marine setting in which an evaporite sequence is being deposited. Sea water recharge ensures replenishment of the solution and sustained mineral precipitation from the brine. Intense evaporation results in marked density stratification and accumulation of highly saline, dense brines in the bottom waters. In environments where there is a constriction at the entrance of the basin, there is little opportunity for this dense brine to discharge back into the deep parts of the open ocean and, hence, the high solubility

components of the solution remain within the basin to eventually precipitate out as bitterns once the solubility limits are exceeded. Bitterns do not accumulate if reflux of the dense bottom water brines to the open ocean occurs because the high solubility components of the solution are then flushed out of the basin. In the latter case, if the bitterns do not precipitate, the evaporite mineral assemblage may be truncated or incomplete and the deposit less diverse in terms of the commodities it could exploit. Figure 5.26b illustrates the typical mineral paragenetic sequence for evaporite deposits as a function of precipitation from sea water, the composition of which is shown on the left-hand side of the figure. The proportion of sea water remaining after evaporation (as a function of a 1000 litre volume) and its increasing density are also shown in relation to the precipitation sequence.

The formation of the giant Phanerozoic saline deposits is a contentious issue, mainly because modern analogues do not exist and ancient evaporite sequences are generally altered and tectonically modified. The evaporite deposits on the floor of the Mediterranean Sea are now believed to have formed in a sabkha environment, implying that much of the Mediterranean basin was subaerial and desiccated during late Miocene times (Kendall and Harwood 1996). Evaporitic minerals which form in a sabkha environment precipitate from groundwater brines that evaporate in the capillary zone immediately above the water table. Recharge of the brine occurs either by periodic flooding or by sea water seepage into the sabkha basin. A good example of a complete evaporate deposit formed as part of a Phanerozoic saline giant is the Boulby Mine in the Zechstein of northeast England (Box 5.3).

Box 5.3 Evaporite Deposits: The Boulby Mine (Na and K) in Cleveland, UK and Salar de Atacama (K and Li) in Chile

Boulby
The majority of the UK's salt and potash, for the de-icing of roads and fertilizer manufacture respectively, comes from the Boulby Mine in the northeast of England. Underground operations commenced in 1973 to depths in excess of 1000 m and extending for a considerable distance offshore beneath the North Sea (Woods 1979). The various salt layers are hosted within shallow marine sediments of late Permian age, in an environment that formed when the Zechstein Sea covered large parts of northern Europe, and at a time when the continent was at lower latitudes than at present. Approximately 1100 m below surface an upper halite (NaCl) layer is encountered which passes abruptly into an underlying anhydrite ($CaSO_4$) bed (Figure 5.3.1), the latter containing lenses of sylvite (KCl). These are in turn underlain by a dolomitic shale unit that contains minor polyhalite ($K_2Ca_2Mg(SO_4)_4 \cdot 2H_2O$) and carnallite ($KMgCl_3 \cdot 6H_2O$), followed by the principal ore horizons, termed the Boulby Potash and Boulby Halite Formations. Mining activities focus on the latter two horizons to produce halite for the de-icing of roads and sylvinite (a mixture of halite and sylvite) for the manufacture of fertilizers. Interestingly, the Boulby Potash unit is strongly deformed and is characterized by a strong gneissic fabric and contorted layering, features that contrast with salt layers above and below that are relatively undisturbed. This is a halokinetic feature and is related to the preferential mobility under stress of the more potassium rich salts. Talbot et al. (1982) have suggested that deformation of the Boulby Potash unit accompanied dehydration of gypsum to anhydrite in the underlying Billingham Main Anhydrite Formation (Figure 5.3.1), with the resultant egress of water into the salt horizons. Both halite and sylvite are markedly softened in contact with water, but with sylvite, being the more mobile under these conditions, deforming more readily.

Approximately 150 m below the Boulby Potash Formation, a well developed layer of polyhalite is encountered – this unit is now being exploited as a natural multi-nutrient fertilizer that has the advantage of not requiring additional processing and can be applied directly to certain crops to bolster their K, Ca, Mg, and S requirements. Marine evaporites, such as at Boulby, representing the so-called saline-giants, can therefore provide both sodium and potassium salts, as well as sulfate minerals, that are the basic feed stock for the global fertilizer industry.

Box 5.3 (Continued)

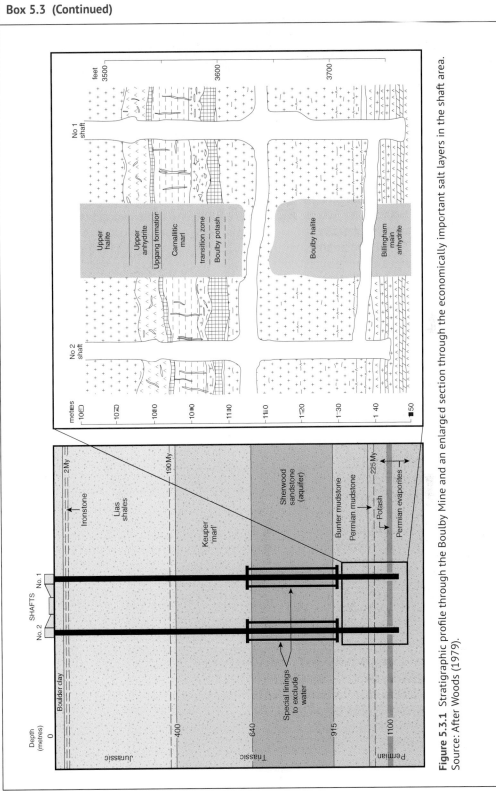

Figure 5.3.1 Stratigraphic profile through the Boulby Mine and an enlarged section through the economically important salt layers in the shaft area. Source: After Woods (1979).

Box 5.3 (Continued)

Salar de Atacama

Historically, most of the world's lithium has been produced from pegmatites that were mined to extract lithium-bearing minerals such as lepidolite, spodumene and petalite. Pegmatite deposits tend to be small and costly to mine, with the Li being relatively difficult to extract from its host minerals. With the advent of new lithium-ion rechargeable battery technologies and increased demand for electric cars, the demand for Li has grown, and much of this is being supplied from intracontinental lacustrine evaporites which have larger resources and produce a LiCl-bearing aqueous solution that is relatively simple and cost-effective to beneficiate.

The arid, high-altitude intermontane regions of the Andes mountain belt host large endorheic (or closed) basins that are characterized by limited ingress of riverine waters and high rates of evaporation, so that they are typically floored by salt pans – locally referred to as *salars*. The Salar de Atacama, in Chile's Atacama Desert, is one of the largest in the region and is the site of a major operation whereby Li-rich brines are pumped from local groundwater aquifers into surface evaporation ponds (Figure 5.3.2). The brine evaporates over a period of several months leaving behind a precipitate of various K and Mg salts, as well as borates and an enriched Li brine. The Li bearing solution is extracted and then reacted with soda ash to form $LiCO_3$.

Parts of the Salar de Atacama are characterized by subsurface brines with unusually high Li contents (Figure 5.3.3). The Li, together with K, Mg, and B, is thought to have been derived from the weathering of Andean volcanic rocks to the east of the salar, with additional Na, Ca, Cl, and SO_4 originating from the dissolution of an older salt bed that diapirically intrudes these volcanic rocks (Warren 2010). The subsurface brines are extracted via shallow wells and pumped into evaporation ponds where a complex operation of brine recycling and solubility control results in the production of KCl and KSO_4 precipitates, in addition to the lithium-enriched brine. Interestingly, lithium does not typically precipitate as an evaporitic mineral and tends to remain in solution in the brine, although at one location, Zabuye Lake in Tibet, a natural Li_2CO_3 precipitate is known to form (Warren 2010). The high solute content of the source brines, together with very high evaporation rates and extreme aridity, ensure that the Salar de Atacama operation is economically viable.

Figure 5.3.2 Aerial view of the evaporation ponds on the Salar de Atacama, Chile. Source: Photo by Ivan Alvarado, courtesy of Thomson Reuters.

Box 5.3 (Continued)

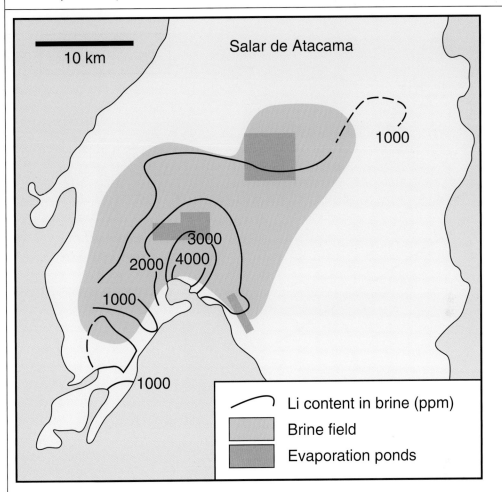

Figure 5.3.3 Distribution of the Li content of subsurface brines in the Salar de Atacama, Chile. Source: After Kesler et al. (2012).

Most of the world's Li producing salt pans occur in hyperarid regions underlain by basins that formed in active or recently-active orogenic belts. The salars of Chile, Argentina, and Bolivia are located on the Altiplano along the crest of the Andean convergent margin, while the USA-based brine evaporation works at Clayton Valley, Nevada and the Salton Sea, California, occur in valleys formed within extensional grabens (Bradley et al. 2013). In all cases it would appear that active faulting promotes the recharge of aquifers and the movement of saline groundwaters within the basin. In addition, because salt compacts readily to an essentially impervious rock, the strata along which saline brines flow need to be shallow (typically <50 m depth) so that adequate porosity is maintained to allow recharge as the fluids are extracted. Finally, fertile Li-rich brines only form when groundwaters come into contact with a leachable form of lithium, such as in felsic tuffs or rocks altered to hectorite (a Li-bearing smectite clay). A schematic illustration of lithium sources and pathways, as applicable to evaporitic extraction of lithium from brines in hyperaridic terranes, is presented in Figure 5.3.4.

(Continued)

Box 5.3 (Continued)

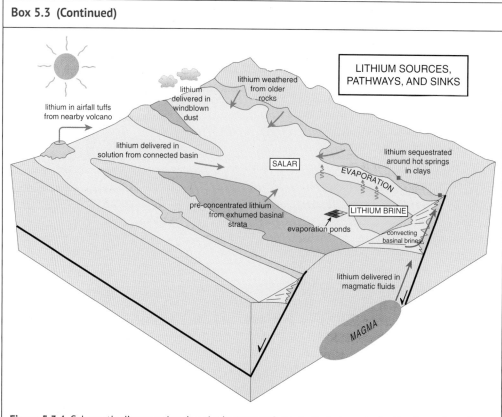

Figure 5.3.4 Schematic diagram showing the important features that characterize the formation of evaporitic lithium deposits in tectonically-active, endorheic basins. Source: After Bradley et al. (2013).

5.4 Fossil Fuels – Oil/Gas Formation and Coalification

A fossil fuel is formed from the altered remains of organic matter (plant and animal) trapped in sedimentary rock. The hydrocarbon compounds that form during the burial and degradation of organic material retain a significant proportion of the chemical energy imparted into the original living organism by the sun. This energy is harnessed in many different ways by burning or combusting the fuel. The main fossil fuels are coal and petroleum, the latter being a general term that encompasses crude oil and natural gas (mainly methane – CH_4), as well as solid hydrocarbons.

Although most ore deposit books do not include descriptions of fossil fuel occurrences, it is felt that any text dealing with ore-forming *processes* would be incomplete without brief

discussion about the accumulation of hydrocarbons in the Earth's crust. There is an enormous volume of published literature on the origin and nature of fossil fuel deposits and, consequently, the following section presents only brief overviews relevant to oil/gas formation and coalification processes. Mention is also made of unconventional deposits such as shale gas, tar sands, oil shales, and gas-hydrates, all of which represent potential fuel resources of the future.

5.4.1 Basic Principles

The global carbon cycle is multifaceted, involving both circulation (in the form of gaseous CO_2 and CH_4) and storage (as reduced carbon or carbonate minerals) in reservoirs such as the deep oceans and sedimentary rocks. Organic processes (mainly photosynthesis) ensure that a small proportion of the carbon in the global cycle is combined with hydrogen and oxygen to form the molecular building blocks of the various biota that inhabit the Earth's surface. The areas of greatest organic productivity on Earth are represented on the continents by tropical vegetative zones, and in the oceans by areas of cold current upwelling. As a first approximation, therefore, it is logical to expect that fossil fuels are most likely to have formed in these two environments. Indeed, coals develop mainly in continental settings, from accumulation of vegetation in either humid, tropical swamp environments or, as in the case of younger coal seams, more temperate to polar regions. Oil and gas deposits, by contrast, develop mainly from the accumulation of phyto- and zooplankton in lacustrine and shallow marine settings.

Hydrocarbons are organic compounds composed of hydrogen and carbon; they are Group 14 hydrides and have a propensity to catenate (i.e. the linking together of C atoms by covalent bonding to form chains or rings) which is why there are so many organic compounds in nature and why hydrocarbons form the basis of organic life. Methane, for example, is the name given to carbon tetrahydride. Hydrocarbons are not to be confused with carbohydrates that are molecules comprising carbon, hydrogen, and oxygen with a simple empirical formula $C_m(H_2O)_n$ – essentially they are hydrates of carbon. Carbohydrates are also referred to as saccharides, a group comprising sugars, starch, and cellulose.

When organisms die, they decompose by bacterial decay and/or oxidation, and rapidly break down to relatively simple molecular constituents such as CO_2 and CH_4. Such a process would not be conducive to the formation of fossil fuels, which requires preservation of much of the complex organic hydrocarbon material. Suitable source rocks within which fossil fuels accumulate must, therefore, have formed in reducing environments where sedimentation rates are neither too slow (in which case oxidation and molecular breakdown might occur) nor too rapid (in which case dilution of the total organic matter content of the rock would occur).

The prevailing theory for the origin of both petroleum and coal is based on the notion that organisms are buried during the sedimentary process and then subjected to a series of alteration stages as pressure and temperature increase. As shown in Figure 5.27, progressive burial of phytoplankton results in the early liberation of CO_2 and H_2O and formation of kerogen (a collective name for sedimentary organic matter that is not soluble in organic solvents and has a polymer like structure; Bjørlykke 1989). As the kerogen "matures" with increasing temperature, the long-chain covalent bonds that typify organic molecules are progressively broken to form lower molecular weight compounds. At around 100–120 °C and burial depths of 3–4 km (depending on the geothermal gradient), a liquid hydrocarbon fraction develops which can then migrate from the source rock. This interval is known as the "oil window." With further burial and cracking of molecular bonds, significant volumes of gas (mainly methane – CH_4) develop, which are also amenable to migration. The solid residue remaining in the sediment is referred to as kerogen, but with progressive burial it devolatilizes further and evolves to a composition approaching pure carbon (graphite). By contrast, when humic, land-derived vegetation is buried, little or no liquid oil is formed, although significant volumes of gas will, again, be generated. The solid residue that remains in this case is more voluminous and compacts to form coal seams. The nature of the solid coal residue changes with progressive burial, from peat and lignite at shallow depths (less than about 500 m), to bituminous coal and subsequently anthracitic coal (at depths of around 5000 m). The calorific value (i.e. the amount of energy produced on combustion) of the coal increases with maturity or burial depth.

5.4.2 Oil and Gas Formation (Conventional)

All living organisms are made up of relatively few molecular building blocks that have apparently changed little over geologic time. The

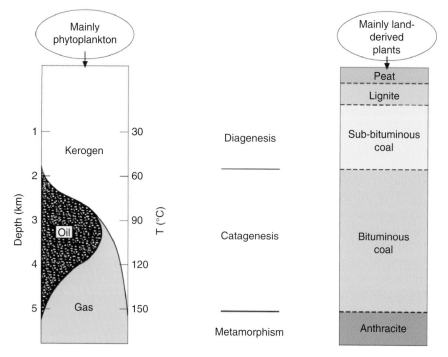

Figure 5.27 Simplified scheme illustrating the formation of oil, gas, and coal by the progressive burial of different types of mainly vegetative matter. Source: After Hunt (1979).

main biological building blocks involved in petroleum formation include:

- Carbohydrates and related substances – these are mono- and polysaccharide polymers and include the sugars, chitin, and lignin, the latter being a major precursor to coal.
- Proteins – these are high molecular weight amino acid polymers and represent one of the most important constituents of life processes.
- Lipids, which are represented in animal fats and vegetable oils and are abundant in crude oils.
- Lignin and other substances such as resins and pigments are of lesser importance but also occur variably in both plant and animal matter.

Plants comprise mainly carbohydrates (40–70%), although the higher, woody forms also contain substantial lignin for strength. Phytoplankton contains around 20% protein. Animals, by contrast, are made up mainly of proteins (55–70%) with lesser carbohydrates and lipids but no lignin (Hunt 1979). The

Table 5.1 Average chemical compositions (in wt%) of the main organic building blocks compared to those of petroleum and a typical coal.

	C	H	O	N	S
Carbohydrates	44	6	50	—	—
Lignin	63	5	31.6	0.3	0.1
Proteins	53	7	22	17	1
Lipids	76	12	12	—	—
Petroleum	85	13	0.5	0.5	1
Coal	70	5	23	1	1

Source: After Hunt (1979).

average chemical composition of these various organic building blocks is shown in Table 5.1.

Petroleum is mainly made up of hydrocarbons of the alkane (or paraffin) and naphthene groups, the latter also referred to either as cycloparaffins or cycloalkenes. Alkanes are saturated (with respect to hydrogen) compounds that are stable and typically have an empirical formula C_nH_{2n+2}. The alkanes methane (CH_4), ethane (C_2H_6), propane (C_3H_8), and butane (C_4H_{10}) are common ingredients of petroleum, and are gaseous at STP, whereas alkanes with higher carbon numbers (i.e. from pentane upwards) are generally liquids. Naphthenes are less stable but are also saturated and have a ring structure (or a series of rings) with the formula C_nH_{2n}. Naphthenes such as cyclopentane (C_5H_{10}) and cyclohexane (C_6H_{12}), for example, are abundant ingredients in liquid petroleum (oils). Less abundant ingredients include the aromatic hydrocarbons which are unsaturated ring compounds, the simplest of which have an empirical formula C_nH_{2n-6}, such as benzene (C_6H_6). Finally, asphaltenes are high molecular weight, complex hydrocarbons that commonly contain N, S, and O in their structure – they are typically viscous or

solid at STP and commonly form as a result of biodegradation of oils or pressure drop.

As the original organic constituents are buried, they are subjected to increases in pressure and temperature, resulting in systematic changes that are divided into three stages, termed diagenesis, catagenesis, and metamorphism. This progression, and its application to the burial of organic matter in sediments, together with the organic processes and changes that occur in each, is summarized in Figure 5.28. Diagenesis refers to the early biological and chemical changes that occur in organic-rich sediments prior to the pronounced temperature-dependent effects of later reactions (i.e. at less than about

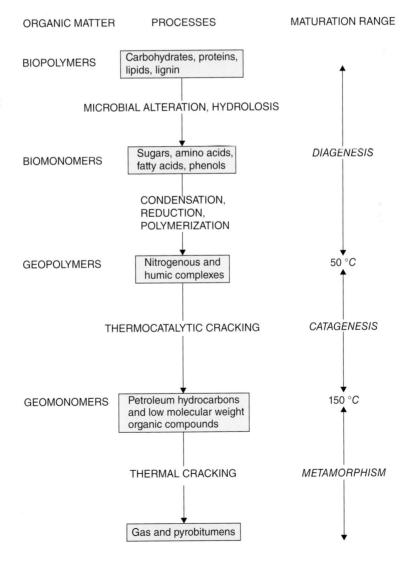

Figure 5.28 Summary of the stages and processes involved in the transformation of organic matter during burial to form oil and gas. Source: After Hunt (1979).

50 °C). During this stage the biopolymers of living organisms undergo a wide variety of low-temperature reactions. Put simply, they are either oxidized or attacked by microbes and converted into less complex molecules which may, in turn, react and condense to form more complex, high molecular weight geopolymers that are the precursors to kerogen. Biogenic reactions occurring during diagenesis also produce significant amounts of gas, termed biogenic gas, which typically escapes into the atmosphere or water column and is not retained in the sediment. Catagenesis occurs between about 50 and 150 °C and is the most important stage as far as petroleum generation is concerned. During this stage, temperature plays an important role in catalyzing reactions, the majority of which result in the formation of light hydrocarbons from high molecular weight kerogen by the breaking of carbon–carbon bonds. During this process, known as *thermal cracking*, a complex organic molecule such as a paraffin or alkane will split and form two smaller molecules (a different paraffin and an alkene or olefin). Each of the products contains a carbon atom, with an outer orbital electron from the original covalent pairing that is now shared. The shared

electron means that the resultant alkane and alkene each contain a "free radical" that makes them reactive and amenable to further breakdown. Alternatively, reactions may occur by *catalytic cracking*, where a carbon atom in one of the resultant molecules takes both electrons, thereby becoming a Lewis acid or double electron acceptor. The molecule that loses the electron pair is called a carbonium ion and with its net positive charge is also amenable to further decomposition. For example, alkanes subjected to catalytic cracking could yield gaseous products such as ethane or butane (Hunt 1979). Catalytic cracking tends to be the dominant process in petroleum generation up to about 120 °C, but at higher temperatures thermal cracking becomes increasingly important.

The integrated effects of time and temperature have important implications for petroleum generation. Temperature and time are inversely related in terms of petroleum productivity so that, for example, a given quantity of oil formed at 110 °C over 25 million years would require 100 million years to form at 90 °C (Hunt 1979). This relationship is quantified on a plot of time versus temperature in Figure 5.29, where data for selected petroleum-bearing basins are also shown.

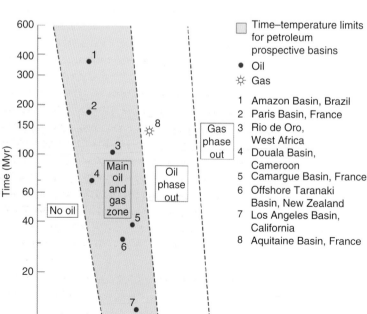

☐ Time–temperature limits for petroleum prospective basins
● Oil
☀ Gas

1 Amazon Basin, Brazil
2 Paris Basin, France
3 Rio de Oro, West Africa
4 Douala Basin, Cameroon
5 Camargue Basin, France
6 Offshore Taranaki Basin, New Zealand
7 Los Angeles Basin, California
8 Aquitaine Basin, France

Figure 5.29 Graph showing the relationships between temperature and time with respect to the generation of oil and gas in sedimentary basins. The shaded area refers to the optimum range of conditions for petroleum generation and the points are actual examples of petroleum-bearing basins from several locations around the world. Source: After Connan (1974).

The graph can be very useful in petroleum exploration as it predicts, for example, that oil would not be formed in a young basin with low geothermal gradient because the threshold for efficient liquid hydrocarbon generation might not yet have been reached. Likewise, little oil is likely to be preserved in an old, hot basin as it would all have been destroyed during metamorphism. The time–temperature evolution of sedimentary basins is, therefore, a critical factor in petroleum exploration. As a general rule, oil resources are sought either in young, hot basins or in old, cold ones, as identified by the relevant field in Figure 5.29.

5.4.2.1 Source Rock Considerations and Organic Maturation

Most of the organic matter on Earth can be classified into two major types, namely *sapropelic*, which refers to the decomposition products of microscopic plants such as phytoplankton, and *humic*, which refers essentially to the maturation products of macroscopic land plants. Sapropelic organic matter has H/C ratios in the range 1.3–1.7, whereas humic matter has lower H/C ratios around 0.9. These compositional differences have led to a more rigorous classification of kerogen types that has relevance to the generation of fossil fuels. The classification was originally devised by Tissot et al. (1974) in terms of the so-called Van Krevelen diagram, a plot of H/C against O/C (Van Krevelen 1961; see inset in Figure 5.30). A more recent version of the Tissot classification is described in Cornford (1998) using a plot of hydrogen index versus oxygen index (Figure 5.30), two parameters that are analogous to H/C and O/C, respectively, but can be obtained directly by Rock-Eval step-wise pyrolysis and chromatography (Bjørlykke 2015).

The classification scheme recognizes three main kerogen types, with a fourth type having little or no relevance to petroleum generation. Type I kerogen (or liptinite with high H/C) is derived from algal material, rich in lipids and deposited typically in lacustrine and lagoonal environments. On maturation Type I kerogen yields mainly oil – oil yields are high but liptinites are relatively uncommon. Type II kerogen, or exenite, by contrast is abundant and is obtained from the breakdown of phytoplankton, plant spores, pollens, exines, and resins deposited in marine environments. It yields both oil and gas with moderate yields. The Jurassic source rock of the North Sea oil deposits, the Kimmeridge Clay Formation, represents an example of petroleum derivation from Type II kerogen. Type III kerogen (or vitrinite with low H/C and high O/C) is derived from humic land plants rich in lignin and cellulose, and on maturation yields only gas. As mentioned previously, this is the material from which coals are formed. Inertinite, or Type IV kerogen, as the name implies yields neither oil nor gas and comprises mainly carbon and oxygen. The arrows in Figure 5.30 indicate the compositional trends of different kerogens with progressive burial and maturation, and all coalesce toward the origin, represented by pure carbon (graphite). In considerations of source it should be noted that oil-prone Type I and II kerogens will yield not only oil, but also both "wet" (i.e. with a high condensable fraction) and "dry" gas (mainly methane). Gas-prone kerogen (usually Type III) will produce mainly dry gas.

Modern analytical and modeling techniques allow the quantitative estimation of gas and oil yields to be calculated from a source rock undergoing progressive burial. A summary of hydrocarbon yields from two different source rocks is shown in Figure 5.31. Figure 5.31a shows the quantities of petroleum products derived from a source rock containing 1% total organic carbon (TOC) as Type II kerogen – the analysis distinguishes between the production of condensates comprising high (C_{15+}) and low (C_2–C_8) molecular weight oils, and a dominantly methane gas component. Diagenetic stages of maturation produce mainly CO_2 and H_2O that are essentially lost to the surface. With progressive burial, oil and gas is generated, with peak oil production taking place at around 150 °C and burial depths of 3000–4000 m. With greater burial depths and at higher temperatures, the main product is

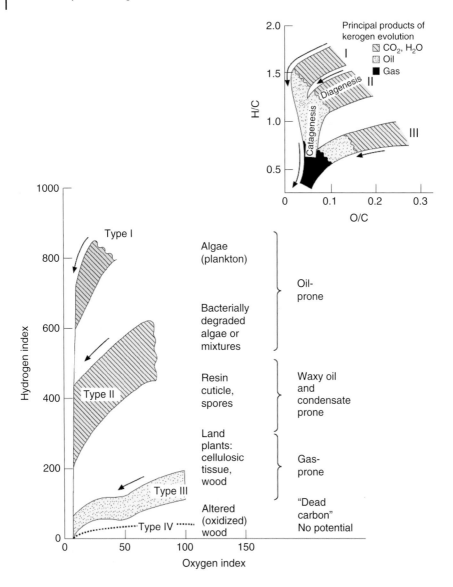

Figure 5.30 Classification scheme for kerogen types based on the hydrogen index and oxygen index. Source: After Cornford (1998). Inset diagram shows an older classification of kerogen on the basis of H/C and O/C atomic ratios. Source: After Tissot and Welte (1984).

dry gas or methane. Figure 5.31b is the equivalent maturation scenario for a source rock sediment comprising (Type III) kerogen that is only gas-prone, which initially yields only a minor amount of wet gas (ethane, propane, butane, etc.) and mainly dry gas (methane), leaving a coal residue. The latter process is relevant to the generation of coal-bed methane resources.

Source rock considerations are important during petroleum exploration. The origin of organic matter is obviously a pointer to the depositional environment of the sedimentary rock and, therefore, also useful in placing constraints on the nature of oil/gas migration and entrapment. Conventional petroleum deposits are formed when hydrocarbons from the source "kitchen" migrate

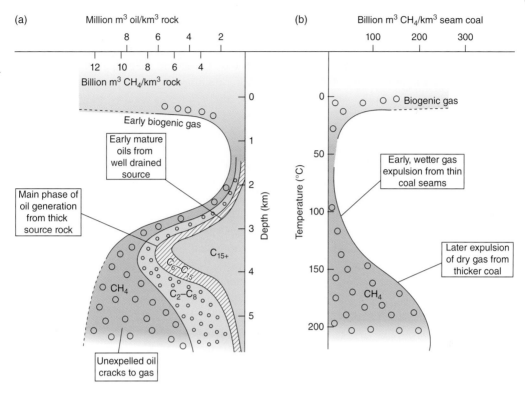

Figure 5.31 Maturity curves showing (a) the cumulative generation of oil types and methane gas from sediment source rock originally containing a 1% total organic carbon (TOC) content of Type II oil-prone kerogen; and (b) the generation of wet and dry gas only from a 1 km³ sediment volume of gas-prone kerogen which ultimately yields a coal deposit. Source: After Cornford (1998).

and concentrate into a trap-site known as the reservoir – unconventional oil and gas deposits (described in Section 5.4.4) typically form when the hydrocarbons do not migrate and the source rock is also the reservoir.

5.4.2.2 Petroleum Migration and Reservoir Considerations

One of the features of oil and gas formation is the fact that petroleum originates in a fine grained source rock and then migrates into more permeable, or coarser grained, reservoir sediments. Knowledge of petroleum migration and entrapment processes is obviously important to both the exploration and exploitation stages of the industry. Although the processes are complex, they can now be accurately evaluated in terms of fluid hydraulic principles and sophisticated computer modeling techniques.

Petroleum engineers differentiate between primary migration, which accounts for the movement of oil and gas out of the source rock during compaction, and secondary migration, which describes flow within the permeable reservoir, as well as the segregation of oil and gas. As the organic source rock is lithified during diagenesis, water (usually a low to moderate salinity brine) is expelled from the sediment to form a basinal fluid (see Chapter 3). In the early catagenic stages of hydrocarbon maturation, therefore, oil and gas exist in the presence of water. The actual mechanisms of hydrocarbon migration and the role of water in this process are not completely understood. Oil is unlikely to migrate in aqueous solution since hydrocarbons generally have low solubilities in water (e.g. 3 ppm for pentane, 24 ppm for methane, 1800 ppm for benzene, all at room temperature; Bjørlykke 1989). However, small hydrocarbon molecules dissolve more readily than large ones and methane (CH_4), for example, becomes fairly soluble in water as pressure increases (about 7500 ppm at 6000 m depth; Hunt 1979). Methane and other low molecular weight compounds can, therefore, be dissolved at depth and then released as the aqueous solution rises to the surface and depressurizes. Major amounts of hydrocarbon liquids and high molecular weight compounds, however, are unlikely to migrate by this mechanism. At catagenic pressures and temperatures, hydrocarbon gases themselves will increasingly dissolve oils and in some source rocks there is evidence that primary migration may be achieved by gas phase dissolution. During the advanced stages of catagenesis, temperatures and pressures are such that many hydrocarbons exist close to their critical points, and density contrasts between liquid and vapor are minimized (see Section 2.2 of Chapter 2). The segregation of oil from gas, therefore, mainly reflects the decrease in temperature that accompanies migration of petroleum products into reservoirs that are well away from the sites of oil/gas generation. The progressive separation of liquid from vapor and the formation of a low density, buoyant gas phase and a more dense, viscous oil, is a process that generally accompanies secondary migration.

Although dissolution mechanisms are likely to play a role in petroleum flow, they cannot account for the huge oil accumulations observed in the major oil producing regions of the world. Two other mechanisms are considered to be more important during primary migration of hydrocarbons, namely oil phase migration and diffusion. The movement of crude oil out of the source rock is initiated once compaction has driven off much of the bulk pore water in the sediment. Initially a source rock will usually contain a relatively high proportion of pore water relative to oil and in such a case the hydrocarbon will be unable to move. This is mainly because the flow of oil is impeded by the presence of water which, because of its lower surface tension, tends to "water-wet" the sediment pore spaces. In such a situation, oil globules floating in the water-dominated pore spaces cannot generate the capillary forces required to initiate migration (Figure 5.32a). As compaction progresses, however, the bulk of the pore water is driven off and the little remaining water is structurally bound on clay particles. With reduced rock porosity, oil now occupies enough of the pore volume to be subjected to capillary forces and start flowing through the rock (Figure 5.32b). Once "oil-saturated" pathways are established in the rock, oil migration becomes feasible because flow is not impeded by the presence of water. In the less common situation when pore spaces are initially "oil-wet" because of very high initial organic/hydrocarbon contents, then porosity reduction is not necessary to initiate the flow of oil through the source rock.

Another way in which hydrocarbon molecules could migrate in the primary environment is by diffusion along an activity or free energy gradient (Hunt 1979). Figure 5.32c illustrates, at a nanometer scale, the arrangement of structured water tetrahedra along the edge of a clay particle. Hydrocarbon molecules or aggregates in the sediment pore space will typically be encased by a jacket of water molecules to form hydrate or clathrate compounds. Interaction of the hydrocarbon clathrate with structured water would, from a thermodynamic viewpoint, require energy to facilitate the breakdown of one or both molecules. A diffusional energy gradient is, therefore, set up that promotes migration of the hydrocarbon clathrate toward lower free energy and, therefore, away from the structured water clay mineral edge. This type of diffusion translates, at a geological scale, to movement of hydrocarbons from finer grained shales toward coarser sands, which is similar to capillary action and accords with natural observations. Although not directly involved as the transporting agent in primary migration, it is evident that water is implicated in all the mechanisms involved.

Figure 5.32 Explanation of oil-phase primary migration. (a) "Water saturated" pore spaces in a sedimentary source rock in which little or no oil migration takes place because oil capillarity is impeded by the presence of water-wet pores which block the pore throats. (b) "Oil saturated" pore spaces form after compaction of the sediment has driven off much of the bulk pore water, which facilitates capillary-related oil migration by removing pore throat blocking free water. Source: After Bjørlykke (1989). (c) Simplified illustration, at a nanometer scale, of a hypothetical pore space and the setting in which diffusional migration of hydrocarbons takes place away from structural water-bound mineral grains. Source: After Hunt (1979).

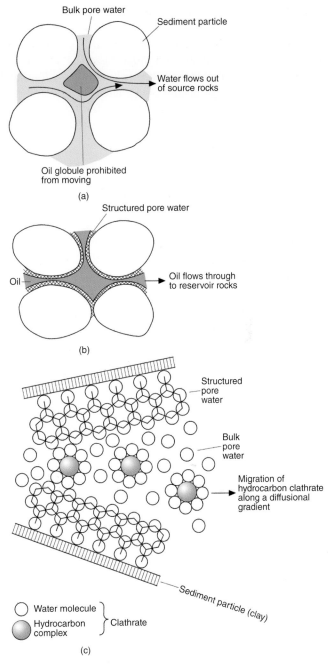

In the case of low organic content source rocks generating mainly gas, hydrocarbon migration could occur by diffusion, in aqueous solution, or directly as the gas phase. In high organic content, oil-prone, source rocks, however, hydrocarbons will migrate predominantly as the oil phase (Hunt 1979).

The efficiencies of primary hydrocarbon migration are variable and depend on a number of factors. Studies have shown that the amount of petroleum generated and the efficiency of its expulsion increase with depth and temperature. Extraction efficiencies are higher for saturated hydrocarbons than for aromatics, and petroleum generation is much more volu-minous from Type II kerogen bearing source rocks than it is from Type III kerogen equiv-alents. Furthermore, petroleum products are generally expelled from their source rocks only once the volume of hydrocarbons has matched or exceeded the volume of pore space in the source rock, and subsequent to water having migrated from the system (Leythauser et al. 1984; Mackenzie et al. 1987).

At a scale of tens or hundreds of kilometers, where secondary migration is important, the main factor that controls hydrocarbon flow is the existence of a pressure differential in the host rock. Particularly relevant to petroleum migration are the concepts of fluid overpres-sure (described in more detail in Section 3.3.2 of Chapter 3) and buoyancy pressure. If the removal of pore water is impeded by low permeability in, for example a fine grained shale, then compaction will be retarded and fluid pressures will increase to values above normal hydrostatic pressures. Fine and coarse sediment will expel pore waters at different rates during burial and will, therefore, com-pact along different fluid pressure gradients. Overpressured fluids will tend to exist in the finer grained sediments. Figure 5.33a shows actual measurements of fluid pressures in a series of drill wells from the Orinoco Basin and the hydrostatic overpressures that occur in the poorly drained shale units relative to the well drained sandstone. This information allows the migration pathways to be identified and shows that the main sandstone aquifer can be fed by fluids from both above and below. Several factors actually control the stresses that cause hydrostatic fluid overpressures in rocks and these include rapid sediment depo-sition, thermal expansion of fluids, tectonic compression, and the actual generation of low density oil and gas in the source rock.

Buoyancy pressure is another factor that controls secondary migration of petroleum. Buoyancy pressure arises from the fact that oil and water have different densi-ties – consequently, a water column overlain by an oil column in a homogeneous sediment will exhibit different pressure gradients, as illustrated in Figure 5.33b. Because water is denser than oil, fluid pressures in the water column will increase more rapidly with depth than those in the oil column. The magnitude of buoyancy pressure is a function of the height of the less-dense fluid column, as shown in Eq. (5.11) (from Gluyas and Swarbrick 2004);

$$P_b = gh(\delta_w - \delta_o) \qquad (5.11)$$

where P_b is the buoyancy pressure, g is the acceleration due to gravity, h is the height of the less-dense fluid column and δ_w and δ_o are the densities of water and oil respectively. As a consequence of the buoyancy effect, petroleum will tend to rise in a sedimentary sequence. The rate at which a fluid will flow through a porous medium is expressed in terms of Darcy's Law, determined experimentally by Henry Darcy in 1856, and then modified by Morris Muskat (Eq. (5.12); Muskat 1937) to make it more generally applicable as a tool to quantify fluid flow rates as a function of vis-cosity, permeability, and pressure differentials in natural petroleum forming environments:

$$Q = -kA(P_x - P_y)/\mu L \qquad (5.12)$$

where Q is the fluid discharge rate, k is the per-meability of the porous medium, A is the cross sectional area of fluid flow, $(P_x - P_y)$ is the pres-sure drop along the flow path, μ is the viscosity and L is the flow path length. It is considera-tions such as these that allow petroleum scien-tists to track petroleum migration pathways as a guide to where oil and gas resources might be located.

5.4.2.3 Entrapment of Oil and Gas

Primary migration of hydrocarbons typically occurs over short distances (hundreds of meters or less) and is constrained by the prox-imity of the first available aquifer to the source rock. Secondary migration, by contrast, occurs over tens, and occasionally even hundreds, of kilometers and is only constrained by the presence of a trap which allows the petroleum

Figure 5.33 (a) Actual measurements of excess hydrostatic pressures (i.e. overpressure) in three drill wells from the Orinoco basin and the anticipated fluid flow lines in the sequence of alternating shale and sandstone. Source: After Hunt (1979). (b) Schematic diagram explaining the concept of buoyancy pressure in a petroleum reservoir beneath a seal. The reservoir comprises mainly water below the oil-water contact (OWC) and oil above it. Fluid pressure gradients differ because water and oil have different densities. The maximum buoyancy pressure (P_b) is in the oil immediately beneath the seal. Source: Modified after Gluyas and Swarbrick (2004).

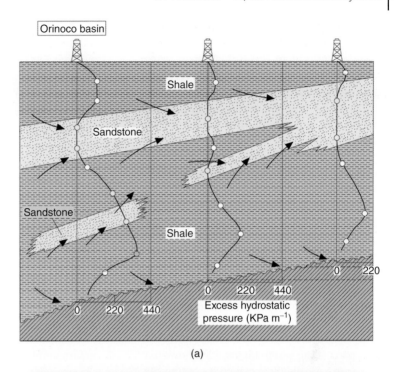

(a)

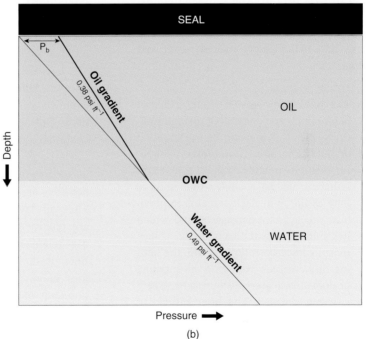

(b)

to accumulate in a constrained volume of rock over a fixed period of time. Hydrocarbon traps are critical to the formation of viable oil and gas deposits and take the form of any geological feature that either reduces permeability of the reservoir or provides a physical barrier that impedes the further migration of fluid. Evaporite layers, for example, may be good hydrocarbon traps as they are laterally extensive and have virtually zero permeability because of the ability of halite to deform plastically even at moderate temperatures. Many of the giant oil fields of the Arabian (Persian) Gulf are trapped by evaporite sequences (Box 5.4), as are some of the North Sea gas deposits. In general, however, structural features such as faults and anticlines tend to be the dominant hydrocarbon traps. This section briefly outlines some of the geological scenarios that represent potential traps for hydrocarbon deposits in various parts of the world. Figure 5.34 shows that there are basically three categories of trap site, in order of decreasing importance: structural, stratigraphic, and a less common miscellaneous class that includes hydrodynamic and asphalt traps.

(i) Structural traps

Structural traps result from sediment deformation and generally provide physical barriers that prevent the continuation of fluid flow along an aquifer. A fault, for example, might juxtapose a reservoir sediment against a shaley unit and, as long as the fault itself remains impervious, will act as a barrage behind which hydrocarbons will accumulate. Since most hydrocarbons flow up-dip, the folding of a reservoir rock into an anticline or dome-like feature also provides a very efficient trap site. The huge oilfield at Prudhoe Bay in Alaska represents an example where petroleum is trapped by a combination of both folding and faulting of the reservoir sediments. The scale of hydrocarbon accumulation in this setting can be estimated by considering the closure volume of the structure and also the spill point, beyond which oil will continue its migration into another aquifer system (Figure 5.34a-iii).

A good example of an oilfield where entrapment of petroleum has occurred in thrust-related anticlines is the Painter Reservoir field in the Wyoming–Utah thrust belt, USA. Figure 5.35a shows a section through this field where it is apparent that oil and gas are trapped in anticlines within the Nuggett sandstone unit, the latter having been buckled by second order thrusts that are spawned off the major Absaroka Thrust below the reservoir (Frank et al. 1982; Gluyas and Swarbrick 2004).

Salt domes and diapirs also represent effective petroleum trap sites and are important in many of the major petroleum basins of the world. When salt layers underlie more dense rock strata, the resultant gravitational instability causes the salt to pierce the overlying beds creating ideal trap sites for hydrocarbons. The lateral and upward migration of salt layers, or *halokinesis*, is a result of density contrasts between a salt layer (with a density of approximately $2.2\,g\,cm^{-3}$ and comprising predominantly halite and sylvite) and overlying clastic or carbonate sediments (where, on compaction, densities are typically 2.5 to $2.7\,g\,cm^{-3}$). The salt layer is gravitationally buoyant relative to the overlying strata and, over relatively short periods of geological time, it domes upwards to eventually form diapiric features that can penetrate hundreds of meters into the overburden (Figures 5.34aii and 5.35b). A wide variety of structural trap sites results from the plastic deformation of salt layers and these have been widely implicated in the pooling of oil and gas in, for example, the Gulf Coast plays of the USA and also in the Zagros mountains of Iran.

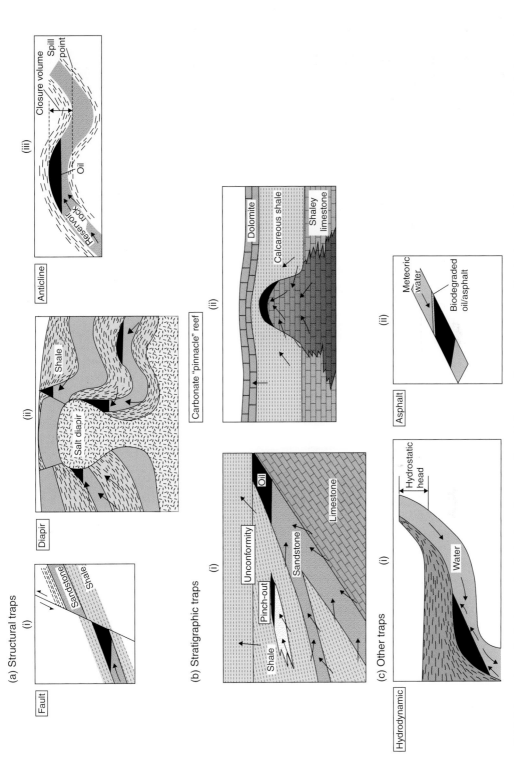

Figure 5.34 Schematic geological scenarios for hydrocarbon trap sites. (a) Structural traps represented by faults, diapiric features, and anticlinal or dome like structures. (b) Stratigraphic traps represented by unconformities, pinch-outs, and carbonate "pinnacle" reefs. (c) Other features, such as hydrodynamic and asphalt traps. Source: After Hunt (1979), Bjørlykke (1989).

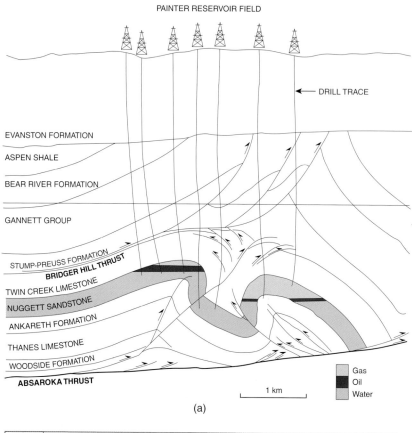

(a)

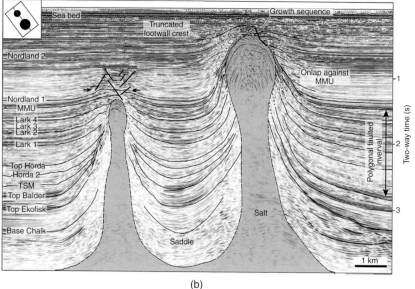

(b)

Figure 5.35 Examples of structural traps for petroleum. (a) Cross section through the Painter Reservoir (Wyoming) showing petroleum that has pooled in anticlines formed in response to secondary thrusts that have spawned off the principal Absaroka Thrust below the oil fields. Source: Modified after Gluyas and Swarbrick (2004). (b) Interpreted seismic profile showing the Pierce salt diapirs in the east Central Graben of the North Sea oilfield. Source: After Carruthers et al. (2013), image courtesy of Joe Cartwright. These features, and many others like them, create potential petroleum trap sites in sedimentary basins undergoing halokinesis.

(ii) Stratigraphic traps

Stratigraphic traps arise entirely from sedimentological features and can be represented by features such as sediment pinch-out zones, where permeability is reduced as one sediment facies grades into another, or unconformities where the reservoir rock sub-outcrops against a unit of reduced flow capability (Figure 5.34b-i). The Athabasca tar sands (see Section 5.4.4 below) are believed to have resulted from migration and entrapment of hydrocarbons along a major unconformity in the early Paleogene period. In addition to sandstones, carbonate rocks represent important reservoirs because the high solubility of minerals such as calcite promotes secondary porosity and migration of hydrocarbons. Sedimentary features such as limestone "pinnacle" reefs (Figure 5.34b-ii), are important trap sites in carbonate sequences, especially in the giant Middle East deposits.

The geometry of sedimentary units as well as the nature of lateral facies changes and diagenesis all play important roles in pinch-out stratigraphic traps. A good example of petroleum entrapment by stratigraphic pinch-out is the giant East Texas oilfield, located in upper Cretaceous sediments of the Woodbine Group (Ambrose et al. 2009). The East Texas oil field, hosted in Woodbine sandstones, is the largest in the contiguous USA and has produced more than 5 billion barrels of oil – it was discovered by wildcat drilling in the early 1930s (Dolson 2016). Although explorers understood the concept of structural traps at this time, stratigraphic traps were generally difficult to identify and the discovery was entirely serendipitous. The Woodbine sandstones of the East Texas Basin were sourced from the north and deposited into a fluvio-deltaic setting that progressively thins toward the east against the Sabine dome (Figure 5.36). The latter uplift occurred continuously during Woodbine deposition such that upper sequences are progressively cut out toward the east and truncated by a regional unconformity, the latter overlain by a limestone unit (the Austin Chalk Group) that acted as an impervious seal against further secondary migration. The angular unconformity responsible for trapping the huge volumes of oil in the East Texas field is very subtle and low-relief, making this type of stratigraphic trap difficult to recognize during exploration.

(iii) Other traps

Other trap sites of lesser importance include sites where oil migration is diverted by a strong flow of groundwater which, as explained previously, will water-wet the sediment pore spaces and impede the flow of hydrocarbons. These are known as hydrodynamic traps and result from the shear stresses set up at the oil–water contact in aquifers where strong water flow is occurring. Groundwater will also interact chemically with reservoir oil, resulting in oxidation, biodegradation, and the formation of a bituminous asphalt layer at the interface. This layer, known as an asphalt trap, may itself act as a barrier beneath which hydrocarbons may accumulate (Figure 5.34c-ii).

5.4.3 Coalification Processes

The majority of the world's energy requirements are still obtained from the burning of coal and lignite, a fact that has major deleterious implications for pollution and global warming. The two largest producers of coal in the world are China and the USA, although Australia ranks top as an exporter of high quality coal.

As mentioned previously, most coal is derived from Type III kerogen and generally represents the in situ accumulation of land-derived vegetation subjected to alteration

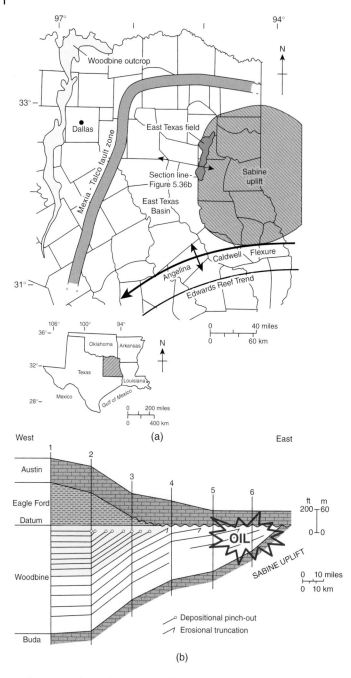

Figure 5.36 (a) Location of the East Texas oil field in the context of the principal structural features of the East Texas basin. (b) Schematic section (not to scale) of a sequence stratigraphic reconstruction of the Woodbine Group showing the sandstone sequences that host petroleum and the pinch-out of these units against a regional unconformity. Source: Modified after Ambrose et al. (2009).

and compaction. Figure 5.27 shows that coal is derived from a peat precursor and, with burial or coalification, is progressively transformed to a series of products ranging from lignite and sub-bituminous coal (known as "brown coals"), to bituminous coal and finally anthracite (the "hard coals"). The formation of peat is an essential initial step in the process and involves restrained biochemical degradation of plant matter without rampant oxidative and bacterial decay. Coalification follows once peat is covered by overburden and subjected to an increase in temperature with progressive burial.

Box 5.4 Fossil Fuels – Oil and Gas: The Arabian (Persian) Gulf, Middle East

---

The Arabian (or Persian) Gulf Basin contains well over half of the conventionally recoverable oil reserves of the world, and also huge reserves of natural gas. This extraordinarily rich basin extends over a length of more than 2000 km from Oman in the south to Syria and southeast Turkey in the north (Figure 5.4.1). Production has been derived from some 250 reservoirs in the region, of which more than 80% are in Jurassic–Cretaceous sediments. The Middle East is unique for the size of its individual deposits, with 14 fields having recoverable reserves in excess of 10 billion barrels (Shannon and Naylor 1989). These include such giants as the Ghawar field in Saudi Arabia and Burgan field in Kuwait (both with more than 70 billion barrels of oil) and the Rumaila and Kirkuk fields in Iraq, each with around 15–20 billion barrels of oil (Tiratsoo 1984). Many of the oil fields in the Gulf region are also associated with substantial gas resources. In addition, however, some very large unassociated gas fields also occur in the region, such as the North Dome field off Qatar. These gas deposits occur in the Permian Khuff limestone and are, therefore, older and underlie the main oil deposits. Unassociated gas deposits in many other parts of the world appear to be related to coal measures of Carboniferous age. There are several reasons why the Middle East is so well endowed in hydrocarbon resources and these are briefly discussed below.

The oil fields of the Arabian Gulf are located in late Mesozoic to early Paleogene reservoir sediments that were deposited on a continental shelf forming off the northeastern margin of the Arabian portion of Pangea. A significant proportion of these sediments are limestones, indicating deposition at equatorial paleolatitudes in the Tethyan seaway (North 1985; Glennie 2000). As Pangea fragmented the Arabian land mass became part of a downgoing slab involved in subduction to the north and east. Subduction ceased in Pliocene times with the formation of the Zagros mountains in present day Iran, and this hydrocarbon-rich basin was preserved largely intact. It is likely that other parts of the Tethyan seaway, probably equally well endowed, were consumed by subduction in Cenozoic times, perhaps accounting for the fact that the western and eastern extremities of the seaway contain fewer oil and gas deposits.

The tectonic and sedimentological regimes active in this environment produced at least three major source rock sediments in the Arabian foreland, along the axis of the present day basin and in the orogenic belt of Iran. Source rocks are all marine in origin and developed over the entire extent of the Arabian platform. Oil and gas are hosted in a variety of different lithologies and the principal reservoirs become progressively younger from southwest to northeast. The thick carbonate and sandstone reservoirs that formed in the basin are also characterized by a remarkable lack of intergranular cements and a high degree of primary porosity, making them very effective for the migration and ultimate concentration of hydrocarbons. This also contributes to good recovery rates and the fact that only a few wells need to be drilled in order to exploit a given field. Periodic uplift and subaerial exposure also resulted in the deposition of numerous evaporitic horizons that make good seals for oil and gas entrapment. Sedimentation in the Jurassic and Cretaceous periods was characterized by a marine transgression over much of the Arabian platform. Initially widespread carbonate sedimentation occurred, and this was followed by increasing aridity and evaporite formation. In detail, sea level fluctuations and climate changes gave rise to complex interfingering of lithologies (Shannon and Naylor 1989), as illustrated in the environmental interpretations for the region during mid-Jurassic and mid-Cretaceous times (Figure 5.4.2).

(Continued)

Box 5.4 (Continued)

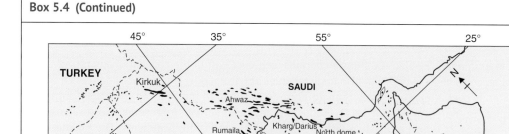

Figure 5.4.1 Location and distribution of some of the major oil and gas fields in the Arabian Gulf. Source: After Shannon and Naylor (1989).

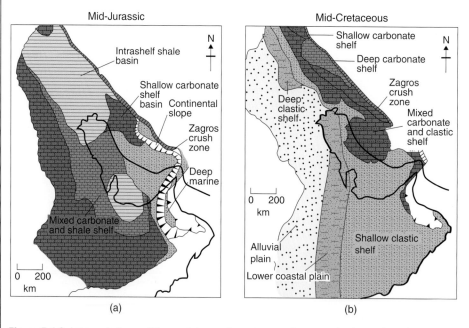

Figure 5.4.2 Interpretations of the evolving sedimentary environment in the region of the Arabian Gulf during (a) mid-Jurassic and (b) mid-Cretaceous times. Source: After Shannon and Naylor (1989).

The structural development of the region was also conducive to the formation of large and effective trap sites. Three types of structural traps are recognized in the Arabian Gulf Basin. These include large, but gently dipping, anticlinal warps which caused oil entrapment mainly in Jurassic limestones (e.g. Ghawar in Saudi Arabia); diapiric salt dome structures arising from basement evaporitic sequences which deform Cretaceous and Jurassic limestones (e.g. Burgan

---

Box 5.4 (Continued)

in Kuwait); and steeply dipping anticlines in the foothills of the Pliocene mountain building episode in Iraq and Iran (e.g. Kirkuk in Iraq). The presence of thick evaporitic sequences is considered to be one of the most important features of the region because the salt horizons are believed to have absorbed much of the shortening that accompanied subduction to the north and east of the basin. This had the effect of preserving the large gently-dipping anticlinal structural traps that characterize the southwestern portions of the region and accounts for the huge sizes (up to 100 km long) of many of the oil fields (Shannon and Naylor 1989). In addition to the presence of salt, some of the other geological features of the Middle East that have made it so productive for oil and gas include a long history of marine sedimentation at relatively low paleolatitudes, numerous transgressive and regressive sedimentary cycles resulting in a close association between source and reservoir rocks, and the occurrence of widespread seals, in the form of evaporitic sequences, that prevent leakage of the oil and gas.

Coal is a sedimentary rock (Figure 5.37) that contains more than 50% by weight of carbonaceous material and can readily be burnt (McCabe 1984). It forms by the compaction of peat in an environment represented by a well vegetated land surface that is saturated with water. This environment is loosely termed a "swamp" but should not necessarily imply a tropical or equatorial climate as it is now well known that many coals have accumulated in cold climates at mid- to high latitudes. The greatest concentrations of peat are presently accumulating in Russia (with about 60% of the world's resources; McCabe 1984) at latitudes of around 50–70 °N. High rainfall is not an environmental prerequisite, although abundant free-standing water promotes plant growth and also submerges dead vegetation, retarding the rate of decomposition. Furthermore, swampy environments often contain waters that are both anaerobic and acidic, which promotes the preservation of organic material by minimizing oxidation and destroying bacteria. Tropical rain forests seldom form peat accumulations because the high temperatures promote rapid oxidation and decay of organic material. The best conditions for peat accumulation reflect a balance between organic productivity and decay, with the added prerequisite of low sediment input necessary to keep dilution of the peat by clastic material to a minimum.

5.4.3.1 Coal Characteristics

Coal is a heterogeneous sedimentary rock that reflects both the different sedimentological regimes within which peat forms and the varied vegetation types from which it is derived. Peat deposits are known to have formed in many settings including lacustrine, fluvial, deltaic, and beach-related (Galloway and Hobday 1983; McCabe 1984; Selley 1988). Peat and coal deposits, therefore, exhibit a wide range of shapes, and are also characterized by the facies variations typical of any sedimentary rock. Figure 5.38 illustrates the stratigraphic variations one might expect to see in peat, as a function of vegetation type and fluctuating water levels.

The vegetation shown in Figure 5.38 applies specifically to the Everglades in Florida, where marine transgression results in lateral facies changes in a swamp environment and deposition of different peats that originate from a variety of plant types. Coals that form by compaction of peat will also, therefore, exhibit marked variations in composition and vegetative make-up. This characteristic has led to a classification of coals, the main elements of which are shown in Table 5.2. In the same way that a rock is made up of minerals, a coal is made up of "macerals," which are components of plant tissue, or their degraded products, altered during compaction and diagenesis.

Figure 5.37 Bituminous coal seam from the Witbank area, South Africa, hosted in carbonaceous shales and deltaic sandstones of the Permian Ecca Group. Source: Photograph courtesy of Bruce Cairncross.

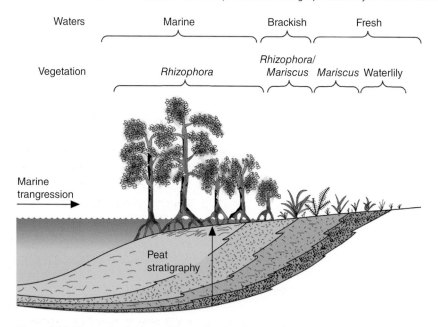

Figure 5.38 Generalized scheme illustrating the nature of peat stratigraphy in terms of vegetation type and transgressive/regressive effects in a swamp environment. Source: After Spackman et al. (1976).

The main maceral groups are *vitrinite* (also called huminite in brown coals), made up of woody material (branches, roots, leaves, etc.), and *exinite* (also called liptinite), which comprises spores, cuticles, waxes, resins, and algae. A third grouping is also recognized, termed *inertinite*, which comprises mainly the remains of oxidized plant material and fungal remains. Each maceral group is further subdivided into individual macerals, but listing

Table 5.2 Coal classification in terms of maceral groups and lithotypes.

Maceral group		Lithotype		
Vitrinite (or huminite)	Cell walls of vascular plants	Vitrain	Vitrinite + <20% exinite	Humic coals
		Clarain	Layers of all three maceral groups	
Exinite (or liptinite)	Spores, cuticles, waxes, resins, and algae	Durain	Mainly inertinite and exinite	
		Fusain	Fusinite	
Inertinite	Oxidized and burnt (fusinite) plant material	Cannel coal	Spore-dominated exinite	Sapropelic coals
		Boghead coal	Algae-dominated exinite	

of these is beyond the scope of this book and can be obtained in more detailed texts, such as Stach (1975). Different combinations of maceral groups identify the coal lithotypes (i.e. analogous to a rock type) and these are also listed in Table 5.2.

The majority of coals are humic (i.e. made up of macroscopic plant parts), are peat derived, and can be subdivided into four lithotypes. These are vitrain, clarain, durain, and fusain, whose maceral compositions are shown in Table 5.2. The coal lithotypes are important in determining its properties, in particular factors of economic importance such as calorific value and liquefaction potential. In general, the highly reflective, more combustive coals are made up of vitrain, whereas the duller, less combustive coals comprise essentially durains. It should be noted, however, that coal lithotypes should not be confused with coal rank, which separates low calorific value brown coals from high heat capacity hard coals, essentially on the basis of degree of compaction or burial depth.

A small proportion of coals are sapropelic and formed largely from microscopic plant remains such as spores and algae (exinite). The sapropelic lithotypes, also referred to as cannel and boghead coals (Table 5.2), did not form from peat. They can be found overlying humic coals, forming in muddy environments

as the swamp was flooded and buried, and occasionally are linked to the formation of oil shales (see Section 5.4.4 below).

In addition to the maceral-based classification, coals can also be subdivided in terms of their chemical composition, specifically their carbon, hydrogen, and oxygen contents. Figure 5.39 is a ternary plot of normalized C, H, and O in which the compositional fields of peat and the coals are shown in relation to those for oil and oil shales. As peat is progressively compacted and coalification commences, the carbon content increases markedly, whereas hydrogen remains fairly constant and oxygen decreases. Hydrogen only starts to diminish noticeably once the hard coals start to develop. The chemical changes that define coalification are accompanied by the production of CO_2 and CH_4 gas. The source rocks for some of the gas fields in the North Sea, for example, are believed to be underlying Westphalian coal beds.

Another important property of coal, but one which is unrelated to its organic or maceral characteristics, is ash content. As peat forms it inevitably incorporates inorganic, clastic detritus from a variety of sources. During coalification this detritus is diagenetically altered to form various authigenic mineral components such as quartz, carbonate, sulfide, and clay minerals. Once the coal is combusted

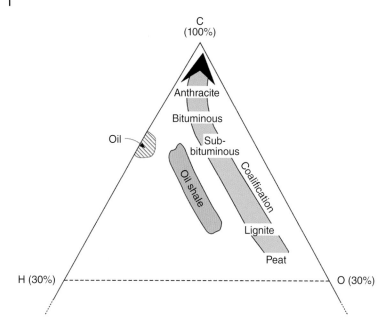

Figure 5.39 Ternary C–H–O plot showing the compositional trend accompanying coalification and the comparison between coals and oil and oil shale compositions. Source: After Forsman and Hunt (1958).

these components remain behind to make up its ash content. Low ash content coals are obviously economically advantageous as this parameter translates favorably into higher calorific values, better grindability, and less deleterious environmental effects such as sulfur emission.

5.4.3.2 A Note Concerning Formation of Economically Viable Coals

The formation of economically viable coals is determined by processes active during both initial peat accumulation and later coalification. In the latter case bituminous coals and anthracite are the high rank products of thermal maturation and burial, and are characterized by the highest calorific values and lowest volatile contents. Progressive compaction will transform most humic coals to these products, which, if extensive and thick enough, will likely be economically viable. In addition to burial, some anthracites are known to be the products of tectonic deformation or devolatilization by nearby igneous intrusions, a good example of the former being the Pennsylvanian coals of the USA, which grade laterally into anthracites as they approach the Appalachian fold belt. Controls

on the formation of economically viable coals during initial peat accumulation are, however, more difficult to constrain. Deltas have traditionally been regarded as the optimal environment for development of viable coal seams (Figure 5.37), although this concept is now questioned because the high clastic sediment input in this setting generally results in peat dilution and lower quality, high ash coals (McCabe 1984). Environmental models now favor the notion that optimal peat accumulation represents a discrete event that was temporally distinct from those of underlying and overlying clastic deposition. The notion of an environment where peat accumulation is optimized, but not diluted by clastic input, is best accommodated by the concept of raised and/or floating swamps. Floating swamps refer to buoyant peats that develop on the surface of shallow lakes to eventually accumulate and accrete on the lake margins (Figure 5.40). Raised swamps are seen as the final product in a continuum of processes which extend from initial development of shallow lakes with floating peat, followed by formation of well-vegetated, low-lying swamps, and eventually to raised swamps as accretion of organic

Figure 5.40 Environmental model for peat accumulation, showing the evolutionary development of swamp types and the formation of raised swamps in which extensive, thick peats, devoid of major sediment input, form. Source: After Romanov (1968), McCabe (1984).

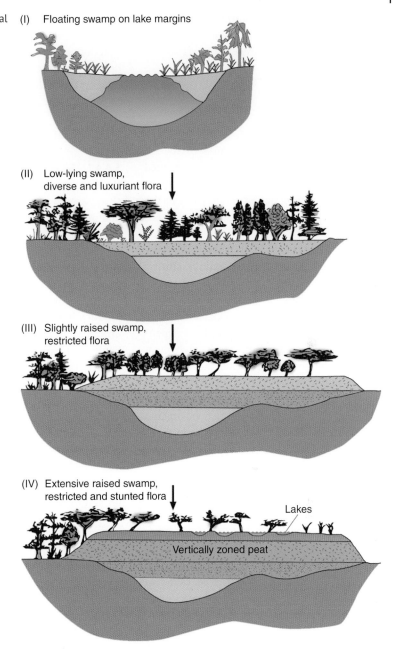

(I) Floating swamp on lake margins

(II) Low-lying swamp, diverse and luxuriant flora

(III) Slightly raised swamp, restricted flora

(IV) Extensive raised swamp, restricted and stunted flora

Lakes

Vertically zoned peat

material eventually sculpts a landscape that is elevated with respect to the sediment substrate. The latter environment is obviously one in which clastic sediment input is minimized. This evolutionary sequence, illustrated in Figure 5.40, serves as a model for the depositional environment in which optimal formation of thick peat, and subsequently economic coal horizons, might form. Raised swamps can form in deltaic environments, but the above model applies specifically to those settings in which sediment input is minimized and the resultant coals are low ash in character. Examples of raised swamps include the

Klang-Langat delta in Malaysia and areas of Sarawak where peat accumulation of more than $2\,mm\,yr^{-1}$ is occurring (McCabe 1984).

5.4.4 Unconventional Hydrocarbons – Shale Gas, Oil Shales, and Tar Sands

Since around 2005 the dynamics of the oil and gas industry in the USA has changed dramatically with the rapid development of shale gas extraction across many parts of the country. Access to huge, previously inaccessible, oil and gas resources has been facilitated by new drilling techniques based essentially on horizontal coring and hydraulic fracturing (colloquially termed "fracking") of petroleum source rocks. These innovative technologies, albeit controversial, have unlocked substantial resources that have reduced the US's dependence on petroleum imports (Stephenson 2015). Although organic matter accumulates in most sediments, it is shale or mud rock that typically contains the highest hydrocarbon contents. Mud rocks that are rich in organic matter are called sapropelites and represent the most common source rocks generating petroleum products on maturation. Conventional oil and gas resources involve the migration of hydrocarbons out of the source kitchen into more porous reservoir rocks, as described in Section 5.4.2 above. Unconventional petroleum plays, however, are hosted in shales that have matured to produce either oil or gas, or both, but the source rocks are so impervious that the hydrocarbons are essentially unable to move away. Drilling a single, or even multiple, vertical holes into an oil/gas shale horizon will not allow the petroleum to be extracted because rock permeabilities are extremely low. The development of techniques whereby horizontal drilling into a flat-lying oil/gas shale from an originally vertical drill hole, and the subsequent creation of multiple, sub-vertical, hydraulically-induced fractures along the drilled out section will, however, artificially enhance permeability so that oil and gas can be extracted. It is the *lack of migration* of petroleum from a source rock that defines the viability of an unconventional deposit, and it is the development of innovative drilling technologies that has allowed access to new resources that may, if necessary, have the potential to out-produce all pre-existing conventional output.

5.4.4.1 Shale Gas and Oil Shales

Routine commercial production of gas from shale is now widespread in parts of the USA and the process, though controversial, is likely to start in other parts of the world too. The description of petroleum bearing shales is perhaps best undertaken by referring to an example, such as the Eagle Ford Formation in southwest Texas, which has the potential to produce both gas and oil from a sedimentary unit that is both source and reservoir. The Eagle Ford Formation is a carbonate-rich shale (marl) of Cretaceous age – the lower parts of the sequence comprise organic rich sediments, with TOC contents of around 6%, that were deposited in a markedly anoxic environment (Bryndzia and Braunsdorf 2014). The marl outcrops in an arc extending through San Antonio and Austin, and dips toward the southeast becoming progressively more deeply buried in the same direction (Figure 5.41). With progressive burial and heating the rocks also become more mature, with the oil window developing down-dip at depths of around 4000 ft, and wet gas and dry gas maturation occurring at systematically greater depths.

The Eagle Ford Formation is a highly favored target for both shale gas and oil shale resources, and has already proven viable for both commodities – the host rock marl is also conducive to hydraulic fracturing because its properties and composition provide a harder and more brittle rock than more typical clay-rich shales (Stephenson 2015).

5.4.4.2 Tar Sands (or Oil Sands)

Tar sands are often confused with oil shales, but they are in fact quite different. Tar sands

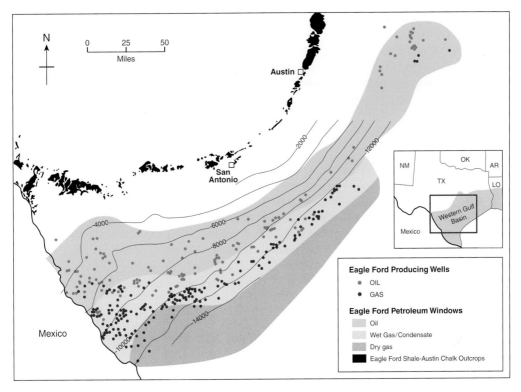

Figure 5.41 Map showing the outcrop of the Eagle Ford Formation and its progressive burial shown as contour lines marking the depth (in feet) from the surface to the top of the formation. The rocks become progressively more mature with depth toward the southeast – the petroleum windows for oil, wet gas, and dry gas are shown as zones of progressive burial and maturation. Producing oil and gas wells from the Eagle Ford shale are also shown. Source: After Stephenson (2015).

are sandstones within which bitumen or asphalt occurs. The bitumen in tar sands is a solid or semi-solid hydrocarbon either derived by biodegradation of crude oils, or forming directly as a high molecular weight hydrocarbon without ever having been a light oil. Many of the world's large tar sand deposits occur in foreland basins where oils accumulate along the cool basin flanks and are subjected to severe biodegradation. Biodegradation is the process whereby microorganisms (usually bacteria, but also yeast and fungi) with a metabolic affinity for hydrocarbons degrade oils under both aerobic and anaerobic conditions. Biodegradation occurs at temperatures below about 80 °C, this being the temperature above which these microorganisms can no longer survive. The processes by which oil

is biodegraded to bitumen is convoluted and several microorganisms and metabolic pathways are implicated. One important process is that by which bacteria and archaea combine syntrophically to biodegrade oil. Bacteria, responsible for the oxidation of alkanes to form H_2 and CO_2, coexist with archaea that utilize the H_2 produced to reduce CO_2 to CH_4. This methanogenic biodegradation of oil to gas leaves behind a residue of heavier molecular weight bitumen (Larter and Head 2014).

In certain instances, tar sands that are near the surface can be extracted and processed to produce oil and a range of other hydrocarbon by-products. The best known deposits are the Athabasca tar sands of northern Alberta, Canada, where bitumen is formed in a variety of sandstone reservoirs of late Devonian to

Cretaceous age (Larter and Head 2014). The resource potential of Athabascan tar sands is vast, with some estimates indicating volumes comparable to the rest of the world's remaining reserves of conventional petroleum (Bjørlykke 2015). The principal tar sand reservoir is the McMurray Formation, a shallow, fluvio-deltaic sandstone reservoir into which oil migrated from older source rocks during the Laramide orogeny. Uplift of the reservoir in Eocene times resulted in widespread biodegradation of the oil to form a viscous bitumen residue. Currently, near-surface bitumen is extracted through highly mechanized, open-cast operations that excavate the tar sand and then separate the hydrocarbon using a process of hot water washing and flotation. Deeper tar sand deposits in Alberta are extracted in situ by heating the bitumen with steam that is pumped into the reservoir along horizontal wells – heating lowers the viscosity of the bitumen which can then be pumped out conventionally through parallel wells. This process is known as *steam assisted gravity drainage* (or SAGD).

As with oil shales, though, technological and environmental difficulties provide major obstacles to more widespread production from this resource at present. Both tar sands and oil shales might, however, represent significant petroleum resources in the future.

5.4.5 Gas Hydrates

A fairly recent discovery in the permafrost regions of the world, as well as in ocean sediments sampled during the Deep Sea Drilling Program, is the presence of vast resources of hydrocarbon locked up as frozen gas hydrates (Kennicutt et al. 1993). Gas hydrates (or clathrates) are crystalline aqueous compounds that form at low temperatures when the ice lattice expands to accommodate a variety of gaseous molecules, the most important of which, for the purposes of this discussion, is methane (CH_4). A methane hydrate forms when sufficient CH_4 is present in water, at the appropriate pressure and temperature,

to form a solid compound with the ideal composition $4CH_4 \cdot 23H_2O$ (i.e. the unit cell comprises 46 water molecules with up to 8 methane molecules). Methane hydrates are now known to occur in both continental and oceanic sedimentary basins, with the latter holding by far the major proportion of known resources. The stability of methane hydrate is constrained to a narrow range of pressure and temperature (Figure 5.42a) and its preservation is, therefore, restricted to sedimentary environments where temperatures are cold ($<2\,°C$) and pressures moderate (>50 atm or around 500 m water depth). Continental shelves, polar regions, and deep continental lakes and seas are the preferred sedimentary environments where methane hydrates accumulate. Other gases, including CO_2, H_2S, and C_2H_6, also form gas hydrates with a similar structure. Gas hydrates represent an important natural sink for greenhouse gases and their preservation may be disconcertingly dependent on future trends of global warming.

It is of interest to consider the conditions under which methane hydrates form in the ocean basins of the world. Figure 5.42a shows a phase diagram with the H_2O ice–water and CH_4 hydrate–gas boundaries marked. If sufficient methane and pore water are available methane hydrates will stabilize in ocean floor sediment where temperatures are around $3–4\,°C$ at pressures of about 50 atm (equivalent to about 500 m water depth). Gas hydrates will melt if either the temperature increases or the pressure decreases. The thickness of the methane hydrate stability zone in oceanic sediments will depend on the pressure (an increase in which has the effect of stabilizing hydrates to lower temperatures) and the geothermal gradient, the latter dictating the rate at which sediment is heated with progressive burial. For a constant geothermal gradient, the thickness of the methane hydrate layer will increase directly as a function of water depth. Figure 5.42b shows this graphically for a typical ocean–sediment profile on the continental shelf. It illustrates the initiation of gas hydrate

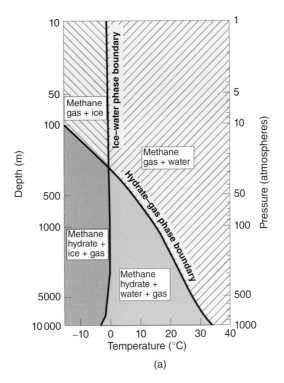

(a)

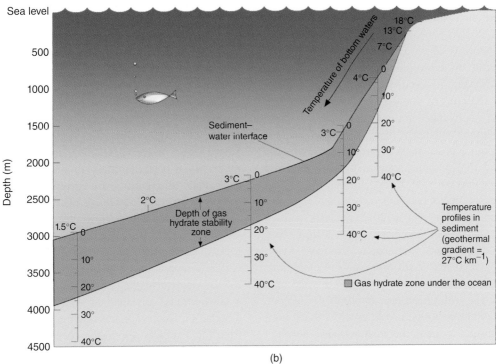

(b)

Figure 5.42 (a) Phase diagram illustrating the regions of methane hydrate stability under most natural conditions in the near-surface environment. Source: After Kvenvolden and McMenamin (1980). (b) Profile across a typical ocean–sediment interface in a continental margin setting showing the progressive increase in the width of the gas hydrate stability zone in the ocean sediment with increasing depth of sea water. Source: After Kvenvolden (1988).

formation in sediment under about 500 m of water and the progressive increase in thickness of the hydrate stability zone with depth. Under 3000 m of water a methane hydrate zone almost 1000 m thick could form if the ingredients were present. This particular profile was calculated for a geothermal gradient of $27\,^{\circ}C\,km^{-1}$, but it would be thinner if the host sediments were subjected to higher geothermal gradients.

The formation of methane hydrates on the Earth's surface is ultimately determined, not only by the specific P–T conditions required to stabilize them, but also by the availability of methane and water in the sedimentary host environment. Significant volumes of methane in particular are needed in the host sediments, be they continental or marine, in order for viable gas hydrate deposits to form. Most gas hydrates in oceanic settings are dominated by pure methane ($>99\%$ CH_4) which is generated essentially in situ by the microbial reduction of CO_2 to CH_4 (Kvenvolden 1995). Microbial methanogenesis is a complex process and one such pathway has already been described in Section 5.4.4, whereby tar (bitumen) formation is attributed to the biodegradation of oil by syntrophic bacteria and archaea. However, in certain settings, such as parts of the Gulf of Mexico and the Caspian Sea, the gas hydrates are impure (i.e. they contain methane together with significant proportions of other gases such as propane and ethane) and the methane appears to have migrated to its present site, having formed by thermal degradation of heavier hydrocarbons at greater depths. Some gas hydrate deposits contain methane that has been derived from both microbial and thermogenic sources (Kvenvolden 1995) – both processes are ubiquitous in hydrocarbon source rocks and this accounts for the widespread occurrence of gas hydrate methane deposits in cold environments, both subaquatic and in the polar regions.

On a global scale, the volume of natural gas present in methane hydrate reservoirs was originally thought to be substantially greater than the total, presently defined, fossil fuel reserve. However, it is now known that Natural Gas Hydrate (NGH) deposits require very specific P–T conditions of formation and the resource is substantially lower than previously thought – currently estimated at around 1.4 to 1.7×10^5 trillion cubic feet (Milkov 2004). NGH deposits are defined as those that are potentially economically viable to extract – one cubic meter of solid methane hydrate could contain as much as $160\,m^3$ of methane (Ruppel 2011), so they do, clearly, represent a significant energy resource for the future. Although NGH resources could exceed the total conventional natural gas reserves of the world, the technical and economic difficulties of exploiting gas from this resource remains highly challenging and for this reason it is unlikely that NGH will supplant conventional gas extraction procedures for some time to come. Nevertheless, several countries have embarked upon research and test programs aimed at identifying resources and extractive technologies for future methane hydrate exploitation. In 2013, for example, Japanese researchers drilled into an offshore gas hydrate deposit located in ocean sediments beneath several hundred meters of sea water in the Nankai Trough, and were successful in extracting methane after depressurizing the host clathrate. A more easily accessible resource has also been located in permafrost hosted gas hydrates along the Alaskan North Slope, where some 85 trillion cubic feet of methane has been identified. Production tests to evaluate this resource are currently underway and will take advantage of the existing petroleum extractive infrastructure in place around Prudhoe Bay (Ruppel 2011). Finally, Russian producers are reported to have inadvertently tapped into a NGH resource during the commercial extraction of methane from a conventional gas field at Messoyakha in Siberia. It is believed that extraction of free gas from this field was responsible for the depressurization of an overlying NGH layer which, upon dissociation, contributed additional methane to the conventional production. The actual proportion of methane contributed from the gas hydrate component, however, remains uncertain (Collett 1992).

5.5 Summary

Sedimentation is a fundamental geological process that takes place over much of the Earth's surface, in response to the pattern of global tectonic cycles. A number of very important mineral and fossil fuel commodities are concentrated during the formation of sedimentary rocks. Placer processes are important during clastic sedimentation and deposits form by the sorting of light from heavy particles. A number of hydrodynamic mechanisms, such as settling, entrainment, shear sorting, and transport sorting, are responsible for the concentration of different commodities in a variety of sedimentary micro- and meso-environments. At a larger scale, placer deposits form mainly in fluvial, deltaic, and beach-related environments and include concentrations of gold, diamonds, tin, zirconium, and titanium.

Chemical sedimentation, where dissolved components precipitate out of solution from sea water, also gives rise to a wide range of important natural resources. Precipitation of ferric iron from sea water is one of the principal mechanisms for the formation of iron ore deposits. BIFs formed in a variety of oceanic settings, mainly in the Paleoproterozoic and prior to the onset of significant levels of oxygen in the atmosphere. Ironstone deposits tend to form mainly in Phanerozoic times. Other types of metal concentration linked to syn-sedimentary and early diagenetic chemical processes occur in carbonaceous black shales (Ni, Co, Cr, Cu, Mn, Zn, Ag, etc.) and in Mn nodules on the ocean floor. Accumulation of phosphorus and the development of phosphorites in ocean settings is a process that is linked both to direct chemical precipitation from sea water and to biological mediation. The upwelling of cold currents onto the continental shelf, and the associated biological productivity, are processes implicated in the formation of most of the world's important phosphate deposits, the latter important for fertilizers and global food production. Chemical sediments formed by high rates of evaporation of sea water and lacustrine brines host most of the world's Na, K, Li, borate, nitrate, and sulfate resources.

Fossil fuels, in the form of oil, natural gas, and coal, are the most strategic and valuable natural resource as they provide the world with most of its combustion-derived energy. Oil and natural gas are derived in the early stages of sedimentation by biodegradation of planktonic organisms in marine environments to form oil-prone kerogen and biogenic gas. Further burial results in subsequent chemical and thermal modification of kerogen to produce wet gas and a liquid condensate (oil), typically at burial depths of 3000–4000 m and temperatures of 100–150 °C. Further burial results only in the production of more dry gas. Burial of vascular land plants gives rise initially to the formation of peat and the production of mainly dry gas. Further compaction of peat gives rise to coals, the rank of which is a function of progressive burial. Other forms of fossil fuel, possibly important for the future but largely unexploited at present for technological and environmental reasons, include oil shales and tar sands, as well as methane hydrates in permafrost regions and ocean floor sediments.

Further Reading

A large volume of literature is available in the fields of sedimentology and fossil fuels, but less so with respect to chemical sedimentation and metallogeny. The following books and reviews are subdivided in accordance with the main topics covered in this chapter.

Sedimentology and Placer Processes

Allen, P.A. (1997). *Earth Surface Processes.* Oxford: Blackwell Science 404 pp.

Carling, P.A. and Breakspear, R.M.D. (2006). Placer formation in gravel-bedded rivers: a review. *Ore Geology Reviews* 28: 377–401.

Leeder, M. (1999). *Sedimentology and Sedimentary Basins: From Turbulence to Tectonics.* Oxford: Blackwell Science 592 pp.

Slingerland, R. and Smith, N.D. (1986). Occurrence and formation of water-laid placers. *Annual Reviews of Earth and Planetary Sciences* 14: 113–147.

Chemical Sedimentation and Ore Formation

Maynard, J.B. (1983). *Geochemistry of Sedimentary Ore Deposits.* Springer-Verlag 305 pp.

Mel'nik, Y.P. (1982). *Precambrian Banded Iron Formations.* New York: Elsevier 310 pp.

Melvin, J.L. (ed.) (1991). *Developments in Sedimentology, Evaporites, Petroleum and Mineral Resources,* vol. 50. New York: Elsevier Chapter 4.

Parnell, J., Ye, L., and Chen, C. (1990). *Sediment-Hosted Mineral Deposits.* Special Publication, 11, International Association of Sedimentologists. Oxford: Blackwell Scientific Publications 227 pp.

Warren, J.K. (2006). *Evaporites: Sediments, Resources and Hydrocarbons.* Springer 1035 pp.

Young, T.P. and Gordon Taylor, W.E. (1989). *Phanerozoic Ironstones.* Special Publication 46. London: The Geological Society of London 257 pp.

Fossil Fuels

Bjørlykke, K. (ed.) (2015). *Petroleum Geoscience – From Sedimentary Environments to Rock Physics,* 2e. Springer 662 pp.

Dolson, J. (2016). *Understanding Oil and Gas Shows and Seals in the Search for Hydrocarbons.* Springer 485 pp.

Engel, M.H. and Mack, S.A. (1993). *Organic Geochemistry: Principles and Applications.* New York: Plenum Press 861 pp.

Glennie, K.W. (1998). *Petroleum Geology of the North Sea: Basic Concepts and Recent Advances.* Oxford: Blackwell Science 636 pp.

Gluyas, J. and Swarbrick, R. (2004). *Petroleum Geoscience.* Oxford: Blackwell Science 359 pp.

Levorsen, A.I. (1967). *Geology of Petroleum,* 2e. W.H. Freeman and Company 723 pp.

North, F.K. (1985). *Petroleum Geology.* London: Allen & Unwin 607 pp.

Selley, R.C. (1998). *Elements of Petroleum Geology,* 2e. New York: Academic Press 470 pp.

Thomas, L. (1992). *Handbook of Practical Coal Geology.* New York: Wiley.

Ward, C.R. (1984). *Coal Geology and Coal Technology.* Oxford: Blackwell Science 345 pp.

Part IV

Global Tectonics and Metallogeny

6

Ore Deposits in a Global Tectonic Context

TOPICS
Patterns in the distribution of mineral deposits
Continental growth and the supercontinent cycle
Geological processes and metallogenesis
evolution of the hydrosphere and atmosphere
global heat production and mantle temperature
global tectonic trends and mantle convection
continental freeboard and eustatic sea-level changes
Metallogeny through time
the Archean Eon
the Proterozoic Eon
the Phanerozoic Eon
Plate tectonic settings and ore deposits – a summary

6.1 Introduction

Most of the world's great mineral districts and deposits are the products of a fortuitous superposition of geological processes that resulted in the formation and preservation of anomalous concentration of metals, usually over short periods of geologic time. The preceding chapters have shown that the formation of ore deposits is, nevertheless, related to much the same sort of processes that gave rise to the formation of normal igneous, sedimentary, and metamorphic rocks in the Earth's crust. The fact that a close relationship exists between rock-forming and ore-forming processes means that metallogeny must be relevant to understanding the nature of crustal evolution through geologic time (Groves and Bierlein 2007; Bradley 2011). Conversely, crust-forming processes and the global plate tectonic paradigm have become indispensable to the broader understanding of how ore deposits form, and no modern economic geologist can practice successfully without an appreciation of the geodynamic evolution of the continents and oceans. Consequently, this chapter summarizes some of the recent thinking that relates the formation of ore deposits to global tectonics and continental evolution. There are essentially two ways of doing this. One is to chart continental evolution and place ore deposits into a secular and tectonic framework (e.g. Meyer 1981, 1988; Veizer et al. 1989; Barley and Groves 1992; Windley 1995; Groves and Bierlein 2007; Bradley 2011; Cawood and Hawkesworth 2015). The other is to empirically describe ore deposits in the context of the tectonic environment and host rocks in which they occur (e.g. Mitchell and Garson 1981; Hutchison 1983; Sawkins

Introduction to Ore-Forming Processes, Second Edition. Laurence Robb.
© 2021 John Wiley & Sons Ltd. Published 2021 by John Wiley & Sons Ltd.

1990). The latter approach has been covered in considerable detail in Sawkins's (1990) *Metal Deposits in Relation to Plate Tectonics*, and readers are referred to this source for more detailed information. The former approach is a more difficult undertaking and requires a thorough knowledge of the evolution of continents with time, a topic that becomes progressively less well constrained – and more controversial – the further back in Earth history one looks. There are, however, significant benefits to be gained by understanding ore formation in terms of the secular evolution of the Earth's crust and atmosphere. This is the approach that has been adopted below, even though it is necessarily cursory and in places speculative.

6.2 Patterns in the Distribution of Mineral Deposits

It has long been recognized that mineral deposits are not randomly distributed, either in time or in space, and that broad patterns exist when relating deposit types to crustal evolution and global tectonic setting. The original compilations of ore deposit type as a function of geological time by Meyer (1981, 1988) have been re-examined by Veizer et al. (1989), Barley and Groves (1992), Kerrich et al. (2005), Groves and Bierlein (2007), and Bradley (2011) and reveal consistent patterns. Figure 6.1, for example, distinguishes between the secular distribution of metal deposits formed in *orogenic* settings and those formed in *anorogenic* environments and in continental basins. Although this compilation is a "first-order" generalization, it is nevertheless clear that deposit types classified as orogenic (including lode-gold or orogenic gold ores, volcanogenic massive base metal sulfides, and the porphyry-epithermal family of base and precious metal deposits) were preferentially formed, or better preserved, either in the late stages of the Archean Eon (between 3000 and 2500 Ma) or in the Phanerozoic Eon (between 541 Ma and the present day). The Proterozoic Eon, between 2500 and 541 Ma,

preserves fewer of these types of deposits. By contrast, the majority of metal deposits that are associated with so-called "anorogenic" magmatism (such as anorthosite hosted Ti deposits and the Olympic Dam and Kiruna type iron oxide–copper–gold [IOCG] ores), as well as sediment hosted ores (such as sedimentary exhalative [SEDEX] type Pb–Zn deposits and stratiform sediment-hosted Cu deposits), are preferentially hosted in Proterozoic rocks (Figure 6.1). An explanation for this observation, as discussed below (see Sections 6.4.3 and 6.5), is that ore formation might be closely linked to the so-called "supercontinent cycle," (Figure 6.2) which describes the broad scale amalgamation and dispersal of the major continental fragments with time (Rogers and Santosh 2004; Bradley 2011; Nance et al. 2014; Cawood and Hawkesworth 2015). It is, for example, evident that peaks in the production of anorogenic and continental sediment-hosted metal deposits (Figure 6.1, lower histogram) coincide with periods of crustal stability and the existence of large, continental amalgamations such as Nuna (also called Columbia) in the Mesoproterozoic, Rodinia in the Neoproterozoic, and Pangea in the early Mesozoic. Periods of large-scale continental fragmentation, by contrast, appear to be orogenically more active and give rise to a different suite of ore deposit types. This pattern is demonstrated in more detail in Figure 6.2 which shows the size of various ore deposit types as a function of time, but subdivided according to whether orogenic processes were predominantly convergent or divergent – reflecting, respectively, the pattern of supercontinent assembly and dispersal (Groves and Bierlein 2007; Cawood and Hawkesworth 2015). The distribution of ore deposit types over time reveals a broad association with the periodicity of supercontinent assembly and breakup, with fewer deposits forming in periods of supercontinent stasis. Porphyry and epithermal ore forming systems, orogenic Au and Mississippi Valley type (MVT) Pb–Zn deposits are typically associated

Figure 6.1 Distribution of ore deposit styles, classified according to orogenic (upper set of histograms) and anorogenic/continental sediment-hosted metal deposits (lower set of histograms) as a function of time. The length of each histogram bar is an estimate of the proportion of ore formed over a 50 million year interval relative to the global resource for that deposit type. Periods of supercontinent amalgamation are shown as Pangea, Rodinia, and Nuna, with decreasing levels of confidence further back in time. Source: The diagram is originally after Meyer (1988) and modified after Barley and Groves (1992).

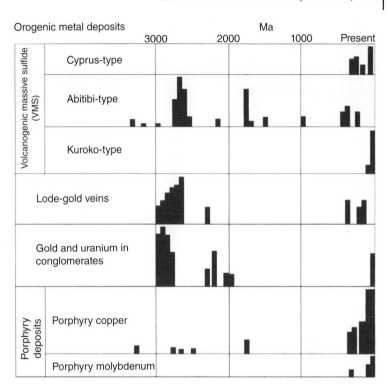

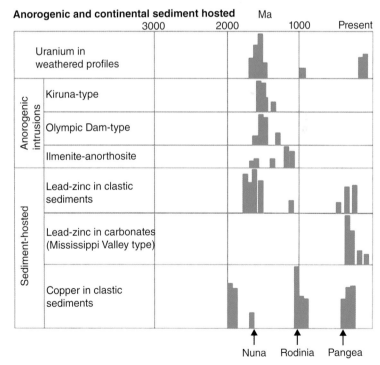

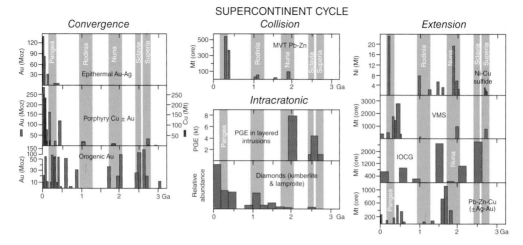

Figure 6.2 Temporal distribution of ore deposit types as function of the assembly and dispersal of supercontinents. Source: After Cawood and Hawkesworth (2015). Periods of supercontinent amalgamation in this diagram are shown as Pangea, Rodinia, Nuna, and Sclavia/Superia, with decreasing levels of confidence further back in time.

with convergent tectonic regimes (Figure 6.2) and mineralized districts tend to form close to craton margins and paleo-sutures. By contrast, the volcanogenic massive sulfide (VMS)-SEDEX family of base metal deposits, as well as some magmatic Ni–Cu and IOCG ores, are preferentially linked to divergent plate margins, whereas diamondiferous kimberlite and platinum group metal-rich layered mafic intrusions show a closer relationship to intracratonic settings. These patterns are discussed in more detail in Section 6.5.

It should be emphasized at this stage that the relationships observed in Figures 6.1 and 6.2 could also be due, at least in part, to a preservation factor. The existence of numerous arc-related orogenic deposits in the late Archean, for example, reflects the preservation of greenstone belts in stable shield areas of the world. In Proterozoic times, by contrast, the preponderance of collisional orogenies and buoyant crust might have uplifted and eroded similar mineralized arcs so that many of the related near-surface ores were destroyed. The products of young Cenozoic orogenies are also preserved because many of them have not yet suffered collision and uplift and, therefore, remain intact.

6.3 Continental Growth and the Supercontinent Cycle

6.3.1 Estimations of Continental Growth Rates

The relationships illustrated in Figures 6.1 and 6.2 suggest that the patterns of continental growth with time are likely to be important to any consideration of continental assembly and dispersal and, therefore, metallogeny. Estimating these patterns, however, is problematic because of the difficulties in calculating the rates of recycling (or destruction) of continental material relative to new crust formation. The availability, globally, of numerous age determinations of detrital zircons, however, has yielded a database that reflects the formation and preservation of continental crust as a function of time (Figure 6.3a) which is, in turn, useful in assessing the secular patterns of crustal growth and accretion (Hawkesworth et al. 2010, 2016).

It is apparent that detrital zircon ages define a series of well-defined peaks that coincide broadly with the ages of supercontinents (Figure 6.3a), the latter defined as the periodic assembly of most of the Earth's continental

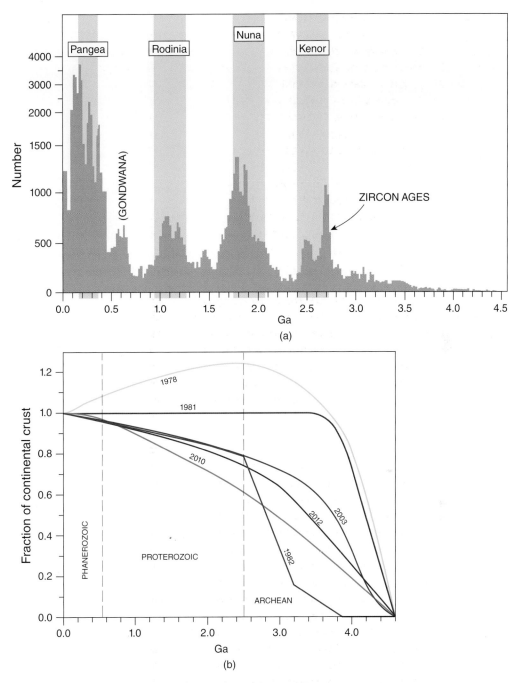

Figure 6.3 (a) Histogram showing peaks of detrital zircon ages over time and the coincidence of those peaks with the tenure of supercontinents. Source: After Hawkesworth et al. (2016). (b) Models for crustal growth rates showing the change in ideas over time. Source: After Korenaga (2013) – early models (1978, 1981, and 1982) reflect ideas largely based on crustal distribution; later models (2003, 2010, and 2012) were based on isotopic constraints. The curve marked 2012 (after Dhuime et al. 2012) proposes a decrease in crustal growth rate due to increased crustal destruction rates from around 3.0 Ga and is considered to indicate the onset of conventional plate tectonics at that time.

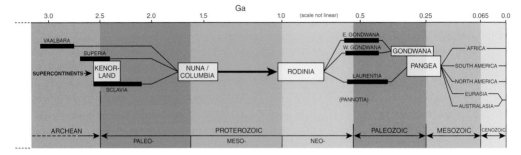

Figure 6.4 Schematic diagram showing the timing of amalgamation and dispersal of supercontinents. A number of different names have been allocated to the various supercontinent cycles and the ones adopted here are from Hawkesworth et al. (2016).

fragments into a single, essentially contiguous, landmass. At least four cycles of supercontinent formation have been recognized over the past 3 billion years (Figure 6.4 and Section 6.3.2), excluding Gondwana which in certain compilations is considered, not as a separate supercontinent, but as part of the eventual amalgamation that resulted in Pangea. The supercontinents depicted here are – from youngest to oldest – Pangea (which existed c. 0.35–0.15 Ga), Rodinia (c. 1.25–0.95 Ga), Nuna (or Columbia at c. 2.1–1.7 Ga), and Kenorland (c. 2.8–2.5 Ga). In Figure 6.3a, the zircon age peaks do not necessarily imply that the growth of continental crust was marked by pulses of magmatic activity, but rather reflect periods of enhanced preservation of crust relative to a global crustal growth rate that was relatively constant – it follows that preservation of crust was best achieved during periods of lithospheric stability and supercontinent stasis (Hawkesworth et al. 2010). The zircon age peaks are, thus, considered to be a proxy for the timing and periodicity of supercontinent amalgamation and dispersal.

Crustal growth rates, expressed as the fraction of crust formed at any given time relative to the present day volume, are difficult to quantify and a variety of models have been proposed (Figure 6.3b). Early models advocated contrasting scenarios in which crustal growth was either very rapid early on and then slowed, or took place in spurts at various periods, notably the late Archaean. More recent

models, based on isotopic data that reflect the relative proportions of juvenile versus recycled crust, suggest that a more rapid pace of growth occurred in the Archaean up to 3.0 Ga, with a lower but more constant rate thereafter (McCulloch and Bennett 1994; Dhuime et al. 2012). The change in crustal growth at 3.0 Ga is taken as an indication of the onset of global plate tectonics, after which a marked increase in crustal destruction rates occurred. In this model it is also envisaged that some 70% of the present day volume of crust had formed by 3.0 Ga, but that a substantial proportion of this was recycled during the late Archaean and Proterozoic, accounting for the relatively small volumes of preserved older continental material seen today.

It is interesting to note that the Phanerozoic Eon appears to have been a period of time that was particularly productive in terms of ore deposit formation, suggesting enhanced crustal growth. The latter is probably not the case but rather supports the notion of preferential preservation of younger ore-forming environments compared to older sequences which become either progressively more eroded, or buried, with time. The patterns of crustal growth and supercontinent cyclicity play a fundamental role in the formation and distribution of ore deposit types over geological time – these processes are now sufficiently well understood to allow the subdivision of Earth history into five geodynamic episodes (after Hawkesworth et al. 2016), namely:

STAGE 1 – HADEAN (>4.0 Ga)

differentiation of the core-mantle system; early crust comprised largely of a mafic magma "ocean"; minor continental crust suggested by rare, ancient detrital zircons; chemical sedimentation likely.

STAGE 2 – PRE-PLATE TECTONIC REGIME (4.0–3.0 Ga)

mantle upwelling; surface gravitationally unstable; formation of greenstone belts and TTG (tonalite-trondhjemite-granodiorite) magmatism.

STAGE 3 – EARLY PLATE TECTONICS (3.0–1.7 Ga)

cratonic stabilization; thickening of continental crust; initial formation of potassic granitoids; initiation of "hot" subduction; collisional orogenics developed; shallow slab break-off; amalgamation of early continental landmasses; intracratonic clastic sedimentation.

STAGE 4 – EARTH'S MIDDLE AGE (1.7–0.75 Ga)

lithospheric stability; "anorogenic" magmatism (such as "massif" type anorthosites); thick, buoyant continental crust; amalgamation of long-lived supercontinents.

STAGE 5 – MODERN PLATE TECTONICS (0.75 Ga to present)

existence of large, composite plates; extensive "cold" subduction and arc formation; thick, strong crust; high-P metamorphic assemblages.

The definition of these stages undoubtedly warrants refinement and further study and, in particular, the debate as to how plate tectonics evolved over time, remains contentious (Korenaga 2013). However, the pattern of crustal evolution is now sufficiently well understood to provide a better understanding of supercontinent cyclicity and metallogenesis.

6.3.2 Supercontinent Cycles

The existence of supercontinents and the cycles within which continental fragments coalesce and disperse is fundamental to all Earth processes, including metallogeny. The topic of supercontinent cyclicity is fundamentally important to understanding the entire Earth system – early attempts to discern these cycles (Windley 1995; Rogers 1996; Unrug 1997) are continuously being improved as more precise paleomagnetic and geochronological studies provide better constraints to the documentation of apparent polar wander paths. The pattern of crustal evolution in the Phanerozoic Eon is well understood, involving the progressive amalgamation of large continental fragments to form the supercontinent of Pangea. The latter existed during Permian and Triassic times and was subsequently dispersed to form the present day continental geography (Figure 6.4). A relatively high degree of confidence marks the reconstruction of continental configuration during the Phanerozoic Eon and geoscientists generally agree on the paleogeography of landmasses such as Gondwana, Laurentia, and Pangea. As one goes back in time into the Precambrian, however, the situation becomes progressively more uncertain and disagreement often accompanies the reconstruction of supercontinental geometry. There are many reasons for this, including the difficulties of acquiring and accurately dating apparent polar wander paths, the high rates of destruction of continental and oceanic crust, and the progressive deformation and burial of crust with time.

Despite these difficulties, progress has been made in reconstructing continental paleogeography back through geological time although, as mentioned, the nature of supercontinent cyclicity in the Proterozoic and Archean Eons remains tenuous. The historical development of the concept of supercontinent cyclicity is described by Nance and Murphy (2013) and Figure 6.4 presents a scheme suggesting the existence of four supercontinents since approximately 3.0 Ga. The following discussion is based on this fourfold subdivision, from oldest to youngest:

6.3.2.1 Kenorland

The oldest amalgamated, but now dispersed, continental landmass, originally referred to as *Ur* (Rogers 1996), was suggested to have comprised the ancient Kaapvaal and Pilbara cratons of southern Africa and Western Australia, respectively, as well as parts of India and Antarctica. Ur is now generally referred to as *Vaalbara* (Button 1979; Cheney 1996; Wingate 1998; Evans et al. 2000), the evidence for which stems largely from the remarkably similar basement and early supracrustal geology of the Kaapvaal and Pilbara cratons, as well as a well constrained overlapping paleomagnetic pole reconstruction at 2.78–2.71 Ga (De Kock et al. 2009). Vaalbara is believed to have coalesced from about 3000 Ma but had broken up by c. 2100 Ma, prior to emplacement of the Bushveld Complex on the Kaapvaal Craton. Other continental amalgamations, in the form of *Superia* and *Sclavia* (Figure 6.4), are also suggested to have existed in Neoarchean and Paleoproterozoic times (Bleeker 2003; Nance et al. 2014) and it is the amalgamation of these fragments that is believed to have resulted in the existence of an early supercontinent called *Kenorland* (Williams et al. 1991). The evidence for Kenorland (or simply Kenor) is conjectural and it may not, in fact, have been a supercontinent at all since it is not clear that Vaalbara was part of this amalgamation in late Archean times.

6.3.2.2 Nuna (also referred to as Columbia)

The Mesoproterozoic Era is believed to have included a more robust supercontinent referred to as either *Nuna* (an acronym for **N**orthern **Eu**rope and **N**orth **A**merica) or *Columbia* (Piper 1976; Hoffman 1988; Park 1995). The evidence for Nuna is again based on comparative geology and observed through alignment and synchroneity of features such as mafic dyke swarms, suture zones, and orogenic belts. There are approximately 35 known fragments of Archean-aged continents preserved on Earth at present (Bleeker 2003) and these would appear to have been largely cohesive at

around 1.8–1.7 Ga, to form what might have been the first genuine supercontinent, Nuna. The existence of numerous compressional orogenic regimes in the period 2.1–1.7 Ga (such as Limpopo, Eburnian, Trans-Hudson, Yavapai, Kola, Svecofennian, Akitkan, Rondonian, Arunta, etc.) provide the evidence that amalgamation of Archean continental fragments was near global in its extent at this time. Reconstructions of Nuna at 1740 Ma and 1590 Ma are shown in Figure 6.5, attesting to the long-lived duration of this supercontinent. Nuna is believed to have consolidated and grown over a considerable period of Earth's "Middle Age," creating lithospheric stability for around 400 million years (between c. 1.8 and 1.4 Ga) through late Paleoproterozoic into Mesoproterozoic times (Gower et al. 1990; Zhao et al. 2004; Zhang et al. 2012; Evans and Mitchell 2011). Equally widespread rifting along continent margins is evidenced by the development of major sedimentary sequences such as the Belt-Purcell Supergroup in Laurentia and the Telemark Supergroup in Baltica, and marks the beginning of Nuna breakup from around 1.5 Ga. Fragmentation of Nuna was also accompanied by extensive anorogenic, anorthositic, and granite-rhyolite magmatism, and then terminated by the emplacement of voluminous mafic dyke swarms such as the 1.27 Ga Mackenzie dykes in Laurentia (Gaál 1987, 1990; LeCheminant and Heaman 1989).

6.3.2.3 Rodinia

The supercontinent cycle repeated itself during the latter stages of the Proterozoic Eon. There are indications that a Neoproterozoic supercontinent, now referred to as *Rodinia* (McMenamin and McMenamin 1990), started to assemble from about 1200 Ma, and then dispersed again by about 700 Ma. As with Nuna, it is the global extent of compressional orogenic belts active in the period 1400–1000 Ma – the so-called "Grenvillian" belts that include examples such as the Sunsas of Brazil, Kibaran of central Africa and Sveconorwegian of Scandinavia – that attest to the re-amalgamation

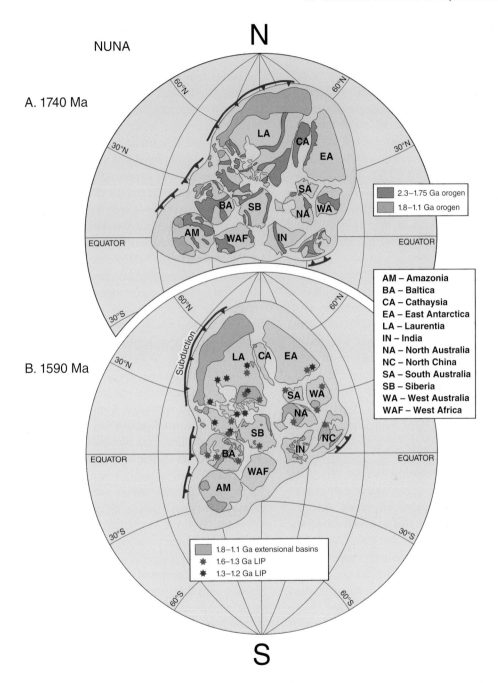

Figure 6.5 Reconstructions of the mainly Paleoproterozoic supercontinent Nuna at two periods, 1740 and 1590 Ma, suggesting a long-lived duration of this entity. Source: After Zhang et al. (2012). Breakup and dispersal of Nuna took place largely in Mesoproterozoic times.

of dispersed Nuna continents by the early Neoproterozoic (Murphy and Nance 1991; Meert and Torsvik 2003; Nance et al. 2014). The position of the Congo Craton within Rodinia remains problematic with some workers suggesting that it was entirely separate during peak stasis (Scotese 2009). Rodinia, like Nuna, appears to have been a long-lived entity providing lithospheric stability between approximately 1.0 and 0.7 Ga. The supercontinent broke up episodically over a protracted period that may have exceeded 200 million years. Episodic breakup may have included rifting of Laurentia (the North American Craton) from East Gondwana at 800–700 Ma to form the Panthalassic Ocean (Figure 6.6a), and Laurentia from West Gondwana somewhat later at around 600–550 Ma (Hoffman 1991; Dalziel et al. 2000; Meert and Torsvik 2003; Scotese 2009). The breakup and dispersal of Rodinia coincided with the advent of catastrophic climate change and dramatic biological diversification associated with the so-called "Snowball Earth" events of the Cryogenian and Ediacaran Periods of the Neoproterozoic Era.

6.3.2.4 Pangea

The onset of the Paleozoic Era was characterized initially by the convoluted rearrangement of Rodinian continental fragments. Like the Grenvillian orogenies that assembled Rodinia, it was the globally extensive Pan-African orogenies that were responsible for coalescence at the onset of the Phanerozoic. Several workers have suggested that the two major Rodinian fragments, essentially West and East Gondwana, rotated apart and then re-collided whilst at the same time consuming the errant Congo Craton to form a transient supercontinent that has been called "*Pannotia*" (Figures 6.4 and 6.6b; Powell 1995; Scotese 2009). The name Pannotia is not widely recognized and some workers refer to the entity as "Greater Gondwana," regarding it as a stepping stone to the eventual construction of *Pangea* in Mesozoic times. Pannotia broke up almost as soon as it had assembled, with Laurentia and

Baltica rifting away from Gondwana to form the Iapetus and Tornquist Seas in Cambrian times. By the early Paleozoic Era, Rodinia (or Pannotia) had broken up into 4 large continents, Gondwana, Laurentia, Baltica, and Siberia, with further fragmentation (to form smaller continental masses such as Avalonia, Cathaysia, and Cimmeria) taking place in the mid- and late-Paleozoic (Scotese 2009).

The most recent supercontinent, Pangea, coalesced from the fragments of Rodinia, and assembled as Laurasia (a combination of Laurentia and Eurasia) and Gondwana re-united by progressive subduction of the Rheic Ocean in late-Paleozoic and Mesozoic times. The geological, paleontological, and paleomagnetic evidence for the existence of a combined landmass in Permian–Triassic times is robust, but the details of how it was assembled are complex (Powell et al. 1993; Murphy et al. 2009; Torsvik and Cocks 2013). In late Devonian times, at 370 Ma, the large continental landmass of Gondwana was centered on the South Pole and separated from Laurasia by the Rheic Ocean (Figure 6.7a). Collision of Laurasia and Gondwana occurred in late Carboniferous times (c. 320 Ma) to form Pangea, that initially existed largely in the southern hemisphere. Pangea drifted northwards as a cohesive block such that in the early Mesozoic, at 250 Ma (Figure 6.7b), significant portions of the supercontinent lay at equatorial latitudes. This explains why significant accumulation of lignite and coal occurred in Permian times and also explains the formation of extensive evaporitic sequences and petroleum source rocks in marginal marine settings at, and subsequent to, this era (Scotese 2009; Torsvik and Cocks 2013).

Pangea was not a long-lived entity and it started to breakup already in the early Mesozoic, soon after Neotethys started to form, and the Panjal Traps were emplaced in northern India at 290 Ma. Major fragmentation, however, only commenced along the fossilized Rheic suture when Laurasia separated from West Gondwana at c. 195 Ma, to

Figure 6.6 (a) Reconstruction of Rodinia at 750 Ma, prior to the onset of breakup. (b) Reconstruction of Pannotia (or Greater Gondwana) at 545 Ma. Source: After Nance et al. (2014).

(a) RODINIA
- early to mid-Neoproterozoic (ca. 725 Ma.)

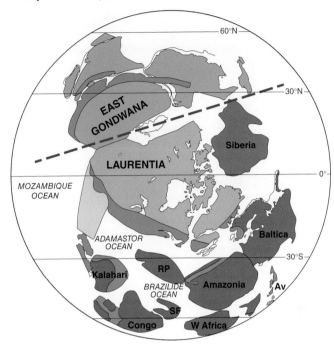

(b) PANNOTIA
- end Neoproterozoic (ca. 545 Ma.)

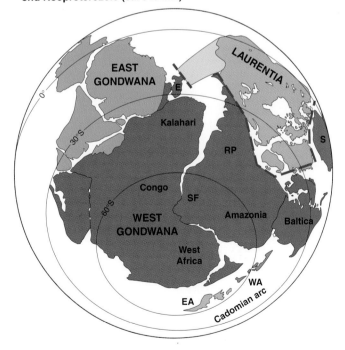

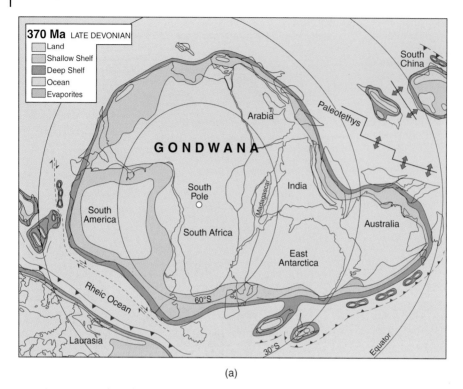

(a)

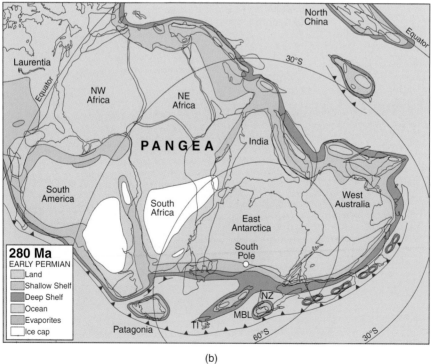

(b)

Figure 6.7 (a) Continental paleogeography in the late Devonian Period, at 370 Ma. Gondwana was a very large continental landmass, but not a supercontinent, centered on the South Pole. (b) Reconstruction of Pangea at 280 Ma. Source: After Torsvik and Cocks (2013).

form the incipient Atlantic Ocean. Further breakup separated West Gondwana from East Gondwana (to form the Indian Ocean) as the Neotethyan Ocean was consumed from 170 Ma onwards. Pangea breakup was catalyzed by the upwelling of major plumes that impacted the base of the lithosphere, resulting initially in the Central Atlantic Magmatic Province (CAMP) at 201 Ma and then the Karoo large igneous province (LIP) at 183 Ma (Torsvik and Cocks 2013). Pangea has now dispersed, and its remnants occur as five major continental landmasses that form the present day geographic configuration (Figure 6.4). Some continents are currently undergoing reassembly, with collision of India and Asia, and the partial amalgamation of Africa and Europe, having taken place over the past 50 million years.

6.4 Geological Processes and Metallogenesis

The review by Barley and Groves (1992) suggests that there were several fundamental geological processes that influenced the pattern of global metallogeny. These include factors such as the evolution of the hydrosphere–atmosphere, a secular decrease in global heat production, the nature of mantle convection, eustatic sea level changes, and, as discussed in Section 6.3, the supercontinent cycle and long-term global tectonic trends. These are briefly discussed below, with an explanation of how and why they influence the temporal and geographic distribution of ore deposit types.

6.4.1 Evolution of the Hydrosphere and Atmosphere

The changing budget of O_2, and to a lesser degree CO_2, in the Earth's atmosphere with time has played a role in the formation of ore deposits, especially those related to redox processes and to the weathering and erosion of continental crust. During the Archean, the atmosphere contained very little free molecular oxygen and what little did exist was the result of inorganic photodissociation of water vapor. A reduced atmosphere in the Archean helps to explain many of the features of ore formation at that time, including the widespread mobility of Fe^{2+} and development of Algoma and Superior type banded iron-formations, as well as the preservation of detrital grains of uraninite and pyrite in sedimentary sequences such as those of the Witwatersrand and Huronian basins. The transition from the Archean to the Proterozoic Eon, at 2500 Ma, broadly coincides with the beginnings of an increase in atmospheric oxygen (Figure 6.8), with the Great Oxidation Event (GOE) having now been dated at c. 2.3 Ga (Bekker et al. 2004; see Section 5.3.1). This event is possibly related to the evolution of primitive life and the widespread development of cyanobacteria capable of photosynthetically producing oxygen. An increase in partial pressures of O_2 in the atmosphere at this time can also be estimated from the mobility of Fe^{2+} in paleosols (the remnants of weathering horizons or soils preserved in the rock record; Rye and Holland 1998).

The interval between about 2500 and 2000 Ma is known as the period of "oxy-atmoinversion" and is widely believed to coincide with a significant rise in atmospheric oxygen levels. It is also the period *after* which development of Algoma and Superior type banded iron-formations (BIFs) and related bedded manganese deposits generally ceased (although some BIFs may have formed soon after the GOE – see discussion in Section 5.3.1). Another significant increase in atmospheric oxygen levels coincided with the end of the Proterozoic Eon, when macroscopic, multicellular life forms (Metazoan fauna which require oxygenated respiration) proliferated. In addition, the early Phanerozoic is also the time when vascular plants evolved and this, too, would have contributed to the increase in atmospheric oxygen levels (Figure 6.8). In

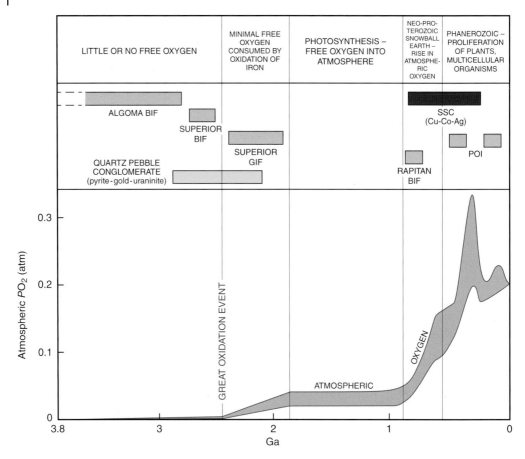

Figure 6.8 Estimated rise of atmospheric oxygen over the past 3.8 Gyr on the basis of five evolutionary stages, after Holland (2006). Also shown are the approximate age ranges over which ore deposit types sensitive to atmospheric oxygen levels and redox processes formed (see Chapter 5). BIF – banded iron-formation; GIF – granular iron-formation; SSC – sediment-hosted stratiform copper; POI – Phanerozoic oolitic iron.

contrast with the net addition of O_2 in the atmosphere, CO_2 contents appear to have progressively decreased with time. The high levels of CO_2 in the early atmosphere were a product of extensive volcanism and outgassing during the Archean, but levels have dropped steadily, but episodically, as a result of carbonate deposition in sediments. The evolution of the Earth's atmosphere and, in particular, the global O_2 and CO_2 budgets, are essentially non-recurrent and irreversible and contribute to the time-specific occurrence of ore deposit types such as banded iron-formations and Witwatersrand type placers (Figure 6.8).

6.4.2 Secular Decrease in Global Heat Production and Mantle Temperature

The production of heat from the Earth, mainly from the decay of long-lived radioisotopes such as U, Th, and K, was significantly greater (by a factor of two to three times) in the Archean than it is today. This pattern, like that for atmospheric evolution, is also non-recurrent and irreversible. The thermal evolution of the Earth is a balance between the radiogenic heat produced in the mantle and crust and that lost by convection in the mantle. Figure 6.9 shows the decline in global heat production, as well

Figure 6.9 Schematic diagram showing the decline in global heat production and mantle temperature with time, and the matching decline in production of komatiite volcanism. Also shown is a crustal growth curve and the period of enhanced mafic or "norite" magmatism in the late Archean and early Proterozoic. Source: Information after Hall and Hughes (1993), Herzberg et al. (2010).

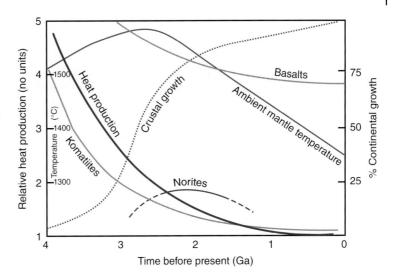

as the modeled changes in mantle temperature over time, all in relation to a crustal growth curve similar to that discussed above. There are several consequences for crust formation of the secular decrease in heat production. One of the most obvious is that the Archean mantle was hotter (by about 250 °C; Windley 1995; Herzberg et al. 2010) than at present and this could have led to an increase in the volume of melting at that time. A hotter mantle might also account for production of the high-Mg komatiites that typify many Archean greenstone belts and which are rare in younger oceanic settings (Figure 6.9). Higher heat production thus provides an explanation for the preferential occurrence, in the Archean, of komatiite-related magmatic Ni–Cu sulfide deposits, such as those at Kambalda, Western Australia and in Zimbabwe. Another consequence of higher heat flow in the Archean is the suggestion that convective overturn in the mantle would have been more rapid, with consequences for ocean crust residence times and the very nature of plate tectonics at the time (Windley 1995; Cawood et al. 2018). Archean crust is also considered to have been dominated by basaltic material with the oceanic crust itself being thicker than in later periods, a view supported by suggestions that ophiolites older than 1 Ga are more than double the thickness

of their younger equivalents (Moores 1993). As another consequence, Archean oceans may have been shallower than those at present, resulting in a reduced continental freeboard at that time (see Section 6.4.4).

6.4.3 Long-Term Global Tectonic Trends and Mantle Convection

As discussed in Section 6.3, the broad-scale amalgamation and dispersal of continental fragments with time, referred to as the "supercontinent cycle," is a feature of Earth history that has implications for virtually every aspect of global evolution, including metallogeny. Figures 6.1 and 6.2 show that broad temporal trends in ore formation reflect the contrast between episodes of orogeny and periods of lithospheric stability. In detail, however, it is the supercontinent cycle that controls the many other factors more directly related to secular metallogenic trends. Unlike the patterns imposed by atmospheric evolution and heat production, the supercontinent cycle is recurrent and cyclical in that continental fragments have amalgamated and dispersed several times through Earth history. This cyclicity helps explain why some ore deposit types, or at least the processes that give rise to them, are recognizably repetitive despite the preservation bias of ore deposits in younger rocks. However, although the supercontinent cycle is recurrent, the pattern of crustal evolution is also unidirectional and related to decreasing mantle temperature and increasing thickness and rigidity of the lithosphere over time. Consequently, crust forming processes have evolved such that those we see in operation today, and during the Phanerozoic Eon, are different from events that dominated earlier periods of Earth history.

Features such as decreasing mantle temperatures over time impact on factors such as mantle viscosity, lithospheric rigidity and the nature and scale of convection in the mantle. These in turn relate to the pattern of crustal evolution and the transition from an early, hot Earth characterized by crustal TTG plutonism and greenstone belts, to a cooler Earth comprising a thicker continental lithosphere and a regime of globally linked, rigid, plate boundaries. Archean crust was likely underlain by a hotter, more rapidly convecting mantle than the present day, such that it was too weak to be subducted in a rigid fashion – conventional plate tectonics may not, therefore, have been possible during the development of the early Earth. These observations are central to the

fundamental question of when plate tectonics commenced and what impact this had on metallogeny (Cawood and Hawkesworth 2015; Cawood et al. 2018).

Figure 6.10 illustrates a model for the nature of lithospheric changes over time. The diagram illustrates a pre-3200 Ma "stagnant lid" regime in which a relatively thin, immobile crust comprising continental islands of TTG type crust and oceanic material represented by greenstone belts is all underlain by hot mantle undergoing small-scale convection. Formation of magma and new crust during stagnant lid processes occurred not so much by "subduction" in the sense of conventional plate tectonics, but by processes, variably termed foundering, sagduction, subcretion, or simply "drip" tectonics (Nebel et al. 2018) – this regime envisages that a basalt-dominated stagnant lid is periodically over-turned back into the mantle with accompanying partial melt events. The change to a more rigid, and yet more mobile, lithosphere is believed to have been transitional and occurred between c. 3200 and 2500 Ma (Bédard 2018; Cawood et al. 2018). This transitional period may have been characterized by the beginnings of true subduction,

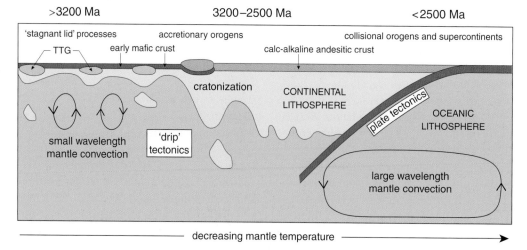

Figure 6.10 Model for the secular evolution of the lithosphere as a function of decreasing mantle temperature and evolving convection dynamics. Tectonic processes are dominated by a stagnant lid regime during early Archean times, but conventional plate tectonics ensued from around 3200–2500 Ma as the continental lithosphere became thicker and more rigid. Source: Modified after Cawood et al. (2018).

but in an episodic rather than sustained fashion. Subduction thickened the continental crust, typically via the onset of potassic granite magmatism, resulting in the differentiation of well-defined continental and oceanic domains – it was subsequent to this, and after 2500 Ma, that conventional plate tectonics became the dominant process controlling lithospheric evolution (Figure 6.10). The transition from "drip" style tectonics to conventional plate tectonics equates to the evolution from a stagnant lid regime to one in which rigid plates are continuously subducted along their margins to form a globally linked system. Sustainability of conventional plate tectonics is ensured by the dominant slab-pull force of the lower subducting plate resulting in continuous introduction of dense ocean crust into the mantle and the return of arc related magmas inboard of trench systems. The onset of sustainable plate tectonics in the period 3200–2500 Ma is supported geologically by features such as the appearance of eclogitic inclusions in diamond at around 3.0 Ga (Shirey and Richardson 2011), the appearance of potassic granites and dyke swarms in many parts of the world at around the same time, and geometric changes from "dome and keel" morphologies of TTG-greenstone belt assemblages in the early Archaean (e.g. Pilbara and Barberton at 3.5–3.2 Ga) to more linear arrays later on (e.g. Yilgarn and Superior at 2.7 Ga) (Figure 6.12).

A detailed understanding of how the patterns of crustal evolution, plate tectonic onset and the supercontinent cycle can be used in the study of ore deposits is likely to remain a profitable area of future research. Some of the concepts that are relevant to both global tectonics and the pattern of metallogenic evolution are discussed in more detail in Section 6.5.

6.4.4 Eustatic Sea Level Changes and "Continental Freeboard"

Continental freeboard, or the relative elevation of the continental land masses with respect to sea level or the geoid, is effectively a measure of the area of the exposed continents compared to that covered by the oceans. Continental freeboard is reduced when the average ocean depth becomes shallower, since this causes a marine transgression and flooding of the continental shelf. This situation is linked to events of large-scale continental dispersal, which in turn reflect enhanced creation of oceanic crust, and active magmatism and uplift along mid-ocean ridges. By contrast, an increase in the area of the continents occurs when the oceans are deepened and marine regression exposes the continental shelf – this situation tends to be associated with periods of tectonic stability and muted mid-ocean ridge magmatic activity (Worsley et al. 1984; Nance et al. 1986). The cyclicity of continental freeboard and ocean depth in the Phanerozoic Eon is well documented (Figure 6.11) and shows that a period of maximum continental exposure and ocean lowstand (i.e. deep oceans) coincides with the existence of the Pangean supercontinent in Permian–Triassic times. The nature and duration of continental dispersion and amalgamation in the Phanerozoic Eon, and their projection into the future, are also shown in Figure 6.11. A similar period of ocean lowstand may also have applied with respect to the amalgamation of an earlier supercontinent, Rodinia, at approximately 1000 Ma (see Section 6.5.2 below). From a metallogenic viewpoint, continental freeboard has its biggest influence on the nature and preservation of sediments forming on the continental shelves. Marine transgressions flood the continental shelf, preserving the sediments within which deposits such as heavy mineral beach placers, SEDEX Pb–Zn ores, banded iron-formations, bedded manganese deposits, and phosphorites might have formed. Oceanic lowstand and exposure of continental margins, on the other hand, results in shelf erosion and possible destruction of any ore deposits present in the sedimentary sequence.

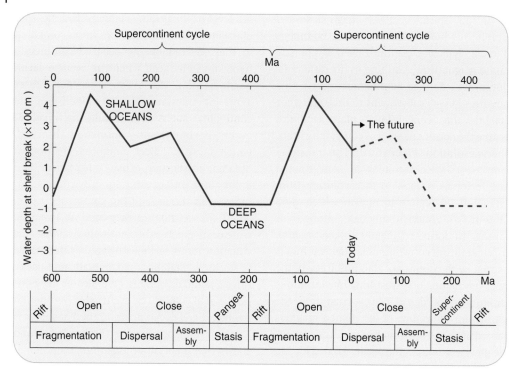

Figure 6.11 Pattern of supercontinent cycles, reflecting Pangean assembly and fragmentation during the Phanerozoic Eon (and into the future), and showing water depth of the continental shelf. Source: After Nance et al. (1986).

6.5 Metallogeny Through Time

The following sections summarize the nature of mineral deposits in terms of Earth's geodynamic evolution (see Section 6.3), but are subdivided and described on the basis of the three major geological time periods (or Eons), namely Archean, Proterozoic, and Phanerozoic.

6.5.1 The Archean Eon

Given the relatively poor degree of preservation and exposure of Archean continental crust it is virtually impossible to reconstruct the paleogeography of these ancient fragments with any certainty. Although it is unlikely that there were any supercontinents in Archean times (or that they will ever be recognized), there are nevertheless intriguing geological similarities between some Archaean cratons that are now spatially separated (Kröner and Layer 1992). The Kaapvaal Craton in southern Africa and the Pilbara Craton in Western Australia, for example, are remarkably similar in terms of lithostratigraphy and age patterns, and it has been suggested that these two fragments were joined in the late Archean. The existence of a "Vaalbara" continent was proposed by Button (1979) and Cheney (1996) and considered to have been in existence from late Archean through early Proterozoic times – its existence has more recently been supported by paleomagnetic data (De Kock et al. 2009). Two other Archean continental assemblages have been proposed (Bleeker 2003), namely Sclavia (possibly comprising cratons such as Slave, Zimbabwe, Wyoming, and Dharwar) and Superia (possibly comprising cratons such as Superior, Karelia, Hearne, and Yilgarn). It has also been suggested (Williams et al. 1991) that

all three of these continents might have coalesced at around 2500 Ma to form a supercontinent called Kenorland (Figure 6.4), although
this configuration is largely conjectural. With
this in mind, a discussion of Archean metallogeny is better undertaken on the basis of
chronostratigraphic subdivisions:

6.5.1.1 The Hadean (>4000 Ma) and Eoarchean (>3600 Ma) stages

The Hadean Era refers to that period of Earth
history for which there is very little evidence
in the rock record and which is nominally
pre-Archean. It was a time of global differentiation and accretion, as well as intense
meteorite bombardment. The Hadean was previously regarded as existing prior to 3800 Ma
(Harland 1989) although the growing evidence
(from U–Pb zircon dating) for rock remnants
at around 4000 Ma suggests that the latter date
is perhaps a more accurate reflection of the
Hadean boundary (Windley 1995). The lack of
any meaningful preservation of Hadean crust
anywhere on the face of the Earth is a feature
generally attributed to widespread destruction
of this ancient material, either by intense
meteorite bombardment or by crustal overturn
associated with a turbulent, rapidly convecting
mantle, or both. There is also evidence to
suggest that the early atmosphere and ocean
formed only at the end of the Hadean era, once
the main period of accretion and meteorite
bombardment had terminated (Kasting 1993).
De Wit and Hynes (1995) have suggested that
the Hadean Earth was also characterized by
loss of heat direct to the atmosphere, in contrast to later periods of time when heat loss
is largely buffered by a liquid hydrosphere.
The implications for metallogenesis are that
sedimentation and hydrothermal processes
are likely to have been inconsequential in the
Hadean, and any ore deposits that did form
were, therefore, probably igneous in character.
It is conceivable, for example, that oxide and
sulfide mineral segregations accumulated from
anorthositic and basaltic magmas at this time.
The only preserved record of such rocks within

reach of humankind at present is, however,
likely to be on the Moon.

The Eoarchean refers to the dawn of Archean
time and to rocks formed prior to 3600 Ma,
although for the purposes of this discussion it is considered to extend between 4000
and 3600 Ma. A well preserved section of
Archean crust that falls into this time bracket
is the 3800 Ma Isua supracrustal belt and
associated Itsaq (previously called Amitsoq)
gneisses of western Greenland. The Isua belt
comprises mafic and felsic metavolcanics,
as well as metasediments, and resembles
younger greenstone belts from elsewhere
in the world. Although only 4×30 km in
dimension, the Isua belt contains a major
chert–magnetite banded iron-formation
component as well as minor occurrences
of copper–iron sulfides in banded amphibo-
lites and in iron-formation (Appel 1983). The
largest iron-formation contains an estimated
2 billion tons of ore at a grade of 32% Fe.
Scheelite mineralization has also been found
in both amphibolite and calc–silicate rocks of
the Isua belt, an association which suggests
a submarine-exhalative origin. The coexistence of banded iron-formations and incipient
VMS style mineralization points to sea-floor
processes involving circulation of seawater
through oceanic crust. Although the zones of
known mineralization in the Isua belt are sub-
economic, at 3800 million years old they represent the oldest known ore deposits on Earth.

6.5.1.2 The Paleo-, Meso-, and Neoarchean stages (3600–2500 Ma)

The main stage of Archean crustal evolution
took place over an extended duration of more
than 1000 million years, during which time
some geological processes were probably not
dissimilar to those of today. It is nevertheless
apparent that this period witnessed a major
transition from a stagnant lid and drip tectonic
regime to one in which conventional plate tectonics was active (see Section 6.4.3). In terms
of crustal evolution there appears to be some
consensus around the existence of substantive

continental landmasses in Archean times, and as mentioned previously, an early Vaalbara continent and later Superia and Sclavia continents are thought to have existed through the Meso- and Neoarchean Eras (Bleeker 2003).

The existence of Vaalbara receives some support from the similarities that exist in the nature and ages of Archean greenstone belts and supracrustal sequences on the Pilbara and Kaapvaal cratons (Cheney 1996; Martin et al. 1998a), a feature that is especially striking when comparing the huge Superior-type banded iron-formations of the two regions. In addition, it was long thought that the Archean Witwatersrand basin on the Kaapvaal Craton was a unique occurrence, but exploration in Western Australia, has recently revealed the existence of sedimentary sequences of similar age that also appear to be well endowed with gold mineralization. The existence of Vaalbara in the late Archean is further supported by the similarities between voluminous, rift-related basaltic volcanism emplaced at 2780–2710 Ma on both cratons, in the form of the Ventersdorp (Kaapvaal) and Fortescue (Pilbara) lavas (De Kock et al. 2012).

In a metallogenic context, Archean crustal evolution can be viewed in terms of a two stage model which suggests that early "shield" formation, in which amalgamation of pre-3200 Ma TTG-greenstone crust was followed by "cratonization," where processes akin to modern plate tectonics, such as subduction and continent collision, occurred post-3200 Ma (de Wit et al. 1992). The characteristics and relevant ore-forming processes of these two stages are illustrated in Figure 6.12. In this model two principal stages, termed "*intra-oceanic shield formation*" (pre-3100 Ma) and "*continental craton formation*" (between about 3100 and 2500 Ma) are distinguished.

6.5.1.3 Shield formation (pre-3100 Ma)

This period conforms approximately with the Paleoarchean era and is typified by amalgamation of oceanic crust with incipient continental crust in the form of TTG plutons (de Wit et al.

1992; Choukroune et al. 1997). This stage of Archean crustal evolution is characterized by the development of early continental shield areas comprising highly deformed (oceanic) greenstone remnants occurring as remnants within extensive TTG terranes, typically referred to as "dome and keel" structures (Figure 6.12a). It is also characterized by processes that are generally believed to have pre-dated the onset of conventional plate tectonics (see Section 6.4.3).

The styles of mineralization that formed during this stage of Earth's evolution are limited and best exemplified by the deposits previously described for the Isua belt of west Greenland. Algoma type banded iron-formations are a common component of early greenstone belt assemblages and reflect the low oxygen levels of the atmosphere and the abundance of ferrous iron in sea water, sourced from exhalative activity on the ocean floor. The existence of shallow oceans at this time and the likelihood of exhalative hydrothermal processes on the sea floor would have resulted in the formation of volcanogenic massive base metal sulfide deposits, although few examples are preserved. An exception is the well preserved Big Stubby VMS deposit, in the 3460 Ma Warrawoona Group metavolcanics of the Pilbara Craton in Western Australia (Barley 1992).

6.5.1.4 Cratonization (c. 3100–2500 Ma)

This stage of Archean crust formation coincides broadly with the Meso- and Neoarchean eras and is illustrated in Figure 6.12b. It is suggested to be the most prolific period of crustal growth in Earth history (Figure 6.3b) and is also a time of major global mineralization. The processes active at this time were not unlike those taking place later in Earth history and involved incipient plate subduction, arc magmatism, continent collision and rifting, and cratonic sedimentation. In Figure 6.12b two sketches are presented. An earlier sub-stage illustrates consecutive accretion of island arcs onto a previously formed continental shield and stabilization of the latter by intrusion

SHIELD FORMATION (3600–3100 Ma)

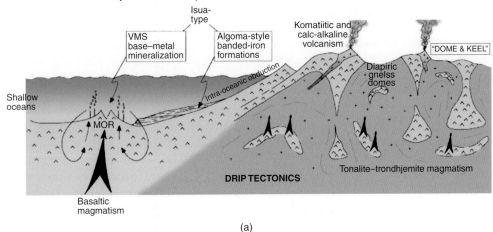

(a)

CRATONIZATION (3100–2500 Ma)

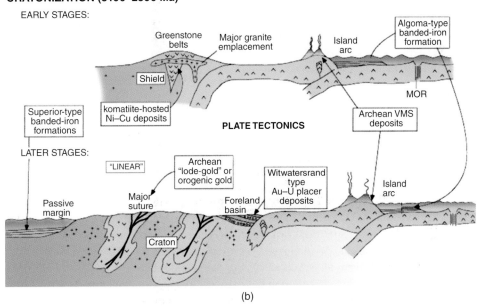

(b)

Figure 6.12 (a) Early stage of Archean crustal evolution, termed intra-oceanic shield formation, occurring in the interval 3600–3100 Ma. Mineralization in this period seems to have been restricted mainly to ocean floor exhalative processes and formation of Algoma-type banded iron-formations. (b) Later stage of Archean crustal evolution, termed intra- and inter-continental craton formation, occurring between 3100 and 2500 Ma. Mineralization processes were varied and resulted in the formation of several important ore deposit types, including lode- or orogenic gold, VMS Cu–Zn, paleoplacer Au–U, Algoma- and Superior-type iron ores and komatiite-hosted Ni–Cu deposits. Source: After de Wit et al. (1992).

of large granite batholiths. A later sub-stage envisages the existence of Archean cratons consisting of numerous terranes, of linear disposition, each bordered by major suture zones (possibly the fossilized sites of subduction or arc collision) and flanked by both active and passive margins. Sites of intracratonic sedimentation are also envisaged in this stage of development. This scenario is consistent with the geometry of Archean crust as observed, for example, in the Superior Province of Canada (Choukroune et al. 1997) and in the Yilgarn Craton of Western Australia.

From a metallogenic viewpoint, this stage of Archean crustal evolution gave rise to a wide variety of ore-forming processes (Figure 6.12b). Arc-related volcanism associated with plate subduction contributed large volumes of magma to the accreting terranes of the time. Well mineralized examples of continental crust formed in the period 3100–2500 Ma are represented by the granite–greenstone terranes of the Superior Province of Canada, as well as the Yilgarn and Zimbabwe cratons. Greenstone belts are hosts to numerous important VMS Cu–Zn ore bodies, such as those at Kidd Creek and Noranda in the Abitibi greenstone belt of the Superior Province. Off-shore, in more distal environments, chemical sedimentation gave rise to Algoma type banded iron-formations, examples of which include the Adams and Sherman deposits, also in the Abitibi greenstone belt. Greenstone belts formed at this time also contain komatiitic basalts that, under conditions favorable for magma mixing and contamination, exsolved immiscible Ni–Cu–Fe sulfide fractions to form deposits such as Kambalda in Western Australia and Trojan in Zimbabwe. During and soon after periods of compressive deformation, major suture zones became the focus of hydrothermal fluid flow derived from either metamorphic devolatilization or late-orogenic magmatism. This resulted in the formation of the varied but common styles of orogenic gold mineralization that are typical of most Meso- and Neoarchean granite–greenstone terranes worldwide. Examples include important deposits such as the Golden Mile, Kalgoorlie district of Western Australia, the Hollinger–McIntyre deposits of the Abitibi greenstone belt, the Sheba–Fairview deposits of the Barberton greenstone belt, and the Freda–Rebecca mine in Zimbabwe. Interestingly, it is rocks of the late Archean (2800–2500 Ma) that are preferentially mineralized with respect to orogenic gold deposits.

Early intracratonic styles of sedimentation, often in a foreland basinal setting, gave rise to concentrations of gold and uraninite represented by the Witwatersrand basin in South Africa. At least some of this mineralization is placer in origin and was derived by eroding a fertile Archean hinterland. The passive margins to these early continents would have developed stable platformal settings onto which laterally extensive Superior type banded iron-formations would have been deposited. A very significant period for deposition of iron ores, such as those of the Hamersley and Transvaal basins of Western Australia and South Africa respectively, as well as the Mesabi range of Minnesota, seems to have been around the Archean–Proterozoic boundary at 2500 Ma, by which time the crust was both thick and rigid.

6.5.2 The Proterozoic Eon

The period of time around 2500 Ma represents a major transition in the nature of crustal evolution, involving changes in the volume and composition of the continents, tectonic regimes, and atmospheric make-up. It is also clear from secular metallogenic patterns (Figure 6.1) that these evolutionary changes affected ore-forming processes and characteristics. The Proterozoic Eon spans a vast period of geological time, from 2500 to 541 Ma, including the period between 1800 and 1000 Ma that was marked by a relative paucity of orogenic and/or magmatic-hydrothermal deposit types, but abundant ores hosted in intracontinental sedimentary basins and anorogenic igneous complexes (Figure 6.1). The reasons for this pattern are multifaceted, but, as a generalization, are related to a higher degree of continental stability and the existence of long-lived supercontinents such as Nuna and Rodinia (Zhang et al. 2012; Nance et al. 2014). A substantial volume of continental crust must have been in existence by the beginning of the Proterozoic Eon (Figure 6.3b) and, as described in Section 6.3.2, it is widely held that the first true supercontinent, Nuna, came into existence during this time.

Although geological processes may not have been significantly different from those of subsequent periods of Earth history, the interval between 1800 and 1000 Ma is typified by ore deposits related to anorogenic magmatism and intracontinental basin deposition (Figure 6.1). The metallogenic characteristics of the three Proterozoic eras are discussed in more detail below.

6.5.2.1 The Paleoproterozoic Era (2500–1600 Ma)

From a metallogenic viewpoint, the period of Earth history between 2500 and 1600 Ma is significant because of the major changes that occurred to the atmosphere, especially the rise in atmospheric oxygen levels at around 2300 Ma (Figure 6.8). Prior to this time, the major oxygen sink was the reduced deep ocean where any photosynthetically produced free oxygen was consumed by the oxidation of volcanic gases, carbon, and ferrous iron. In this environment banded iron-formations, as well as bedded manganese ores, developed, as is evident from the widespread preservation of both Algoma and Superior type iron deposits. The increase in ferric/ferrous iron ratio in the surface environment that accompanied oxyatmoinversion at 2300 Ma, and the accompanying depletion in the soluble iron content of the oceans at around this time, resulted in fewer BIFs forming after this time (Figure 6.8). The stability of easily oxidizable minerals such as uraninite and pyrite is also to a certain extent dependent on atmospheric oxygen levels and it is, therefore, relevant that major Witwatersrand-type placer deposits did not form after about 2000 Ma. Besides these non-recurrent changes, which only affected redox-sensitive ore-forming processes, the types of mineral deposit that formed in the Paleoproterozoic reflect the patterns of crustal evolution of the time, as discussed below.

Metallogenic patterns during the Paleoproterozoic Era were dominated by orogenic

processes prevailing during the breakup of Superia and Sclavia and the assembly of Nuna between c. 2100 and 1800 Ma (Section 6.3.2 and Figure 6.5). The breakup of Superia was accompanied, between 2000 and 1700 Ma, by plate motions that gave rise to the Trans-Hudson, Yavapai–Mazatzal and Svecofennian orogenies – these formed new crust within which VMS Cu–Zn deposits such as Flin Flon in Canada, Jerome, Arizona, and the Skellefte (Sweden)–Lokken (Norway) ores of Scandinavia are preserved. Elsewhere on Nuna, in Zimbabwe for example, rifting at around 2575 Ma (Oberthür et al. 2002) gave rise to intrusion of the Great Dyke, with its significant Cr and platinum group element (PGE) reserves. At 2055 Ma on the Kaapvaal craton, the enormous Bushveld complex with its world-class PGE, Cr, and Fe–Ti–V reserves was emplaced, as was the Phalaborwa alkaline complex with its contained Cu–P–Fe–REE mineralization. In West Africa the period between 2100 and 1900 Ma saw the development of substantial juvenile crust during the Eburnean orogeny, accompanied by the formation of extensive orogenic gold mineralization. The Australian Barramundi and southern African Kheis orogenies are events which contributed to the growth of Nuna and, in the latter region, for example, gave rise to the Haib porphyry Cu deposit in Namibia and small MVT-type Pb–Zn deposits along the western edge of the Kaapvaal craton.

The amalgamation of Nuna was followed by a long period of cratonic stability that resulted in the deposition, between 1800 and 1500 Ma, of marginal marine sedimentary basins that host the world-class SEDEX Pb–Zn ores of eastern Australia (Mount Isa, Broken Hill, and McArthur River) and South Africa (Aggeneys and Gamsberg). Deposition of the Athabasca basin in Canada formed at around 1700 Ma – it should be noted, however, that the very rich uranium ores in the latter are epigenetic and probably formed during later episodes of fluid flow between 1500 and 1000 Ma (Hecht and Cuney 2000).

Anorogenic magmatism was widespread in Nuna times – in South Australia, for example, the 1590 Ma Roxby Downs granite–rhyolite complex (Johnson and Cross 1995), host to the enormous magmatic-hydrothermal Olympic Dam iron oxide–Cu–Au-U deposit, was emplaced. Anorogenic granite magmatism at 1880 Ma may also have given rise to the later stages of IOCG style mineralization (such as Estrela; Volp 2005) in the giant deposits of Carajas, Brazil.

6.5.2.2 The Mesoproterozoic Era (1600–1000 Ma)

The dispersed fragments of Nuna eventually started to re-assemble as Rodinia in the Mesoproterozoic, as evidenced globally by the existence of the Grenvillian orogenies between 1400 and 1000 Ma. The events that created Rodinia have preserved few world-class ore deposits – the reasons for this are enigmatic but possibly related, in part, to the construction of buoyant continents and the erosive removal of mineralized upper crust. The deposits that did form in the Mesoproterozoic are often termed "anorogenic" as they are not typically related to subduction processes. In South Africa, for example, mineralization at this time is represented by the magmatic Cu–sulfide ores associated with deep-seated mafic intrusions of the Okiep copper district in the 1220–1030 Ma Namaqualand belt (Robb et al. 1999). Elsewhere, large volumes of anorogenic magmatism, especially in the period 1500–1000 Ma, provided the host rocks to a number of mineral deposits in a belt stretching from southern California through Labrador into Scandinavia (Windley 1995). For example, gabbro–anorthosite intrusions host the large magmatic Fe–Ti (ilmenite) ore bodies of the Marcy massif in the Adirondacks and the c. 1060 Ma Lac Tio deposit in Quebec, as well as the 920 Ma Tellnes deposit of SW Norway (Charlier et al. 2015). The same belt also contains alkali granite–rhyolite complexes which give rise to Fe–Au–REE resources such as those of the c. 1480 Ma St

Francois mountains of Missouri (Denison et al. 1984). This basement was eroded to form the sediments of the 1440 Ma Belt Basin in the northwest USA, host to the Sullivan Pb–Zn SEDEX deposit in British Columbia, Canada. In addition, intracontinental rifting at around 1100 Ma in ancestral Rodinia gave rise to the formation of the 2000 km long Keweenawan mid-continental rift, stretching from Michigan to Kansas and filled with a thick sequence of bimodal basalt–rhyolite volcanics overlain by rift sediments. The latter form the host rocks to the stratiform Cu–Ag White Pine deposit in Michigan.

6.5.2.3 The Neoproterozoic Era (1000–541 Ma)

The Neoproterozoic commenced with the formation, from around 1000 Ma, of the supercontinent Rodinia. Like Nuna, Rodinia was long-lived and only started to partially fragment after a static period of more than 250 million years (see Section 6.3.2). Substantial parts of what had been Rodinia then reconvened toward the end of the Proterozoic (at 541 Ma) to form the very substantial Gondwanan landmass (or Pannotia) during the Pan-African orogeny (Figure 6.6b). At this stage the amalgamation of east and west Gondwana had taken place and this large continental mass was situated at polar latitudes and across the Iapetus seaway from an equatorially located Laurentia.

In detail, the assembly of Gondwana was long-lived and polyphase. The time between about 750 and 550 Ma was characterized by the development of at least two, and in places possibly four, major ice ages, one or more of which was near global in cover and extended to equatorial latitudes. The concept of the Neoproterozoic "Snowball Earth" (Harland 1965; Hoffman et al. 1998) has important implications for understanding climate change and, especially, for the proliferation and diversification of organic life at the Precambrian–Cambrian boundary. Global glaciations also have implications for the nature and formation of ore deposits in the Neoproterozoic Era.

The major ore deposits of the Neoproterozoic reflect the conditions of Gondwanan continental stability, as well as the periods of near global ice cover and attendant anoxia, that prevailed at this time. The extensively developed ironstone ores of northwest Canada and South Australia, associated with the 750–725 Ma Rapitan and Sturtian glaciogenic rocks respectively, are considered to be the result of the buildup of ferrous iron derived from offshore hydrothermal vents in the reduced ocean waters that accompanied the development of vast continental and oceanic ice sheets at this time. Receding glaciers and a return to more oxidizing conditions would have resulted in conversion of labile ferrous iron to insoluble ferric iron and precipitation of the latter from the ocean water column, together with clastic and glaciomarine detritus to form the ironstones (Lottermoser and Ashley 2000). Even more oxidizing conditions would also have resulted in the precipitation of manganese oxides or carbonates in the succession.

In a similar vein, the Precambrian–Cambrian boundary at around 540 Ma is also characterized by the first major global phosphogenic event that resulted in the development of substantial deposits of phosphatic sedimentary rock (phosphorites) in several parts of the world (Cook and Shergold 1984; Filippelli 2008). As with sedimentary iron ores, phosphorites reflect the upwelling of deep, nutrient-rich ocean waters onto shallow continental shelves with the syn-sedimentary precipitation of carbonate–fluorapatite onto the shelf floor. Although the actual formation of phosphate ores is a complex process (see Chapter 5), there is compelling evidence to suggest that the onset of phosphogenesis at 540 Ma was related to climatic conditions prevailing toward the end of the Neoproterozoic. Many phosphorite deposits worldwide immediately overlie glaciogenic sediments, suggesting that upwelling of phosphorus-rich ocean waters was promoted by overturn of a stagnant ocean

during the widespread blanketing of sea ice associated with a Snowball Earth scenario. The Precambrian–Cambrian phosphogenic event also coincides with the proliferation and diversification of organic life and it is pertinent that a significant proportion of organisms that evolved at this time developed calcium phosphate skeletal structures. Phosphorus is, in addition, a universal nutrient (unlike oxygen) and its concentration in the oceans at the end of the Proterozoic, and in the Cambrian, may also be linked to the evolution of organic life.

The formation of the extensive, stratiform Cu–Co clastic sediment hosted ores of the Central African Copperbelt is also considered to have formed in an environment influenced by the Snowball Earth. The host Katangan sediments were deposited on a fertile Paleoproterozoic basement (Rainaud et al. 2005) in what was initially an intracontinental rift, the development of which overlapped with both the Sturtian and Marinoan glacial events. The Grand and Petit Conglomerats of the Katangan sequence, for example, represent glaciogenic sediments capped by carbonates which are correlated, respectively, with the Sturtian and Marinoan events (Windley 1995; Master et al. 2005). Influx of oxide-soluble Cu and Co, perhaps derived from the local basement, might have occurred as diagenetic fluids migrated along growth faults and through the basin during the postglacial stages of deposition. Precipitation of ore sulfides would have occurred when the metal-charged oxidized fluids encountered reduced sediments or fluids (Greyling et al. 2005).

The proliferation and diversification of organic life in late Neoproterozoic times also resulted in the first substantial development of highly carbonaceous sediments. A good example of these are the Hormuz sediments that flank the eastern Arabian shield and which represent some of the source rocks for the vast Mesozoic oil and gas fields of the Arabian Gulf (Chapter 5, Box 5.3; Windley 1995). The Neoproterozoic is notable for the paucity of orogenic type deposits (Figure 6.1). Even the extensive Pan-African orogenic belts, representing the suturing of continental fragments during Gondwanan assembly, are deficient in, for example, world-class VMS deposits, the latter being relatively abundant in the Paleoproterozoic orogens of the world. A few small examples do nevertheless occur, such as the deposits of the Matchless amphibolite belt in Namibia, Bleïda in Morocco, and the Ducktown, Tennessee deposit (Titley 1993). The global deficiency of these ore types in the Neoproterozoic perhaps reflects a lack of preservation.

6.5.3 The Phanerozoic Eon

By comparison with earlier eons, crustal evolution during the Phanerozoic is well understood and there is a large measure of confidence that accompanies the interpretation of tectonic, chemical, and biological processes over the past 540 million years. The latter half of this period, the Mesozoic and Cenozoic eras, appears to have been accompanied by a prolific development of mineral deposits, perhaps unprecedented since the late Archean bonanza, although this may be enhanced by favorable preservation. Ore-forming processes and their relationships to the pattern of Earth evolution in the Phanerozoic are also reasonably well understood and additional details regarding some of the topics discussed below can be found in Nance et al. (1986), Larson (1991), Barley and Groves (1992), Titley (1993), Kerrich and Cassidy (1994), Windley (1995), Barley et al. (1998), Groves and Bierlein (2007), and Cawood and Hawkesworth (2015).

A significant proportion of the Phanerozoic Eon was characterized by geological processes that reflect a supercontinent cycle, namely the sequence of events that saw the dispersal of Pannotia and Gondwana in the early Paleozoic, followed by amalgamation of most existing continental material to form the Pangean supercontinent, by early Mesozoic times. The remainder of the Phanerozoic has witnessed the start of another cycle, involving the dispersal of Pangea to form the present day continental geography. It seems likely that the dispersal of Pangea is now well advanced (Nance et al. 1986) and that current plate movements will result in the reassembly of a significant proportion of the continents over the next 100–200 million years (Figure 6.11). Continued continental amalgamation in the future will extend processes currently taking place, such as the collision of the Indo-Australian plate (previously part of east Gondwana) with the combined Baltica–Siberia (now called the Eurasian plate) and closure of the Mediterranean Sea.

As previously mentioned, breakout of Laurentia from Rodinia and dispersal of other continental fragments at around 725 Ma (Figure 6.4 and Figure 6.6) eventually led to the assembly of the precursors to Gondwana (or Pannotia), by the end of the Proterozoic Eon. Dispersal of Pannotia commenced in Cambrian–Ordovician times (Figure 6.13a) and may have been facilitated by the earlier development of a superplume (Larson 1991), evidence for which is seen in the formation of major dyke swarms at 650–580 Ma (Torsvik et al. 1996) in parts of what became Gondwana. The Iapetus Ocean formed as Laurentia drifted away from Baltica and Gondwana in early Paleozoic times. The pattern of continental fragmentation was accompanied by a decrease in continental freeboard, sea-level highstand and marine transgression (Figure 6.11). This was followed by progressive basin closure, terrane accretion, and granitoid magmatism during the period extending from Devonian through Carboniferous times (about 420–300 Ma). This phase of collision and continental assembly commenced with the rapid consumption of Iapetus as Baltica and Avalonia (i.e. a small terrane comprising England, Wales, southern Ireland, and eastern Newfoundland) moved toward lower latitudes and collided with Laurentia in late Silurian times (Figure 6.13b). The orogenies that reflect this collision phase are referred to as the Caledonian in Scandinavia and Scotland/Ireland and

Appalachian in the eastern USA. Continental amalgamation continued as Gondwana drifted northwards toward lower latitudes, consuming the Rheic Ocean during the Devonian and Carboniferous, and eventually colliding with Laurentia–Avalonia–Baltica in the late Carboniferous–Permian periods. These complex and polyphase collisions are referred to as the Variscan and Hercynian orogenies (the terms are essentially synonymous) in present day Europe, and the Alleghanian orogeny in the eastern USA. Continental amalgamation continued into the Permian period with the accretion of Siberia to Baltica along the Urals suture, after which the Pangean supercontinent had essentially formed (Figure 6.13c).

The breakup of Pangea commenced soon after the time of its maximum coalescence in the Triassic at around 230 ± 5 Ma (Veevers 1989). It was, therefore, a short-lived supercontinent compared to Rodinia, which appears to have endured during much of the early Neoproterozoic. One reason for this might have been the development of a superplume, or plumes, that existed for some 70 million years, in late Carboniferous and Permian times (320–250 Ma), an interval that also coincided with a protracted period of constant reversed magnetic polarity (Larson 1991). The effects are seen geologically by increased production of oceanic crust and global volcanism (e.g. the outpouring of the Siberian continental flood basalt province and also the Central Atlantic Magmatic Province), increased atmospheric greenhouse gas production (with associated warming and organic proliferation), and enhanced deposition of petroleum and coal bearing sediments (Figure 6.14). The first landmasses to actually drift apart, more than 70 million years after initial plume-related rifting, were Laurentia from Gondwana in the Jurassic at around 180 Ma. West and East Gondwana also started to split at this time, with the development of the proto-Indian Ocean and outpouring of the Karoo igneous province. The opening of the Atlantic Ocean was particularly prevalent in the early Cretaceous (around

130 Ma) and this extensional phase was also marked by continental flood basalt volcanism of the Etendeka–Parana provinces in Namibia and Brazil respectively. Continued dispersal of Pangea was stimulated by a second superplume event in the mid-Cretaceous between 120 and 80 Ma (Figures 6.13d and 6.14). The evidence for this event is again seen in a 40 million year period of constant, normal magnetic polarity, as well as the expected rises in oceanic crust production, eustatic sea level, atmospheric temperatures, organic productivity, and black shale deposition. The large oceanic plateaus of the western Pacific (e.g. Ontong–Java) are a magmatic reflection of this mid-Cretaceous event, as is the development of significant global oil reserves related to the surges in nutrient supply and organic productivity on the voluminous continental shelves formed by the rise in sea level (Larson 1991). In post-plume times the late Cretaceous saw the final separation of Indo-Australia from Antarctica (Figure 6.13e), propelling the former northwards and resulting in continent–continent collision with Eurasia and the onset of the Himalayan orogeny in the early Cenozoic. The Alpine orogeny of southern Europe was more or less coeval with its Himalayan counterpart and resulted from the collision of the African and Arabian plates with Eurasia after consumption of the Tethyan Ocean.

An important component of Pangean breakup in the Mesozoic–Cenozoic is reflected in the development of new crust that accompanied the Cordilleran and Andean orogenies along the western margins of North and South America. This huge and complex orogenic belt, extending along the ocean–continent interface of western Pangea (Figure 6.13c), occurred in response to subduction and translocation of the original Panthalassan Ocean (now the eastern Pacific) beneath Laurentia (now North America), the Cocos plate (now central America), and segmented west Gondwana (now South America). An island arc lay to the west of North America during Pangean times and this was accreted onto the continental margin, together

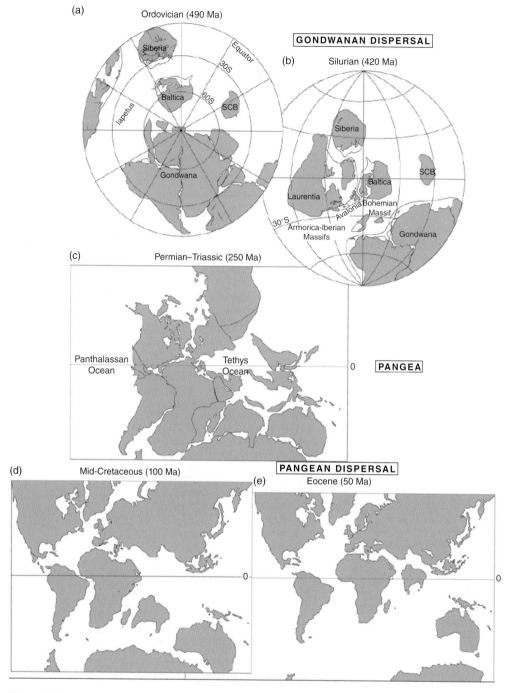

Figure 6.13 The paleogeographic patterns of Gondwanan and Pangean dispersal in the Phanerozoic Eon. (a) and (b) Two reconstructions in the early Ordovician and Silurian. Source: Modified after Torsvik et al. (1996). (c) A reconstruction of Pangea at its peak amalgamation in the Permian–Triassic. (d) and (e) Two instants of Pangean dispersal in the mid-Cretaceous and Eocene. Source: After Windley (1995).

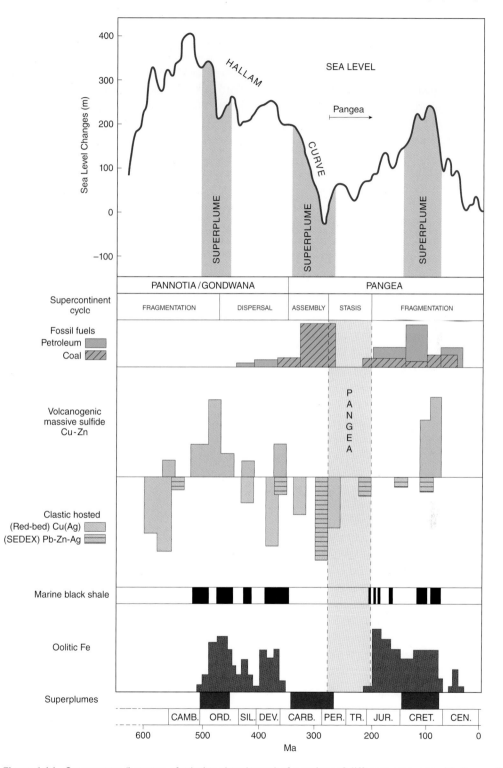

Figure 6.14 Occurrences (in terms of relative abundances) of a variety of different ore types with time and with respect to the Phanerozoic supercontinent cycle. Source: Modified after Titley (1993), with information from Nance et al. (1986), Larson (1991), Hallam (1992), and Barley et al. (2005).

with numerous other exotic terranes, during the consumption of several thousand kilometers of the Panthalassan Ocean beneath North America in the Mesozoic Era. The enormous continent-floored magmatic arc that resulted from this subduction gave rise to the 130–80 Ma granite batholiths and felsic volcanic rocks (such as the Sierra Nevada batholith) of the Cordillera. Continued subduction, albeit at a shallower angle than previously, and crustal thickening during the early Cenozoic, gave rise to in-board extension and exhumation of metamorphic core complexes. Ongoing subduction during the Laramide orogeny (80–40 Ma) continued to feed the magmatic arc and numerous, metallogenically important, I-type granite batholiths were emplaced at this time. As subduction waned in the late Eocene–Oligocene, the magmatic arc migrated westwards, resulting in continued calc–alkaline magmatism. Thermal collapse commenced in the Neogene and the resulting crustal extension ultimately led to the formation of the Basin and Range province. In South America, Andean evolution was analogous, but less complex, with that further north, and was also characterized by a smaller degree of accretion and crustal extension. Subduction of the Nazca plate beneath a Paleozoic sedimentary margin, in the late Cretaceous, gave rise to a volcanic arc, behind which back-arc sedimentation took place (Lamb et al. 1997). This arc was the site of protracted magmatism that built the Western Cordillera and gave rise to the volcanic edifices which host the important porphyry and epithermal styles of polymetallic mineralization in Peru, Bolivia, and Chile. Continued subduction transferred compressional stresses in-board and created a thin-skin fold and thrust belt in what is now referred to as the Eastern Cordillera. In the late Oligocene, crust was shortened even more and uplift gave rise to rapid erosion of the mountain belt to form a thick sedimentary plateau of dominantly Miocene-aged gravel and red-bed sequences known as the Altiplano. Subduction-related volcanism continued throughout this period in the Western Cordillera, which was further elevated to form the high Andes and then eroded to contribute sediment into the Altiplano basin as well as westwards onto the coastal plain. Further to the east, magmatism commenced in the early Miocene as a result of convective removal of the basal lithosphere and crustal melting beneath the Altiplano (Lamb et al. 1997). The buoyancy and elevated profile of the Andes is maintained by present day underthrusting of the Brazilian shield beneath the Eastern Cordillera.

6.5.3.1 Phanerozoic Tectonic Cycles and Metallogeny

The relatively well defined global tectonic cycles of the Phanerozoic Eon, summarized above, are also clearly reflected in secular metallogenic trends. Titley (1993) noted that the distribution of stratabound ores (i.e. VMS, clastic sediment hosted Pb–Zn [SEDEX], and sediment-hosted Cu [red-bed] deposit types) could be related to the tectonic cycles of Pangean amalgamation and breakup (Figure 6.14). Preferential development of VMS deposits, for example, appears to be associated with periods in the supercontinent cycle of elevated sea level (highstand) associated with continental dispersal, namely, *after* Gondwana breakup in the early Paleozoic and in post-Pangean Mesozoic times. This association was considered to reflect the processes of rifting, enhanced ocean crust production, and hydrothermal exhalation that accompany the fragmentation of continents. Similar patterns are also evident in the accumulation of organic-rich shales and oolitic ironstone ores, which preferentially occur in the same two intervals, namely after Gondwana breakup in the Ordovician–Devonian and in post-Pangean Jurassic–Cretaceous times (Figure 6.14). In these cases, increased exhalative activity, carbon dioxide production, global warming, and organic productivity are interrelated processes that result in suitable conditions for black shale and ironstone precipitation in the oceans. In contrast, clastic sediment hosted base metal ores of the SEDEX Pb–Zn–Ag and red-bed Cu types tend to form at different stages of the tectonic cycle. These ores are preferentially developed *at the time* of maximum continental amalgamation and stasis, namely in Gondwana during the Precambrian–Cambrian transition and in Pangea during the late Paleozoic–early Mesozoic (Figure 6.14). In terms of processes, these ore types appear to have formed preferentially at sites of intracratonic rifting and during sea-level lowstand (Titley 1993).

Two of the Phanerozoic superplume events described above are also related in time to periods of enhanced organic productivity and formation of fossil fuels (Figure 6.14). The superplume events appear to be linked to periods of enhanced volcanic activity, greenhouse gas emission, and global warming, which, in turn, stimulated organic productivity. The mid-Cretaceous superplume, active between 120 and 80 Ma, coincides with the formation of voluminous organic-rich shales deposited largely in the low-latitude Tethyan seaway. These shales are thought to be genetically linked to some 60% of the world's oil reserves (Larson 1991). The late Carboniferous–Permian superplume is similarly associated with deposition of a large proportion of the world's coal reserves, formed between 320 and 250 Ma, again at a time when organic productivity increased and tropical, swampy conditions prevailed along flooded continental margins associated with the corresponding sea-level rise. Continental hotspots, or beheaded plumes, are also possibly related to the preferential development of alkaline and kimberlitic magmatism and reflected in the enhanced igneous activity of mid-Cretaceous times and again in the Cenozoic. Many of these intrusions, important for Cu–REE–P–Fe–Au mineralization and also diamonds, are found on old, stable cratons and located along ancient lineaments that might have been reactivated during extension and crustal thinning associated with hotspot activity.

6.5.3.2 Time-Bound and Regional Aspects of Phanerozoic Metallogeny

A summary of ore deposit trends as a function of the major Phanerozoic tectonic events reveals that the metallogenic inventory, especially of convergent margin lithosphere, appears to increase with time and with each tectonic cycle (Barley et al. 1998). The distribution of ore deposit types also has a regional character to it and, as mentioned previously, may reflect the extent of erosion in different regions and the preservation potential of ores.

The large-scale development of new crust along the convergent Mesozoic–Cenozoic plate margins of the Americas did not occur on the same scale during the Paleozoic dispersal stage. Hence, a cyclic pattern in the development of orogenic magmatic-hydrothermal ores is not as evident as it is for stratabound deposits, where controls on ore formation are different. There are, for example, few known porphyry Cu–Mo type deposits of Paleozoic age, although sub-economic examples are recognized along convergent margins such as the Caledonides of Scotland. Orogenic gold mineralization is also typically related to the late stages of collisional orogens. Cretaceous examples of this style of gold mineralization are evident in California, British Columbia, China, and New Zealand and could conceivably be related to the enhanced thermal regime, and magmatism, related to the superplume breakout at this time. It has also been suggested that these gold provinces might rather have been linked to areas where mid-ocean ridge, as opposed to normal oceanic crust, was subducted. Scenarios in which hot crust was subducted, compared to periods when normal subduction of cold oceanic crust prevailed, have been suggested as an explanation for the anomalous thermal conditions required for episodic gold mineralization (Haeussler et al. 1995).

The amalgamation stage of Pangea commenced with the collision of Avalonia–Baltica with Laurentia during the polyphase Paleozoic Caledonian–Appalachian orogenies. VMS deposits formed in response to these processes in, for example, the Bathurst–Newcastle district of New Brunswick, the Buchans area of Newfoundland, and Scandinavia. New Brunswick also has examples of Devonian aged granites, some of which are Sn–W–Mo–F mineralized. The carbonate-hosted MVT Pb–Zn deposits of the southeastern USA are also related to the Appalachian orogeny and formed when the thrust front expelled basinal fluids into the carbonate platform on the continental margin, giving rise to epigenetic hydrothermal mineralization. The Variscan orogeny, formed when Gondwana collided with an already amalgamated Laurentia–Baltica–Avalonia, is also characterized by carbonate-hosted ore bodies. Some of these, particularly the Irish deposits at Navan, are transitional in style between MVT and SEDEX-type Pb–Zn deposits, but are also related to compression-related, gravity- and tectonic-driven, basinal fluid flow. The Variscan orogeny of Europe is particularly well known for its association with widespread, 300–275 Myr old, S-type Sn–W–U granites. These are well mineralized in many of the historic European mining districts, such as Cornwall in southwest England (South Crofty, St Just, etc.), the Massif Central of France (Bellezane), the Bohemian massif of the Czech Republic, and Portugal (Panasqueira). Other deposits associated with the Variscan of Europe include the Devonian aged Rammelsberg and Meggen SEDEX-type deposits of Germany and the numerous VMS deposits of the Iberian Pyrite Belt (Rio Tinto, Neves Corvo, etc.). The brief interval in the Permian–Triassic periods of relative stability that accompanied the existence of the Pangean supercontinent was one in which few mineral deposits formed (Figure 6.14). An exception to this was the stratiform red-bed hosted Cu–Ag ores of the Permian Kupferschiefer in Poland and Germany.

Mineralization associated with post-Pangean breakup and orogenesis is both voluminous and widespread. Exceptions to this would appear to be the products of continent–continent collision, such as regions affected by the Himalayan orogeny, which appear to be devoid of any world-class ore deposits. The Karakoram mountains of Pakistan, for example, contain rich gemstone deposits (such as the rubies of the Hunza area), but otherwise only minor MVT and vein-related precious and base metal type ores, as well as a few small hydrothermal uranium deposits, occur (Windley 1995). More significant stratabound and ophiolite hosted styles of mineralization are contained in rocks caught up in the Himalayan orogeny, but the ore-forming processes probably predate the latter. Regions affected by the Alpine orogeny are sporadically mineralized, although certain districts may be well endowed. MVT-type Pb–Zn deposits occur in mid-Triassic limestones of the eastern Alps and north Africa. The Carpathian arc of southeastern Europe contains porphyry Cu–Mo styles of mineralization, such as at Sar Cheshmeh in Iran and Bor in Serbia, and also contains significant potential for epithermal gold mineralization. Important ophiolite-hosted VMS Cu–Zn and chromite mineralization occurs in numerous obducted remnants of ocean floor, in places such as Cyprus (Troodos), Albania and Turkey. Again, however, these deposits are likely related to ore-forming processes that predate the Alpine orogeny.

The Andean and Cordilleran orogens of the western Americas contain the greatest concentration of metals on Earth, and are pervasively mineralized from one end to the other. During the Jurassic–Cretaceous period of terrane accretion in North America, small VMS-type Cu–Zn deposits were formed, as were the "mother-lode" hydrothermal gold systems, hosted in obducted oceanic material, in California, British Columbia, and Mexico. Mesozoic and early Cenozoic plutonism gave rise to the numerous world-class porphyry Cu–Mo deposits of the USA and Canada, as well as related skarn and polymetallic epithermal vein deposits, such as the Bingham system in Utah. Thermal collapse and the development of Basin and Range extension gave rise to the Eocene (c. 40 Ma), hydrothermal, sediment-hosted Carlin-type Au deposits of northeast Nevada, followed by Miocene volcanism and development of epithermal Au–Ag mineralization such as the Comstock Lode of Virginia City, Nevada. Similarly, in South America a wide variety of mineralization styles are associated with Mesozoic–Cenozoic orogeny. During the Jurassic–Cretaceous period island arc volcanism and associated hydrothermal activity gave rise to the Fe oxide–phosphate ores now preserved in the western, coastal portions of the belt. The axis of subduction related magmatism commenced in the west during the late Triassic and early Jurassic and then migrated eastwards through the Cretaceous and early Cenozoic (Sillitoe 1976). In the Eastern Cordillera, however, minor magmatism occurred early in the Mesozoic, but most of the magmatism took place in the Miocene. The enormous porphyry copper deposits of Chile (such as Chuquicamata and El Teniente) and Peru (Morococha and Toquepala) formed in response to diachronous episodes of subduction-related magmatism. Many of the important porphyry copper deposits in Peru are Paleocene in age, whereas in northern Chile they are Eocene–Oligocene and, further south, Miocene–Pliocene (Sillitoe 1976). East of the porphyry copper belt a broad zone of vein-related and skarn type Cu–Pb–Zn–Ag mineralization occurs, spatially related to Eocene–Oligocene calc–alkaline plutons. These deposits are most prolific in Peru (Antamina and Cerro de Pasco) but the belt continues southwards into Bolivia and Argentina. Finally, a distinct zone of Sn–W–Ag–Bi mineralization occurs mainly in the Eastern Cordillera of Bolivia and southern Peru, in the elbow of the central Andes. These ores occur as both vein-types associated mainly with Mesozoic intrusions, and disseminated porphyry type deposits (such as Llallagua and Potosi in Bolivia) associated with Oligocene–Miocene S-type granites.

Also important with respect to mineralization associated with post-Pangean orogenesis is the extensive chain of island arcs in the northern and western Pacific. The northern and western Pacific is predominantly characterized by collisions of oceanic crust (as opposed to the ocean–continent collisions of the western Americas) which formed island arcs and built new crust on oceanic (simatic) basement rather than on continental (sialic) basement. There are several types of island arc, each with a particular metallogenic character. These include intra-oceanic arcs (such as Tonga, New Hebrides, and the Solomons), island arcs separated from continental crust by a narrow back-arc sea (such as Japan), and those built directly against a continent (such as Java–Sumatra). Soon after initiation of ocean–ocean collision in the western circum-Pacific region, early, relatively mafic (andesitic–dacitic), stages of calc–alkaline magmatism resulted in porphyry Cu–Au deposit formation, examples of which include Grasberg, Indonesia, and Bougainville in Papua New Guinea. Besshi-type VMS deposits also developed in back-arc settings where andesite–dacite volcanism and ocean floor exhalative activity occurred synchronously with deep water sedimentation. During the later stages of arc construction and calc–alkaline magmatism, dacite–rhyolite volcanism occurred resulting in the development of Kuroko-type VMS deposits, examples of which are known from the Miocene-aged Green Tuff belt of Japan. Also important, in geologically more recent times, is the formation of large epithermal Au–Ag deposits. Several world-class ore deposits of this type occur in the western circum-Pacific region, such as Baguio and Lepanto in the Philippines, Hishikari in Japan, Ladolam, and Porghera in Papua New Guinea, and Emperor in Fiji. Porphyry Cu–Au styles of mineralization are known to occur beneath such epithermal deposits, such as at Baguio and Ladolam.

As a final comment, it is interesting to note that the Paleozoic Central Asian orogenic belts are characterized by many of the features, both temporal and tectonic, of the younger

Andean and Cordilleran orogens. Still somewhat underexplored, these terranes probably represent particularly attractive metallotects for future mineral exploration.

6.6 Plate Tectonic Settings and Ore Deposits – A Summary

The description of ore deposits in the context of their plate tectonic setting has been carried out in considerable detail in numerous publications (Mitchell and Garson 1981; Tarling 1981; Hutchison 1983; Sawkins 1990; Groves and Bierlein 2007). A useful diagrammatic summary is presented in Mitchell and Garson (1981), from which Figures 6.15 and 6.16 are extracted. The latter two diagrams serve as a useful overview of the major plate-related tectonic settings and the ore deposit types associated with each.

6.6.1 Extensional Settings

Incipient rifting of stable continental crust is represented in Figure 6.15a, where thinning and extension may be related to hotspot activity. Magmatism is often localized along craton margins and preserved lineaments or sutures and is alkaline or ultrapotassic in character (to form kimberlites and lamproites). Anorogenic granites such as those of the Bushveld Complex (Sn, W, Mo, Cu, F, etc.), alkaline intrusions such as Phalaborwa (Cu–Fe–P–U–REE) in South Africa and Mount Weld (REE–Y–Nb–Ta–P–Zr) in Western Australia, as well as kimberlites (diamonds), represent ore deposit types formed in this setting. Intracontinental rifts can host SEDEX-type Pb–Zn–Ba–Ag deposits as well as stratiform sediment-hosted Cu–Co deposits. As continental rifting extends to the point that incipient oceans begin to open (such as the Red Sea; Figure 6.15b), basaltic volcanism marks the site of a mid-ocean ridge and this site is also accompanied by exhalative hydrothermal activity and VMS deposit

formation. Such settings also provide the environments for chemical sedimentation and precipitation of banded iron-formations and manganiferous sediments. Continental platforms commonly host organic accumulations that, on catagenesis, give rise to oil deposits. Carbonate sedimentation ultimately provides the rocks which host MVT deposits, although the hydrothermal processes that give rise to these epigenetic Pb–Zn ores are typically associated with circulation during compressional stages of orogeny. Mid-ocean ridges are the culmination of extensional processes (Figure 6.15c). Exhalative activity at these sites gives rise to "black-smoker" vents (such as those at 21°N on the East Pacific Rise) that provide the environments for the formation of Cyprus type VMS deposits. The basalts that form at mid-ocean ridges also undergo fractional crystallization at sub-volcanic depths to form podiform chromite deposits as well as Cu–Ni–PGE sulfide segregations.

6.6.2 Compressional Settings

Andean type collisional margins are both widespread and important as metallotects and are represented in Figure 6.16a. These are the sites of the great porphyry Cu–Mo provinces of the world, while inboard of the arc significant Sn–W granitoid-hosted mineralization also occurs. The volcanic regions above the porphyry systems are also the sites of epithermal precious metal mineralization. A similar tectonic setting can exist between two slabs of oceanic crust, as represented by the island arc environment in Figure 6.16b. Porphyry Cu–Au deposits occasionally occur associated with the early stages of magmatism in these settings, whereas the later, more evolved calc–alkaline magmatism gives rise to Kuroko-type VMS deposits. Back-arc basins represent the sites of Besshi type VMS deposition. Arc–arc collision in the back-arc environment can also result in the preservation of obducted oceanic spreading centers within which podiform Cr and sulfide segregations

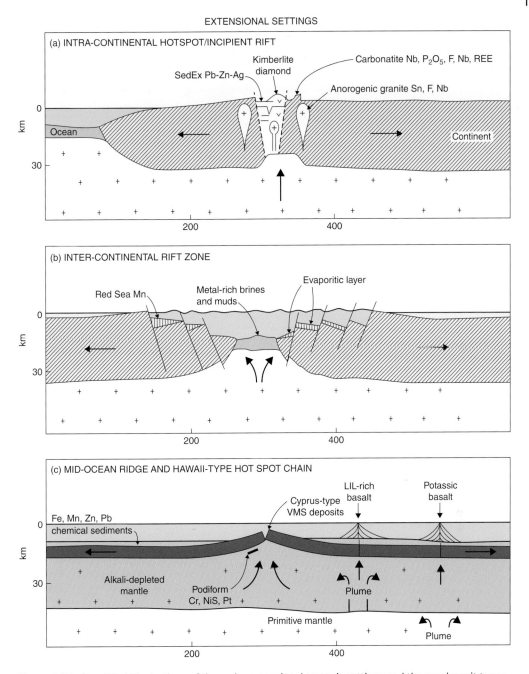

Figure 6.15 Simplified illustrations of the major extensional tectonic settings and the ore deposit types associated with each. Source: Modified after Mitchell and Garson (1981).

might be preserved. In a Japanese-style setting, where the island arc develops fairly close to a continent (Figure 6.16c), marginal sedimentary basins are floored by oceanic crust. This setting typically hosts Besshi- and Cyprus-type VMS deposits. As the arc and continent accrete, ophiolite obduction can occur, and felsic magmatism may give rise to

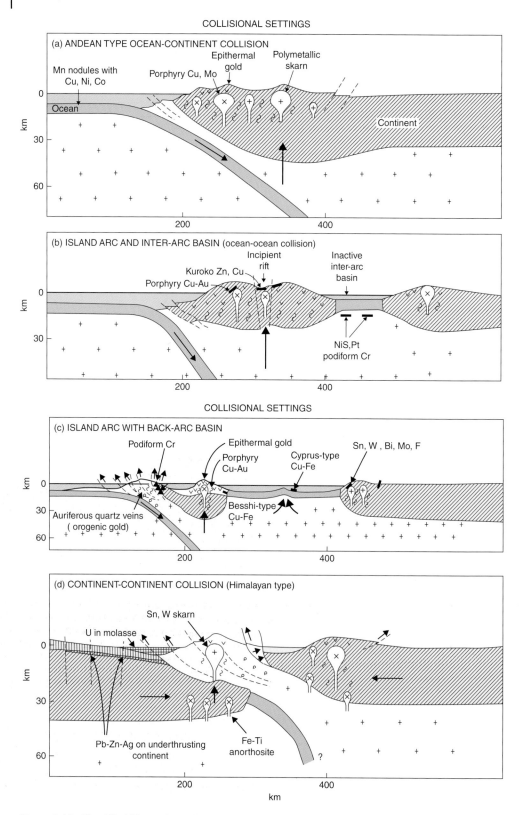

Figure 6.16 Simplified illustrations of the major compressional tectonic settings and the ore deposit types associated with each. Source: Modified after Mitchell and Garson (1981).

large-ion lithophile element mineralization. Ultimately, the oceanic crust is totally consumed to form a zone of continent–continent collision (Figure 6.16d). Modern examples such as the Himalayas and Alps do not appear to be significantly mineralized, but this may be an expression of insufficient exhumation of mineralized zones. Older examples preserve Sn–W–U mineralization in S-type granites, whereas orogeny-driven fluids give rise to orogenic vein-related lode Au systems and MVT Pb–Zn deposits in suitably preserved platformal sediments.

6.7 Summary

Broad trends emerge when relating ore deposit types to geologic time and global tectonic setting. The mechanisms by which the crust has evolved, and the amalgamation and dispersal of continents over time, have played critical roles in the formation of all deposit types. Although the processes are obscure in the early periods of Earth history, it appears that primitive continents started to form after the late heavy meteorite bombardment, from about 4000 Ma. Only a limited range of ore deposit types, mainly related to ocean floor hydrothermal processes, appears to have developed at this stage. As more evolved continents formed from around 3000 Ma and conventional plate tectonic processes were entrenched, the range of ore deposit types broadened, although many of the near-surface variants may not be preserved due to erosion. The Neoarchean Era, especially around 2700 Ma, was a period of intense global orogenesis and ore deposits formed at this time tend to be arc-related and magmatic-hydrothermal to hydrothermal in nature, not unlike those typifying the latter part of the Phanerozoic Eon. Although global tectonic processes continued unabated in the Proterozoic Eon, and a number of major crust-forming orogenies occurred, this extensive period of time appears to have been characterized by longer periods of tectonic quiescence and continental stability – possibly related to the long-lived existence of supercontinents such as Nuna and Rodinia. Consequently, deposits reflect hydrothermal and sedimentary ore-forming processes in intracratonic settings or along passive margins. Likewise, magmatism is more commonly anorogenic in nature and is associated with an entirely different suite of metal deposits than those found in arc-related provinces.

Global tectonic processes are much better understood in the Phanerozoic Eon and the distribution of continental land masses is well documented in terms of the supercontinent cycle. Certain deposit types (such as VMS Cu–Zn ores) are preferentially associated with breakup and dispersal stages of the cycle, such as Gondwana in the early Paleozoic and Pangea in the Mesozoic Eras. Other ores (such as SEDEX Pb–Zn and sediment-hosted Cu deposit types) exhibit a complementary association in terms of the supercontinent cycle and are linked with peak continental amalgamation and stasis. The biggest concentration of metals on the Earth's surface is, however, linked to subduction along the western margins of Pangea, commencing in the Mesozoic and extending through the Cenozoic Era. The present day eastern Pacific Ocean has been, and still is being, consumed beneath North and South America and has given rise to a large variety of world-class deposit types, of which the porphyry and epithermal ores are the most prolific. The skew in the distribution of these and other deposit types is, however, also related to the enhanced preservation of young ore bodies that have not yet been eroded or consumed at a plate margin.

Further Reading

Bradley, D.C. (2011). Secular trends in the geological record and the supercontinent cycle. *Earth-Science Reviews* 108: 16–33.

Cawood, P.A. and Hawkesworth, C.J. (2015). Temporal relations between mineral deposits and global tectonic cycles. In: *Ore Deposits in an Evolving Earth* (eds. G.R.T. Jenkin, P.A.J. Lusty, I. McDonald, et al.), 9–21. Geological Society of London. Special Publication, 393.

Goldfarb, R.J., Groves, D.I., and Gardoll, S. (2001). Orogenic gold and geologic time: a global synthesis. *Ore Geology Reviews* 18: 1–75.

Groves, D.I. and Bierlein, F.P. (2007). Geodynamic settings of mineral deposit systems. *Journal of the Geological Society* 164: 19–30.

Hawkesworth, C.J., Dhuime, B., Pietranik, A.B. et al. (2010). The generation and evolution of the continental crust. *Journal of the Geological Society* 167: 229–248.

Hawkesworth, C.J., Dhuime, B., and Cawood, P.A. (2016). Tectonics and crustal evolution. *GSA Today* 26 (9): 4–11. https://doi.org/10.1130/GSATG272A.1.

Hutchison, C.S. (1983). *Economic Deposits and Their Tectonic Setting*. London: Macmillan Press, 365 pp.

Kesler, S. (1997). Metallogenic evolution of convergent margins: selected ore deposit models. *Ore Geology Reviews* 12: 153–171.

Korenaga, J. (2013). Initiation and evolution of plate tectonics on Earth: theories and observations. *Annual Review of Earth and Planetary Sciences* 41: 117–151.

Pattrick, R.A.D. and Polya, D.A. (1993). *Mineralization in the British Isles*. London: Chapman and Hall, 499 pp.

Sawkins, F.J. (1990). *Metal Deposits in Relation to Plate Tectonics*, 2e. New York: Springer-Verlag, 461 pp.

Solomon, M., Groves, D.I., and Jacques, A.L. (2000). *The Geology and Origin of Australia's Mineral Deposits*. Hobart: University of Tasmania and University of Western Australia, 1002 pp.

Torsvik, T.H. and Cocks, L.R.M. (2013). Gondwana from top to base in space and time. *Gondwana Research* 24: 999–1030.

Windley, B.F. (1995). *The Evolving Continents*, 3e, 526. New York: Wiley.

References

Ague, J.J. (2014). Fluid flow in the deep crust. In: *Treatise on Geochemistry*, 2e, vol. 13 (eds. H.D. Holland and K.K. Turekian), 203–247. Oxford: Elsevier.

Ague, J.J. and Brimhall, G.H. (1989). Geochemical modeling of steady state fluid flow and chemical reactions during supergene enrichment of porphyry copper deposits. *Economic Geology* 84: 506–528.

Aiglsperger, T.H. (2015). Mineralogy and geochemistry of the platinum group elements (PGE), rare earth elements (REE) and scandium in nickel laterites. Doctoral Thesis. University of Barcelona.

Albarede, F. (1996). *Introduction to Geochemical Modeling*. Cambridge University Press, 543 pp.

Alderton, D.H.M. (1993). Mineralization associated with the Cornubian Granite Batholith. In: *Mineralization in the British Isles* (eds. R.A.D. Pattrick and D.A. Polya), 270–354. Chapman & Hall.

Allen, P.A. (1997). *Earth Surface Processes*. Blackwell Science, 404 pp.

Alpers, C.N. and Brimhall, G.H. (1989). Paleohydrological evolution and geochemical dynamics of cumulative metal supergene enrichment at La Escondida, Atacama Desert, northern Chile. *Economic Geology* 84: 229–255.

Ambrose, W.A., Hentz, T.F., Bonnaffé, F. et al. (2009). Sequence-stratigraphic controls on complex reservoir architecture of highstand fluvial-dominated deltaic and lowstand valley-fill deposits in the upper Cretaceous (Cenomanian) Woodbine Group, East Texas field: regional and local perspectives. *AAPG Bulletin* 93 (2): 231–269.

Anderson, G.M. (1975). Precipitation of Mississippi Valley type ores. *Economic Geology* 70: 937–942.

Anderson, G.M. and Burnham, C.W. (1965). The solubility of quartz in supercritical water. *American Journal of Science* 263: 494–511.

Anderson, J.C.Ø., Rasmussen, H., Nielsen, T.F.D., and Ronsbo, J.G. (1998). The Triple Group and the Platinova gold and palladium reefs in the Skaergaard Intrusion: stratigraphic and petrographic relations. *Economic Geology* 91: 488–509.

Annels, A.E. (1984). The geotectonic environment of Zambian copper-cobalt mineralization. *Journal of the Geological Society* 141: 279–289.

Annels, A.E. (1989). Ore genesis in the Zambian copperbelt with particular reference to the northern sector of the Chambishi basin. In: *Sediment-Hosted Stratiform Copper Deposits*, Special Paper, vol. 36 (eds. R.W. Boyle, A.C. Brown, C.W. Jefferson, et al.), 427–452. Geological Association of Canada.

Appel, P.W.U. (1983). Mineral occurrences in the 3.6 Ga old Isua supracrustal belt, west Greenland. In: *Iron Formation: Facts and Problems*, Developments in Precambrian Geology, vol. 6 (eds. A.F. Trendall and R.C. Morris), 593–603. Elsevier.

Arribas, A., Hedenquist, J.W., Itaya, T. et al. (1995). Contemporaneous formation of adjacent porphyry and epithermal Cu–Au

Introduction to Ore-Forming Processes, Second Edition. Laurence Robb.
© 2021 John Wiley & Sons Ltd. Published 2021 by John Wiley & Sons Ltd.

deposits over 300 ka in northern Luzon, Philippines. *Geology* 23: 337–340.

Atkinson, D. and Baker, D.J. (1986). Geology of the MacTung tungsten skarn deposit. In: *Mineral Deposits of the Northern Canadian Cordillera, Yukon–North Eastern British Columbia* (eds. J.G. Abbott and R.J.W. Turner). Geological Survey of Canada, Open File 2169, 279 pp.

Audétat, A., Gunther, D., and Heinrich, C.A. (1998). Formation of a magmatic-hydrothermal ore deposit: insights with LA-ICP-MS analysis of fluid inclusions. *Science* 279: 2091–2094.

Audétat, A., Gunther, D., and Heinrich, C.A. (2000). Causes for large scale metal zonation around mineralized plutons; fluid inclusion LA-ICP-MS evidence from the Mole Granite, Australia. *Economic Geology* 95: 1563–1581.

Audétat, A., Petke, T., Heinrich, C.A., and Bodnar, R.J. (2008). The composition of magmatic-hydrothermal fluids in barren and mineralized intrusions. *Economic Geology* 103: 877–908.

Bagnold, R.A. (1941). *The Physics of Blown Sand and Desert Dunes*. Methuen, 265 pp.

Baker, T., Van Achterburg, E., Ryan, C.G., and Lang, J.R. (2004). Composition and evolution of ore fluids in a magmatic-hydrothermal skarn deposit. *Geology* 32: 117–120.

Baldridge, W.S., McGetchin, T.R., and Frey, F.A. (1973). Magmatic evolution of Hekla, Iceland. *Contributions to Mineralogy and Petrology* 42: 245–258.

Ballard, R.D. (1977). Notes on a major oceanographic find. *Oceanus* 20 (3): 35–44.

Ballhaus, C. (1998). Origin of podiform chromite deposits by magma mingling. *Earth and Planetary Science Letters* 156: 185–193.

Ballhaus, C. and Sylvester, P. (2000). Noble metal enrichment processes in the Merensky Reef, Bushveld Complex. *Journal of Petrology* 41: 545–561.

Banks, D.A., Yardley, B.W.D., Campbell, A.R., and Jarvis, K.E. (1994). REE composition of an aqueous magmatic fluid: a fluid inclusion study from the Capitan Pluton, New Mexico, USA. *Chemical Geology* 113: 259–272.

Bao, Z. and Zhao, Z. (2008). Geochemistry of mineralization with exchangeable REY in the weathering crusts of granitic rocks in South China. *Ore Geology Reviews* 33: 519–535.

Barley, M.E. (1992). A review of Archean volcanic-hosted massive sulfide and sulfate mineralization in Western Australia. *Economic Geology* 87: 855–872.

Barley, M.E. and Groves, D.I. (1992). Supercontinent cycles and distribution of metal deposits through time. *Geology* 20: 291–294.

Barley, M.E., Pickard, A.L., and Sylvester, P.J. (1997). Emplacement of a large igneous province as a possible cause of banded iron formation 2.45 billion years ago. *Nature* 385: 55–58.

Barley, M.E., Krapez, B., Groves, D.I., and Kerrich, R. (1998). The late Archaean bonanza: metallogenic and environmental consequences of the interaction between mantle plumes, lithospheric tectonics and global cyclicity. *Precambrian Research* 91: 65–90.

Barley, M.E., Pickard, A.L., Hagemann, S.G., and Folkert, S.L. (1999). Hydrothermal origin for the 2 billion year old Mount Tom Price giant iron ore deposit, Hamersley Province, Western Australia. *Mineralium Deposita* 34: 784–789.

Barley, M.E., Bekker, A., and Krapez, B. (2005). Late Archean to Early Proterozoic global tectonics, environmental change and the rise of atmospheric oxygen. *Earth Planetary Science Letters* 238: 156–171.

Barnes, H.L. (1967). *Geochemistry of Hydrothermal Ore Deposits*. Holt, Rinehart and Winston, 670 pp.

Barnes, H.L. (1975). Zoning of ore deposits: types and causes. *Transactions of the Royal Society Edinburgh* 69: 295–311.

Barnes, H.L. (1979a). *Geochemistry of Hydrothermal Ore Deposits*, 2e. Wiley, 798 pp.

Barnes, H.L. (1979b). Solubilities of ore minerals. In: *Geochemistry of Hydrothermal Ore Deposits*, 2e (ed. H.L. Barnes), 404–460. Wiley.

Barnes, H.L. (1997). *Geochemistry of Hydrothermal Ore Deposits*, 3e. Wiley, 972 pp.

Barnes, S.-J. and Francis, D. (1995). The distribution of the platinum-group elements, nickel, copper, and gold in the Muskox Layered Intrusion, Northwest Territories, Canada. *Economic Geology* 90: 135–154.

Barnes, S.-J. and Maier, W.D. (2002). Platinum group elements and microstructures of normal Merensky Reef from Impala Platinum Mines, Bushveld Complex. *Journal of Petrology* 43: 103–128.

Barnes, S.-J., von Achterbergh, E., Makovicky, E., and Li, C. (2001). Proton microprobe results for the partitioning of platinum group elements between monosulphide solid solution and sulphide liquid. *South African Journal of Geology* 104: 275–286.

Barnicoat, A.C., Henderson, I.H.C., Knipe, R.J. et al. (1997). Hydrothermal gold mineralization in the Witwatersrand basin. *Nature* 386: 820–824.

Barshad, I. (1966). The effect of a variation in precipitation on the nature of clay mineral formation in soils from acid and basic igneous rocks. *Proceedings of the International Clay Conference* 1: 167–173.

Barton, M.D. (1996). Granitic magmatism and metallogeny of southwestern North America. *Transactions of the Royal Society Edinburgh: Earth Sciences* 87: 261–280.

Bates, R.L. and Jackson, J.A. (1987). *Glossary of Geology*. American Geological Institute, 788 pp.

Battey, M.H. and Pring, A. (1997). *Mineralogy for Students*. Longman, 363 pp.

Bavinton, O. (1981). The nature of sulfidic metasediments at Kambalda and their broad relationships with associated ultramafic rocks and nickel ores. *Economic Geology* 76: 1606–1628.

Beales, F.W. and Jackson, S.A. (1966). Precipitation of lead-zinc ores in carbonate reservoirs as illustrated in Pine Point ore field, Canada. *Transactions of the Institute of Mining Metallurgy B* 75: 278–285.

Bédard, J.H. (2018). Stagnant lids and mantle overturns: implications for Archaean tectonics, magmagenesis, crustal growth, mantle evolution and the start of plate tectonics. *Geoscience Frontiers* 9: 19–49.

Behrens, H. and Gaillard, F. (2006). Geochemical aspects of melts: volatiles and redox behaviour. *Elements* 2 (5): 275–280.

Bekker, A., Holland, H.D., Wang, P.-L. et al. (2004). Dating the rise of atmospheric oxygen. *Nature* 427: 117–120.

Bekker, A., Slack, J.F., Planavsky, N. et al. (2010). Iron formation: the sedimentary product of a complex interplay among mantle, tectonic, oceanic and biospheric processes. *Economic Geology* 105: 467–509.

Bekker, A., Planavsky, N., Krapez, B. et al. (2014). Iron formations: their origins and implications for ancient seawater chemistry. In: *Treatise on Geochemistry*, 2e, vol. 13 (eds. H.D. Holland and K.K. Turekian), 561–628. Oxford: Elsevier.

Bell, K., Kjarsgaard, B.A., and Simonetti, A. (1999). Carbonatites – into the twenty-first century. *Journal of Petrology* 39: 1839–1845.

Bentor, Y.K. (1980). Phosphorites – the unsolved problems. In: *Marine Phosphorites*, vol. 29 (ed. Y.K. Bentor), 3–18. Society of Economic Palaeontologists and Mineralogists, Special Publication,.

Berner, R.A. (1971). *Principles of Chemical Sedimentology*. McGraw-Hill, 240 pp.

Berner, E.K. and Berner, R.A. (1987). *The Global Water Cycle: Geochemistry and Environment*. Prentice Hall.

Berry, L.G., Mason, B., and Dietrich, R.V. (1983). *Mineralogy*. W.H. Freeman and Co, 561 pp.

Best, M.G. (2003). *Igneous and Metamorphic Petrology*, 2e. Blackwell Publishing, 729 pp.

Best, J.L. and Brayshaw, A.C. (1985). Flow separation – a physical process for the concentration of heavy minerals within alluvial channels. *Journal of the Geological Society* 142: 747–755.

Beukes, N.J. and Cairncross, B. (1991). A lithostratigraphic-sedimentological reference profile for the late Archaean Mozaan Group,

Pongola Sequence: application to sequence stratigraphy and correlation with the Witwatersrand Supergroup, South Africa. *South African Journal of Geology* 94: 44–49.

Beukes, N.J. and Gutzmer, J. (2008). Origin and paleoenvironmental significance of major iron formations at the Archean-Paleoproterozoic boundary. In: *Banded Iron Formation-Related High-Grade Iron Ore*, Reviews in Economic Geology, vol. 15 (eds. S. Hagemann, C. Rosière, J. Gutzmer and N.J. Beukes), 5–47. Society of Economic Geologists.

Bierlein, F.P. and Crowe, D.E. (2000). Phanerozoic orogenic lode gold deposits. In: *Gold in 2000*, Reviews in Economic Geology, vol. 13 (eds. S. Hagemann and P.E. Brown), 103–140. Society of Economic Geologists.

Birch, G.F. (1980). A model of penecontemporaneous phosphatization by diagenetic and authigenic mechanisms from the western margin of Southern Africa. In: *Marine Phosphorites*, vol. 29 (ed. Y.K. Bentor), 79–100. Society of Economic Palaeontologists and Mineralogists, Special Publication,.

Bird, D.K., Brooks, C.K., Gannicott, R.A., and Turner, T.A. (1991). A gold bearing horizon in the Skaergaard Intrusion, east Greenland. *Economic Geology* 86: 1083–1092.

Bischoff, J.L. (1969). Red Sea geothermal brine deposits: their mineralogy, geochemistry and genesis. In: *Hot Brines and Recent Heavy Metal Deposits in the Red Sea* (eds. E.T. Degens and D.A. Ross), 368–401. Springer-Verlag.

Bjørlykke, K. (1989). *Sedimentology and Petroleum Geology*. Springer-Verlag, 363 pp.

Bjørlykke, K. (1994). Fluid-flow processes and diagenesis in sedimentary basins. In: *Geofluids: Origin, Migration and Evolution of Fluids in Sedimentary Basins*, vol. 78 (ed. J. Parnell), 127–140. Geological Society, Special Publication.

Bjørlykke, K. (ed.) (2015). *Petroleum Geoscience – From Sedimentary Environments to Rock Physics*, 2e. Springer, 662 pp.

Bland, W. and Rolls, D. (1998). *Weathering: an Introduction to the Scientific Principles*. Arnold, 271 pp.

Bleeker, W. (2003). The late Archean record: a puzzle in ca. 35 pieces. *Lithos* 71: 99–134.

Blevin, P.L. and Chappell, B.W. (1992). The role of magma sources, oxidation states and fractionation in determining the granite metallogeny of eastern Australia. *Transactions of the Royal Society Edinburgh: Earth Sciences* 83: 305–316.

Blevin, P.L., Chappell, B.W., and Allen, C.M. (1996). Intrusive metallogenic provinces in eastern Australia based on granite source and composition. *Transactions of the Royal Society Edinburgh: Earth Sciences* 87: 281–290.

Blundy, J., Mavrogenes, J., Tattitch, B. et al. (2015). Generation of porphyry copper deposits by gas-brine reaction in volcanic arcs. *Nature Geoscience* 8: 235–240.

Bodnar, R., Reynolds, T.J., and Kuehn, C.A. (1985). Fluid inclusion systematics in epithermal systems. In: *Geology and Geochemistry of Epithermal Systems*, Reviews in Economic Geology, vol. 2 (eds. B.R. Berger and P.M. Bethke), 73–97. Society of Economic Geologists.

Bodnar, R.J., Lecumberri-Sanchez, P., Moncada, D., and Steele-MacInnis, M. (2014). Fluid inclusions in hydrothermal ore deposits. In: *Treatise on Geochemistry*, 2e, vol. 13 (eds. H.D. Holland and K.K. Turekian), 120–142. Oxford: Elsevier.

Boer, R., Meyer, F.M., Robb, L.J. et al. (1995). Mesothermal-type mineralization in the Sabie–Pilgrim's Rest Goldfield, South Africa. *Economic Geology* 90 (4): 860–876.

Bottinga, Y. and Javoy, M. (1990). The degassing of Hawaiian tholeiite. *Bulletin of Volcanology* 53: 73–85.

Boudreau, A.E. and Meurer, W.P. (1999). Chromatographic separation of the platinum-group elements, gold, base metals and sulphur during degassing of a compacting and solidifying igneous crystal pile. *Contributions to Mineralogy and Petrology* 134: 174–185.

Bowles, J.F.W. (1986). The development of platinum-group minerals in laterites. *Economic Geology* 81: 1276–1285.

Bowles, J.F.W., Gize, A.P., Vaughan, D.J., and Norris, S.J. (1994). Development of platinum-group minerals in laterites – initial comparison of organic and inorganic controls. *Transactions of the Institution of Mining Metallurgy* 103: B53–B56.

Boxer, G.L., Lorenz, V., and Smith, C.B. (1989). The geology and volcanology of the Argyle (AK1) Lamproite Diatreme, Western Australia. *Geological Society of Australia, Special Publication* 14: 140–152.

Boyce, A.J., Fallick, A.E., Little, C.T.S. et al. (1999). A hydrothermal vent tube worm in the Ballynoe barite deposit, Silvermines, Ireland: implications for timing and ore genesis. In: *Mineral Deposits: Processes to Processing* (eds. C. Stanley, A.H. Rankin, R.J. Bodnar, et al.), 825–827. Balkema.

Bradbury, M.H. and Baeyens, B. (2002). Sorption of Eu on Na- and Ca-montmorillonites: experimental investigations and modelling with cation exchange and surface complexation. *Geochimica et Cosmochimica Acta* 66 (13): 2325–2334.

Bradley, D.C. (2011). Secular trends in the geological record and the supercontinent cycle. *Earth-Science Reviews* 108: 16–33.

Bradley, D., Munk L., Jochens, H., Hynek, S. and Labay, K. (2013) A preliminary deposit model for lithium brines. *U.S Geological Survey, Open File Report 2013-1006*, 6pp.

Brathwaite, R.L. and Faure, K. (2002). The Waihi epithermal gold-silver-base metal sulfide-quartz vein system, New Zealand. *Economic Geology* 97: 269–290.

Brooker, R.A., Kohn, S.C., Holloway, J.R., and McMillan, P.F. (2001). Structural controls on the solubility of CO_2 in silicate melts. Part 1: bulk solubility data. *Chemical Geology* 174: 225–239.

Brown, P.E. (1998). Fluid inclusion modeling for hydrothermal systems. In: *Techniques in Hydrothermal Ore Deposits Geology*, Reviews in Economic Geology, vol. 10 (eds. J.P. Richards and P.B. Larson), 151–171. Society of Economic Geologists.

Brugger, J., Etschmann, B., Grosse, C. et al. (2013). Can biological toxicity drive the contrasting behaviour of platinum and gold in surface environments? *Chemical Geology* 343: 99–110.

Bryndzia, L.T. and Braunsdorf, N.R. (2014). From source rock to reservoir: the evolution of self-sourced unconventional resource plays. *Elements* 10: 271–276.

Buntin, T.J., Grandstaff, D.E., Ulmer, G.C., and Gold, D.P. (1985). A pilot study of geochemical and redox relationships between potholes and adjacent normal Merensky Reef of the Bushveld Complex. *Economic Geology* 80: 975–987.

Burnham, C.W. (1967). Hydrothermal fluids in the magmatic stage. In: *Geochemistry of Hydrothermal Ore Deposits* (ed. H.L. Barnes), 34–76. Holt, Rinehart and Winston.

Burnham, C.W. (1979). Magmas and hydrothermal fluids. In: *Geochemistry of Hydrothermal Ore Deposits*, 2e (ed. H.L. Barnes), 71–136. Wiley.

Burnham, C.W. (1997). Magmas and hydrothermal fluids. In: *Geochemistry of Hydrothermal Ore Deposits*, 3e (ed. H.L. Barnes), 63–123. Wiley.

Burnham, C.W. and Ohmoto, H. (1980). Late-stage processes of felsic magmatism. *Mining Geology*, Special Issue 8: 1–11.

Burrows, D.R. and Spooner, E.T.C. (1987). Generation of a magmatic H_2O-CO_2 fluid enriched in Mo, Au and W within an Archaean sodic granodiorite stock, Mink Lake, northwestern Ontario. *Economic Geology* 82: 1931–1957.

Butt, C.R.M. and Cluzel, D. (2013). Nickel laterite ore deposits: weathered serpentinites. *Elements* 9: 123–128.

Butt, C.R.M., Lintern, M.J., and Anand, R.R. (2000). Evolution of regoliths and landscapes in deeply weathered terrain – implications for geochemical exploration. *Ore Geology Reviews* 16: 167–183.

Button, A. (1979). Transvaal and Hamersley basins—review of basin development and mineral deposits. *Minerals Science and Engineering.* 8 (4): 262–293.

Cairns-Smith, A.G. (1978). Precambrian solution photochemistry, inverse segregation and banded iron-formation. *Nature* 276: 807–808.

Campbell, I.H. and Naldrett, A.J. (1979). The influence of silicate:sulphide ratios on the geochemistry of magmatic sulphide deposits. *Economic Geology* 74: 2248–2253.

Campbell, I.H., Naldrett, A.J., and Barnes, S.J. (1983). A model for the origin of the platinum-rich sulfide horizons in the Bushveld and Stillwater Complexes. *Journal of Petrology* 24: 133–165.

Camus, F. and Dilles, J.H. (2001). A special issue devoted to porphyry copper deposits of northern Chile: preface. *Economic Geology* 96: 233–237.

Candela, P.A. (1989a). Felsic magmas, volatiles and metallogenesis. In: *Ore Deposition Associated with Magmas*, Reviews in Economic Geology, vol. 4 (eds. J.A. Whitney and A.J. Naldrett), 223–233. Society of Economic Geology.

Candela, P.A. (1989b). Magmatic ore-forming fluids: thermodynamic and mass transfer calculations of metal concentrations. In: *Ore Deposition Associated with Magmas*, Reviews in Economic Geology, vol. 4 (eds. J.A. Whitney and A.J. Naldrett), 203–221. Society of Economic Geology.

Candela, P.A. (1991). Physics of aqueous phase evolution in plutonic environments. *American Mineralogist* 76: 1081–1091.

Candela, P.A. (1992). Controls on ore metal ratios in granite related ore systems: an experimental and computational approach. *Transactions of the Royal Society Edinburgh: Earth Sciences* 83: 317–326.

Candela, P.A. (1997). A review of shallow, ore-related granites: textures, volatiles and ore metals. *Journal of Petrology* 38: 1619–1633.

Candela, P.A. and Holland, H.D. (1984). The partitioning of copper and molybdenum between silicate melts and aqueous fluids. *Geochimica et Cosmochimica Acta* 48: 373–380.

Candela, P.A. and Holland, H.D. (1986). A mass transfer model for copper and molybdenum in magmatic hydrothermal systems: the origin of porphyry-type copper deposits. *Economic Geology* 81: 1–18.

Candela, P.A. and Piccoli, P.M. (1995). Model ore-metal partitioning from melts into vapor and vapor-brine mixtures. In: *Magmas, Fluids and Ore Deposits*, vol. 23 (ed. J.F.H. Thompson), 101–128. Mineralogical Association of Canada, Short Course.

Canfield, D.E., Poulton, S.W., Knoll, A.H. et al. (2008). Ferruginous conditions dominated later Neoproterozoic deep-water chemistry. *Science* 15: 949–952.

Carling, P.A. and Breakspear, R.M.D. (2006). Placer formation in gravel-bedded rivers: a review. *Ore Geology Reviews* 28: 377–401.

Carlisle, D. (1983). Concentration of uranium and vanadium in calcretes and gypcretes. In: *Residual Deposits: Surface Related Weathering Processes and Materials* (ed. R.C.L. Wilson), 185–195. The Geological Society of London/Blackwell Scientific Publications.

Carr, H.W., Kruger, F.J., Groves, D.I., and Cawthorn, R.G. (1999). The petrogenesis of Merensky Reef potholes at the Western Platinum Mine, Bushveld Complex: Sr-isotopic evidence for the synmagmatic deformation. *Mineralium Deposita* 34: 335–347.

Carruthers, D., Cartwright, J., Jackson, M.P.A., and Schutjens, P. (2013). Origin and timing of layer-bound radial faulting around North Sea salt stocks: new insights into the evolving stress state around rising diapirs. *Marine and Petroleum Geology* 48: 130–148.

Cathles, L.M. (1981). Fluid flow and genesis of hydrothermal ore deposits. *Economic Geology* 75: 424–457.

Cawood, P.A. and Hawkesworth, C.J. (2015). Temporal relations between mineral deposits and global tectonic cycles. In: *Ore Deposits in an Evolving Earth*, vol. 339 (eds. G.R.T. Jenkin, P.A.J. Lusty, I. McDonald, et al.), 9–21. London, Special Publication: Geological Society.

Cawood, P.A., Hawkesworth, C.J., Pisarevsky, S.A. et al. (2018). Geological archive of the onset of plate tectonics. *Philosophical Transactions Royal Society A* 376 http://dx.doi.org/10.1098/rsta.2017.0405.

Cawthorn, R.G. (1999). Platinum group element mineralization in the Bushveld Complex – a critical reassessment of geochemical models. *South African Journal of Geology* 102: 268–281.

Cawthorn, R.G. (2002). The role of magma mixing in the genesis of PGE mineralization in the Bushveld Complex: thermodynamic calculations and new interpretations – a discussion. *Economic Geology* 97: 663–667.

Cawthorn, R.G. and McCarthy, T.S. (1985). Incompatible trace element behavior in the Bushveld Complex. *Economic Geology* 80: 1016–1026.

Cawthorn, R.G., Harrison, C., and Kruger, F.J. (2000). Discordant ultramafic pegmatoidal pipes in the Bushveld Complex. *Contributions to Mineralogy and Petrology* 140: 119–133.

Cawthorn, R.G., Lee, C.A., Schouwstra, R.P., and Mellowship, P. (2002). Relationship between PGE and PGM in the Bushveld Complex. *Canadian Mineralogist* 40: 311–328.

Cerny, P. (1991). Fertile granites of Precambrian rare element pegmatite fields: is geochemistry controlled by tectonic setting or source lithologies? *Precambrian Research* 51: 429–468.

Cerny, P. and Meintzer, R.E. (1988). Fertile granites in the Archean and Proterozoic fields of rare-element pegmatites: crustal environment, geochemistry and petrogenetic relationships. In: *Recent Advances in the Geology of Granite Related Mineral Deposits*, vol. 39 (eds. R.P. Taylor and D.F. Strong), 170–207. Canadian Institution of Mining, Special Volume.

Chan, M.A. and Kocurek, G. (1988). Complexities in eolian and marine interactions: processes and eustatic controls on erg development. *Sedimentary Geology* 56: 283–300.

Chappell, B.W. and White, A.J.R. (1974). Two contrasting granite types. *Pacific Geology* 8: 173–174.

Charlier, B., Namur, O., Malpas, S. et al. (2010). Origin of the giant Allard Lake ilmenite deposit (Canada) by fractional crystallization, multiple magma pulses and mixing. *Lithos* 117: 119–134.

Charlier, B., Namur, O., Bolle, O. et al. (2015). Fe-Ti-V-P ore deposits associated with Proterozoic massif-type anorthosites and related rocks. *Earth-Science Reviews* 141: 56–81.

Chávez, W.X. (2000) Supergene oxidation of copper deposits: zoning and distribution of copper oxide minerals. *Society of Economic Geologists Newsletter*, 41.

Cheney, E.S. (1996). Sequence stratigraphy and plate tectonic significance of the Transvaal succession of southern Africa and its equivalent in Western Australia. *Precambrian Research* 79: 3–24.

Choukroune, P., Ludden, J.N., Chardon, D. et al. (1997). Archaean crustal growth and tectonic processes: a comparison of the Superior Province, Canada and the Dharwar craton, India. In: *Orogeny Through Time*, vol. 121 (eds. J.-P. Burg and M. Ford), 63–98. Geological Society, Special Publication.

Clark, A.H., Farrar, E., Kontak, D.J. et al. (1990). Geologic and geochronologic constraints on the metallogenic evolution of the Andes of southeastern Peru. *Economic Geology* 85: 1520–1583.

Clemens, J.D. and Vielzeuf, D. (1987). Constraints on melting and magma production in the crust. *Earth and Planetary Science Letters* 86: 287–306.

Clemmey, H. (1985). Sedimentary ore deposits. In: *Sedimentology: Recent Developments and Applied Aspects* (eds. P.J. Brenchley and B.P.J. Williams), 229–247. Geological Society London/Blackwell Scientific Publications.

Cloke, P.L. and Kelly, W.C. (1964). Solubility of gold under inorganic supergene conditions. *Economic Geology* 59: 259–270.

Coetzee, F. (1965). Distribution and grain size of gold, uraninite, pyrite and certain other heavy minerals In gold-bearing reefs of the Witwatersrand basin. *Transactions of the Geological Society of South Africa* 68: 61–68.

Coetzee, J. and Twist, D. (1989). Disseminated tin mineralization in the roof of the Bushveld granite pluton at the Zaaiplaats Mine, with implications for the genesis of magmatic-hydrothermal tin systems. *Economic Geology* 84: 85–102.

Collett, T.S. (1992). Potential of gas hydrates outlined. *Oil and Gas Journal* 22: 84–87.

Colvine, A.C. (1989). An empirical model for the formation of Archean gold deposits: products of final cratonization of the Superior Province, Canada. In: *The Geology of Gold Deposits: The Perspective in 1988*, Economic Geology Monograph, vol. 6 (eds. R.R. Keays, W.R.H. Ramsay and D.I. Groves), 37–53. The Economic Geology Publishing Company.

Comer, J.B. (1974). Genesis of Jamaican bauxite. *Economic Geology* 69: 1251–1264.

Connan, J. (1974). Time–temperature relations in oil genesis. *Bulletin of the American Association of Petroleum Geologists* 58: 2516–2521.

Constantinou, G. and Govett, G.J.S. (1973). Geology, geochemistry and genesis of Cyprus sulfide deposits. *Economic Geology* 68: 843–858.

Cook, P.J. and McElhinny, W. (1979). A reevaluation of the spatial and temporal distribution of sedimentary phosphate deposits in the light of plate tectonics. *Economic Geology* 74: 315–330.

Cook, P.J. and Shergold, J.H. (1984). Phosphorus, phosphorites and skeletal evolution at the Precambrian-Cambrian boundary. *Nature* 308: 2331–2337.

Cooke, D.R. and Simmons, S.F. (2000). Characteristics and genesis of epithermal gold deposits. In: *Gold in 2000*, Reviews in Economic Geology, vol. 13 (eds. S. Hagemann and P.E. Brown), 221–244. Society of Economic Geologists.

Coppin, F., Berger, G., Bauer, A. et al. (2002). Sorption of lanthanides on smectite and kaolinite. *Chemical Geology* 182: 57–68.

Corbett, G.J. and Leach, T.M. (1998). *Southwest Pacific Rim Gold–copper Systems: Structure, Alteration and Mineralization*, vol. 6. Society of Economic Geologists, Special Publication, 237 pp.

Cornell, D.H. and Schütte, S.S. (1995). A volcanic-exhalative origin for the World's largest (Kalahari) manganese field. *Mineralium Deposita* 30: 146–151.

Cornford, C. (1998). Source rocks and hydrocarbons of the North Sea. In: *Petroleum Geology of the North Sea: Basic Concepts and Recent Advances* (ed. K.W. Glennie). Blackwell Science, 636 pp.

Cox, K.G., Bell, J.D., and Pankhurst, R.J. (1979). *The Interpretation of Igneous Rocks*. George Allen & Unwin, 450 pp.

Cox, S.F., Knackstedt, M.A., and Braun, J. (2001). Principles of structural control on permeability and fluid flow in hydrothermal systems. In: *Structural Controls on Ore Genesis*, Reviews in Economic Geology, vol. 14 (eds. J.P. Richards and R.M. Tosdal), 1–24. Society of Economic Geologists.

Craig, H. (1961). Isotopic variations in meteoric waters. *Science* 133: 1702–1703.

Craig, J.R. and Vaughan, D.J. (1994). *Ore Microscopy and Ore Petrography*. Wiley, 434 pp.

Craig, J.R., Vaughan, D.J., and Skinner, B.J. (1996). *Resources of the Earth: Origin, Use and Environmental Impact*. Prentice Hall, 472 pp.

Creaser, R.A. and Cooper, J.A. (1993). U-Pb geochronology of middle Proterozoic felsic magmatism surrounding the Olympic Dam Cu-U-Au-Ag and Moonta Cu-Au-Ag deposits, South Australia. *Economic Geology* 88: 186–197.

Dalziel, I.W.D., Mosher, S., and Gahagan, L.M. (2000). Laurentia–Kalahari collision and the assembley of Rodinia. *Journal of Geology* 108: 499–513.

Davies, G. and Tredoux, M. (1985). The platinum group element and gold contents of the marginal rocks and sills of the Bushveld Complex. *Economic Geology* 80: 838–848.

Davis, D.W. (2008). Sub-million year resolution of Precambrian igneous events by thermal extraction – thermal ionization mass spectrometer Pb dating of zircon: application to crystallization of the Sudbury impact melt sheet. *Geology* 36 (5): 383–386.

De Kock, M.O., Evans, D.A.D., and Beukes, N.J. (2009). Validating the existence of Vaalbara in the Neoarchean. *Precambrian Research* 174 (1): 145–154.

De Kock, M.O., Beukes, N.J., and Armstrong, R.A. (2012). New SHRIMP U-Pb zircon ages from the Hartswater Group, South Africa: implications for correlations of the Neoarchean Ventersdorp Supergroup on the Kaapvaal craton and with the Fortescue Group on the Pilbara craton. *Precambrian Research* 204–205: 66–74.

De Voto, R.H. (1978). *Uranium Geology and Exploration*. Colorado School of Mines, 188 pp.

de Wit, M.J. and Hynes, A. (1995). The onset of interaction between the hydrosphere and oceanic crust, and the origin of the first continental lithosphere. In: *Early Precambrian Processes*, vol. 95 (eds. M.P. Coward and A.C. Ries), 1–9. Geological Society, Special Publication.

de Wit, M.J., de Ronde, C.E.J., Tredoux, M. et al. (1992). Formation of an Archaean continent. *Nature* 357: 553–562.

Deer, W.A., Howie, R.A., and Zussman, J. (1982). *Rock Forming Minerals*, 2e, vol. 1–5. Longman.

Delaney, J.R. and Cosens, B.A. (1982). Boiling and metal deposition in submarine hydrothermal systems. *Marine Technology Society Journal* 16 (3): 62–66.

Denison, R. E., Lidiak, E.G., Bickford, M.E. and Kisvarsanyi, E.B. (1984). Geology and geochemistry of the Precambrian rocks in the Central Interior Region of the United States. *Geological Survey Professional Paper 1241-C*, 20 pp.

Dhuime, B., Hawkesworth, C.J., Cawood, P.A., and Storey, C.D. (2012). A change in the geodynamics of continental growth 3 billion years ago. *Science* 335: 1334–1336.

Dill, H.G. (2015). Supergene alteration of ore deposits: from nature to humans. *Elements* 11 (5): 311–316.

Dingwell, D.B., Holtz, F., and Behrens, H. (1997). The solubility of H_2O in peralkaline and peraluminous granitic melts. *American Mineralogist* 82: 3–4.

Dolson, J. (2016). *Understanding Oil and Gas Shows and Seals in the Search for Hydrocarbons*. Springer 485 pp.

Drennan, G.R., Cathelineau, M., Boiron, M.-C. et al. (1999). Characteristics of post-depositional fluids in the Witwatersrand Basin. *Mineralogy and Petrology* 66: 83–109.

Drew, L.J., Qingrun, M., and Weijun, S. (1990). The Bayan Obo Fe-REE-Nb deposits, Inner Mongolia, China. *Lithos* 26: 46–65.

Duane, M.J. and de Wit, M.J. (1988). Pb–Zn ore deposits of the northern Caledonides: products of continentalscale fluid mixing and tectonic expulsion during tectonic collision. *Geology* 16: 999–1002.

Durrance, E.M. and Bristow, C.W. (1986). Kaolinisation and isostatic readjustment in southwest England. *Proceedings of the Ussher Society* 6: 318–322.

Eddy, C.A., Dilek, Y., Hurst, S., and Moores, E.M. (1998). Seamount formation and associated caldera complex and hydrothermal mineralization in ancient oceanic crust, Troodos ophiolite (Cyprus). *Tectonophysics* 292: 189–210.

Eidel, J.J. (1991). Basin analysis for the mineral industry. In: *Sedimentary and Diagenetic Mineral Deposits: A Basin Analysis Approach to Exploration*, Reviews in Economic Geology, vol. 5 (eds. E.R. Force, J.J. Eidel and J.B. Maynard), 1–15. Society of Economic Geologists.

Einaudi, M. (1982). Description of skarns associated with porphyry copper plutons: southwestern North America. In: *Advances in Geology of the Porphyry Copper Deposits: Southwestern North America* (ed. S.R. Titley), 139–183. University of Arizona Press.

Einaudi, M. (2000). Mineral resources: assets and liabilities. In: *Earth Systems: Processes and Issues* (ed. W.G. Ernst), 346–372. Cambridge University Press.

Einaudi, M., Meinert, L.D., and Newberry, R.J. (1981). Skarn deposits. *Economic Geology* 75: 317–391.

Eldridge, C.S., Compston, W., Williams, I.S. et al. (1995). Applications of the SHRIMP I Ion Microprobe to the understanding of processes and timing of diamond formation. *Economic Geology* 90: 271–280.

Emmons, W.H. (1936) Hypogene zoning in metalliferous lodes. *Report 1 of the 16th International Geological Congress*, pp. 417–432.

England, P. and Houseman, G. (1984). On the geodynamic setting of kimberlite genesis. *Earth and Planetary Science Letters* 67: 109–122.

England, G.L., Rasmussen, B., Krapez, B., and Groves, D.I. (2001). The origin of uraninite, bitumen nodules and carbon seams in Witwatersrand gold–uranium–pyrite ore deposits, based on a Permo-Triassic analogue. *Economic Geology* 96: 1907–1920.

Eriksson, S.C. (1989). Palaborwa: a saga of magmatism, metasomatism and miscibility. In: *Carbonotites: Genesis and Evolution* (ed. J.D. Bell), 221–254. Unwin Hyman.

Eugster, H.P. (1980). Geochemistry of evaporitic lacustrine deposits. *Annual Reviews Earth and Planetary Sciences* 8: 35–63.

Evans, D.A.D. and Mitchell, R.N. (2011). Assembly and break-up of the core of Paleoproterozoic-Mesoproterozoic supercontinent Nuna. *Geology* 39 (5): 443–446.

Evans, D.A.D., Martin, D.McB., Nelson, D.R., Powell, C.McA. and Wingate, M.T.D. (2000) The Vaalbara hypothesis reviewed. Extended Abstract. 31st International Geological Congress, Rio de Janeiro, Brazil (6–17 August 2000).

Exley, C.S. (1976). Observations on the formation of kaolinite in the St Austell granite, Cornwall. *Clay Minerals* 11: 51–63.

Fallick, A.E., Ashton, J.H., Boyce, A.J. et al. (2001). Bacteria were responsible for the magnitude of the world-class hydrothermal base metal sulphide ore body at Navan, Ireland. *Economic Geology* 96: 885–890.

Ferguson, J. and Currie, K.L. (1971). Evidence of liquid immiscibility in alkaline ultrabasic dikes at Callander Bay, Ontario. *Journal of Petrology* 12: 561–585.

Field, M., Gibson, T.A., Wilkes, J. et al. (1997). The geology of the Orapa A/K1 kimberlite, Botswana: further insight into the emplacement of kimberlite pipes. *Russian Geology and Geophysics* 38 (1): 261–276.

Filippelli, G.M. (2008). The global phosphorus cycle: past, present, and future. *Elements* 4: 89–95.

Fischer, B.J., Martel, E. and Falck, H. (2018). Geology of the Mactung tungsten skarn and area – review and 2016 field observations. *Northwest Territories Geological Survey, NWT Open File 2018-02*, 84 pp.

Fleischer, V.D. (1984). Discovery, geology and genesis of copper-cobalt mineralization at Chambishi Southeast prospect, Zambia. *Precambrian Research* 25: 119–133.

Fleischer, V.D., Garlick, W.G., and Haldane, R. (1976). Geology of the Zambian copper belt. In: *Handbook of Strata-bound and Stratiform Ore Deposits: II Regional Studies and Specific Deposits*, vol. 6 (ed. K.H. Wolff), 223–352. Elsevier.

Force, E.C. (1991a). Geology of titanium deposits. *Bulletin Geological Society of America* 129: 19–37.

Force E.C. (1991b) Placer deposits in shoreline-related sands of Quaternary Age. *United States Geological Survey, Special Publication*.

Force, E.C. and Maynard, J.B. (1991). Manganese: syngenetic deposits on the margins of anoxic basins. In: *Sedimentary and Diagenetic Mineral Deposits: A Basin Analysis Approach to Exploration*, Reviews in Economic Geology, vol. 5 (eds. E.R. Force, J.J. Eidel and J.B. Maynard), 147–157. Society of Economic Geologists.

Force, E.C., Eidel, J.J., and Maynard, J.B. (1991). *Sedimentary and Diagenetic Mineral Deposits: A Basin Analysis Approach to Exploration*, Reviews in Economic Geology, vol. 5. Society of Economic Geologists, 216 pp.

Forsman, J.P. and Hunt, J.M. (1958). Insoluble organic matter (kerogen) in sedimentary rocks. *Geochimica et Cosmochimica Acta* 15: 170–182.

Frakes, L.A. and Bolton, B.R. (1992). Effects of ocean chemistry, sea level and climate on the formation of primary sedimentary manganese deposits. *Economic Geology* 87: 1207–1217.

Francheteau, J., Needham, H.D., Choukroune, P. et al. (1979). Massive deep-sea sulphide ore deposits discovered on the East Pacific Rise. *Nature* 277: 523–528.

Frank, J.R., Cluff, S., and Bauman, J.E. (1982). Painter reservoir, East Painter reservoir and Clear Creek Fields, Unita County, Wyoming. In: *Geologic Studies of the Cordilleran Thrust Belt* (ed. R.B. Powers), 601–618. Colorado: Rocky Mountain Association of Geologists.

Franklin, J.M. (1993). Volcanic-associated massive sulfide deposits. In: *Mineral Deposit Modeling*, Special paper, vol. 40 (eds. R.V. Kirham, W.D. Sinclair, R.I. Thorpe, et al.), 315–334. Geological Association of Canada.

Franklin, J.M., Lydon, J.W., and Sangster, D.F. (1981). Volcanic-associated massive sulfide deposits. *Economic Geology* 75: 485–627.

Franklin, J.M., Gibson, H.L., Jonasson, I.R., and Galley, A.G. (2005). Volcanogenic massive sulfide deposits. In: *One Hundredth Anniversary Volume* (eds. J.W. Hedenquist, J.F.H. Thompson, R.J. Goldfarb and J.P. Richards), 523–560. Society of Economic Geologists.

Freitsch, R. (1978). On the origin of iron ores of the Kiruna type. *Economic Geology* 73: 478–485.

Friedman, G.M., Sanders, J.E., and Kopaska-Merkel, D.C. (1992). *Principles of Sedimentary Deposits*. Macmillan, 717 pp.

Frimmel, H.E. (1997). Detrital origin of hydrothermal Witwatersrand gold: a review. *Terra Nova* 9 (4): 192–197.

Frimmel, H.E. and Minter, W.E.L. (2002). Recent developments concerning the geological history and genesis of the Witwatersrand gold deposits, South Africa. *Society of Economic Geologists*, Special Publication 9: 17–45.

Frimmel, H., Hallbauer, D.K., and Gartz, V.H. (1999). Gold mobilizing fluids in the Witwatersrand basin: composition and possible sources. *Mineralogy and Petrology* 66: 55–81.

Fuchs, W.A. and Rose, A.W. (1974). The geochemical behaviour of platinum and palladium in the weathering cycle in the Stillwater Complex, Montana. *Economic Geology* 69: 332–346.

Fyfe, W.S. (2000). The life support system: toward Earth sense. In: *Earth Systems: Processes and Issues* (ed. W.G. Ernst), 506–515. Cambridge University Press.

Gaál, G. (1987). An outline of the Precambrian evolution of the Baltic Shield. *Precambrian Research* 35: 15–52.

Gaál, G. (1990). Tectonic styles of Early Proterozoic ore deposition in the Fennoscandian Shield. *Precambrian Research* 46: 83–114.

Galley, A.G., Hannington, M.D., and Jonasson, I. (2007). Volcanogenic massive sulfide deposits. In: *Mineral Deposits of Canada* (ed. W.D. Goodfellow), 141–162. Geological Association of Canada Special Publication 5.

Galloway, W.E. and Hobday, D.K. (1983). *Terrigeneous Clastic Depositional Systems*. Springer-Verlag, 423 pp.

Gardiner, N.J., Robb, L.J., Morley, C.K. et al. (2016). The tectonic and metallogenic framework of Myanmar: a Tethyan mineral system. *Ore Geology Reviews* 79: 26–45. https://doi.org/10.1016/j.oregeorev.2016.04.02.

Garlick, W.G. (1982). Erosion of the folded copper-rich arenite filling of a rolled-up algal mat, Mufulira, Zambia. *Economic Geology* 77: 1934–1950.

Garrels, R.M. and Christ, C.L. (1965). *Solutions, Minerals and Equilibria*. Harper and Row, 450 pp.

Garven, G. and Raffensperger (1997). Hydrogeology and geochemistry of ore genesis in sedimentary basins. In: *Geochemistry of Hydrothermal Ore Deposits*, 3e (ed. H.L. Barnes), 125–190. Wiley.

Garven, G., Ge, S., Person, M.A., and Sverjensky, D.A. (1993). Genesis of stratabound ore deposits in the midcontinent basins of North America. *American Journal of Science* 293: 497–568.

Gernon, T.M., Fontana, G., Field, M. et al. (2009). Pyroclastic flow deposits from a kimberlite eruption: the Orapa south crater, Botswana. *Lithos* 112S: 566–578.

Gibson, R.L. and Reimold, W.U. (1999). The significance of the Vredefort dome for the thermal and structural evolution of the Witwatersrand Basin, South Africa. *Mineralogy and Petrology* 66: 5–23.

Giggenbach, W. (1997). The origin and evolution of fluids in magmatic-hydrothermal systems. In: *Geochemistry of Hydrothermal Ore Deposits*, 3e (ed. H.L. Barnes), 737–796. Wiley.

Giggenbach, W.F. and Soto, R.C. (1992). Isotopic and chemical composition of water and steam discharges from volcanic-magmatic-hydrothermal systems of the Guanacaste Geothermal Province, Costa Rica. *Applied Geochemistry* 7: 309–332.

Gill, R. (1996). *Chemical Fundamentals of Geology*. Chapman & Hall, 290 pp.

Gilmour, P. (1976). Some transitional types of mineral deposits in volcanic and sedimentary rocks. In: *Handbook of Strata-bound and Stratiform Ore Deposits*, vol. 1 (ed. K.H. Wolf), 111–160. Elsevier.

Giordano, T.H. (1994). Metal transport in ore fluids by organic ligand complexation. In: *Organic Acids in Geological Processes* (eds. E.D. Pittman and M.D. Lewan), 319–354. Springer-Verlag.

Giordano, T.H. (2002). Transport of Pb and Zn by carboxylate complexes in basinal brine fluids and related petroleum-field brines at 100°C: the influence of pH and oxygen fugacity. *Geochemical Transactions* 3 (8): 56–72.

Giordano, T.H. and Kharaka, Y.K. (1994). Organic ligand distribution and speciation in sedimentary basin brines, diagenetic fluids and related ore solutions. In: *Geofluids: Origin, Migration and Evolution of Fluids in Sedimentary Basins*, vol. 78 (ed. J. Parnell), 175–202. Geological Society, Special Publication.

Gize, A.P. (1999). Organic alteration in hydrothermal sulfide ore deposits. *Economic Geology* 94: 967–980.

Gize, A.P. and Hoering, T.C. (1980). The organic matter in Mississippi Valley-type deposits. *Carnegie Institute of Washington Yearbook* 79: 384–388.

Glennie, K.W. (2000). Cretaceous tectonic evolution of Arabia's eastern plate margin: a tale of two oceans. *SEPM*, Special Publication 69: 9–20.

Gluyas, J. and Swarbrick, R. (2004). *Petroleum Geoscience*. Oxford: Blackwell Science, 359 pp.

Goldhaber, M.B., Reynolds, R.L., and Rye, R.O. (1978). Origin of a south-Texas roll-type uranium deposit: II. Sulfide petrology and sulfur isotope studies. *Economic Geology* 73: 1690–1705.

Goldhaber, M.B., Reynolds, R.L., and Rye, R.O. (1983). Role of fluid mixing and fault-related sulfide in the origin of the Ray Point uranium district, south Texas. *Economic Geology* 78: 1043–1063.

Golightly, J.P. (1981). Nickeliferous laterite deposits. *Economic Geology* 75: 710–735.

Goodfellow, W.D., Lydon, J.W., and Turner, R.J.W. (1993). Geology and genesis of stratiform sediment-hosted (SedEx) zinc-lead-silver sulfide deposits. In: *Mineral Deposit Modeling*, Special paper, vol. 40 (eds. R.V. Kirham, W.D. Sinclair, R.I. Thorpe, et al.), 201–252. Geological Association of Canada.

Gower, C.F., Ryan, A.B., and Rivers, T. (1990). Mid-Proterozoic Laurentia-Baltica: an overview of its geological evolution and a summary of the contributions made by this volume. In: *Mid-Proterozoic Laurentia-Baltica*, Special Paper, vol. 38 (eds. C.F. Gower, T.

Rivers and A.B. Ryan), 1–20. Geological Association Canada.

Gradstein, F.M., Ogg, J.G., and Smith, A.G. (2004). *A Geologic Time Scale 2004.* Cambridge University Press.

Gradstein, F.M., Ogg, J.G., Schmitz, M., and Ogg, G. (2012). *The Geologic Time Scale 2012.* Elsevier, 1176 pp.

Grainger, C.J., Groves, D.I., and Costa, C.H.C. (2002). The epigenetic sediment-hosted Serra Pelada Au-PGE deposit and its potential genetic association with Fe oxide Cu-Au mineralization within the Carajás Mineral Province, Amazon Craton, Brazil. In: *Tectonics and Metallogeny of the Tethyan Orogenic Belt* (ed. J.P. Richards), 47–64. Society of Economic Geologists, Special Publication Number 19.

Greyling, L.N., Robb, L.J., Master, S., and Boiron, M.C. (2005). The nature of early mineralising fluids associated with the Chambishi deposit, Zambian Copperbelt. *Journal of African Earth Sciences*. Special Issue 42: 159–172.

Grinenko, L.N. (1985). Sources of sulphur of the nickeliferous and barren gabbro-dolerite intrusions of the northwest Siberian platform. *International Geological Reviews* 27: 695–708.

Gross, G.A. (1980). A classification of iron-formation based on depositional environments. *Canadian Mineralogist* 18: 215–222.

Gross, G.A., Gower, C.F. and Lefebure, D.V. (1997) Magmatic Ti–Fe–V oxide deposits. *Geological Fieldwork 1997*, British Columbia Ministry of Employment and Investment Paper 1998-1, pp. 24J-1–24J-3.

Groves, D.I. (1993). The crustal continuum model for late-Archaean lode-gold deposits of the Yilgarn block, Western Australia. *Mineralium Deposita* 28: 366–374.

Groves, D.I. and Bierlein, F.P. (2007). Geodynamic settings of mineral deposit systems. *Journal of the Geological Society* 164: 19–30.

Groves, D.I. and McCarthy, T.S. (1978). Fractional crystallization and the origin of tin deposits in granitoids. *Mineralium Deposita* 13: 11–26.

Groves, D.I., Goldfarb, R.J., Gebre-Mariam, M. et al. (1998). Orogenic gold deposits: a proposed classification in the context of their crustal distribution and relationship to other gold deposit types. *Ore Geology Reviews* 13: 7–27.

Guilbert, J.M. and Park, C.F. (1986). *The Geology of Ore Deposits*. W.H. Freeman and Co, 985 pp.

Gustafson, L.B. and Williams, N. (1981). Sediment-hosted stratiform deposits of copper, lead and zinc. *Economic Geology* 75: 139–178.

Haeussler, P.J., Bradley, D., Goldfarb, R. et al. (1995). Link between ridge subduction and gold mineralization in southern Alaska. *Geology* 23: 995–998.

Hagemann, S.G. and Cassidy, K.F. (2000). Archean orogenic lode gold deposits. In: *Gold in 2000*, Reviews in Economic Geology, vol. 13 (eds. S. Hagemann and P.E. Brown), 9–68. Society of Economic Geologists.

Hagemann, S.G., Angerer, T., Duuring, P. et al. (2016). BIF-hosted iron mineral system: a review. *Ore Geology Reviews* 76: 317–359.

Haggerty, S.E. (1989). Mantle metasomes and the kinship between carbonatites and kimberlites. In: *Carbonatites Genesis and Evolution* (ed. J.D. Bell). Unwin Hyman, 618 pp.

Haggerty, S.E. (1999). A diamond trilogy: superplumes, supercontinents and supernovae. *Science* 285: 851–860.

Hall, A. (1996). *Igneous Petrology*. Longman, 551 pp.

Hall, R.P. and Hughes, D.J. (1993). Early Precambrian crustal development: changing styles of mafic magmatism. *Journal of the Geological Society* 150: 625–635.

Hallam, A. (1992). *Phanerozoic Sea-Level Changes*. Columbia University Press, 266 pp.

Hannah, J.L. and Stein, H.J. (1990). Magmatic and hydrothermal processes in ore-bearing systems. In: *Ore-bearing Granite Systems; Petrogenesis and Mineralizing Processes*, Special Paper, vol. 246 (eds. H.J. Stein and J.L. Hannah), 1–10. Geological Society of America.

Hannington, M.D. (2014). Volcanogenic massive sulfide deposits. In: *Treatise on Geochemistry*,

2e, vol. 13 (eds. H.D. Holland and K.K. Turekian), 463–488. Oxford: Elsevier.

Hanor, J.S. (1979). The sedimentary genesis of hydrothermal fluids. In: *Geochemistry of Hydrothermal Ore Deposits* (ed. H.L. Barnes), 137–172. Wiley-Interscience.

Hanor, J.S. (1994). Origin of saline fluids in sedimentary basins. In: *Geofluids: Origin, Migration and Evolution of Fluids in Sedimentary Basins*, vol. 78 (ed. J. Parnell), 151–174. Geological Society, Special Publication.

Hardisty, J. (1994). Beach and nearshore sediment transport. In: *Sediment Transport and Depositional Processes* (ed. K. Pye), 219–255. Blackwell Scientific.

Harland, W.B. (1965). Critical evidence for a great infra-Cambrian glaciation. *Geologische Rundschau* 54: 45–61.

Harland, W.B. (1989). *A Geologic Time Scale*. Cambridge University Press, 263 pp.

Harmer, R.E. and Gittins, J. (1998). The case for primary mantle-derived carbonatite magma. *Journal of Petrology* 39: 1895–1903.

Hartleb, J.W.O. (1988). The Langer Heinrich uranium deposit: Southwest Africa/Namibia. *Ore Geology Reviews* 3: 277–287.

Hawkesworth, C.J., Dhuime, B., Pietranik, A.B. et al. (2010). The generation and evolution of the continental crust. *Journal Geological Society London* 167: 229–248.

Hawkesworth, C.J., Cawood, P.A., and Dhuime, B. (2016). Tectonics and crustal evolution. *GSA Today* 26 (9): 4–11.

Haynes, D.W. (1986a). Stratiform copper deposits hosted by low-energy sediments: I. Timing of sulfide precipitation – an hypothesis. *Economic Geology* 81: 250–265.

Haynes, D.W. (1986b). Stratiform copper deposits hosted by low-energy sediments: II. Nature of source rocks and composition of metal-transporting water. *Economic Geology* 81: 266–280.

Haynes, D.W., Cross, K.C., Bills, R.T., and Reed, M.H. (1995). Olympic Dam ore genesis: a fluid-mixing model. *Economic Geology* 90: 281–307.

Heath, G.R. (1981). Ferromanganese nodules of the deep sea. *Economic Geology* 75: 736–765.

Hecht, L. and Cuney, M. (2000). Hydrothermal alteration of monazite in the Precambrian crystalline basement of the Athabasca Basin (Saskatchewan, Canada); implications for the formation of unconformity-related uranium deposits. *Mineralium Deposita* 35: 791–795.

Heckenmüller, M., Narita, D. and Klepper, G. (2014). Global availability of phosphorus and its implications for global food supply: an economic overview. Kiel Institute for the World Economy, Working Paper 1897.

Hedenquist, J.W. and Aoki, M. (1991). Meteoric interaction with magmatic discharges in Japan and the significance for mineralization. *Geology* 14: 1041–1044.

Hedenquist, J.W., Simmons, S.F., Giggenbach, W.F., and Eldridge, C.S. (1993). White Island, New Zealand, volcanic-hydrothermal system represents the geochemical environment of high-sulfidation Cu and Au ore deposition. *Geology* 21: 731–734.

Hedenquist, J.W., Izawa, E., White, N.C. et al. (1994). Geology, geochemistry and origin of high sulfidation Cu-Au mineralization in the Nansatsu district, Japan. *Economic Geology* 89: 1–30.

Hedenquist, J.W., Izawa, E., Arribas, A., and White, N.C. (1996). *Epithermal gold deposits: styles, characteristics and exploration*, Resource Geology Special Publication, vol. 1. Society of Resource Geology, 17 pp.

Hedenquist, J.W., Arribas, A., and Reynolds, T.J. (1998). Evolution of an intrusion-centred hydrothermal system: far Southeast-Lepanto porphyry – epithermal Cu–Au deposits, Phillipines. *Economic Geology* 93: 373–404.

Hedenquist, J.W., Arribas, R.A., and Gonzalez, U.E. (2000). Exploration for epithermal gold deposits. In: *Gold in 2000*, Reviews in Economic Geology, vol. 13 (eds. S. Hagemann and P.E. Brown), 245–277. Society of Economic Geologists.

Hein, J.R. and Koschinsky, A. (2014). Deep-ocean ferromanganese crusts and nodules. In: *Treatise on Geochemistry*, 2e, vol.

13 (eds. H.D. Holland and K.K. Turekian), 273–291. Oxford: Elsevier.

Heinrich, C.A., Ryan, C.G., Mernagh, T.P., and Eadington, P.J. (1992). Segregation of ore metals between magmatic brine and vapour: a fluid inclusion study using PIXE microanalysis. *Economic Geology* 87: 1566–1583.

Heinrich, C.A., Gunther, D., Audetat, A. et al. (1999). Metal fractionation between magmatic brine and vapor, determined by microanalysis of fluid inclusions. *Geology* 27: 755–758.

Hekinian, H. and Fouquet, Y. (1985). Volcanism and metallogenesis of axial and off-axial structures on the East Pacific Rise near 13°N. *Economic Geology* 80: 221–249.

Helgeson, H. (1964). *Complexing and Hydrothermal Ore Deposition*. Pergamon Press, 128 pp.

Helmy, H.M., Ballhaus, C., Fonseca, R.O.C. et al. (2013). Noble metal nanoclusters and nanoparticles precede mineral formation in magmatic sulphide melts. *Nature Communications* 4: 2405. https://doi.org/10.1038/ncomms3405.

Herzberg, C., Condie, K., and Korenaga, J. (2010). Thermal history of the Earth and its petrological expression. *Earth and Planetary Science Letters* 292: 79–88.

Herzig, P.M. and Hannington, M.D. (1995). Polymetallic massive sulfides at the modern seafloor; a review. *Ore Geology Reviews* 10: 95–115.

Hiemstra, S.A. (1979). The role of collectors in the formation of the platinum deposits in the Bushveld complex. In: *Nickel-Sulfide and Platinum-group-element Deposits*, vol. 17 (ed. A.J. Naldrett), 469–482. Canadian Mineralogist.

Hildebrand, R.S. (1986). Kiruna-type deposits: their origin and relationship to intermediate subvolcanic plutons in the Great Bear Magmatic Zone, Northwest Canada. *Economic Geology* 81: 640–659.

Hoffman, P.F. (1988). United plates of America, the birth of a craton: early Proterozoic assembly and growth of Laurentia. *Annual Reviews of Earth and Planetary Science* 16: 543–603.

Hoffman, P.F. (1991). Did the breakout of Laurentia turn Gondwanaland inside-out? *Science* 252: 1409–1411.

Hoffman, P.F., Kaufman, A.J., Halverson, G.P., and Schrag, D.P. (1998). A Neoproterozoic snowball Earth. *Science* 281: 1342–1346.

Hofstra, A. and Cline, J. (2000). Characteristics and models for Carlin type gold deposits. In: *Gold in 2000*, Reviews in Economic Geology, vol. 13 (eds. S. Hagemann and P.E. Brown), 163–220. Society of Economic Geologists.

Holland, H.D. (1972). Granites, solutions and base metal deposits. *Economic Geology* 67: 281–301.

Holland, H.D. (1984). *The Chemical Evolution of the Atmosphere and the Oceans*. Princeton University Press, 598 pp.

Holland, H.D. (2006). The oxygenation of the atmosphere and oceans. *Philosophical Transactions Royal Society* B361: 903–915.

Holloway, J.R. (1987). Igneous fluids. In: *Thermodynamic Modelling of Geologic Materials: Minerals, Fluids and Melts*, Reviews in Mineralogy, vol. 17 (eds. I.S.E. Carmichael and H.P. Eugster), 211–234. Mineralogical Society of America.

Holwell, D.A. and Keays, R.R. (2014). The formation of low-volume, high-tenor magmatic PGE-Au sulphide mineralization in closed systems: evidence from precious and base metal geochemistry of the Platinova Reef, Skaergaard Intrusion, East Greenland. *Economic Geology* 109: 387–406.

Holzheid, A., Sylvester, P., O'Neill, H.S.C. et al. (2000). Evidence for a late chondritic veneer in the Earth's mantle from high-pressure partitioning of palladium and platinum. *Nature* 406: 396–399.

Hughes, M.G., Keene, J.B., and Joseph, R.G. (2000). Hydraulic sorting of heavy mineral grains by swash on a medium-sand beach. *Journal of Sedimentary Research* 70: 994–1004.

Humphreys, M.C.S. (2011). Silicate liquid immiscibility within the crystal mush: evidence from Ti in plagioclase from the

Skaergaard Intrusion. *Journal of Petrology* 52: 147–174.

Hunt, J.M. (1979). *Petroleum Geochemistry and Geology.* W.H. Freeman and Company, 617 pp.

Huppert, H.H. and Sparks, R.S.J. (1980). Restrictions on the compositions of mid-ocean ridge basalts; a fluid dynamical investigation. *Nature* 286: 46–48.

Hutchinson, R.W. and Searle, D.L. (1971). *Stratabound pyrite deposits in Cyprus and relation to other sulphide ores*, vol. 3, 198–205. Society of Mining Geologists Japan , Special Publication.

Hutchison, C.S. (1983). *Economic Deposits and Their Tectonic Setting.* Macmillan Press, 365 pp.

Hyndman, D.W. (1981). Controls on source and depth of emplacement of granitic magma. *Geology* 9: 244–249.

Irvine, T.N. (1977). Origin of chromitite layers in the Muskox Intrusion and other stratiform intrusions: a new interpretation. *Geology* 5: 273–277.

Irvine, T.N., Keith, D.W., and Todd, S.G. (1983). The J-M platinum–palladium reef of the Stillwater Complex, Montana: II. origin by double-diffusive convective magma mixing and implications for the Bushveld Complex. *Economic Geology* 78: 1287–1334.

Irvine, T.N., Anderson, J.C.Ø., and Brooks, K. (1998). Included blocks (and blocks within blocks) in the Skaergaard Intrusion: geologic relations and the origins of rhythmic modally graded layers. *Bulletin of the Geological Society of America* 110: 1398–1447.

Ishihara, S. (1977). The magnetite-series and ilmenite-series granitic rocks. *Mining Geology* 26: 293–305.

Ishihara, S. (1978). Metallogenesis in the Japanese island arc system. *Journal of the Geological Society* 135: 389–406.

Ishihara, S. (1981). The granitoid series and mineralization. *Economic Geology* 75: 458–484.

Ishihara, S. and Takenouchi, S. (1980). *Granitic Magmatism and Related Mineralization*, vol. 8.

Society of Mining Geologists of Japan, Special Issue, 247 pp.

Isley, A.E. (1995). Hydrothermal plumes and the delivery of iron to banded iron formation. *The Journal of Geology* 103: 169–185.

Isley, A.E. and Abbott, D.H. (1999). Plume-related mafic volcanism and the deposition of banded iron-formation. *Journal of Geophysical Research* 104 (B7): 15461–15477.

Issa Filho, A.I., Riffel, B.F. and de Faria Sousa, C.A. (2014). Some aspects of the mineralogy of CBMM niobium deposit and mining and pyrochlore ore processing, Araxá, MG, Brazil. www.cbmm.com

Ixer, R.A. (1990). *Atlas of Opaque and Ore Minerals in Their Associations.* Open University Press, 208 pp.

Izawa, E., Etho, J., Misuzu, H. et al. (2001). Hishikari gold mineralization: a case study of the Hosen No. 1 vein hosted by basement Shimanto rocks, southern Kyushu, Japan. In: *Epithermal Gold Mineralization and Modern Analogues, Kyushu, Japan*, Guidebook Series, vol. 34 (eds. C.A. Feebrey, T. Hayashi and S. Taguchi), 21–30. Society of Economic Geologists.

Jackson, N.J., Willis-Richards, J., Manning, D.A.C., and Sams, M. (1989). Evolution of the Cornubian Ore Field, southwest England: part II. Mineral deposits and ore-forming processes. *Economic Geology* 84: 1101–1133.

Jahns, R.H. and Burnham, C.W. (1969). Experimental studies of pegmatite genesis: I. A model for the derivation and crystallization of granitic pegmatites. *Economic Geology* 64: 843–864.

James, H.L. (1954). Sedimentary facies of iron-formation. *Economic Geology* 49: 235–293.

Jean, G.E. and Bancroft, G.M. (1985). An XPS and SEM study of gold deposition at low temperatures on sulphide mineral surfaces: concentration of gold by adsorption/reduction. *Geochimica et Cosmochimica Acta* 49: 979–987.

Jean, G.E. and Bancroft, G.M. (1986). Heavy metal adsorption by sulphide mineral surfaces. *Geochimica et Cosmochimica Acta* 50: 1455–1463.

Johnson, J.P. and Cross, K.C. (1991). Geochronological and Sm–Nd isotopic constraints on the genesis of the Olympic Dam Cu–U–Au–Ag deposit, South Australia. In: *Source, Transport and Deposition of Metals* (ed. M. Pagel), 395–400. Balkema.

Johnson, J.P. and Cross, K.C. (1995). U-Pb geochronological constraints on the genesis of the Olympic Dam Cu-U-Au-Ag deposit, South Australia. *Economic Geology* 90 (5): 1046–1063.

Johnson, J.P. and McCulloch, M.T. (1995). Sources of mineralizing fluids for the Olympic Dam deposit (South Australia): Sm–Nd isotopic constraints. *Chemical Geology* 121: 177–199.

Jones, H.D. and Kesler, S.E. (1992). Fluid inclusion gas chemistry in east Tennessee Mississippi Valley-type districts: evidence for immiscibility and implications for depositional mechanisms. *Geochimica et Cosmochimica Acta* 56: 137–154.

Jones, A.P., Genge, M., and Carmody, L. (2013). Carbonate melts and carbonatites. *Reviews in Mineralogy and Geochemistry* 75: 289–322.

Jowett, E.C., Rydzewski, A., and Jowett, R.J. (1987). The Kupferschiefer Cu–Ag deposits in Poland: a re-appraisal of the evidence of their origin and presentation of a new genetic model. *Canadian Journal of Earth Sciences* 24: 2016–2037.

Jugo, P.J. (2009). Sulfur content at sulphide saturation in oxidized magmas. *Geology* 37: 415–418.

Jugo, P.J., Luth, R.W., and Richards, J.P. (2005). An experimental study of the sulphur content in basaltic melts saturated with immiscible sulphide or sulphate liquids at 1300°C and 1.0 GPa. *Journal of Petrology* 46: 783–798.

Kasting, J.F. (1993). Evolution of the Earth's atmosphere and hydrosphere. In: *Organic Geochemistry* (eds. M.H. Engel and S.A. Macko). Plenum Press, 821 pp.

Kazakov, A.V. (1937). The phosphorite facies and the genesis of phosphorites. *U.S.S.R. Transactions of the Scientific Institute of Fertilizers and Insecto-Fungicides* 42: 95–113.

Kazakov, A.V. (1939). Phosphatic facies: origin of phosphorites and geological factors of the formation of deposits. *U.S.S.R. Transactions of the Scientific Institute of Fertilizers and Insecto-Fungicides* 45: 1–108.

Kearey, P. (ed.) (1993). *The Encyclopedia of the Solid Earth Sciences*. Blackwell Scientific Publications, 713 pp.

Kearey, P. and Vine, F. (1996). *Tectonics*. Blackwell Science, 333 pp.

Kendall, A.C. and Harwood, G.M. (1996). Marine evaporites; arid shorelines and basins. In: *Sedimentary Environments, Processes, Facies and Stratigraphy* (ed. H.G. Reading), 281–324. Blackwell Science.

Kennicutt, M.C., Brooks, J.M., and Cox, H.B. (1993). The origin and distribution of gas hydrates in marine sediments. In: *Organic Geochemistry* (eds. M.H. Engel and S.A. Macko). Plenum Press, 861 pp.

Keppler, H. and Wyllie, P.J. (1991). Partitioning of Cu, Sn, Mo, W, U, and Th between melt and aqueous fluid in the systems haplogranite–H_2O–HCl and haplogranite–H_2O–HF. *Contributions to Mineralogy and Petrology* 109: 139–150.

Kerrich, R. and Cassidy, K.F. (1994). Temporal relationships of lode-gold mineralization to accretion, magmatism, metamorphism and deformation – Archean to present: a review. *Ore Geology Reviews* 9: 263–310.

Kerrich, R., Goldfarb, R.J., and Richards, J.P. (2005). Metallogenic provinces in an evolving geodynamic framework. In: *One Hundredth Anniversary Volume* (eds. J.W. Hedenquist, J.F.H. Thompson, R.J. Goldfarb and J.P. Richards), 1097–1136. Society of Economic Geologists.

Kesler, S.E., Gruber, P.W., Medina, P.A. et al. (2012). Global lithium resources – relative importance of pegmatite, brine and other deposits. *Ore Geology Reviews* 48: 55–69.

Key, R., Liyungu, A.K., Njamu, F.M. et al. (2001). The western arm of the Lufilian arc, NW Zambia and its potential for Cu mineralization. *Journal of African Earth Sciences* 33: 503–528.

Kharaka, Y.K. and Hanor, J.S. (2014). Deep fluids in sedimentary basins. In: *Treatise on Geochemistry*, 2e, vol. 13 (eds. H.D. Holland and K.K. Turekian), 472–515. Oxford: Elsevier.

Kilinc, I.A. and Burnham, C.W. (1972). Partitioning of chloride between a silicate melt and coexisting aqueous phase from 2 to 8 kilobars. *Economic Geology* 67: 231–235.

Kimura, K., Lewis, R.S., and Anders, S. (1974). Distribution of gold and rhenium between nickel-iron and silicate melts; implications for the abundance of siderophile elements on the Earth and Moon. *Geochimica et Cosmochimica Acta* 38: 683–701.

Kinloch, E.D. (1982). Regional trends in the platinum group mineralogy of the Critical Zone of the Bushveld Complex, South Africa. *Economic Geology* 77: 1328–1347.

Kinnaird, J.A. (1987). Hydrothermal alteration and mineralization of the Nigerian anorogenic ring complexes, with special reference to the Saiya-Shokobo Complex. PhD thesis. University of St. Andrews.

Kinnaird, J.A., Kruger, F.J., Nex, P.A.M., and Cawthorn, R.G. (2002). Chromitite formation: a key to understanding processes of platinum enrichment. In: *21st Century of Pt–Pd Deposits: Current and Future Potential* (eds. I. McDonald, A.G. Gunn and H.M. Prichard), 23–35. Institute of Mining Metallurgy.

Kirk, J., Ruiz, J., Chesley, J. et al. (2001). A detrital model for the origin of gold and sulfides in the Witwatersrand basin based on Re–Os isotopes. *Geochimica et Cosmochimica Acta* 65: 2149–2159.

Kirk, J., Ruiz, J., Chesley, J. et al. (2002). A major Archean, gold- and crust-forming event in the Kaapvaal Craton, South Africa. *Science* 297: 1856–1858.

Kirkham, R.V. (1989). Distribution, settings and genesis of sediment-hosted stratiform copper deposits. In: *Sediment-Hosted Stratiform Copper Deposits*, Special Paper, vol. 36 (eds. R.W. Boyle, A.C. Brown, C.W. Jefferson, et al.), 3–38. Geological Association of Canada.

Kirkham, R.V. and Roscoe, S.M. (1993). Atmospheric evolution and ore deposit formation. *Resource Geology*, Special Issue 15: 1–17.

Kirschvink, J.L. and Hagadorn, J.W. (2000). A grand unified theory of biomineralization. In: *The Biomineralization of Nano- and Microstructures* (ed. E. Bauerlein), 139–150. Wiley-VCH Verlag GmbH.

Kisvarsanyi, G. (1977). The role of the Precambrian igneous basement in the formation of stratabound lead-zinc-copper deposits in southeast Missouri. *Economic Geology* 72: 435–442.

Klappa, C.F. (1983). A process-response model for the formation of pedogenic calcretes. In: *Residual Deposits: Surface Related Weathering Processes and Materials*. Special Publication, vol. 11 (ed. R.C.L. Wilson), 211–220. Geological Society of London.

Klein, C. and Beukes, N.J. (1993). Sedimentology and geochemistry of the glaciogenic late Proterozoic Rapitan Iron-Formation in Canada. *Economic Geology* 88: 542–565.

Klemd, R., Hirdes, W., Olesch, M., and Oberthür, T. (1993). Fluid inclusions in quartz-pebbles of the gold-bearing Tarkwaian conglomerates of Ghana as guides to their provenance area. *Mineralium Deposita* 28: 334–343.

Kneeshaw, M. (2002). *Guide to the Geology of the Hamersley and Northeast Pilbara Iron Ore Provinces. Handbook*. BHP Billiton, 34 pp.

Knipe, S.W., Foster, R.P., and Stanley, C.J. (1992). Role of sulphide surfaces in sorption of precious metals from hydrothermal fluids. *Transactions of the Institute of Mining Metallurgy Section B* 101: B83–B88.

Kocurek, G.A. (1996). Desert aeolian systems. In: *Sedimentary Environments, Processes, Facies and Stratigraphy* (ed. H.G. Reading), 125–153. Blackwell Science.

Kojima, S., Takeda, S., and Kogita, S. (1994). Chemical factors controlling the solubility of uraninite and their significance in the genesis of unconformity related uranium deposits. *Mineralium Deposita* 29: 353–560.

Komar, P.D. (1989). Physical processes of waves and currents and the formation of marine placers. *Reviews in Aquatic Sciences* 1 (3): 393–423.

Komar, P.D. and Wang, C. (1984). Processes of selective grain transport and the formation of placers on beaches. *Journal of Geology* 92: 637–655.

Konhauser, K.O. (1998). Diversity of bacterial iron mineralization. *Earth-Science Reviews* 43: 91–121.

Konhauser, K.O. (2003). *Introduction to Geomicrobiology*. Blackwell Publishing.

Konhauser, K.O. and Ferris, F.G. (1996). Diversity of iron and silica precipitation by microbial mats in hydrothermals waters, Iceland: implications for Precambrian iron-formations. *Geology* 24: 323–326.

Konhauser, K.O., Kappler, A., and Roden, E.E. (2011). Iron in microbial metabolisms. *Elements* 7: 89–93.

Korenaga, J. (2013). Initiation and evolution of plate tectonics on Earth: theories and observations. *Annual Reviews Earth & Planetary Sciences* 41: 117–151.

Krauskopf, K.B. and Bird, D.K. (1995). *Introduction to Geochemistry*. McGraw-Hill, 647 pp.

Krogh, T.E., Davis, D.W., and Corfu, F. (1984). Precise U-Pb zircon and baddeleyite ages for the Sudbury area. In: *The Geology and Ore Deposits of the Sudbury Structure*, vol. 1 (eds. E.G. Pye, A.J. Naldrett and P.E. Giblin), 431–446. Ontario Geological Survey, Special Publication.

Kröner, A. and Layer, P.W. (1992). Crust formation and plate motion in the Early Archean. *Science* 256: 1405–1411.

Kruger, F.J. (1994). The Sr-isotopic stratigraphy of the western Bushveld Complex. *South African Journal of Geology* 97: 393–398.

Kruger, F.J. and Marsh, J.S. (1982). Significance of $^{87}Sr/ ^{86}Sr$ ratios in the Merensky cyclic unit of the Bushveld igneous Complex, South Africa. *Nature* 298: 53–55.

Kruger, F.J. and Marsh, J.S. (1985). The mineralogy, petrology, and origin of the Merensky Cyclic Unit in the western Bushveld Complex. *Economic Geology* 80: 958–974.

Kushiro, I. (1980). Viscosity, density and structure of silicate melts at high pressure and their petrologic implications. In: *Physics of Magmatic Processes* (ed. R.B. Hargraves), 93–120. Princeton University Press.

Kutina, J., Telupil, A., Adam, J., and Pacesova, M. (1967). On the vertical extent of ore deposition of the Prbram ore veins. *Transactions of the Institution of Mining and Metallurgy, Section B: Applied Earth Science* 76: 732.

Kvenvolden, K.A. (1988). Methane hydrate: a major reservoir of carbon in the shallow geosphere. *Chemical Geology* 71: 41–51.

Kvenvolden, K.A. (1995). A review of the geochemistry of methane in natural gas hydrate. *Organic Geochemistry* 23: 997–1008.

Kvenvolden, K.A. and McMenamin, M.A. (1980) Hydrates of natural gas: a review of their geologic occurrence. *United States Geological Survey, Circular 825*.

Kynicky, J., Smith, J.P., and Xu, C. (2012). Diversity of rare earth deposits: the key example of China. *Elements* 8 (5): 361–367.

LaBerge, G.L. (1973). Possible biologic origin of Precambrian iron-formations. *Economic Geology* 68: 1098–1109.

Lamb, S., Hoke, L., Kennan, L., and Dewey, J. (1997). Cenozoic evolution of the central Andes in Bolivia and northern Chile. In: *Orogeny Through Time,* vol. 121 (eds. J.-P. Burg and M. Ford), 237–264. The Geological Society, Special Publication.

Langmuir, D. (1978). Uranium solution–mineral equilibria at low temperatures with applications to sedimentary ore deposits. In: *Uranium Deposits, Their Mineralogy and Origin*, Short Course Handbook, vol. 3 (ed. M.M. Kimberley), 17–55. Mineralogical Association of Canada.

Langmuir, C.H. (1989). Geochemical consequences of in situ crystallization. *Nature* 340: 199–205.

Large, R.R. (1992). Australian volcanic-hosted massive sulfide deposits: features, styles and genetic models. *Economic Geology* 87: 471–510.

Larsen, K.G. (1977). Sedimentology of the Bonneterre Formation, Southeast Missouri. *Economic Geology* 72: 408–419.

Larson, R.L. (1991). Geological consequences of superplumes. *Geology* 19: 963–966.

Larter, S.R. and Head, I.M. (2014). Oil sands and heavy oil: origin and exploitation. *Elements* 10: 277–284.

Le Bas, M.J. (1987). Nephelinites and carbonatites. In: *Alkaline Igneous Rocks* (eds. J.G. Fitton and B.G.J. Upton), 53–83. Geological Society of London, Special Publication, 30.

Le Fort, P. (1975). Himalayas: the collided range. Present knowledge of the continental arc. *American Journal Science* 275A: 1–44.

Leach, D.L., Bradley, D., Lewchuk, M.T. et al. (2001). Mississippi Valley-type lead-zinc deposits through geologic time: implications from recent age-dating research. *Mineralium Deposita* 36: 711–740.

LeCheminant, A.N. and Heaman, L.M. (1989). Mackenzie igneous events, Canada; middle Proterozoic hotspot magmatism associated with ocean opening. *Earth and Planetary Science Letters* 96: 38–48.

Lee, C.A. and Butcher, A.R. (1990). Cyclicity in the Sr isotope stratigraphy through the Merensky and Bastard reef units, Atok section, eastern Bushveld Complex. *Economic Geology* 85: 877–883.

Leeder, M. (1999). *Sedimentology and Sedimentary Basins from Turbulence to Tectonics*. Blackwell Science, 592 pp.

Lesher, C.M. (1989). Komatiite-hosted nickel sulfide deposits. *Reviews in Economic Geology* 4: 45–101.

Leventhal, J. (1993). Metals in black shales. In: *Organic Geochemistry* (eds. M.H. Engel and S.A. Macko), 581–592. Plenum Press.

Leythauser, D., Radke, M., and Schaefer, R.G. (1984). Efficiency of petroleum expulsion from shale source rocks. *Nature* 311: 745–748. https://doi.org/10.1038/311745a0.

Li, C. and Naldrett, A.J. (1994). A numerical model for the compositional variations of Sudbury sulfide ores and its application to exploration. *Economic Geology* 89: 1599–1607.

Li, C., Maier, W.D., and de Waal, S.A. (2001a). Magmatic Ni–Cu versus PGE deposits: contrasting genetic controls and exploration implications. *South African Journal of Geology* 104: 309–318.

Li, C., Maier, W.D., and De Waal, S. (2001b). The role of magma mixing in the genesis of PGE mineralization in the Bushveld Complex; thermodynamic calculations and new interpretations. *Economic Geology* 96: 653–662.

Lightfoot, P.C. (2016). *Nickel Sulphide Ores and Impact Melts: Origin of the Sudbury Igneous Complex*. Elsevier, 680 pp.

Lightfoot, P.C., Keays, R.R., and Doherty, W. (2001). Chemical evolution and origin of nickel sufide mineralization in the Sudbury Igneous Complex, Ontario, Canada. *Economic Geology* 96: 1855–1875.

Lindgren, W. (1933). *Mineral Deposits*. McGraw-Hill, 930 pp.

Ling, M.-X., Liu, Y.-L., Williams, I.S. et al. (2013). Formation of the World's largest REE deposit through protracted fluxing of carbonatite by subduction-derived fluids. *Scientific Reports* 3: 1776. https://doi.org/10.1038/srep01776.

Linnen, R., Trueman, D.L., and Burt, R. (2014). Tantalum and niobium. In: *Critical Metals Handbook* (ed. G. Gunn), 361–384. Wiley/AGU.

Lipin, B.R. (1993). Pressure increases, the formation of chromite seams, and the development of the Ultramafic Series in the Stillwater Complex, Montana. *Journal of Petrology* 34: 955–976.

Liu, K., Cruzan, J., and Saykally, R.J. (1996). Water clusters. *Science* 271: 929–933.

London, D. (1990). Internal differentiation of rare-element pegmatites: a synthesis of recent research. In: *Ore-bearing Granite Systems; Petrogenesis and Mineralizing processes*,

Special Paper, vol. 246 (eds. H.J. Stein and J.L. Hannah), 35–50. Geological Society of America.

London, D. (1992). The application of experimental petrology to the genesis and crystallization of granitic pegmatites. *Canadian Mineralogist* 30: 499–540.

London, D. (1996). Granitic pegmatites. *Transactions of the Royal Society Edinburgh: Earth Sciences* 87: 305–319.

London, D. (2005). Granitic pegmatites: an assessment of current concepts and directions for the future. *Lithos* 80: 281–303.

Lottermoser, B.G. and Ashley, P.M. (2000). Geochemistry, petrology and origin of Neoproterozoic ironstones in the eastern part of the Adelaide Geosyncline, South Australia. *Precambrian Research* 101: 49–67.

Lowenstam, H.A. (1981). Minerals formed by organisms. *Science* 211: 1126–1131.

Lowenstern, J.B. (1994). Dissolved volatile concentrations in an ore-forming magma. *Geology* 22: 893–896.

Lowenstern, J.B. (2001). Carbon dioxide in magmas and implications for hydrothermal systems. *Mineralium Deposita* 36: 490–502.

Lowenstern, J.B., Mahood, G.A., Rivers, M.L., and Sutton, S.R. (1991). Evidence for extreme partitioning of copper into a magmatic vapor phase. *Science* 252: 1405–1408.

Lydon, J.W. (1988). Volcanogenic massive sulphide deposits. Part 2: Genetic models. *Geoscience Canada* 15: 43–65.

Lynn, M.D., Wipplinger, P.E., and Wilson, M.G.C. (1998). Diamonds. In: *The Mineral Resources of South Africa*. Handbook, 16 (eds. C.R. Anhaeusser and M.G.C. Wilson), 232–258. Council for Geoscience.

MacKenzie, F.T. (1975). Sedimentary cycling and the evolution of sea water. In: *Chemical Oceanography* (eds. J.P. Riley and G. Skirrow), 309–364. Academic Press.

Mackenzie, A.S., Price, I., Leythauser, D. et al. (1987). The expulsion of petroleum from Kimmeridge Clay source-rocks in the area of the Brae Oilfield, UK continental shelf. In: *Petroleum Geology of North West Europe* (eds.

J. Brooks and K. Glennie), 865–877. London: Graham & Trotman.

MacLean, W.H. (1969). Liquidus phase relationships in the FeS-FeO-Fe$_3$O$_4$-SiO$_2$ system, and their application in geology. *Economic Geology* 64: 865–884.

Maier, W.D., Karykowski, B.T., and Yang, S.-H. (2016). Formation of transgressive anorthosite seams in the Bushveld Complex via tectonically induced mobilization of plagioclase-rich crystal mushes. *Geoscience Frontiers* 7: 875–889.

Mann, A.W. (1984). Mobility of gold and silver in lateritic weathering profiles: some observations from Western Australia. *Economic Geology* 79: 38–49.

Mann, A.W. and Deutscher, R.L. (1978). Genesis principles for the precipitation of carnotite in calcrete drainages in Western Australia. *Economic Geology* 73: 1724–1737.

Manning, D.A.C. and Henderson, P. (1984). The behaviour of tungsten in granitic melt–vapour systems. *Contributions to Mineralogy and Petrology* 76: 257–262.

Mariano, A.N. (1989). Nature of economic mineralization in carbonatites and related rocks. In: *Carbonatites: Genesis and Evolution* (ed. J.D. Bell). Unwyn Hyman, 618 pp.

Martin, D.M., Clendenin, C.W., Krapez, B., and McNaughton, N.J. (1998a). Tectonic and geochronological constraints on late Archaean and Palaeoproterozoic stratigraphic correlation within and between the Kaapvaal and Pilbara cratons. *Journal of the Geological Society* 155: 311–322.

Martin, D.M., Li, Z.X., Nemchin, A.A., and Powell, C.M. (1998b). A pre-2.2 Ga age for the giant hematite ores of the Hamersley Province, Australia? *Economic Geology* 93: 1064–1090.

Martin, L.H.J., Schmidt, M.W., Mattsson, H.B., and Guenther, D. (2013). Element partitioning between immiscible carbonatite and silicate melts for dry and H$_2$O-bearing systems at 1-3 GPa. *Journal of Petrology* 54: 2301–2338.

Master, S., Rainaud, C., Armstrong, R.A. et al. (2005). Provenance ages of the Neoproterozoic

Katanga Supergroup (Central African Copperbelt), with implications for basin evolution. *Journal African Earth Sciences* 42: 41–60.

Mathez, E.A. (1989a). Vapor associated with mafic magma and controls on its composition. In: *Ore Deposition Associated with Magmas*, Reviews in Economic Geology, vol. 4 (eds. J.A. Whitney and A.J. Naldrett), 21–31. Society of Economic Geology.

Mathez, E.A. (1989b). Interactions involving fluids in the Stillwater and Bushveld Complexes: observations from the rocks. In: *Ore Deposition Associated with Magmas*, Reviews in Economic Geology, vol. 4 (eds. J.A. Whitney and A.J. Naldrett), 167–179. Society of Economic Geology.

Mathez, E.A. (1999). On factors controlling the concentrations of platinum group elements in layered intrusions and chromitites. In: *Dynamic Processes in Magmatic Ore Deposits and Their Application to Mineral Exploration*. Short course notes,, vol. 13 (eds. R.R. Keays, C.M. Lesher, P.C. Lightfoot and C.E.G. Farrow), 251–285. Geological Society Canada.

Maynard, J.B. (1991). Uranium: syngenetic to diagenetic deposits in foreland basins. In: *Sedimentary and Diagenetic Mineral Deposits: A Basin Analysis Approach to Exploration*, Reviews in Economic Geology, vol. 5 (eds. E.R. Force, J.J. Eidel and J.B. Maynard), 187–197. Society of Economic Geologists.

Maynard, J.B. (2014). Manganiferous sediments, rocks and ores. In: *Treatise on Geochemistry*, 2e, vol. 13 (eds. H.D. Holland and K.K. Turekian), 327–349. Oxford: Elsevier.

McBirney, A.R. (1985). Further considerations of double-diffusive stratification and layering in the Skaergaard Intrusion (Greenland). *Journal of Petrology* 26: 993–1001.

McBirney, A.R. and Noyes, R.M. (1979). Crystallization and layering of the Skaergaard Intrusion. *Journal of Petrology* 20: 487–554.

McCabe, P.J. (1984). Depositional environments of coal and coal-bearing strata. In: *Special Publication 7, International Association of Sedimentologists* (eds. R.A. Rahmani and R.M.

Flores), 13–42. Blackwell Scientific Publications.

McCarthy, T.S. and Hasty, R.A. (1976). Trace element distribution patterns and their relationship to the crystallization of granitic melts. *Geochimica et Cosmochimica Acta* 40: 1351–1358.

McCarthy, T.S., Cawthorn, R.G., Wright, C.J., and McIver, J.R. (1985). Mineral layering in the Bushveld Complex: implications of Cr abundances in magnetitite and intervening silicate-rich layers. *Economic Geology* 80: 1062–1074.

McCuaig, T.C. and Kerrich, R. (1998). P–T–t deformation-fluid characteristics of lode gold deposits: evidence from alteration systematics. *Ore Geology Reviews* 12: 381–453.

McCulloch, M.T. and Bennett, V.C. (1994). Progressive growth of the Earth's continental crust and depleted mantle: geochemical constraints. *Geochimica et Cosmochimica Acta* 58: 4717–4738.

McInnes, B.I.A., McBride, J.S., Evans, N.J. et al. (1999). Osmium isotope constraints on ore metal recycling in subduction zones. *Science* 286: 512–516.

McKibben, M.A. and Hardie, L.A. (1997). Ore-forming brines in active continental rifts. In: *Geochemistry of Hydrothermal Ore Deposits* (ed. H.L. Barnes), 877–930. Wiley Interscience.

McKibben, M.A., Andes, J.P., and Williams, A.E. (1988). Active ore formation at a brine interface in metamorphosed deltaic lacustrine sediments: the Salton Sea geothermal system, California. *Economic Geology* 83: 511–523.

McMenamin, M.A.S. and McMenamin, D.L.S. (1990). *The Emergence of Animals: The Cambrian Breakthrough*. Columbia University Press, 217 pp.

Meert, J.G. and Torsvik, T.H. (2003). The making and unmaking of a supercontinent: Rodinia revisited. *Tectonophysics* 375: 261–288.

Meinert, L.D. (1992). Skarns and skarn deposits. *Geoscience Canada* 19: 145–162.

Meinert, L.D. (2000). Gold in skarns related to epizonal intrusions. In: *Gold in 2000*, Reviews in Economic Geology, vol. 13 (eds. S. Hagemann and P.E. Brown), 347–375. Society of Economic Geologists.

Meland, N. and Norman, J.O. (1966). Transport velocities of single particles in bed-load motion. *Geografisca Annaler* 48A: 165–182.

Merkle, R. (1992). Platinum-group minerals in the middle group of chromitite layers at Marikana, western Bushveld complex: indications for collection mechanism and postmagmatic modification. *Canadian Journal of Earth Sciences* 29: 209–221.

Metcalfe, R., Rochelle, C.A., Savage, D., and Higgo, J.W. (1994). Fluid–rock interactions during continental red-bed diagenesis: implications for theoretical models of mineralization in sedimentary basins. In: *Geofluids: Origin, Migration and Evolution of Fluids in Sedimentary Basins*, vol. 78 (ed. J. Parnell), 301–324. Geological Society, Special Publication.

Meyer, C. (1981). Ore-forming processes in geologic history. *Economic Geology* 75: 6–41.

Meyer, C. (1988). Ore deposits as guides to the geologic history of the Earth. *Annual Reviews in Earth Planetary Science* 16: 147–171.

Meyer, C. and Hemley (1967). Wall rock alteration. In: *Geochemistry of Hydrothermal Ore Deposits* (ed. H.L. Barnes), 166–235. Holt, Rinehart and Winston.

Meyer, F.M., Happel, U., Hausberg, J., and Wiechowski, A. (2002). The geometry and anatomy of the Los Pijiguaos bauxite deposit, Venezuela. *Ore Geology Reviews* 20: 27–54.

Meyers, P.A., Pratt, L.M., and Nagy, B. (1992). Introduction to geochemistry of metalliferous black shales. *Chemical Geology* 99: vii–xi.

Miall, A.D. (1978). Fluvial sedimentology: an historical review. In: *Fluvial Sedimentology*, vol. 5 (ed. A.D. Miall), 1–47. Canadian Society for Petroleum Geology, Memoir.

Migdisov, A.A. and Williams-Jones, A.E. (2013). A predictive model for metal transport of silver chloride by aqueous vapour in ore-forming magmatic-hydrothermal systems. *Geochimica Cosmochimica Acta* 104: 123–135.

Mikucki, E.J. (1998). Hydrothermal transport and depositional processes in Archean lode-gold systems: a review. *Ore Geology Reviews* 13: 307–321.

Milési, J.P., Ledru, P., Marcoux, E. et al. (2002). The Jacobina Paleoproterozoic gold-bearing conglomerates, Bahia, Brazil: a "hydrothermal shear–reservoir" model. *Ore Geology Reviews* 19: 95–136.

Milkov, A.V. (2004). Global estimates of hydrate-bound gas in marine sediments: how much is really out there? *Earth-Science Reviews* 66 (3–4): 183–197.

Minter, W.E.L. (1978). A sedimentological synthesis of placer gold, uranium and pyrite concentrations in Proterozoic Witwatersrand deposits. In: *Fluvial Sedimentology*, vol. 5 (ed. A.D. Miall), 801–829. Canadian Society for Petroleum Geology, Memoir.

Misra, K.C. (2000). *Understanding Mineral Deposits*. Kluwer Academic Publishers, 845 pp.

Mitchell, A.H.G. and Garson, M.S. (1981). *Mineral Deposits and Global Tectonic Settings*. Academic Press, 457 pp.

Mitchell, A.H.G. and Leach, T.M. (1991). *Epithermal Gold in the Phillipines: Island Arc Metallogenesis, Geothermal Systems and Geology*. Academic Press, 457 pp.

Miyano, T. and Beukes, N.J. (1987). Physicochemical environments for the formation of quartz-free manganese oxide ores from the early Proterozoic Hotazel formation, Kalahari manganese field, South Africa. *Economic Geology* 82: 706–718.

Moldoveanu, G. and Papangelakis, V.G. (2012). Recovery of rare earth elements adsorbed on clay minerals. *Hydrometallurgy* 117–118: 71–78.

Moore, D.W., Young, L.E., Modene, J.S., and Plahuta, J.T. (1986). Geologic setting and genesis of the Red Dog zinc-lead-silver deposit, Western Brooks range, Alaska. *Economic Geology* 81: 1696–1727.

Moores, E.M. (1993). Neoproterozoic oceanic crustal thinning, emergence of continents, and the origin of the Phanerozoic ecosystem: a model. *Geology* 21: 5–8.

Morris, R.C. (1985). Genesis of iron ore in banded iron-formation by supergene and supergene-metamorphic processes – a conceptual model. In: *Handbook of Strata-bound and Stratiform Ore Deposits*, vol. 13 (ed. K.H. Wolf), 33–235. Elsevier.

Morris, R.C. and Kneeshaw, M. (2011). Genesis modelling for the Hamersley BIF-hosted iron ores of Western Australia – a critical review. *Australian Journal of Earth Sciences* 58: 417–451.

Morris, R.C., Thornber, M.R., and Ewers, W.E. (1980). Deep-seated iron ores from banded iron formation. *Nature* 288: 250–252.

Mote, T.I., Brimhall, G.H., Tidy-Finch, E. et al. (2001). Application of mass balance modeling of sources and sinks of supergene enrichment to exploration and discovery of the Quebrada Turquesa exotic Cu ore body, El Salvador district, Chile. *Economic Geology* 96: 367–386.

Muir Wood, R. (1994). Earthquakes, strain-cycling and the mobilization of fluids. In: *Geofluids: Origin, Migration and Evolution of Fluids in Sedimentary Basins*, vol. 78 (ed. J. Parnell), 85–98. Geological Society, Special Publication.

Murck, B.W. and Campbell, I.H. (1986). The effects of temperature, oxygen fugacity and melt composition on the behavior of chromium in basic and ultrabasic melts. *Geochimica et Cosmochimica Acta* 50: 1871–1887.

Murphy, J.B. and Nance, R.D. (1991). Supercontinent model for the contrasting character of Late Proterozoic orogenic belts. *Geology* 19: 469–472.

Murphy, J.B., Nance, R.D., and Cawood, P.A. (2009). Contrasting modes of supercontinent formation and the conundrum of Pangea. *Gondwana Research* 15: 408–420.

Muskat, M. (1937). *The Flow of Homogeneous Fluids through Porous Media*. New York: McGraw-Hill.

Mysen, B.O. (1976). The role of volatiles in silicate melts: solubility of carbon dioxide and water in feldspar, pyroxene and feldspathoid

melts to 30 kbar and 1625 °C. *American Journal of Science* 276: 969–996.

Mysen, B.O. (2014). Water-melt interaction in hydrous magmatic systems at high temperature and pressure. *Progress in Earth and Planetary Science* 1 (4) https://doi.org/10.1186/2197-4282-1-4.

Mysen, B.O. and Kushiro, I. (1976). Melting behaviour of peridotite at high pressure. *Eos* 57: 354.

Nagaseki, H. and Hayashi, K. (2004). Vapor–liquid partitioning experiments of ore metals in boiling hydrothermal solutions using synthetic fluid inclusions. Abstract from the Annual Meeting of the Society of Resource Geology, Tokyo (June 2004), p. 17.

Nahon, D.B., Boulangé, B., and Colin, F. (1992). Metallogeny of weathering: an introduction. In: *Weathering, Soils and Paleosols*, Developments in Earth Surface Processes, vol. 2 (eds. I.P. Martini and W. Chesworth), 445–471. Elsevier.

Naldrett, A.J. (1989a). *Magmatic Sulphide Deposits*, Oxford Monographs on Geology and Geophysics. Clarendon Press, 186 pp.

Naldrett, A.J. (1989b). Stratiform PGE deposits in layered intrusions. In: *Ore Deposition Associated with Magmas*, Reviews in Economic Geology, vol. 4 (eds. J.A. Whitney and A.J. Naldrett), 135–165. Society of Economic Geology.

Naldrett, A.J. (1997). Models for the formation of stratabound concentrations of platinum group elements in layered intrusions. In: *Mineral Deposit Modeling*, Special paper, vol. 40 (eds. R.V. Kirham, W.D. Sinclair, R.I. Thorpe, et al.), 373–388. Geological Association of Canada.

Naldrett, A.J. (1999). World-class Ni-Cu-PGE deposits: key factors in their genesis. *Mineralium Deposita* 34: 227–240.

Naldrett, A.J. and Lehmann, J. (1988). Spinel nonstoichiometry as the explanation for Ni-, Cu-, and PGE-enriched sulphides in chromitites. In: *Geo-platinum*, vol. 87 (ed. H.M. Prichard), 93–109.

Naldrett, A.J. and MacDonald, A.J. (1980). Tectonic Setting of Some Ni-Cu Sulphide

Ores: Their Importance in Genesis and Exploration. In: *The Continental Crust and its Mineral Deposits*, Special paper, vol. 20 (ed. D.W. Strangway), 633–657. Geological Association of Canada.

Naldrett, A.J. and von Grünewaldt, G. (1989). Association of platinum-group elements with chromitite in layered intrusions and ophiolite complexes. *Economic Geology* 84: 180–187.

Naldrett, A.J. and Wilson, A.H. (1991). Horizontal and vertical variation in the noble metal distribution in the Great Dyke of Zimbabwe: a model for the origin of the PGE mineralization by fractional segregation of sulfide. *Chemical Geology* 88: 279–300.

Naldrett, A.J., Rao, B.V., and Evensen, N.M. (1986). Contamination at Sudbury and its role in ore formation. In: *Metallogeny of Basic and Ultrabasic Rocks* (eds. M.J. Gallager, R.A. Ixer, C.R. Neary and H.M. Prichard), 75–91. Institution of Mining Metallurgy.

Nami, M. and Ashworth, S.G.E. (1993). *Principles of a Sediment Sorting Model and its Application for Predicting Economic Values in Placer Deposits*, vol. 17, 543–551. International Association Sedimentology, Special Publication.

Nami, M. and James, C.S. (1987). Numerical Simulation of Gold Distribution in the Witwatersrand Placers. In: *Recent Developments in Fluvial Sedimentology*, vol. 39 (eds. F.G. Ethridge, R.M. Flores and M.D. Harvey), 353–370. Society of Economic Palaeontologists and Mineralogists, Special Publication.

Nance, R.D. and Murphy, J.B. (2013). Origins of the supercontinent cycle. *Geoscience Frontiers* 4: 439–448.

Nance, R.D., Worsley, T.R., and Moody, J.B. (1986). Post-Archean biogeochemical cycles and long-term episodicity in tectonic processes. *Geology* 14: 514–518.

Nance, R.D., Murphy, J.B., and Santosh, M. (2014). The supercontinent cycle: a retrospective essay. *Gondwana Research* 25: 4–29.

Nash, J.T., Granger, H.C., and Adams, S.S. (1981). Geology and concepts of genesis of important types of uranium deposits. *Economic Geology* 75: 63–116.

Naslund, H. R. (1976) Liquid immiscibility in the system $KalSi_3O_8$-$NaAlSi_3O_8$-FeO-Fe_2O_3-SiO_2 and its application to natural magmas. *Carnegie Institution of Washington Year Book 75*. pp. 592–597.

Naslund, H.R. (1983). The effect of oxygen fugacity on liquid immiscibility in iron bearing silicate melts. *American Journal of Science* 283: 1034–1059.

Naslund, H.R., Henriquez, F., Nystrom, J.O. et al. (2002). Magmatic iron ores and associated mineralization: examples from the Chilean high Andes and coastal Cordillera. In: *Hydrothermal Iron Oxide-Copper-Gold and Related Deposits: A Global Perspective* (ed. T.M. Porter), 207–226. PGC Publishing.

Nebel, O., Capitanio, F.A., Moyen, J.-F. et al. (2018). When crust comes of age: on the chemical evolution of Archaean, felsic continental crust by crustal drip tectonics. *Philosophical Transactions of the Royal Society A* A376 https://doi.org/10.1098/rsta.2018.0103.

Newton, R.C. and Manning, C.E. (2000). Quartz solubility in H_2O-$NaCl$ and H_2O-CO_2 solutions at deep crust-upper mantle pressures and temperatures: 2-15 kbar and 500-900°C. *Geochimica Cosmochimica Acta* 64: 2993–3005.

Nex, P. (2002) Bifurcating chromitites: analogues to sedimentary structures in the Bushveld Complex, South Africa. *11th Quadrennial IAGOD Symposium and Geocongress 2002: Extended abstracts*, Windhoek, Namibia (22–26 July 2002). Geological Survey of Namibia.

Nex, P.A.M., Kinnaird, J.A., and Oliver, G.J.H. (2001). Petrology, geochemistry and uranium mineralization of post-collisional magmatism around Goanikontes, southern Central Zone, Namibia. *Journal of African Earth Sciences* 33: 481–502.

Nixon, P.H. (1995). The morphology and nature of primary diamondiferous occurrences. *Journal Geochemical Exploration* 53: 41–71.

North, F.K. (1985). *Petroleum Geology*. Allen & Unwin, 607 pp.

Northrop, H.R. and Goldhaber, M.B. (1990). Genesis of the tabular-type vanadium–uranium deposits of the Henry Basin, Utah. *Economic Geology* 85: 215–268.

Norton, S.A. (1973). Laterite and bauxite formation. *Economic Geology* 68: 353–361.

Norton, I. and Cathles, L.M. (1979). Thermal aspects of ore deposition. In: *Geochemistry of Hydrothermal Ore Deposits* (ed. H.L. Barnes), 611–631. Wiley.

Nyström, J.O. and Henríquez, R. (1992). Magmatic features of iron ore of the Kiruna Type in Chile and Sweden: ore textures and magnetite geochemistry. *Economic Geology* 89: 820–839.

Oberthür, T., Davis, D.W., Blenkinsop, T.G., and Hoehndorf, A. (2002). Precise U-Pb mineral ages, Rb-Sr and Sm-Nd systematics for the Great Dyke, Zimbabwe – constraints on crustal evolution and metallogenesis of the Zimbabwe Craton. *Precambrian Research* 113: 293–306.

Odling, N.E. (1997). Fluid flow in fractured rocks at shallow levels in the Earth's crust: an overview. In: *Deformation-enhanced Fluid Transport in the Earth's Crust and Mantle* (ed. M.B. Holness), 289–298. Chapman & Hall.

Ogg, J.G., Ogg, G., and Gradstein, F.M. (2016). *A Concise Geologic Time Scale 2016*. Elsevier, 240 pp.

Ohmoto, H., Mizukami, M., Drummond, S.E. et al. (1983). Chemical processes of Kuroko formation. *Economic Geology Monograph* 5: 570–604.

Oliver, J. (1986). Fluids expelled tectonically from orogenic belts: their role in hydrocarbon migration and other geologic phemomena. *Geology* 14: 99–102.

Oliver, N.H.S. (1996). Review and classification of structural controls on fluid flow during regional metamorphism. *Journal of Metamorphic Geology* 14: 477–492.

Oreskes, N. and Einaudi, M.T. (1990). Origin of rare earth element-enriched hematite breccias at the Olympic Dam Cu-U-Au-Ag deposit, Roxby Downs, South Australia. *Economic Geology* 85: 1–28.

Oreskes, N. and Einaudi, M.T. (1992). Origin of hydrothermal fluids at Olympic Dam: preliminary results from fluid inclusions and stable isotopes. *Economic Geology* 87: 64–90.

Osborn, E.F. (1979). The reaction principle. In: *The Evolution of the Igneous Rocks: Fiftieth Anniversary Perspective* (ed. H.S. Yoder), 133–169. Princeton University Press.

Ossandón, G.C., Freraut, C.R., Gustafson, L.B. et al. (2001). Geology of the Chuquicamata Mine: a progress report. *Economic Geology* 96: 249–270.

Paarlberg, N.L. and Evans, L.L. (1977). Geology of the Fletcher Mine, Viburnum Trend, southeast Missouri. *Economic Geology* 72: 391–407.

Padilla Garza, R.A., Titley, S.R., and Pimental, F.B. (2001). Geology of the Escondida porphyry Cu deposit, Antofagasta region, Chile. *Economic Geology* 96: 307–324.

Park, R.G. (1995). Palaeoproterozoic Laurentia–Baltica relationships: a view from the Lewisian. In: *Early Precambrian Processes*, vol. 95 (eds. M.P. Coward and A.C. Ries), 211–224. Geological Society, Special Publication.

Parnell, J. (ed.) (1994). *Geofluids: Origin, Migration and Evolution of Fluids in Sedimentary Basins*, vol. 78. Geological Society, Special Publication, 372 pp.

Partington, G.A. and Williams, P.J. (2000). Proterozoic lode gold and (iron)-copper-gold deposits: a comparison of Australian and global examples. In: *Gold in 2000*, Reviews in Economic Geology, vol. 13 (eds. S. Hagemann and P.E. Brown), 69–101. Society of Economic Geologists.

Patterson, C. (1999). *Evolution*. Cornell University Press, 166 pp.

Pearson, R.G. (1963). Hard and soft acids and bases. *Journal of the American Chemical Society* 85: 3533–3539.

Perkins, W.G. (1997). Mount Isa lead-zinc ore bodies: replacement lodes in a zoned syndeformational copperlead-zinc system? *Ore Geology Reviews* 12: 61–110.

Peyerl, W. (1982). The influence of the Driekop dunite pipe on the platinum-group mineralogy of the UG-2 chromitite in its vicinity. *Economic Geology* 77: 1432–1438.

Phillips, G.N. (1986). Geology and alteration in the Golden Mile, Kalgoorlie. *Economic Geology* 91: 779–808.

Phillips, G.N. and Evans, K.A. (2004). Role of CO_2 in the formation of gold deposits. *Nature* 429: 860–863.

Phillips, G.N. and Law, J.D.M. (2000). Witwatersrand gold fields: geology, genesis and exploration. *Reviews in Economic Geology* 13: 439–500.

Phillips, G.N., Williams, P.J., and De-Jong, G. (1994). The nature of metamorphic fluids and significance for metal exploration. In: *Geofluids: Origin, Migration and Evolution of Fluids in Sedimentary Basins*, vol. 78 (ed. J. Parnell), 55–68. Geological Society, Special Publication.

Philpotts, A.R. (1967). Origin of certain iron-titanium oxide and apatite rocks. *Economic Geology* 62: 303–315.

Philpotts, A.R. (1982). Compositions of immiscible liquids in volcanic rocks. *Contributions to Mineralogy and Petrology* 80: 201–218.

Piper, J.D. (1976). Palaeomagnetic evidence for a Proterozoic super-continent. In: *A Discussion on Global Tectonics in Proterozoic Times*, vol. 280 (eds. J. Sutton, R.M. Shackleton and J.C. Briden), 469–490. Philosophical Transactions of the Royal Society London.

Piper, D.Z. and Perkins, R.B. (2014). Geochemistry of a marine phosphate deposit: a signpost to phosphogenesis. In: *Treatise on Geochemistry*, 2e, vol. 13 (eds. H.D. Holland and K.K. Turekian), 293–312. Oxford: Elsevier.

Pirajno, F. (1992). *Hydrothermal Mineral Deposits*. Springer-Verlag, 709 pp.

Pitcher, W.S. (1997). *The Nature and Origin of Granite*. Chapman & Hall, 387 pp.

Plimer, I.R. (1978). Proximal and distal stratabound ore deposits. *Mineralium Deposita* 13: 345–353.

Porter, S.M., Meisterfeld, R., and Knoll, A.H. (2003). Vase-shaped micro-fossils from the Neoproterozoic Chuar Group, Grand Canyon: a classification guided by modern testate amoebae. *Journal of Paleontology* 77: 409–429.

Powell, C.M.A. (1995). Are Neoproterozoic glacial deposits preserved on the margins of Laurentia related to the fragmentation of two supercontinents? Comment. *Geology* 23: 1053–1054.

Powell, C.M., Li, Z.X., McElhinny, M.W. et al. (1993). Paleomagnetic constraints on timing of the Neoproterozoic breakup of Rodinia and the Cambrian formation of Gondwana. *Geology* 21: 889–892.

Powell, C.M., Oliver, N.H., Li, Z.X. et al. (1999). Synorogenic hydrothermal origin for the giant Hamersley iron oxide ore bodies. *Geology* 27: 175–178.

Pretorius, D.A. (1976). The stratigraphic, geochronologic, ore-type and geologic-environment sources of mineral wealth in the Republic of South Africa. *Economic Geology* 71: 5–15.

Prevec, S.A., Lightfoot, P.C., and Keays, R.R. (2000). Evolution of the sublayer of the Sudbury Igneous Complex: geochemical, Sm-Nd isotopic and petrologic evidence. *Lithos* 51: 271–292.

Pye, K. (1994). *Sediment Transport and Depositional Processes*. Blackwell Scientific Publications, 397 pp.

Rainaud, C., Armstrong, R.A., Master, S., and Robb, L.J. (1999). A fertile Palaeoproterozoic magmatic arc beneath the Central African Copperbelt. In: *Mineral Deposits: Processes to Processing* (eds. C. Stanley, A.H. Rankin, R.J. Bodnar, et al.), 1427–1430. Balkema.

Rainaud, C., Armstrong, R.A., Master, S. et al. (2002a). Contributions to the geology and mineralization of the Central African Copperbelt: I. Nature and geochronology of the pre-Katangan basement. In: *Economic Geology Research Institute, Information*

Circular 362 (ed. C.R. Anhaeusser), 19–23. University of the Witwatersrand.

Rainaud, C., Master, S., Armstrong, R.A. et al. (2002b). Contributions to the geology and mineralization of the Central African Copperbelt: IV. Monazite U-Pb dating and 40Ar-39Ar thermochronology of metamorphic events during the Lufilian Orogeny. In: *Economic Geology Research Institute, Information Circular 362* (ed. C.R. Anhaeusser), 33–37. University of the Witwatersrand.

Rainaud, C., Master, S., Armstrong, R.A., and Robb, L.J. (2005). Geochronology and nature of the Palaeoproterozoic basement in the Central African Copperbelt (Zambia and the DRC), with regional implications. *Journal African Earth Sciences* 42: 1–31.

Raiswell, R.W., Brimblecombe, P., Dent, D.L., and Liss, P.S. (1980). *Environmental Chemistry: The Earth– Air–Water Factory*. Wiley, 184 pp.

Ramanaidou, E.R. and Wells, M.A. (2014). Sedimentary hosted iron ores. In: *Treatise on Geochemistry*, 2e, vol. 13 (eds. H.D. Holland and K.K. Turekian), 313–355. Oxford: Elsevier.

Reading, H.G. and Collinson, J.D. (1996). Clastic coasts. In: *Sedimentary Environments: Processes, Facies and Stratigraphy* (ed. H.H. Reading), 154–231. Blackwell Science.

Reed, M.H. (1982). Calculation of multicomponent equilibria and reaction processes involving minerals, gases and an aqueous phase. *Geochimica et Cosmochimica Acta* 46: 513–528.

Reed, M.H. (1997). Hydrothermal alteration and its relationship to ore fluid composition. In: *Geochemistry of Hydrothermal Ore Deposits* (ed. H.L. Barnes), 303–366. Wiley.

Reeve, J.S., Cross, K.C., Smith, R.N., and Oreskes, N. (1990). Olympic Dam copper-uranium-gold-silver deposit. In: *Geology of the Mineral Deposits of Australia and Papua New Guinea*. Monograph Series, vol. 14 (ed. F.E. Hughes), 1009–1035. Australasian Institute of Mining and Metallurgy.

Reich, M. and Vasconcelos, P.M. (2015). Geological and economic significance of supergene metal deposits. *Elements* 11 (5): 305–310.

Reid, I. and Frostick, L.E. (1994). Fluvial sediment transport and deposition. In: *Sediment Transport and Depositional Processes* (ed. K. Pye), 89–156. Blackwell Scientific Publications.

Retallack, G. (2010). Lateritization and bauxitization events. *Economic Geology* 105: 655–667.

Reynolds, R.L. and Goldhaber, M.B. (1978). Origin of a south-Texas roll-type uranium deposit: I. Alteration of iron-titanium oxide minerals. *Economic Geology* 73: 1677–1689.

Reynolds, R.L. and Goldhaber, M.B. (1983). Iron disulfide minerals and the genesis of roll-type uranium deposits. *Economic Geology* 78: 105–120.

Rhodes, A.L. and Oreskes, N. (1999). Oxygen isotope compositions of magnetite deposits at El Laco, Chile: evidence of formation from isotopically heavy fluids. In: *Geology and Ore Deposits of the Central Andes*, vol. 7 (ed. B.J. Skinner), 333–351. Society of Economic Geologists, Special Publication.

Richards, J.P., Boyce, A.J., and Pringle, M.S. (2001). Geologic evolution of the Escondida area, northern Chile: a model for spatial and temporal localization of porphyry Cu mineralization. *Economic Geology* 96: 271–306.

Richardson, S.H., Gurney, J.J., Erlank, A.J., and Harris, J.W. (1984). Origin of diamonds in old enriched mantle. *Nature* 310: 198–202.

Rickard, D. (1976). Hydrocarbons associated with lead-zinc ores at Laisvall, Sweden. *Nature* 255: 131–133.

Rickard, D., Willden, M.Y., Marinder, N.E., and Donnelly, T.H. (1979). Studies on the genesis of the Laisvall sandstone lead-zinc deposit, Sweden. *Economic Geology* 74: 1255–1285.

Ridley, J.R. and Diamond, L.W. (2000). Fluid chemistry of orogenic lode gold deposits and implications for genetic models. In: *Gold in 2000*, Reviews in Economic Geology, vol. 13

(eds. S. Hagemann and P.E. Brown), 141–162. Society of Economic Geologists.

Riggs, S.R. (1979). Phosphorite sedimentation in Florida; a model phosphogenic system. In: *An Issue Devoted to Phosphate, Potash and Sulfur*, Economic Geology, vol. 74 (eds. P.F. Howard, J.C. Dunlap, R.J. Hite and A.J. Bodenlos), 285–314.

Rimstidt, J.D. (1979) The kinetics of silica–water reactions. Doctoral Thesis. Pennsylvania State University.

Rimstidt, J.D. (1997). Gangue mineral transport and deposition. In: *Geochemistry of Hydrothermal Ore Deposits* (ed. H.L. Barnes), 487–516. Wiley.

Ripley, E.M. and Al-Jassar, T.J. (1987). Sulfur and oxygen isotope studies of melt-country rock interaction, Babbitt Cu-Ni deposit, Duluth Complex, Minnesota, USA. *Economic Geology* 82: 87–107.

Ripley, E.M. and Li, C. (2013). Sulfide saturation in mafic magmas: is external sulphur required for magmatic Ni-Cu-(PGE) ore genesis? *Economic Geology* 108: 45–58.

Ripley, E.M., Merino, E., Moore, C., and Ortoleva, P. (1985). Mineral zoning in sediment-hosted copper deposits. In: *Handbook of Stratabound and Stratiform Ore Deposits, Volume 6* (ed. K.H. Wolf), 237–260. Elsevier.

Ripley, E.M., Severson, M.J., and Hauck, S.A. (1998). Evidence for sulfide and Fe-Ti-P rich liquid immiscibility in the Duluth Complex, Minnesota. *Economic Geology* 93: 1052–1062.

Riveros, K., Veloso, E., Campos, E. et al. (2014). Magnetic properties related to hydrothermal alteration processes at the Escondida porphyry copper deposit, northern Chile. *Mineralium Deposita* https://doi.org/10.1007/s00126-014-0514-7.

Robb, L.J. and Hayward, C. (2014). Geochemistry of placer gold – a case study of the Witwatersrand deposits. In: *Treatise on Geochemistry*, 2e, vol. 13 (eds. H.D. Holland and K.K. Turekian), 433–461. Oxford: Elsevier.

Robb, L.J., Charlesworth, E.G., Drennan, G.R. et al. (1997). Tectono-metamorphic setting and paragenetic sequence of Au-U mineralization in the Archaean Witwatersrand Basin, South Africa. *Australian Journal of Earth Sciences* 44: 353–371.

Robb, L.J., Armstrong, R.A., and Waters, D.J. (1999). The history of granulite-facies metamorphism and crustal growth from single zircon geochronology: Namaqualand, South Africa. *Journal of Petrology* 40 (12): 1747–1770.

Robb, L.J., Freeman, L.A., and Armstrong, R. (2000). Nature and longevity of hydrothermal fluid flow and mineralization in granites of the Bushveld Complex, South Africa. *Transactions of the Royal Society Edinburgh: Earth Sciences* 91: 269–281.

Robb, L.J., Master, S., Greyling, L. et al. (2002). Contributions to the geology and mineralization of the Central African Copperbelt: V. Speculations regarding the "Snowball Earth" and redox controls on stratabound Cu-Co and Pb-Zn mineralization. In: *Economic Geology Research Institute, Information Circular 362* (ed. C.R. Anhaeusser), 38–42. University of the Witwatersrand.

Roedder, E. (1984). *Fluid Inclusions*, Reviews in Mineralogy, vol. 12. Mineralogical Society of America, 644 pp.

Roedder, E. and Bodnar, R.J. (1980). Geologic pressure determinations from fluid inclusion studies. *Annual Reviews of Earth Planetary Sciences* 8: 263–301.

Rogers, J.J.W. (1996). A history of continents in the past three billion years. *Journal of Geology* 104: 91–107.

Rogers, J.J. and Santosh, M. (2004). *Continents and Supercontinents*. Oxford University Press, 289 pp.

Rollinson, H. (1993). *Using Geochemical Data: Evaluation, Presentation, Interpretation*. Longman, 352 pp.

Romanov, V.V. (1968). *Hydrophysics of bogs*. Jerusalem: Israel Program for Scientific Translations 299 pp.

Rose, A.W. (1976). The effect of cuprous chloride complexes in the origin of red-bed copper and related deposits. *Economic Geology* 71: 1036–1048.

Rose, A.W. and Bianchi-Mosquera, G.C. (1993). Adsorption of Cu, Pb, Zn, Co, Ni, and Ag on goethite and hematite: control on metal mobilization from redbeds into stratiform copper deposits. *Economic Geology* 88: 1226–1236.

Rose, A.W. and Burt, D. (1979). Hydrothermal alteration. In: *Geochemistry of Hydrothermal Ore Deposits* (ed. H.L. Barnes), 173–235. Wiley.

Rowins, S.M., Groves, D.I., McNaughton, M.J. et al. (1997). A reinterpretation of the role of granitoids in the genesis of Neoproterozoic gold mineralization in the Telfer Dome, Western Australia. *Economic Geology* 92: 133–160.

Rudnick, R.L. and Gao, S. (2014). Composition of the continental crust. In: *Treatise on Geochemistry*, 2e, vol. 4 (eds. H.D. Holland and K.K. Turekian), 1–51. Oxford: Elsevier.

Ruppel, C. (2011). Methane hydrates and the future of natural gas. *MITEI Natural Gas Report, Supplementary Paper 4*, 25 pp.

Rye, R. and Holland, H.D. (1998). Paleosols and the evolution of the atmosphere: a critical review. *American Journal of Science* 298: 621–672.

Sallenger, A.H. (1979). Inverse grading and hydraulic equivalence in grain-flow deposits. *Journal of Sedimentary Petrology* 49: 553–562.

Sams, M.S. and Thomas-Betts, A. (1988). Models of convective fluid flow and mineralization in southwest England. *Journal of the Geological Society* 145: 809–817.

Samson, I.M., Williams-Jones, A.E., Ault, K.M. et al. (2008). Source of fluids forming distal Zn-Pb-Ag skarns: evidence from LA-ICP-MS analysis of fluid inclusions from El Mochito, Honduras. *Geology* 36: 947–950.

Sassani, D.C. and Shock, E.L. (1990). Speciation and solubility of palladium in aqueous magmatic-hydrothermal solutions. *Geology* 18: 925–928.

Sato, T. (1972). Behavior of ore forming solutions in sea water. *Mining Geology* 22: 31–42.

Saunders, J.A., Hofstra, A.H., Goldfarb, R.J., and Reed, M.H. (2014). Geochemistry of hydrothermal gold deposits. In: *Treatise on Geochemistry*, 2e, vol. 13 (eds. H.D. Holland and K.K. Turekian), 383–424. Oxford: Elsevier.

Sawkins, F.J. (1990). *Metal Deposits in Relation to Plate Tectonics*. Springer-Verlag, 461 pp.

Sawlowicz, Z. (1992). Primary sulphide mineralization in Cu-Fe-S zones of the Kupferschiefer, Fore-Sudetic monocline, Poland. *Transactions of the Institute of Mining Metallurgy Section B* 101: B1–B8.

Schiffries, C.M. (1982). The petrogenesis of a platiniferous dunite pipe in the Bushveld Complex: infiltration metasomatism by a chloride solution. *Economic Geology* 77: 1439–1453.

Schiffries, C.M. and Rye, D.M. (1990). Stable isotope systematics of the Bushveld Complex: II. Constraints on hydrothermal processes in layered intrusions. *American Journal of Science* 290: 209–245.

Schmid, G. (1994). Ligand Stabilized Giant Metal Clusters and Colloids. In: *Physics and chemistry of metal cluster compounds* (ed. L.J. de Jongh), 107–134. Kluwer.

Schoenberg, R., Kruger, F.J., Nagler, T.F. et al. (1999). PGE enrichment in chromitite layers and the Merensky Reef of the western Bushveld Complex: a Re-Os and Rb-Sr isotope study. *Earth and Planetary Science Letters* 172: 49–64.

Schwartz, A.W. (1971). Phosphate; solubilization and activation on the primitive Earth. In: *Chemical Evolution and the Origin of Life: Molecular Evolution*, vol. 1 (eds. R. Buvet and C. Ponnamperuma), 207–215. Elsevier.

Scoon, R.N. and Teigler, B. (1994). Platinum group element mineralization in the Critical Zone of the western Bushveld Complex: I. Sulfide poor chromitites below the UG-2. *Economic Geology* 89: 1094–1121.

Scotese, C.R. (2009). Late Proterozoic plate tectonics and paleogeography: a tale of two supercontinents, Rodinia and Pannotia. *Geological Society of London Special Publication* 326: 67–83.

Scott, S.D. (1997). Submarine hydrothermal systems and deposits. In: *Geochemistry of Hydrothermal Ore Deposits*, 3e (ed. H.L. Barnes), 797–875. Wiley.

Selley, R.C. (1988). *Applied Sedimentology*. Academic Press, 446 pp.

Seward, T.M. (1981). Metal complex formation in aqueous solutions at elevated temperatures and pressures. In: *Chemistry and Geochemistry of Solutions at High Temperatures and Pressures*, Physics and Chemistry of the Earth, vol. 13 and 14 (eds. D. Rickard and F. Wickman), 113–128.

Seward, T.M. (1991). The hydrothermal geochemistry of gold. In: *Gold Metallogeny and Exploration* (ed. R.P. Foster), 37–62. Blackie.

Seward, T.M. and Barnes, H.L. (1997). Metal transport by hydrothermal ore fluids. In: *Geochemistry of Hydrothermal Ore Deposits* (ed. H.L. Barnes), 435–486. Wiley.

Seward, T.M., Williams-Jones, A.E., and Migdisov, A.A. (2014). The chemistry of metal transport and deposition by ore-forming hydrothermal fluids. In: *Treatise on Geochemistry*, 2e, vol. 13 (eds. H.D. Holland and K.K. Turekian), 29–57. Oxford: Elsevier.

Shannon, P.M. and Naylor, D. (1989). *Petroleum Basin Studies*. Graham & Trotman, 206 pp.

Sheldon, R.P. (1980). Episodicity of phosphate deposition and deep ocean circulation: a hypothesis. In: *Marine Phosphorites*, vol. 29 (ed. Y.K. Bentor), 239–247. Society of Economic Palaeontologists and Mineralogists, Special Publication.

Sheppard, S.M.F. (1977). The Cornubian batholith, southwest England: D/H and $^{18}O/^{16}O$ studies of kaolinite and other alteration minerals. *Journal of the Geological Society* 133: 573–591.

Sheppard, S.M.F., Nielsen, R.L., and Taylor, H.P. (1971). Hydrogen and oxygen isotope ratios in minerals from porphyry copper deposits. *Economic Geology* 66: 515–542.

Shinohara, H. (1994). Exsolution of immiscible vapor and liquid phases from a crystallizing silicate melt: implications for chlorine and metal transport. *Geochimica et Cosmochimica Acta* 58: 5215–5221.

Shinohara, H. and Hedenquist, J.W. (1997). Constraints on magma degassing beneath the Far Southeast Porphyry Cu-Au Deposit, Philippines. *Journal of Petrology* 38: 1741–1752.

Shinohara, H., Iiyama, J.T., and Matso, S. (1989). Partition of chlorine compounds between silicate melt and hydrothermal solutions: I. partition of NaCl-KCl. *Geochimica et Cosmochimica Acta* 53: 2517–2630.

Shirey, S.B. and Richardson, S.H. (2011). Start of the Wilson Cycle at 3 Ga shown by diamonds from subcontinental mantle. *Science* 333: 434–436.

Shuey, R.T. (1975). *Semiconducting Ore Minerals*, Developments in Economic Geology, vol. 4. Elsevier, 364 pp.

Sibson, R.H. (1986). Earthquakes and rock deformation in crustal fault zones. *Annual Reviews in Earth Planetary Sciences* 14: 149–175.

Sibson, R.H. (1987). Earthquake rupturing as a mineralizing agent in hydrothermal systems. *Geology* 15: 701–704.

Sibson, R.H. (1994). Crustal stress, faulting and fluid flow. In: *Geofluids: Origin, Migration and Evolution of Fluids in Sedimentary Basins*, vol. 78 (ed. J. Parnell), 69–84. Geological Society, Special Publication.

Sibson, R.H., Moore, J.M., and Rankin, A.H. (1975). Seismic pumping: a hydrothermal fluid transport mechanism. *Journal of the Geological Society* 131: 653–659.

Sibson, R.H., Robert, F., and Poulsen, K.H. (1988). High angle reverse faults, fluid-pressure cycling, and mesothermal gold-quartz deposits. *Geology* 16: 551–555.

Siehl, A. and Thein, J. (1989). Minette-type ironstones. In: *Phanerozoic Ironstones*, vol. 46 (eds. T.P. Young and W.E.G. Taylor), 175–193. Geological Society London, Special Publication.

Sillitoe, R.H. (1976). Andean mineralization: a model for the metallogeny of unconvergent plate margins. In: *Metallogeny and Plate*

Tectonics, Special Paper, vol. 14 (ed. D.F. Strong), 59–100. Geological Association of Canada.

Sillitoe, R.H. and Burrows, D.R. (2002). New field evidence bearing on the origin of the El Laco magnetite deposit northern Chile. *Economic Geology* 97: 1101–1109.

Sillitoe, R.H. and McKee, E.H. (1996). Age of supergene oxidation and enrichment in the Chilean porphyry copper province. *Economic Geology* 91: 164–179.

Sillitoe, R.H., Folk, R.L., and Saric, N. (1996). Bacteria as mediators of copper sulphide enrichment during weathering. *Science* 272: 1153–1155.

Sillitoe, R.H., Perello, J., Creaser, R.A. et al. (2017). Age of the Zambian Copperbelt. *Mineralium Deposita* 52: 1245–1268.

Simmons, S.F. and Brown, K.L. (2006). Gold in magmatic hydrothermal solutions and the rapid formation of a giant ore deposit. *Science* 314: 288–291.

Simmons, S.F., White, N.C., and John, D.A. (2005). Geological characteristics of epithermal precious and base metal deposits. In: *Economic Geology 100th Anniversary Volume: 1905–2005* (eds. J.W. Hedenquist, J.F.H. Thompson, R.J. Goldfarb, et al.), 485–522. Society of Economic Geologists.

Simonson, B. (2011). Iron ore deposits associated with Precambrian iron formations. *Elements* 7: 119–120.

Skinner, B.J. (1976). A second iron age ahead? *Scientific American* 64: 258–269.

Skinner, B.J. (1997). Hydrothermal mineral deposits: what we do and don't know. In: *Geochemistry of Hydrothermal Ore Deposits* (ed. H.L. Barnes), 1–29. Wiley.

Skinner, B.J. and Peck, D.L. (1969). An immiscible sulfide melt from Hawaii. *Economic Geology Monograph* 4: 310–322.

Skinner, B.J., White, D.E., Rose, H.J., and Mays, R.E. (1969). Sulfides associated with the Salton Sea geothermal brine. *Economic Geology* 62: 316–330.

Slingerland, R. (1984). Role of hydraulic sorting in the origin of fluvial placers. *Journal of Sedimentary Petrology* 54: 137–150.

Slingerland, R. and Smith, N.D. (1986). Occurrence and formation of water-laid placers. *Annual Review of Earth and Planetary Sciences* 14: 133–147.

Smirnov, V.I. (1977). *Ore Deposits of the USSR*, vol. 1–3. Pitman.

Smith, N.D. and Beukes, N.J. (1983). Bar to bank flow convergence zones: a contribution to the origin of alluvial placers. *Economic Geology* 78: 1342–1349.

Smith, M.P. and Henderson, P. (2000). Preliminary fluid inclusion constraints on fluid evolution in the Bayan Obo Fe-REE-Nb deposit, Inner Mongolia, China. *Economic Geology* 95: 1371–1388.

Smith, N.D. and Minter, W.E.L. (1980). Sedimentological controls of gold and uranium in two Witwatersrand paleoplacers. *Economic Geology* 75: 1–14.

Smith, C.B., McCallum, M.E., Coppersmith, H.G., and Eggler, D.H. (1979). Petrochemistry and structure of kimberlites in the Front Range and Laramie Range, Colorado–Wyoming. In: *Kimberlites, Diatremes, and Diamonds: Their Geology, Petrology, and Geochemistry (Proceedings of the International Kimberlite Conference)*, vol. 2 (eds. F.R. Boyd and H.O.A. Meyer), 178–189. American Geophysical Union.

Solomon, M. and Groves, D.I. (1994). *The Geology and Origin of Australia's Mineral Deposits*. Oxford University Press, 951 pp.

Solomon, M., Groves, D.I., and Jaques, A.L. (2000). *The Geology and Origin of Australia's Mineral Deposits*. University of Tasmania/University of Western Australia, 1002 pp.

Southam, G. and Beveridge, T.J. (1994). The in vitro formation of placer gold by bacteria. *Geochimica et Cosmochimica Acta* 58: 4527–4530.

Southam, G. and Saunders, J.A. (2005). The geomicrobiology of ore deposits. In: *Economic Geology 100th Anniversary Volume: 1905–2005* (eds. J.W. Hedenquist, J.F.H. Thompson, R.J. Goldfarb, et al.), 1067–1084. Society of Economic Geologists.

Spackman, W., Davis, A., and Mitchell, G.D. (1976). The fluorescence of liptinite macerals. *Geology Studies* 22: 59–75.

Spooner, E.T.C. (1980). Cu-pyrite mineralization and sea water convection in oceanic crust: the ophiolitic ore deposits of Cyprus. In: *The Continental Crust and its Mineral Deposits*, Special paper, vol. 20 (ed. D.W. Strangway), 685–704. Geological Association of Canada.

Stach, E. (1975). *Textbook of Coal Petrography*. Borntraeger, 428 pp.

Stanton, R.L. (1972). *Ore Petrology*. McGraw-Hill, 771 pp.

Stephenson, M. (2015). *Shale Gas and Fracking – The Science behind the Controversy*. Elsevier, 153 pp.

Stevens, G., Boer, R., and Gibson, R.L. (1997). Metamorphism, fluid flow and gold remobilization in the Witwatersrand Basin: towards a unifying model. *South African Journal of Geology* 100: 363–375.

Stoffell, B., Wilkinson, J.J., and Jeffries, T.E. (2004). Metal transport and deposition in hydrothermal veins revealed by 213 nm UV laser ablation microanalysis of single fluid inclusions. *American Journal of Science* 304: 533–557.

Stolper, E. (1982). The speciation of water in silicate melts. *Geochimica et Cosmochimica Acta* 46: 2609–2620.

Strong, D.F. (1981). Ore deposit models: 5. A model for granophile mineral deposits. *Geoscience Canada* 8: 155–161.

Strong, D.F. (1988). A review and model for granite-related mineral deposits. In: *Recent Advances in the Geology of Granite-related Mineral Deposits*, vol. 39 (eds. R.P. Taylor and D.F. Strong), 424–445. Canadian Institute of Mining Metallurgy, Special Volume.

Sundborg, A. (1956). The River Klaralven: a study of fluvial processes. *Geographic Annals* 38: 127–316.

Susak, N.J. and Crerar, D.A. (1981). Factors controlling mineral zoning in hydrothermal ore deposits. *Economic Geology* 76: 476–482.

Sverjensky, D.A. (1981). The origin of a Mississippi Valley-type deposit in the Viburnum Trend, southeast Missouri. *Economic Geology* 76: 1848–1872.

Sverjensky, D.A. (1986). Genesis of Mississippi Valley-type lead-zinc deposits. *Annual Reviews of Earth and Planetary Sciences* 14: 177–199.

Sverjensky, D.A. (1989). Chemical Evolution of Basinal Brines that form Sediment Hosted Cu-Pb-Zn Deposits. In: *Sediment-Hosted Stratiform Copper Deposits*, Special Paper, vol. 36 (eds. R.W. Boyle, A.C. Brown, C.W. Jefferson, et al.), 499–518. Geological Association of Canada.

Sweeney, M.A., Binda, P.L., and Vaughan, D.J. (1991). Genesis of the ores of the Zambian Copperbelt. *Ore Geology Reviews* 6: 51–76.

Talbot, C.J., Tully, C.P., and Woods, P.J.E. (1982). The structural geology of the Boulby (potash) mine, Cleveland, United Kingdom. *Tectonophysics* 85: 167–204.

Tardy, Y. (1992). Diversity and terminology of lateritic profiles. In: *Weathering, Soils and Paleosols*, Developments in Earth Surface Processes, vol. 2 (eds. I.P. Martini and W. Chesworth), 379–403. Elsevier.

Tardy, Y. and Roquin, C. (1992). Geochemistry and evolution of lateritic landscapes. In: *Weathering, Soils and Paleosols*, Developments in Earth Surface Processes, vol. 2 (eds. I.P. Martini and W. Chesworth), 415 439. Elsevier.

Tarling, D.H. (1981). *Economic Geology and Geotectonics*. Blackwell Scientific, 213 pp.

Taylor, S.R. (1964). Abundance of chemical elements in the continental crust: a new table. *Geochimica et Cosmochimica Acta* 28: 1273–1285.

Taylor, H.P. (1997). O and H isotope relationships in hydrothermal mineral deposits. In: *Geochemistry of Hydrothermal Ore Deposits* (ed. H.L. Barnes), 229–302. Wiley.

Taylor, B.E. (2007). Epithermal gold deposits. In: *Mineral Deposits of Canada: a Synthesis of Major Deposit Types, District Metallogeny, the*

Evolution of Geologic Provinces and Exploration Methods (ed. W.D. Goodfellow), 113–139. Geological Association of Canada, Special Publication 5.

Taylor, K.G. and Konhauser, K.O. (2011). Iron in Earth surface systems: a major player in chemical and biological processes. *Elements* 7: 83–88.

Taylor, J.R. and Wall, V.J. (1992). The behavior of tin in granitoid magmas. *Economic Geology* 87: 403–420.

Taylor, D., Dalstra, H.J., Harding, A.E. et al. (2001). Genesis of high-grade hematite orebodies of the Hamersley Province, Western Australia. *Economic Geology* 96: 837–873.

Tertre, E., Berger, G., Simoni, E. et al. (2006). Eu retention onto clay minerals from 25 to 150 °C: experimental measurements, spectroscopic features and sorption modelling. *Geochimica et Cosmochimica Acta* 70: 4563–4578.

Thomas, R. and Webster, J.D. (2000). Strong tin enrichment in a pegmatite forming melt. *Mineralium Deposita* 35: 570–582.

Thomas, A.V., Bray, C.J., and Spooner, E.T.C. (1988). A discussion of the Jahns–Burnham proposal for the formation of zoned granitic pegmatites using solid– liquid–vapour inclusions from the Tanco pegmatite, SE Manitoba, Canada. *Transactions of the Royal Society Edinburgh: Earth Sciences* 79: 299–315.

Thomas, R., Webster, J.D., and Heinrich, W. (2000). Melt inclusions in pegmatite quartz: complete miscibility between silicate melts and hydrous fluids at low pressure. *Contributions to Mineralogy and Petrology* 139: 394–401.

Thompson, A.J.B. and Thompson, J.F.H. (eds.) (1996). *Atlas of Alteration*, vol. 6. Geological Association of Canada, Special Publication 119 pp.

Thoreson, R.F., Jones, M.E., Breit, F.J. et al. (2000). The geology and gold mineralization of the Twin Creek gold deposits, Humboldt County, Nevada. In: *Geology and Gold Deposits of the Getchell Region, Part II*. Guidebook Series, vol. 332, 175–187. Society of Economic Geologists.

Tiratsoo, E.N. (1984). *Oilfields of the World*. Beaconsfield: Scientific Press, 392 pp.

Tissot, B.P. and Welte, D.H. (1984). *Petroleum Formation and Occurrence*. Springer, 538 pp.

Tissot, B.P., Durand, B., Espistalie, J., and Combaz, A. (1974). Influence of nature and diagenesis of organic matter in formation of petroleum. *American Association of Petroleum Geologists Bulletin* 58: 499–506.

Titley, S.R. (1993). Relationship of stratabound ores with tectonic cycles of the Phanerozoic and Proterozoic. *Precambrian Research* 61: 295–322.

Torsvik, T.H. and Cocks, L.R.M. (2013). Gondwana from top to base in space and time. *Gondwana Research* 24: 999–1030.

Torsvik, T.H., Smethurst, M.A., Meert, J.G. et al. (1996). Continental break-up and collision in the Neoproterozioc and Palaeozoic: a tale of Baltica and Laurentia. *Earth-Science Reviews* 40: 229–258.

Traversa, G., Gomes, C.B., Brotzu, P. et al. (2001). Petrography and mineral chemistry of carbonatites and mica-rich rocks from the Araxa complex – Alto Paranaiba Province, Brazil. *Annals Academia Brasileira de Ciências* 73: 71–98.

Tredoux, M., Lindsay, N.M., Davies, G., and McDonald, I. (1995). The fractionation of platinum group elements in magmatic systems, with the suggestion of a novel causal mechanism. *South African Journal of Geology* 98: 157–167.

Troly, G., Esterle, M., Pelletier, B.G., and Reibell, W. (1979). Nickel deposits in New Caledonia; some factors influencing their formation. In: *International Laterite Symposium* (eds. D.J.I. Evans, R.S. Shoemaker and H. Veltman), 85–117. American Institute of Mining Metallurgical and Petroleum Engineers.

Trudinger, P.A. (1976). Microbiological processes in relation to ore genesis. In: *Handbook of Stratabound and Stratiform Ore Deposits, Volume 2* (ed. K.H. Wolf), 135–190. Elsevier.

Tsikos, H., Beukes, N.J., Harris, C., and Moore, J.M. (2003). Deposition, diagenesis and secondary enrichment of metals in the

Palaeoproterozoic Hotazel iron-formation, Kalahari Manganese Field, South Africa. *Economic Geology* 98: 1449–1462.

Tsikos, H., Matthews, A., Erel, Y., and Moore, J.M. (2010). Iron isotopes constrain biogeochemical cycling of Fe and Mn in a Palaeoproterozoic stratified basin. *Earth & Planetary Science Letters* 298: 125–134.

Tsoar, H. and Pye, K. (1987). Dust transport and the question of desert loess formation. *Sedimentology* 34: 139–153.

Turner, J.S. (1980). A fluid-dynamical model of differentiation and layering in magma chambers. *Nature* 285: 213–215.

Ulmer, G.C. (1969). Experimental investigation on chromite spinels. *Monograph in Economic Geology* 4: 114–131.

Ulrich, T., Günther, D., and Heinrich, C.A. (2001). The evolution of a porphyry Cu-Au deposit based on LA-ICP-MS analysis of fluid inclusions: Bajo de la Alumbrera, Argentina. *Economic Geology* 96: 1743–1774.

Ulrich, T., Günther, D., and Heinrich, C.A. (2002). The evolution of a porphyry Cu-Au deposit, based on LA-ICP-MS analysis of fluid inclusions: Bajo de la Alumbrera, Argentina. *Economic Geology*: 97, 1889–1920, reprinted with corrections.

Unrug, R. (1988). Mineralization controls and source of metals in the Lufilian Fold belt, Shaba (Zaire), Zambia and Angola. *Economic Geology* 83: 1247–1258.

Unrug, R. (1997). Rodinia to Gondwana: the geodynamic map of Gondwana supercontinent assembley. *GSA Today* 7 (1): 1–6.

Van Hinsberg, V.J., Berlo, K., Migdisov, A.A., and Williams-Jones, A.E. (2016). CO_2 – fluxing collapses metal mobility in magmatic vapour. *Geochemical Perspectives Letters* 2: 169–177.

Van Houten, F.B. and Authur, M.A. (1989). Temporal patterns among Phanerozoic oolitic ironstones and oceanic anoxia. In: *Phanerozoic Ironstones*, vol. 46 (eds. T.P. Young and W.E.G. Taylor), 33–49. Geological Society London, Special Publication.

Van Krevelen, D.W. (1961). *Coal*, 1e. New York: Elsevier, 514 pp.

Van Niekerk, A., Vogel, K.R., Slingerland, R.L., and Bridge, J.S. (1992). Routing of heterogeneous sediments over a movable bed: model development. *Journal of Hydraulic Engineering* 118: 246–262.

Van Wyk, J.P. and Pienaar, L.F. (1986). Diamondiferous gravels of the lower Orange River, Namaqualand. In: *Mineral Deposits of Southern Africa, Volume 2* (eds. C.R. Anhaeusser and S. Maske), 2309–2322. Geological Society of South Africa.

Veevers, J.J. (1989). Middle-late Triassic (230±5 Ma) singularity in the stratigraphic and magmatic history of the Pangean heat anomaly. *Geology* 17: 784–787.

Veizer, J., Laznicka, P., and Jansen, S.L. (1989). Mineralization through geologic time: recycling perspective. *American Journal of Science* 289: 484–524.

Viljoen, M.J. and Viljoen, R.P. (1969) The geology and geochemistry of the lower ultramafic unit of the Onverwacht Group and a proposed new class of igneous rock. Upper Mantle Project, collected papers and abstracts presented at the Symposium held in Pretoria in 1969. Geological Society of South Africa, Special Publication No. 2, pp. 55–85.

Vogel, K.R., Van Niekerk, A., Slingerland, R.L., and Bridge, J.S. (1992). Routing of heterogeneous sediments over a movable bed: model verification. *Journal of Hydraulic Engineering* 118: 263–279.

Volp, K.M. (2005). The Estrela copper deposit, Carajas, Brazil: geology and implications of a Proterozoic copper stockwork. In: *Mineral Deposit Research – Meeting the Global Challenge* (eds. J. Mao and F.P. Bierlein), 1085–1088. Berlin: Springer.

Von Grünewaldt, G., Hatton, C.J., Merkle, R.K.W., and Gain, S. (1986). Platinum group element–chromitite associations in the Bushveld Complex. *Economic Geology* 81: 1067–1079.

Wager, L.R. and Brown, M. (1968). *Layered Igneous Rocks*. Oliver & Boyd, 588 pp.

Wager, L.R., Vincent, E.A., and Smales, A.A. (1957). Sulphides in the Skaergaard Intrusion, east Greenland. *Economic Geology* 52: 855–899.

Wall, F. (2014). Rare earth elements. In: *Critical Metals Handbook* (ed. G. Gunn), 312–339. Wiley/AGU.

Walshe J.L., Neumayr P., Petersen K., Halley S., Roache A., and Young, C. (2006). Scale-integrated, architectural and geodynamic controls on alteration and geochemistry of gold systems in the Eastern Goldfields Province, Yilgarn Craton, Project M358. Minerals and Energy Research Institute of Western Australia, Report No. 256, 200 pp.

Wang, J., Tatsumoto, M., Li, X. et al. (1994). A precise ^{232}Th-^{208}Pb chronology of fine-grained monazite: age of the Bayan Obo REE-Fe-Nb ore deposit, China. *Geochimica Cosmochimica Acta* 58: 3155–3169.

Warren, J.K. (2010). Evaporites through time: tectonic, climatic and eustatic controls in marine and non-marine deposits. *Earth-Science Reviews* 98: 217–268.

Watterson, J.R. (1991). Preliminary evidence for the involvement of budding bacteria in the origin of Alaskan placer gold. *Geology* 20: 315–318.

Webb, M. and Rowston, P. (1995). The geophysics of the Ernest Henry Cu-Au deposit (NW) Queensland. *Exploration Geophysics* 26: 51–59.

Webster, J.G. (1986). The solubility of gold and silver in the system Au-Ag-S-O$_2$-H$_2$O at 25 °C and 1 atmosphere. *Geochimica et Cosmochimica Acta* 50: 1837–1845.

Webster, J.D. and Holloway, J.R. (1990). Partitioning of F and Cl Between Magmatic Hydrothermal Fluids and Highly Evolved Granitic Magmas. In: *Ore-bearing Granite Systems; Petrogenesis and Mineralizing Processes*, Special Paper, vol. 246 (eds. H.J. Stein and J.L. Hannah), 21–34. Geological Society of America.

Wedepohl, K.H. (1969). *Handbook of Geochemistry, Volume 2*. Springer-Verlag, loose leaf.

Wendorff, M. (2001). Grey RAT orebodies with Cu-Co in the Katangan copperbelt of the DRC; genetic implications of tectonstratigraphy. In: *Mineral Deposits at the Beginning of the 21st Century: Proceedings of the Joint 6th Biennial SGA-SEG Meeting*, Krakow, Poland (26–29 August 2001). (eds. A. Piestrzynka, S. Speczik, J. Pasava, et al.), 251–254. A.A. Balkema.

Wenk, H-R. and Bulakh, A. (2017). *Minerals – Their Constitution and Origin*, 2e. Cambridge University Press, 621 pp.

White, D.E. (1963). The Salton Sea geothermal brine, an ore-transporting fluid. *Mining Engineering* 15: 60.

Whitney, J.A. (1975). Vapor generation in a quartz monzonite magma: a synthetic model with application to porphyry copper deposits. *Economic Geology* 70: 346–358.

Whitney, J.A. (1989). Origin and evolution of silicic magmas. In: *Ore Deposition Associated with Magmas*, Reviews in Economic Geology, vol. 4 (eds. J.A. Whitney and A.J. Naldrett), 183–201. Society of Economic Geology.

Widler, A.M. and Seward, T.M. (2002). The adsorption of gold(I) hydrosulphide complexes by iron sulphide surfaces. *Geochimica et Cosmochimica Acta* 66: 383–402.

Wilkinson, J.J. (2013). Triggers for the formation of porphyry ore deposits in magmatic arcs. *Nature Geoscience* 6: 917–925.

Williams, H., Hoffman, P.H., Lewry, J.F. et al. (1991). Anatomy of North America: thematic geologic portrayals of the continents. *Tectonophysics* 187: 117–134.

Williams-Jones, A.E. and Heinrich, C.A. (2005). Vapor transport of metals and the formation of magmatic-hydrothermal ore deposits. *Economic Geology* 100: 1287–1312.

Willis-Richards, J. and Jackson, N.J. (1989). Evolution of the Cornubian Ore Field, southwest England: part I. Batholith modeling distribution. *Economic Geology* 84: 1078–1100.

Wilson, A.F. (1984). Origin of quartz-free gold nuggets and supergene gold found in laterites and soils: a review and some new observations. *Australian Journal of Earth Sciences* 31: 303–316.

Windley, B.F. (1995). *The Evolving Continents*. Wiley, 526 pp.

Wingate, M.T.D. (1998). A palaeomagnetic test of the Kaapvaal–Pilbara (Vaalbara) connection at 2.78 Ga. *South African Journal of Geology* 101: 257–274.

Wirth, R., Reid, D., and Schreiber, A. (2013). Nanometer-sized platinum-group minerals (PGM) in base metal sulfides: new evidence for an orthomagmatic origin of the Merensky Reef PGE ore deposit, Bushveld Complex, South Africa. *Canadian Mineralogist* 51: 143–155.

Witt, W.K. (1991). Regional metamorphic controls on alteration associated with gold mineralization in the Eastern Goldfields province, Western Australia: implications for the timing and origin of Archean lode-gold deposits. *Geology* 19: 982–985.

Wood, B.J. and Blundy, J.D. (2002). The effect of H_2O on crystal–melt partitioning of trace elements. *Geochimica et Cosmochimica Acta* 66: 3647–3656.

Wood, B.J. and Fraser, D.G. (1976). *Elementary Thermodynamics for Geologists*. Oxford University Press, 303 pp.

Wood, B.J., Pawley, A., and Frost, D.R. (1996). Water and carbon in the Earth's mantle. *Philosophical Transactions of the Royal Society A* 354: 1495–1511.

Wood, S.A. (1990). The interaction of dissolved platinum with fulvic acid and simple organic acid analogues in aqueous solutions. *Canadian Mineralogist* 28: 665–673.

Wood, S.A. (1996). The role of humic substances in the transport and fixation of metals of economic interest (Au, Pt, Pd, U, V). In: *Organics and Ore Deposits*, Ore Geology Reviews, vol. 11 (ed. T.H. Giordano), 1–31. Elsevier.

Wood, S.A. (1998). Calculation of activity–activity and log fO_2–pH diagrams. In: *Techniques in Hydrothermal Ore Deposits Geology*, Reviews in Economic Geology, vol. 10 (eds. J.P. Richards and P.B. Larson), 81–96. Society of Economic Geologists.

Wood, S.A. and Samson, I.M. (1998). Solubility of ore minerals and complexation of ore metals in hydrothermal solutions. In: *Techniques in Hydrothermal Ore Deposits Geology*, Reviews in Economic Geology, vol. 10 (eds. J.P. Richards and P.B. Larson), 33–80. Society of Economic Geologists.

Wood, S.A. and Vlassopoulos, D. (1990). The dispersion of Pt, Pd and Au in surficial media about two PGE-Cu-Ni prospects in Quebec. *Canadian Mineralogist* 28: 649–663.

Wood, S.A., Mountain, B.W., and Pan, P. (1992). The aqueous geochemistry of Pt, Pa and Au: recent experimental constraints and a re-evaluation of theoretical predictions. *Canadian Mineralogist* 30: 955–982.

Woodcock, L. (2014). Gibbs density surface of water and steam: 2nd debate on the absence of Van Der Waals' "critical point". *Natural Science* 6: 411–432. https://doi.org/10.4236/ns.2014.66041.

Woods, P.J.E. (1979). The geology of the Boulby mine. *Economic Geology* 74: 409–418.

Woolley, A.R. and Kjarsgaard, B.A. (2008) Carbonatite occurrences of the world: map and database. *Geological Survey of Canada, Open File #5796*, 28pp.

Worsley, T.R., Nance, D., and Moody, J.B. (1984). Global tectonics and eustacy for the past 2 billion years. *Marine Geology* 58: 373–400.

Xu, C., Campbell, I.H., Kynicky, J. et al. (2008). Comparison of the Daluxiang,and Maniuping carbonatitic REE deposits with Bayan Obo, China. *Lithos* 106: 12–24.

Xu, C., Wang, L., Song, W., and Wu, M. (2010). Carbonatites in China: a review for genesis and mineralization. *Geoscience Frontiers* 1: 105–114.

Yang, K. and Seccombe, P.K. (1994). Contrasting hydrothermal behavior of platinum group elements of Ir and Pd sub-groups as exemplified by platinum group minerals in Great Serpentine Belt, eastern Australia. *Transactions of the Institute of Mining Metallurgy* 104: B39–B44.

Young, T.P. (1993). Sedimentary iron ores. In: *Mineralization in the British Isles* (eds. R.A.D. Pattrick and D.A. Polya), 446–487. Chapman & Hall.

Young, T.P. and Taylor, G.W.E. (1989). *Phanerozoic Ironstones*, vol. 46. The Geological Society of London, Special Publication, 247 pp.

Zaccerini, F., Garuti, G., and Cawthorn, R.G. (2002). Platinum group minerals in chromitite xenoliths from the Onverwacht and Tweefontein ultramafic pipes, eastern Bushveld Complex, South Africa. *Canadian Mineralogist* 40: 481–497.

Zammit, C.M., Shuster, J.P., Gagen, E.J., and Southam, G. (2015). The geomicrobiology of supergene metal deposits. *Elements* 11 (5): 337–342.

Zhang, S., Li, Z-X., Evans, D.A.D. et al. (2012). Pre-Rodinia supercontinent Nuna shaping up: a global synthesis with new paleomagnetic results from North China. *Earth and Planetary Science Letters* 353–354: 145–155.

Zhao, G., Sun, M., Wilde, S., and Li, S.Z. (2004). A Paleo-Mesoproterozoic supercontinent: assembly, growth and break-up. *Earth-Science Reviews* 67: 91–123.

Index

Note: Page numbers in *italics* indicate figures and tables. 'n' following a page number indicates a footnote.

Introduction to Ore-Forming Processes, Second Edition. Laurence Robb.
© 2021 John Wiley & Sons Ltd. Published 2021 by John Wiley & Sons Ltd.